NEOTROPICAL MIGRATORY BIRDS

NEOTROPICAL MIGRATORY BIRDS

NATURAL HISTORY, DISTRIBUTION, AND POPULATION CHANGE

Richard M. DeGraaf

U.S. Forest Service
Northeastern Forest Experiment Station
Amherst, Massachusetts

John H. Rappole

Smithsonian Institution
National Zoological Park
Conservation and Research Center
Front Royal, Virginia

Comstock Publishing Associates

a division of Cornell University Press

ITHACA AND LONDON

First published 1995 by Cornell University Press

Library of Congress Cataloging-in-Publication Data
DeGraaf, Richard M.
    Neotropical migratory birds : natural history, distribution, and
population change / Richard M. DeGraaf and John H. Rappole.
        p. cm.
    Includes bibliographical references and index.
    ISBN 0-8014-8265-8 (pbk. : alk. paper)
    1. Birds—North America.   2. Birds—Wintering—Latin America.
3. Birds—Migration—North America. 4. Birds—Migration—Latin
America.   I. Rappole, John H.   II. Title.
QL681.D43   1995
598.297—dc20   94-37130

Printed in the United States of America

To Gabriel Charles, late of St. Lucia, W.I., and to

Thomas E. Lovejoy, pioneers in Neotropical conservation

Contents

Acknowledgments

We gratefully acknowledge those who contributed to this effort: William R. Danielson and Christopher J. E. Welsh, U.S. Forest Service, Amherst, Massachusetts, for assistance with literature reviews; Jorge Vega-Rivera and Sharon F. Leathery, Conservation and Research Center, Smithsonian Institution, Front Royal, Virginia, for preparing range maps; Sam Droege and Bruce Peterjohn, U.S. Fish and Wildlife Service, Laurel, Maryland, for providing Breeding Bird Survey data; Nancy Haver, Amherst, Massachusetts, for preparing the map of physiographic regions; and Robert A. Askins, Connecticut College, New London, for his careful review of the manuscript. Liz Pierson painstakingly copyedited the manuscript and, with her knowledge of field ornithology, offered many helpful suggestions. Peter Stangel, National Fish and Wildlife Foundation, Washington, D.C., provided encouragement and background information from the outset. Mary A. Sheremeta, U.S. Forest Service, Amherst, Massachusetts, diligently typed the manuscript. Publication of this book was made possible, in part, by a grant from the Northeastern Forest Experiment Station, U.S. Forest Service.

We thank Princeton University Press for permission to reprint an excerpt from John Terborgh's *Where Have All the Birds Gone?*, copyright © 1989 by Princeton University Press; and Blackwell Scientific Publications, Inc. for permission to reprint, in modified form, figure 2 from R. Dirzo and M. L. Garcia, "Rates of Deforestation in Los Tuxtlas, a Neotropical Area in Southeast Mexico," *Conservation Biology* 6, no. 1 (1992): 84–90.

R. M. D.

J. H. R.

NEOTROPICAL MIGRATORY BIRDS

Introduction

North America has a long and rich ornithological history. The accumulated contributions of both professionals and amateurs have provided a wealth of information on the distribution, abundance, and ecology of the continent's avifauna. The breeding biology of most species has been studied intensively, and many private and public lands have been set aside for bird conservation.

More than half of the species that breed in North America, however, spend only a portion of their life cycle on the breeding grounds. For these migratory species, the breeding season is often brief, just long enough to take advantage of the abundant food resources available in the Nearctic summer. Migration begins soon after the young fledge, and for many species it concludes on another continent, often in a completely different habitat.

This book summarizes the life histories and distributions of 361 species of Neotropical migratory birds, species that for the most part breed in the United States and Canada and migrate to wintering grounds in the Caribbean, Mexico, and southward. Until recently, the winter segment of the life cycle of most of these species was essentially unknown, with even the most basic information on distribution and habitat use poorly documented. Winter was a period when, from a conservation standpoint, migrants were for the most part out of sight and seemingly out of mind. For migratory birds that spend six months or more away from the breeding grounds, however, conservation measures require a thorough understanding of each species' ecology throughout the entire life cycle. For most species of migratory birds, this task is just beginning.

The quest to understand the complexities of the ecology of Neotropical migratory birds has accelerated with a sense of great urgency in recent years as a fortunate result of an unfortunate circumstance—namely, evidence of long-term declines of many species and increased rates of habitat destruction, particularly in the Neotropics.

1

Introduction

The accumulating reports of declining populations of migratory birds have jolted our ornithological and environmental senses. For scientists and bird-watchers alike, the fact that a variety of species—wood-warblers, shorebirds, thrushes, raptors, flycatchers, grosbeaks, and others—are decreasing at alarming rates in the presence of one of the most elaborate conservation infrastructures in the world has served as a clear signal that our geographically biased conservation efforts are not simply inadequate but failing.

The reasons why migratory birds are declining are subject to lively debate among ornithologists and ecologists. Loss of breeding habitat in eastern North America since the late 1940s is a factor for many species. Recent losses of winter habitat are probably increasingly important for others. The combined effects are alarming. Declines were first reported for forest birds, which constitute about a third of Neotropical migratory species. Another third are aquatic birds, including waterfowl, shorebirds, and herons, and old-field, grassland, and shrubland birds complete the list. Declines are occurring among these groups as well. For example, the fastest-declining bird in New England is the Rufous-sided Towhee (Hagan et al. 1992), a bird of brushy fields and forest edges. Vesper and Grasshopper Sparrows, Bobolinks, Eastern Meadowlarks, and Yellow-breasted Chats are dwindling as abandoned farmland reverts to forest. The Upland Sandpiper is virtually absent from New England now and essentially is found only at the largest airports and in other open habitats such as the blueberry barrens in Maine. Habitat changes throughout the northeastern United States are resulting in widespread declines of grassland, old-field, and thicket birds.

As forests reclaim much of the Northeast, a major concern is habitat availability or suitability on the breeding grounds. Many wood-warblers, such as the Ovenbird and Northern Parula, show great habitat specificity, using the interiors of fairly extensive mature deciduous forests during the breeding season, although they may occur in a variety of habitats, even disturbed small tracts, in winter. Other species, such as the Gray-cheeked Thrush and Louisiana Waterthrush, seem to select extensive forest in winter. Pollution, especially from pesticides, is a continuing threat to many wintering aquatic species such as the Black-crowned Night-Heron.

Even in fairly stable North American landscapes where the proportions of forest, agricultural land, and suburban development have remained essentially constant for decades, subtle changes have greatly altered conditions to the benefit of some species and the detriment of others (Terborgh 1989). The availability of waste grain, especially corn, from mechanical harvesting has resulted in population increases and changes in migratory habits of Canada Geese and Mourning Doves and of resident or short-distance migrants such as Common Crows, European Starlings, and Common Grackles. The shift from picked to chopped silage corn in much of the Northeast, in combination with the shift from cultivation to herbicide weed control since the 1960s, has resulted in the elimination of Ring-necked Pheasants from much of the region for the opposite reason—there is no waste grain or cover in the cut-over fields. In many areas the ever-increasing proportion of alfalfa in hay fields has resulted in abandonment by Bobolinks, which prefer grasses.

Within woodlands, housing developments, even those with small lawns and low-maintenance yards, subtly break up contiguous forest tracts and result in population increases of generalist predators such as cats and raccoons. In some areas, high densities of white-tailed deer result in removal of the forest understory in which many migrants such as Black-throated Blue and Black-throated Green Warblers and Rose-breasted Grosbeaks, among others, nest.

The conservation of Neotropical migrants depends on the availability of suitable breeding and wintering habitat as well as migration stopover areas. On the breeding grounds, Neotropical migrants are adapted to a wide range of habitats, from early-successional grasslands and old fields to mature forests. Maintaining the largest blocks of each habitat possible is probably the best conservation strategy. Just as there are species adapted to forest interior, such as the Wood Thrush, Ovenbird, and Cerulean Warbler, there are species adapted to extensive early-successional habitats—Chestnut-sided Warblers, Yellow-breasted Chats, and Eastern Meadowlarks, for example. Both groups contain area-sensitive species that do not occur or may not successfully breed in small patches of otherwise suitable habitat. Bobolinks, for example, rarely nest if the area of contiguous grassland falls much below 25 ha.

Early-successional migrants were probably adapted to breed in natural prairies or in forested areas disturbed by fire or wind. Such events are irregular and often were extensive; populations of species such as Willow and Alder Flycatchers, Chestnut-sided Warblers, Yellow-breasted Chats, and Prairie Warblers depend on extensive areas of disturbance to rapidly build up their populations. They then are able to survive as species by dispersing to search for other suitable breeding areas as their former breeding habitats revert to forest. Throughout much of eastern North America, a shifting mosaic of habitats—from recent burns and vast blowdowns to patchy young forests to extensive mature woodlands—provided breeding habitat for all grassland, thicket, and forest-dwelling migrants over time.

In presettlement times, the Ohio River Valley was apparently a stable, mature forest area, disturbed mainly by the falling of individual trees. To the west and north, the Great Plains and boreal forests burned frequently and over large areas. To the east and southeast, the Atlantic coastal plain was subject to frequent fires and hurricanes, storms that also cut periodic swaths through New England. Eastern North America was not an unbroken expanse of mature forest.

Native eastern prairies were among the first places to be settled or farmed, and fires were strictly controlled. New England and the Middle Atlantic region were rapidly cleared after about 1750. Agriculture has steadily declined since its peak between 1820 and 1840 in New England and much of the Northeast. The opening of the Erie Canal in 1825 started an emigration of New England farmers as the fertile Midwest was opened, and the California gold rush and Civil War led to more land abandonment, hastening the reversion of family farms to forest, a trend that continues today. The dramatic declines of grassland and shrubland birds since the early 1980s show that we are at the end of the supply of early-successional habitats provided by the initial decline of agriculture in the Northeast. Forests increase in age and extent each year. Except in northern Maine, timber cutting on any appreciable scale is uncommon on most of the privately owned northeastern forest. It will take active management to maintain the ephemeral habitats of grassland, old-field, and shrubland birds (and other species) in much of eastern North America. We will have to think in terms of managed landscapes so that habitat management mimics natural disturbances historically caused by wind and fire, and provides successional and mature breeding habitats in large enough areas for the species adapted to them.

Although North American habitats are always changing to some degree, the rates of change are slow compared with those in the Neotropics where more than 300 species of migratory birds winter. The rate of human population growth has placed high demands on the land for food and fiber. Exploitation of the humid tropics is the Western Hemisphere's great ecological disaster of the twentieth century. Biologist John Terborgh's description of this disaster, from his 1989 book *Where Have All the Birds Gone?*, can scarcely be improved:

> In many countries . . . , the most pervasive changes are occurring in conjunction with a massive wave of abusive overexploitation of virgin lands, a wave propelled by a gold-rush mentality oblivious to the basic precepts of renewability, sustainability, and future need. The pattern is uncomfortably reminiscent of the excesses that occurred in our own country during the eighteenth, nineteenth and early twentieth centuries, before the federal government began to regulate the use of renewable resources such as fish, game, forests, soils, and water. The spasm of exploitation that is currently under way in the tropics is being pursued with the same total lack of . . . concern for posterity as occurred here between 1860 and 1920, when the

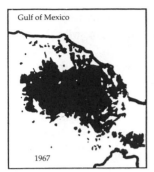  

Figure 1. Loss of rain forest in the Tuxtla Mountain region of southern Veracruz, Mexico (after Dirzo and Garcia 1992). Presettlement forest coverage was 849.6 km². By 1967, 309.6 km² (36%) remained; by 1976, 211.2 km² (25%); and by 1986, 136.1 km² (16%).

> American bison went from being the most abundant large mammal on the continent to the rarest, and when more than 400 million acres of virgin forest east of the Great Plains were felled in a mere sixty years. (P. xiv)

Since 1989, when Terborgh wrote with such urgency, the rate of deforestation in Mexico, Central America, the Caribbean, and tropical South America is annually 16,713,500 ha, or 1,911 ha per hour (United Nations Food and Agricultural Organization 1991: Table 1), and shows no sign of slowing (Fig. 1).

On the wintering grounds, the continuance of secure habitat probably lies, somewhat ironically, in industrialization and in controlled, sustainable commercialization of key resident wildlife. Concentrating people in cities, maintaining production farming on suitable land, and hunting at low densities may allow valuable native species to exist in large enough areas to provide secure habitat for Neotropical migrants and resident wildlife in perpetuity. For a review of issues involved in the maintenance of both biodiversity and sustainable development, see J. G. Robinson and K. H. Redford, *Neotropical Wildlife Use and Conservation* (Univ. of Chicago Press 1991), wherein the case is made that subsistence and commercialization of wildlife for food and other products are essential to the conservation of Neotropical ecosystems and their wildlife.

Although much remains to be learned about the ecology of Neotropical migrants, there is also a need to synthesize existing knowledge about their distribution and habitat requirements. Our purpose in compiling this book is to provide land managers, resource professionals, and birdwatchers with information on the basic habitat requirements and distributions of 361 species of Neotropical migrants that breed regularly in the United States and Canada. Clearly, conservation efforts for these species must be undertaken throughout the birds' ranges rather than just on the breeding grounds. Our goal is to summarize practical information on the abundance, distribution, nesting sites, and habitat needs of Neotropical migrants on their breeding and wintering grounds and to illustrate their complete distribution. Our hope is that this information will foster an appreciation for the complexity of the species' entire life cycles.

The chapter "What Is a Neotropical Migratory Bird?" defines those species that compose the Neotropical migrants. We attempt to place this group into a hemispheric perspective, and we focus on information that we feel is of greatest importance to the birds' management. The next chapter, "Population Change," summarizes existing knowledge on general population trends and presents current information on the factors believed to be contributing to the declines of populations of Neotropical migrants.

These two chapters are followed by the main body of the text: the natural history accounts of 361 species of Neotropical migratory birds. The taxonomic order and English and Latin names of all species follow the American Ornithologists' Union's *Check-list of North American Birds* and supplements. For each species, we provide notes on distribution and a complete range map for the species' distribution in the Western Hemisphere—basic information that until recently has been largely unavailable in usable form to the management community.

The maps depict those parts of the Western Hemisphere where a species occurs regularly as a summer, winter, or permanent resident. Light shading is used for the breeding range, darker shading for the winter range, and black for areas where a species is resident. Range in the West Indies is shown by a circle around the islands on which a species is found: a broken circle for breeding, a solid circle for winter, and two circles for resident. Portions of the range where a species is found only during migration are not shown.

Our primary source of information on breeding and wintering ranges is the American Ornithologists' Union's 1983 *Check-list of North American Birds* and supplements (American Ornithologists' Union 1983, 1985, 1987, 1989, 1991). This source has been supplemented with information from several other books and technical papers (cited in the species accounts), along with the notes of numerous ornithologists. The range maps and range descriptions are not likely to match precisely. The descriptions tend to incorporate places where only one or two records document occurrences, whereas the maps are meant to show places where each species occurs regularly. Despite the excellent work by Olrog (1984), Sick (1985), Meyer de Schauensee (1982), Ridgely and Tudor (1989), and others, vast areas of South America are still poorly known ornithologically, and maps of species' distributions in many parts of this continent should be considered as little more than a first approximation that will be corrected as better information becomes available.

Details on status, breeding habitat, and special habitat requirements are also provided for each species. When available, material on migration and winter habitats is presented. Also, when available, management notes are provided, especially for breeding habitats. Spanish and French common names are given to facilitate use by managers of wintering-ground habitats. Spanish names are taken primarily from the 1989 Spanish translation of Peterson and Chalif's *A Field Guide to Mexican Birds*. Last, a list of basic references for each species is provided for the United States and Canada, Mexico and Central America, South America, and the Caribbean.

Following the species accounts are two appendixes. Appendix A consists of a description of the major habitats used by Neotropical migrants in the Nearctic and Neotropics; six tables of habitat types and the species that breed in them; and a seventh table that lists the winter habitats of each species. Appendix B lists species that are increasing or decreasing at significant rates in various physiographic regions of North America, based on long-term (1966–1994) and short-term (1980–1994) trends gleaned from the Breeding Bird Surveys (BBS) conducted annually since 1966 by the U.S. Fish and Wildlife Service.

What Is a Neotropical Migratory Bird?

The subjects of this book are those bird species that breed in the Nearctic faunal region and winter in the Neotropics. There has been a lively debate regarding what to call members of this group and how to define them. Part of the difficulty lies in finding an accurate definition that does not reflect a regional prejudice. *Nearctic avian breeders that winter in the Neotropics, Neotropical avian species that breed in the Nearctic,* or even *Nearctic-Neotropical migrants* are accurate but cumbersome. *North American migrants, Nearctic migrants, wintering migrants, tropical migrants,* and *long-distance migrants* all appear in the literature, but they fail the accuracy or prejudice test. Hagan and Johnston (1992) argued for the term *Neotropical migrant.* Of course this term includes those species that migrate solely within the Neotropical faunal region as well. Nevertheless, we have chosen to adopt the term *Neotropical migrant,* despite its problems, because it has the benefit of properly placing most Nearctic-breeding species that winter in the Neotropics as part of the Neotropical migration system.

Identifying the species to be included in this assemblage has proved to be even more problematic than choosing a name for them. The research working group of the Partners in Flight program (Finch and Martin 1991) provided a list of Neotropical migrants that included only 255 species. They limited their list to landbirds "generally recognized" as being Neotropical migrants, omitting waterbirds and shorebirds because "other initiatives for the conservation of waterfowl and shorebirds are already under way, and a separate initiative for migrant landbirds was warranted."

In contrast to this somewhat arbitrary definition, we have defined Neotropical migrants on the basis of zoogeographic criteria, namely the location of breeding and wintering areas. This procedure requires a firm knowledge, some of which is still lacking, of the breeding and wintering ranges of Western Hemisphere birds. However, this definition has the advantage of being dependent on measurable, biologically relevant criteria.

The Nearctic and Neotropical faunal regions are two of six broad geographic areas characterized by similarities in animal life (Darlington 1957). The Nearctic region includes North America north of the tropics, and the Neotropical region includes tropical Middle and South America and the West Indies (Mayr 1985). The border between these two faunal regions follows the northern edge of tropical, evergreen forest distribution in Mexico. However, the transition from subtropical to tropical habitat types is tortuous in some areas, and exact boundaries are difficult to establish. To facilitate the identification of species to be included in this book, we chose the Tropic of Cancer as the boundary between the Nearctic and Neotropical regions. This border is reasonable biologically and provides a precise reference point when examining breeding and wintering distributions of birds. For our purposes, then, Neotropical migrants are those "Western Hemisphere species, all or part of whose populations breed north of the Tropic of Cancer and winter south of that line," the definition proposed by Rappole et al. (1983). According to this definition, there are 361 species of Neotropical migratory birds, though our list is probably incomplete because the migratory movements of subtropical species such as White-tailed Hawk, Couch's Kingbird, Sulphur-bellied Flycatcher, and many others are poorly known. Some of these subtropical-tropical species have been recorded as migrants (Thiollay 1977, 1979, Wetmore 1943), and it is likely that many more will be documented to be at least partially migratory as our knowledge of seasonal movements of these species increases (Rappole et al. in press). As in Rappole et al. (1983), we do not include in this book the strictly pelagic species of the Order Procellariiformes (tube-noses) and the Family Stercorariidae (skuas and jaegers).

Our 361 species of Neotropical migrants embrace 38 of the 62 families of birds breeding in the United States and Canada. Some families, such as the Pelecanidae, contain only one or two Neotropical migrants, whereas other families contain several species. The family containing the most by far is the Emberizidae (wood-warblers, tanagers, orioles, and sparrows), which has 95 species—more than 25% of all Neotropical migrants.

The phenomenon of bird migration is a spectacular event that goes unnoticed by many. Even birdwatchers, often engrossed in trying to see as many species as possible, may not realize the remarkable biological feats these birds accomplish. The details have been pieced together for a few species on the basis of thousands of hours of painstaking research. For instance, research on Blackpoll Warblers has shown that within a few weeks after completing the breeding cycle, the birds accumulate enough fat to nearly double their weight—enough to make the nonstop journey over the Atlantic from breeding grounds in northern New England to wintering grounds in Venezuela in 85 hours (Nisbet et al. 1963, Williams and Williams 1978). They arrive at their winter home emaciated and must then compete for food with the many resident species there. How can Blackpoll Warblers, and other Neotropical migrants, accomplish this, competing with highly evolved and specialized tropical residents for resources for half of the year before returning to their boreal breeding grounds in spring?

The answer is that, in all probability, the ancestors of migratory birds were resident in what are now winter ranges. Most migrants are Neotropical birds that move north in the summer to breed. More than three-quarters of Neotropical migratory species are closely related to species that have breeding populations in the Neotropics. Forty-eight percent of all Neotropical migrants have breeding populations in the Neotropics. At least 30 species of Neotropical migrants have overlapping distributions of year-round resident populations and winter resident populations, including such familiar species as the Great Blue Heron, Turkey Vulture, Osprey, Killdeer, Common Tern, Mourning Dove, Vermilion Flycatcher, and Yellow Warbler. These facts make comprehensible the annual invasion of migrants into Neotropical habitats, which already contain diverse communities of resident species. Neotropical migrants originated in the Neotropics and are well-adapted to exploit niches in tropical habitats.

Why did Neotropical species develop the habit of migrating to a northern breeding area? We propose that the explanation is that migration increased breeding success. Intense competition and high predation rates in tropical populations result in low annual productivity. Yet because of the longevity of the adults, the needs of surviving young often exceed the amount of food and space available (Fogden 1972, Willis 1972). For members of those tropical species able to exploit the abundant food resources, especially insects and other invertebrates, and reduced levels of competition and predation in the temperate and boreal regions during summer, much higher production rates of young are possible than with comparable efforts in the Neotropics (Ricklefs 1972, Skutch 1976). Thus migration is a natural result of intense competition and high predation rates in the tropics, abundant food resources in the Nearctic summer, and the mobility of birds.

The development of migration is an ongoing process. The Cattle Egret, originally a bird of southern Europe, Africa, and Asia, invaded South America in the late 1800s (Meyer de Schauensee 1966). The first wanderers arrived in the United States in 1952. The species now occurs widely across the United

States and parts of Canada, with breeding populations occurring as far north as New England and the northern Great Plains. It has become a Neotropical migrant within four decades of its first appearance in the Nearctic zone. For a detailed discussion of the evolution and zoogeography of Neotropical migrants, see Rappole et al. 1983 and Rappole and Tipton 1992.

The preponderance of data from the breeding grounds conveys the impression that the breeding season is the most important portion of the life cycle for Neotropical migrants. However, most Neotropical migrants spend a longer period each year migrating or on the wintering grounds than on the breeding grounds. For example, Smith (1980) calculated that Broad-winged Hawks spend four months on the breeding grounds, six months on the wintering grounds, and two months in transit. Swainson's Hawks divide the year more or less equally among the three periods. Schwartz (1964) recorded individual residence times of seven and six months, respectively, for American Redstarts and Northern Waterthrushes in a city park in northern Venezuela. He estimated that these species were present in northern South America for up to eight months of the year. In a study of 13 common warblers (Parulinae), Keast (1980) found that most spent three months or less breeding in Ontario, two to three months going to and from the wintering grounds, and six to seven months on the wintering grounds. Thus, rather than representing a brief absence from the North American breeding grounds, the migration and wintering periods dominate the life cycles of many Neotropical migrants.

Our knowledge of the distribution of Neotropical migrants during the nonbreeding season is still primitive. Political history, economics, language barriers, and until recently, the lack of field guides, have all postponed study of Neotropical migrants in Latin America (Terborgh 1989). Nonetheless, we still know far more about the birds' distribution during the nonbreeding season than we do about their ecology during this time.

Three published symposia since 1970 have helped to expose and dispel our ecological ignorance regarding these species: Buechner and Buechner 1970, Keast and Morton 1980, and Hagan and Johnston 1992. Many of the papers presented at these symposia specifically addressed Neotropical migrant ecology and conservation, allowing the first real assessment of what these birds do during the nonbreeding season.

The most illuminating revelations in this regard have been the discoveries of the intimate details of the "migrant at home" on its wintering site, dispelling the myth of the migrant as a seasonal interloper in tropical environments, subsisting as a wanderer on temporary resource flushes in marginal habitats (Karr 1976). These revelations began with the pioneering studies of Schwartz (1963, 1964) in which migrant Northern Waterthrushes were found to hold and defend territories, to which they returned in subsequent years, in stable tropical habitats throughout the winter. Van Tyne (1936), Loftin and his collaborators (Loftin 1963a, 1963b, 1967, 1977, Loftin et al. 1966, 1967), Nickell (1968), and many others used extensive banding studies to demonstrate the precise interyear seasonal fidelity of individual migrants of many species to specific tropical wintering sites. Rappole et al. (1989) documented that Wood Thrushes defended small territories (0.5 ha) in rain forest and that birds unable to locate territories in this habitat in Veracruz, Mexico, used second growth as an alternative, moving over large distances and suffering higher mortality rates than their sedentary compatriots holding rain-forest territories. These "floaters" continued to explore rain forest, probing existing occupied areas for weak or absent territory owners until they found a suitable vacant site. Rappole and Warner (1980) discovered that male and female Hooded Warblers held territories in different habitat types in southern Mexico, a phenomenon that has since been documented elsewhere in the Neotropics and for other species of wintering migrants (Holmes et al. 1989, Lopez Ornat and Greenberg 1990, Lynch et al. 1985, Wunderle 1992). Morton et al. (1987) showed that this difference was a matter of choice, at least for Hooded Warblers in Yucatán, Mexico. Staicer (1992) reported some remarkable variations in winter resource-use behavior for migrants on her study sites in dry scrub of Puerto Rico. She found that Northern Parulas and Cape May and Prairie Warblers showed highly variable spacing and resource-use systems, with individuals moving about independently within overlapping home ranges, or occasionally in small cohesive flocks. As in Rappole and Warner's (1980) Veracruz studies, Staicer found the birds to be quite faithful to their wintering sites from one year to the next (48–54% return rates). Holmes and Sherry (1992), in their studies of American Redstarts and Black-throated Blue Warblers, documented higher return rates to winter sites in Jamaica than to breeding areas in New Hampshire.

Eugene S. Morton has published several papers on the subject of the tight interrelationship that exists between some migrants and their tropical habitats. For instance, he describes the curiously interwoven tie between the tropical fruiting tree *Erythrina fusca* and wintering Orchard Orioles in Panamanian forest. The color and shape of the tree's flowers mimic the oriole's colors, attracting orioles to approach and enter the flower to extract nectar and, in the process of moving from flower to flower, to pollinate the plant (Morton 1979). Tennessee Warblers have a similarly intimate relationship with the tropical vine *Combretum fruticosum*. The warbler extracts nectar from the plant's flower, which is perfectly shaped to brush pollen on the bird in the process. Morton (1980b) further suggests that the bird may benefit in providing this pollination service, not only from the nectar but from the bright red pollen, or "war paint," left on the bird's face, which could enhance the bird's ability to defend flowers from other Tennessee Warblers.

George Powell's work on the Resplendent Quetzal has caused a "paradigm shift" in the way we view migrants and tropical residents (Powell et al. in press). Using radio-tracking in Costa Rica, Powell and his co-workers have found that the Resplendent Quetzal, long presumed to be a tropical resident of cloud forests, is in fact a migrant. The bird breeds in cloud forest but migrates 20 to 30 km to a different habitat after the breeding season to exploit fruiting trees. The bird moves to a third habitat later in the season before finally returning to its breeding habitat in cloud forest in time to begin reproduction. Powell's work demonstrates that the difference between an intratropical migrant and one that migrates to the Nearctic is one of degree rather than kind. Fundamentally, they are participants in the same extraordinary process.

Perhaps the most remarkable information about migrants to come from studies in the Neotropics concerns the birds' vulnerability to tropical habitat alteration. As an example, consider the Wood Thrush, a species that, like several other migrants, winters in rain forest and old second growth. In southern Belize, Wood Thrush winter populations are estimated to have declined by only 17% since the time of European settlement. The figure is 76% in northeastern Costa Rica, however, and in southern Veracruz, Mexico, 95% of the Wood Thrush winter population has been extirpated as the rain-forest habitats on which the birds depend have disappeared (Powell et al. 1992, Rappole et al. in press). The dependence of these birds on tropical habitats during the winter makes them extraordinarily vulnerable to the massive environmental changes presently occurring throughout much of the tropics.

Neotropical migrants constitute an internationally shared resource—integral parts of both tropical and temperate ecosystems. Indeed there are no major New World avifaunal communities that do not include migrants. Their population trends are indicators of habitat conditions for many other species. We know many details about the breeding biology and ecology of Neotropical migrants, but we have a great deal to learn about their distribution, ecology, and habitat needs on the wintering grounds. Research is urgently needed to improve our knowledge of these species. We believe that improved management and protection of habitat on the wintering grounds of migratory birds is the greatest current challenge to conservation in the Western Hemisphere.

Population Change

Migratory birds have complex annual cycles in which they must exploit habitats in widely separated geographic regions in order to survive and reproduce. This way of life has considerable potential benefits in terms of maximizing survival and reproduction (Rappole and Tipton 1992, Williams 1958), but it may also expose migrants to a higher likelihood of extinction than resident species are exposed to, because of the multiplicity of environmental risks that migrants confront in an annual cycle.

The vulnerability of migratory birds to multiple pressures in different parts of the Western Hemisphere was recognized by waterfowl conservationists early in the 1900s. The Migratory Bird Treaty (U.S. Congress 1918), to which Mexico, Canada, and the United States are signatories, was an attempt to address the problem of waterfowl population declines caused by mortality factors during different portions of the annual cycle. A broad system of national wildlife refuges was established in the United States principally to protect migratory waterfowl stopover and wintering habitats.

Recognition that other species of migrants might require special protection has been slower to develop. Ecologist William Vogt was one of the first to clearly articulate this concern in a symposium he helped to organize at the Smithsonian Institution in 1966 (Buechner and Buechner 1970). He pointed out that massive environmental changes were in progress in areas used by a large portion of the avifauna of North America as stopover or wintering habitat, and he expressed this view: "The future of the bird life between the Rio Bravo [Rio Grande] and Central South America is definitely dark. It is virtually certain that at no period in history have human populations been expanding with such speed and with such a destructive technology. To continue ignoring the situation is virtually to condemn to near or complete extinction a number of species that depend on the tropical American habitats" (Vogt 1970:12).

14

Many of the experts in Neotropical ornithology who attended this meeting reassured Dr. Vogt that North American migrants did not fall into the category of "species that depend on the tropical American habitats." They maintained that migrants were essentially marginal species in the tropics, exploiting excess resources in disturbed habitats; habitat disturbance and destruction might even favor migrant birds.

The degree to which migratory birds are dependent on stopover or wintering sites for survival is still a subject of debate among researchers. Morel and Bourlière (1962), Karr (1976), and Petit et al. (1992) have taken positions similar to those voiced by many persons participating in the 1966 Smithsonian symposium (Buechner and Buechner 1970), namely, that migrants are marginal species in tropical communities, unable to compete for stable resources with tropical residents, and thus dependent on temporary resource concentrations for survival. Other researchers have argued that migrants are an integral part of and reliant on tropical communities (Rappole and Warner 1980, Rappole et al. 1982, Rappole et al. 1993, Stiles 1980). Still others have stressed how little we know about the relationship between these birds and the communities they inhabit during the nonbreeding portion of the life cycle (Hutto 1988).

Despite the inconclusiveness of this debate, concern for the welfare of migratory species has continued to grow among scientists (Lovejoy 1983, Morton and Greenberg 1989, Pasquier and Morton 1982, Terborgh 1989, 1992) as well as the public at large (Deis 1981, Howe 1983, Steinhart 1984, Wallace 1986).

Evaluation of Evidence of Population Declines

Table 1 shows 28 species of Neotropical migrants that the U.S. Fish and Wildlife Service has classified as endangered or as being of management concern because of documented significant reductions in their populations. The presumptive causes for the declines are quite disparate, ranging from DDT poisoning for the Peregrine Falcon (Newton 1979) to beach-front development for the Gull-billed Tern (Via and Duffy 1992) to brood parasitism by cowbirds for the Black-capped Vireo (Grzybowsky et al. 1986).

Several scientists remain skeptical that reported population declines, especially of forest-dwelling migrants, represent actual threats to Neotropical migrant species (Hutto 1988). James et al. (1992), for instance, point out that apparent overall population declines in certain species of migrants may be the result of regional population fluctuations related to regional breeding conditions rather than to any populationwide phenomenon.

Table 1. Neotropical migrants classified by the U.S. Fish and Wildlife Service as endangered or as being of management concern

Species	Status	Source	Suspected cause
Brown Pelican	Endangered	USFWS 1991	DDT poisoning
American Bittern	Management concern	USFWS 1987	Habitat loss
Least Bittern	Management concern	USFWS 1987	Habitat loss
Reddish Egret	Management concern	USFWS 1987	Habitat loss
White-faced Ibis	Management concern	USFWS 1987	Habitat loss, environmental contaminants
Wood Stork	Endangered	USFWS 1991	Habitat loss
Northern Harrier	Management concern	USFWS 1987	Habitat loss
Peregrine Falcon	Endangered	USFWS 1991	DDT poisoning
Black Rail	Management concern	USFWS 1987	Habitat loss
Whooping Crane	Endangered	USFWS 1991	Hunting, habitat loss
Snowy Plover	Management concern	USFWS 1987	Habitat loss, human disturbance
Piping Plover	Endangered	USFWS 1991	Habitat loss
Eskimo Curlew	Endangered	USFWS 1991	Hunting
Gull-billed Tern	Management concern	USFWS 1987	Habitat loss, human disturbance
Roseate Tern	Management concern	USFWS 1987	Hunting, habitat loss
Least Tern	Endangered	USFWS 1991	Habitat disturbance
Black Tern	Management concern	USFWS 1987	Habitat loss
Olive-sided Flycatcher	Management concern	USFWS 1987	Habitat loss
S.W. Willow Flycatcher [a]	Endangered	USFWS 1992	Cowbirds, habitat loss
Vermilion Flycatcher	Management concern	USFWS 1987	Habitat loss, cowbirds
Loggerhead Shrike	Management concern	USFWS 1987	Habitat loss, human disturbance
Least Bell's Vireo [b]	Endangered	USFWS 1991	Habitat loss, cowbirds
Black-capped Vireo	Endangered	USFWS 1991	Cowbirds, habitat loss
Bachman's Warbler	Endangered	USFWS 1991	Winter habitat loss?
Golden-winged Warbler	Management concern	USFWS 1987	Habitat loss, hybridization
Golden-cheeked Warbler	Endangered	USFWS 1991	Habitat loss
Kirtland's Warbler	Endangered	USFWS 1991	Cowbirds, habitat loss
Cerulean Warbler	Management concern	USFWS 1987	Habitat loss, human disturbance

[a]Southwestern Willow Flycatcher is a subspecies of Willow Flycatcher.
[b]Least Bell's Vireo is a subspecies of Bell's Vireo.

Questions that may help to resolve these disagreements include the following: (1) Are these declines disproportionate when compared with those of resident species of the same guild? (2) Are there indications that the number of species of migrants showing declines is increasing? (3) Are there fundamental causes related to the strategy of migration connecting declines for the entire group, and if so, what can be done?

Table 2 lists another 79 species of Neotropical migrants for which declines have been reported in the literature. Again, these species derive from a variety of breeding habitats, but roughly half of them are forest dwellers. Declines in forest-dwelling species lie at the heart of the debate over causes of migrant conservation problems for several reasons: (1) forest area in the tropics has decreased dramatically in recent years (Sader and Joyce 1988), (2) forest area in the United States has increased during the same period (Powell and Rappole 1986), yet (3) changes to breeding-ground forest habitats have occurred as well (fragmentation, drought, etc.) (Askins et al. 1990), and (4) the dependence of forest-dwelling migrants on tropical forest is not well understood.

The methods used to assess changes in migrant populations fall into five major categories: (1) long-term studies of breeding populations at single sites where no obvious site alteration has occurred; (2) long-term (>20 yr), rangewide breeding bird surveys; (3) short-term (<20 yr), rangewide breeding bird surveys; (4) measurement of populations during migration (transient studies); and (5) regional breeding-ground surveys. Each of these methods presents a somewhat different perspective on the population measurement problem, along with its own potential biases, as discussed below.

Long-Term Studies at One or More Breeding-ground Sites

The first data indicating that populations of forest-dwelling Neotropical migrants might be in the kind of trouble predicted by Vogt (1970) derived from studies performed in and around Washington, D.C. (Briggs and Criswell 1979, Robbins 1979). Since that time, at least 20 long-term studies have been done in several parts of the country (Table 2). The results of most of these studies have been similar: Neotropical migrant populations declined while resident species generally showed no significant change (Askins et al. 1990:2, Briggs and Criswell 1979, Lynch and Whitcomb 1978).

Long-term studies comparing earlier with later counts involve several procedural difficulties: (1) locating a site where good records of breeding bird numbers have been kept, (2) documenting the fact that no significant vegetational changes have occurred at the site, (3) obtaining the information

Table 2. Neotropical migrants for which declines have been reported in the literature

Species	Type of study	Reference
Sharp-shinned Hawk	L	Temple & Temple 1976
Broad-winged Hawk	T	Bednarz et al. 1990, Titus et al. 1990
Swainson's Hawk	R	Risebrough et al. 1989
American Kestrel	T	Bednarz et al. 1990
Upland Sandpiper	R	Carter 1992, Osborne & Peterson 1984
Whimbrel	T	Howe et al. 1989
Sanderling	T	Howe et al. 1989
Short-billed Dowitcher	T	Howe et al. 1989
Franklin's Gull	B, S	Sauer & Droege 1992
Black-billed Cuckoo	R, S	R: Witham & Hunter 1992; S: Sauer & Droege 1992
Yellow-billed Cuckoo	B, L, S	B, S: Sauer & Droege 1992; L: Briggs & Criswell 1979, Robbins 1979
Short-eared Owl	R	Tate 1992
Whip-poor-will	S	Sauer & Droege 1992
Ruby-throated Hummingbird	L	Briggs & Criswell 1979, Robbins 1979
Western Wood-Pewee	L	Sharp 1985
Eastern Wood-Pewee	B, L, S	B, S: Sauer & Droege 1992; L: Briggs & Criswell 1979, Robbins 1979, Serrao 1985
Yellow-bellied Flycatcher	T	Hussell et al. 1992
Acadian Flycatcher	L	Briggs & Criswell 1979, Robbins 1979
Least Flycatcher	L, R, T	L: Ambuel & Temple 1982, Holmes & Sherry 1988; R: Witham & Hunter 1992; T: Hussell et al. 1992
Great Crested Flycatcher	L	Briggs & Criswell 1979, Criswell 1975, Robbins 1979
Scissor-tailed Flycatcher	B	Titus 1990
Barn Swallow	S	Sauer & Droege 1992
Sedge Wren	L	Temple & Temple 1976
Blue-gray Gnatcatcher	L	Ambuel & Temple 1982, Briggs & Criswell 1979
Veery	B, L, S, T	B, S: Sauer & Droege 1992; L: Askins & Philbrick 1987, Briggs & Criswell 1979, Leck et al. 1981, Litwin & Smith 1992, Serrao 1985; T: Hussell et al. 1992
Gray-cheeked Thrush	T	Hussell et al. 1992
Swainson's Thrush	L, R, T	L: Baird 1990, Hall 1984a, Holmes & Sherry 1988, Marshall 1988; R: Holmes & Sherry 1988; T: Hussell et al. 1992
Wood Thrush	B, L, R, S, T	B: Sauer & Droege 1992; L: Askins & Philbrick 1987, Briggs & Criswell 1979, Johnston & Winings 1987, Leck et al. 1981, Robbins 1979, Serrao 1985; R: Holmes & Sherry 1988, Witham & Hunter 1992; S: Howe et al. 1989; T: Hussell et al. 1992

Table 2—*cont.*

Species	Type of study	Reference
Gray Catbird	L, R, T	L: Briggs & Criswell 1979; R: Witham & Hunter 1992; T: Hussell et al. 1992
Cedar Waxwing	R	Witham & Hunter 1992
White-eyed Vireo	B, S	Sauer & Droege 1992
Solitary Vireo	R	Witham & Hunter 1992
Yellow-throated Vireo	L	Ambuel & Temple 1982, Briggs & Criswell 1979, Johnston & Winings 1987, Robbins 1979, Serrao 1985, Temple & Temple 1976
Warbling Vireo	L, R	L: Sharp 1985; R: Witham & Hunter 1992
Red-eyed Vireo	L	Askins & Philbrick 1987, Briggs & Criswell 1979, Johnston & Winings 1987, Kendeigh 1982, Leck et al. 1981, Litwin & Smith 1992, Robbins 1979, Serrao 1985
Tennessee Warbler	S, T	S: Sauer & Droege 1992; T: Jones 1986, Stewart 1987
Orange-crowned Warbler	T	Stewart 1987
Nashville Warbler	T	Hussell et al. 1992, Stewart 1987
Northern Parula	L, S, T	L: Briggs & Criswell 1979, Johnston & Winings 1987, Robbins 1979; S: Sauer & Droege 1992; T: Hagan et al. 1992
Yellow Warbler	L, R	L: Briggs & Criswell 1979; R: Witham & Hunter 1992
Chestnut-sided Warbler	L, R, S, T	L: Litwin & Smith 1992; R: Witham & Hunter 1992; S: Sauer & Droege 1992; T: Stewart 1987
Magnolia Warbler	L, T	L: Baird 1990, Hall 1984a, Litwin & Smith 1992; T: Stewart 1987
Cape May Warbler	S, T	S: Sauer & Droege 1992; T: Hagan et al. 1992, Stewart 1987
Black-throated Blue Warbler	L	Baird 1990
Yellow-rumped Warbler	R, T	R: Witham & Hunter 1992; T: Hagan et al. 1992, Hussell et al. 1992, Stewart 1987
Black-throated Green Warbler	L, S, T	L: Askins & Philbrick 1987, Baird 1990, Litwin & Smith 1992, Serrao 1985; S: Sauer & Droege 1992; T: Stewart 1987
Blackburnian Warbler	L, T	L: Hall 1984b, Holmes & Sherry 1988, Wilcove 1983; T: Stewart 1987
Pine Warbler	R	Witham & Hunter 1992
Prairie Warbler	B	Sauer & Droege 1992
Palm Warbler	T	Stewart 1987
Bay-breasted Warbler	S, T	S: Sauer & Droege 1992; T: Stewart 1987
Blackpoll Warbler	S, T	S: Sauer & Droege 1992; T: Hagan et al. 1992, Stewart 1987
Black-and-white Warbler	L	Askins & Philbrick 1987, Briggs & Criswell 1979, Leck et al. 1981, Litwin & Smith 1992, Robbins 1979, Serrao 1985

19

Table 2—*cont.*

Species	Type of study	Reference
American Redstart	L, R, S, T	L: Ambuel & Temple 1982, Briggs & Criswell 1979, Johnston & Winings 1987, Leck et al. 1981, Litwin & Smith 1992, Robbins 1979, Serrao 1985; R: Witham & Hunter 1992; S: Sauer & Droege 1992; T: Hussell et al. 1992
Worm-eating Warbler	L, S	L: Robbins 1979, Serrao 1985; S: Sauer & Droege 1992
Ovenbird	L, S, T	L: Ambuel & Temple 1982, Askins & Philbrick 1987, Baird 1990, Briggs & Criswell 1979, Johnston & Winings 1987, Leck et al. 1981, Litwin & Smith 1992, Robbins 1979, Serrao 1985; S: Sauer & Droege 1992; T: Hussell et al. 1992
Northern Waterthrush	L, T	L: Litwin & Smith 1992; T: Hussell et al. 1992, Stewart 1987
Louisiana Waterthrush	L	Briggs & Criswell 1979, Robbins 1979
Kentucky Warbler	B, L, S	B, S: Sauer & Droege 1992; L: Briggs & Criswell 1979, Johnston & Winings 1987, Robbins 1979
Common Yellowthroat	L, R, S, T	L: Briggs & Criswell 1979, Sharp 1985; R: Witham & Hunter 1992; S: Sauer & Droege 1992; T: Hagan et al. 1992
Hooded Warbler	L	Askins & Philbrick 1987, Briggs & Criswell 1979, Johnston & Winings 1987, Leck et al. 1981, Litwin & Smith 1992, Robbins 1979, Serrao 1985, Wilcove 1983
Wilson's Warbler	L, T	L: Sharp 1985; T: Hussell et al. 1992, Stewart 1987
Canada Warbler	L, S, T	L: Baird 1990, Litwin & Smith 1992, Wilcove 1983; S: Sauer & Droege 1992; T: Hussell et al. 1992, Stewart 1987
Yellow-breasted Chat	B, L	B: Sauer & Droege 1992; L: Sharp 1985
Scarlet Tanager	L, R, S	L: Ambuel & Temple 1982, Baird 1990, Briggs & Criswell 1979, Litwin & Smith 1992, Robbins 1979; R: Witham & Hunter 1992; S: Sauer & Droege 1992
Rose-breasted Grosbeak	L, R, S, T	L: Baird 1990; R: Witham & Hunter 1992; S: Sauer & Droege 1992; T: Hussell et al. 1992
Black-headed Grosbeak	L	Sharp 1985
Lazuli Bunting	L, S	L: Sharp 1985; S: Sauer & Droege 1992
Indigo Bunting	B, L, R, S	B, S: Sauer & Droege 1992; L: Baird 1990, Briggs & Criswell 1979; R: Witham & Hunter 1992
Painted Bunting	B, S	Sauer & Droege 1992
Dickcissel	B	Sauer & Droege 1992
Rufous-sided Towhee	L, R, T	L: Sharp 1985; R: Witham & Hunter 1992; T: Hagan et al. 1992, Hussell et al. 1992

Table 2—cont.

Species	Type of study	Reference
Chipping Sparrow	T	Hussell et al. 1992
Bobolink	B, R	B: Robbins et al. 1986, Sauer & Droege 1992; R: Bollinger & Gavin 1992
Eastern Meadowlark	B, R	B: Droege 1991; R: Witham & Hunter 1992
Western Meadowlark	B	Droege 1991
Orchard Oriole	B	Sauer & Droege 1992
Baltimore Oriole [a]	R	Ambuel & Temple 1982, Witham & Hunter 1992
Bullock's Oriole [a]	B	Sauer & Droege 1992

Notes: B = long-term (>20 yr), rangewide Breeding Bird Surveys; L = long-term (>20 yr) comparative studies at one or more sites; R = regional studies; S = short-term (<20 yr), rangewide Breeding Bird Surveys; T = counts of birds in transit between breeding and wintering ground.

[a]These are subspecies of the Northern Oriole.

necessary to repeat the methods used to make the early counts, and (4) replicating the original observer's observational skills.

The strength of these studies is in their concordance (Askins et al. 1990:4–5). When only a few studies were done at widely separated sites using quite different methods, the data were not convincing, particularly since they appeared to be largely contradicted by the U.S. Fish and Wildlife Service's national Breeding Bird Survey. However, at least 17 separate studies have now been done showing remarkably similar declines for Neotropical migrants (Table 2), and the Breeding Bird Survey trend analysis for the period 1980–1988 now shows a striking degree of similarity to the results of these long-term, single-site surveys (Table 2).

Rangewide Breeding Bird Surveys

The U.S. Breeding Bird Survey was initiated by the U.S. Fish and Wildlife Service in 1966 and is based on roadside counts performed each summer by large numbers of volunteers covering specified routes through representative habitats (Robbins et al. 1986). Survey procedures are similar to those used in other singing bird counts and are subject to problems regarding basic assumptions (Rappole et al. 1993). Because of these problems and the lack of in-depth habitat information, the counts are not useful as population determinants. However, the counts may be useful as a gross measure of rangewide trends for population increases or decreases. At least it is clear that if a species is absent from a piece of habitat it will not be heard (though the reverse may not be true), so if migrant populations are undergoing serious declines, the survey should eventually reflect those changes, though the effects may be masked at first by floaters joining the breeding population (Rappole et al. 1983:70).

In this review, we have included the Breeding Bird Survey trend information on Neotropical migrants as presented in Sauer and Droege (1992) for 1966–1988, which they refer to as "long-term," along with the trends for 1978–1988, which they refer to as "short-term." This distinction is important because, for several species, the more recent information has shown trends that were not apparent during the first years of the survey (Robbins et al. 1989b). In fact, trends for several species showed increases from 1966 to 1978.

Transient Studies

A few studies have been done to measure Neotropical migrant populations during migration on the basis of repeated sampling of grounded migrants using lines of mist nets or visual surveys (Hussell et al. 1992). Because of the large annual fluctuations at stopover sites, these studies require considerable effort and several years of information to determine any significant trends. Biases include the following: (1) the birds stopping at the site represent a sample drawn from a regional base of unknown size that probably fluctuates from year to year, and (2) weather plays a major role in determining the number of birds that are sampled in a given season.

Because of these problems, one would predict that only a very large positive or negative change would be likely to appear as a consistent trend using this sampling procedure. Small changes would be masked by the annual variation. Despite this characteristic of the data, significant trends have been detected for several species in three different groups of migrants: songbirds (Hagan et al. 1992, Hussell et al. 1992, Jones 1986, Stewart 1987), raptors (Bednarz et al. 1990), and shorebirds (Howe et al. 1989) (Table 2).

Weather radar can also be used to crudely measure the density of migrants past a given point (Gauthreaux 1992). Radar studies cannot distinguish species. Each individual migrant in flight appears as a blip on the screen. The size of the blip depends on the size of the bird and its distance from the radar unit, among other factors. For large numbers of birds in a flock, the blips merge to block out entire areas of the screen. Films of radar screens can be used to document passage densities on nights when large numbers of birds are in transit versus those when few birds are passing. Gauthreaux

(1992:99) documented a remarkable difference in total number of nights on which major migratory flights across the Gulf of Mexico occurred when comparing radar screen films for the period 8 April–15 May for 1965–1967 versus 1987–1989; he found a decrease of almost 50% for the later years.

Regional Breeding-ground Studies

By definition, these studies pertain to only a portion of a species' range. They are therefore subject to forces that may be quite different for the species as a whole. In fact, reported population changes based on regional work can be the opposite of those seen in national trends (James et al. 1992). Nevertheless, regional population studies can provide important insights and potential early warning signs, particularly when the data are derived from a variety of different studies.

As an example, in a study of the Bobolink, Bollinger and Gavin (1992) presented regional Breeding Bird Survey information, intensive single-site studies, and regional habitat assessments to build a convincing case for the existence and causes of Bobolink population declines in the northeastern United States.

Explanations for Observed Population Declines

Declines have now been documented for 107 species of Neotropical migrants (Tables 1 and 2), about a third of the total of all Neotropical migrant species. Some of these declines have been documented by a single study or methodology, whereas others have been documented by several studies using different methodologies. Declines in Wood Thrush populations, for instance, have been reported in seven long-term, single-site studies, one long-term study of transient numbers, two regional studies, and both short- and long-term rangewide breeding bird surveys.

There is as yet no consensus among scientists regarding the existence of a threat to populations of long-distance migrants that winter in the Neotropics (Hagan and Johnston 1992). The declines that have been observed have been attributed to a variety of causes, some of which (e.g., periodic drought cycles) would not have long-term effects on species' populations (Table 3). Part of the difficulty arises from the fact that migrants are exposed to multiple risks, and different species are threatened by different factors.

Reasons that have been presented to explain observed migrant declines are discussed below.

Table 3. Reasons given for apparent declines in populations of Neotropical migrants

Reason	Source[a]
Loss of breeding-ground habitat	USFWS 1987
Habitat fragmentation	
Island biogeography effects	Lynch & Whitcomb 1978
Area effect	Galli et al. 1976
Brood parasitism by cowbirds	Brittingham & Temple 1983
Nest predation	Wilcove 1985
Loss of critical microhabitats	Robbins et al. 1989a
Interspecific competition	Butcher et al. 1981
Successional changes to breeding-ground habitat	Litwin & Smith 1992
Breeding habitat alteration by white-tailed deer	Baird 1990
Contaminant poisoning	Newton 1979
Normal population fluctuation	
Variation in food resources on breeding sites	Holmes et al. 1986
Climatic cycles on the breeding ground	Blake et al. 1992
Procedural biases	
Assumptions	Rappole et al. 1993
Analytical errors	James et al. 1992
Sampling errors	Hutto 1988
Stopover habitat alteration	Howe et al. 1989
Winter habitat alteration	Rappole et al. 1983:73–75

[a]Additional references are given in the text and in the extensive review by Askins et al. (1990).

Loss of Breeding Habitat

Perhaps the single factor that all parties agree will result in declines of Neotropical migrants is loss of breeding habitat. The U.S. Fish and Wildlife Service, in their assessment of migratory bird populations, lists habitat loss as the primary perceived cause of population decline for 24 of the 28 species that they have selected to be "endangered" or "of management concern" (U.S. Fish and Wildlife Service 1987:15). The report does not make a distinction between loss of breeding and nonbreeding habitat, but information in the species accounts makes clear that loss of breeding habitat is generally what is meant.

One can hardly argue with the logic that loss of breeding habitat can result in declining populations. However, loss of breeding habitat is sometimes presented as an explanation when data are equivocal or even nonexistent. For example, in the case of the Least Bell's Vireo, an endangered subspecies, Goldwasser et al. (1980:742) note that whereas loss of breeding habitat has been a factor in the decline of the population, many formerly occupied areas of apparently suitable habitat are now unoccupied.

For some of the species in Table 1 and many of those listed in Table 2, loss of breeding habitat does not provide a sufficient explanation for observed declines, particularly for those species that breed in eastern deciduous forests, a habitat type that has increased substantially since the turn of the century (Birch and Wharton 1982, Brooks and Birch 1988, Powell and Rappole 1986). In fact, it is this reversion of large amounts of farmland to forest that has been cited as a major reason for the disappearance of species dependent on old-field habitat in the eastern United States, such as Bachman's Sparrow (LeGrand and Schneider 1992:307–308).

Habitat Fragmentation

Habitat fragmentation has come to be used as a general term to refer to all of the various effects on species diversity associated with size and degree of isolation of a particular piece of habitat. More attention has been focused on this aspect of migratory bird breeding biology than on any other in recent years. Reviews of the extensive pertinent literature appear in Robbins et al. (1989a), Askins et al. (1990), Finch (1991), and various papers in the symposium volume by Hagan and Johnston (1992).

The principal evident effects of habitat fragmentation are decreases in species richness and decreases in species density (Askins et al. 1990:20). The causes of these changes are not clear, mostly because it is very difficult to limit the number of variables tested at the landscape scale. The following explanations have been proposed to explain the evident effects of habitat fragmentation.

Island Biogeography Theory. Several biologists have theorized that island size and distance from the mainland are somehow related to the number of resident species (Arrenhius 1921, Cain and Castro 1959, Gleason 1922). MacArthur and Wilson (1967) developed a model and theoretical explanation for this phenomenon which states that the number of species in a community is dependent on a balance between the relative rates of colonization and extinction. These rates are affected by the size (area) occupied by the community and its degree of isolation. The model predicts that in large sites with a low degree of isolation, an equilibrium between colonization and extinction will be established that is at the high end in terms of species richness, whereas small, isolated sites will show an equilibrium that is at the low end. Lynch and Whitcomb (1978) hypothesized that island biogeography theory could provide an explanation for the observed declines of breeding Neotropical migrants in forested sites in Maryland and Washington, D.C.

25

Area Effect. Testing island biogeography concepts on Neotropical migrant populations has been difficult, and although no studies have established a strong causal link between degree of isolation and Neotropical migrant species diversity, several studies have indicated that size of the site is an important factor affecting species diversity of migrant breeding populations (Ambuel and Temple 1983, Askins et al. 1990, Freemark and Collins 1992, Galli et al. 1976). Nevertheless, even in these studies it is not clear whether the "area effect" is due to those factors predicted by the model (i.e., random events of colonization and extinction) or to other events associated with reduction in area (e.g., increased predation, increased brood parasitism, decreased microhabitat heterogeneity, etc.) (Askins et al. 1990:23).

A related question is how breeding-bird populations on a site are affected when extensive regional habitat changes occur but the site itself remains apparently unchanged. Askins and Philbrick (1987), for instance, reported that breeding populations of forest-dwelling migrants in New England declined between 1953 and 1976, a period during which regional forest area also declined. However, several of the same species showed increases from 1976 to 1985, when regional forest area increased.

Brood Parasitism by Cowbirds and Nest Predation. In some parts of the United States, rates of brood parasitism by Brown-headed Cowbirds and nest predation are higher for forest-dwelling migrants on small forest patches than on large ones, leading some investigators to propose that the disappearance of migrants from small forest patches could be caused by those factors (Brittingham and Temple 1983, Lynch and Whitcomb 1978, Robinson 1992, Whitcomb 1977, Wilcove 1985). Temple and Cary (1988) formulated a simulation model that predicts eventual abandonment of poor-quality (small) habitat patches because of low reproductive success caused by cowbirds or nest predation.

A decrease in return rates to breeding territories by adult birds is an expected result of reproductive failure. Because of immigration from higher-quality habitats (Pulliam 1988), however, disappearance of a species from these poor-quality sites may not occur unless the species' entire population is declining. As long as a stable population of adults is maintained, there may be enough excess individuals coming north each spring to locate and fill marginal habitats. Only when the entire population is declining should marginal sites begin to appear empty (Wilcove and Terborgh 1984). Robinson (1992), for instance, still had populations of Neotropical migrants on sites in southern Illinois that had existed as fragments for several years despite nest predation and parasitism rates of 70 to 80% and annual return rates averaging 15%. Similarly, Temple and Cary (1988) noted that fragmentation of the habitats used in their study in Wisconsin reached a maximum in the early 1900s; forest patch isolation has actually decreased since that time (Curtis 1956, Temple and Cary 1988:346).

Except in those situations where fragmentation is characteristic of most or all of the breeding habitat for a species, as is the case in Temple and Cary's (1988) model, disappearance of species from marginal habitats may be a result of, rather than a cause of, declining populations (Askins et al. 1990: 26). If this hypothesis is correct, and if size is the best predictor of quality (Askins et al. 1990, Freemark and Collins 1992), then the "area sensitivity" of a species will not be a fixed number (Robbins et al. 1989a); rather, it will increase as the total population decreases until only the largest, best-quality habitats are occupied.

Loss of Critical Microhabitats. On the basis of sampling probability alone, a small piece taken from a larger, heterogeneous whole is unlikely to represent the entire range of heterogeneity present in the whole. Similarly, forest remnants are unlikely to represent the same range of microhabitats that were present in large, unbroken forest (Robbins et al. 1989a). It is logical to conclude that this factor, along with factors that reduce reproductive success, can combine to affect the overall quality of a patch. Although not all patches are of low quality simply because of small size, they are more likely to be of low quality than a larger patch.

Interspecific Competition. Whitcomb (1977) suggested that residents, mostly presumed food "generalists," might be favored over migrant "specialists" in fragmented communities (though see Morse 1971). Butcher et al. (1981) hypothesized that "edge"/"edge-interior" species could have a competitive advantage over "forest interior" species in fragmented habitats, and Askins and Philbrick (1987) found a significant, negative correlation between abundance of these two species groups.

Some authors have argued that habitat fragmentation is a primary cause of migratory bird declines (Butcher et al. 1981, Lynch and Whigham 1984, Robbins 1979, Sherry and Holmes 1992:432), yet several questions remain, including the following:

1. Why have studies discovered declines in the absence of any evidence of recent fragmentation (Briggs and Criswell 1979, Johnston and Winings 1987, Litwin and Smith 1992)?

2. Why have declines been recorded in regions of the United States where fragmentation is a relatively minor phenomenon? Although it is clear that fragmentation is an important conservation problem in some parts of the country (e.g., southern Wisconsin, Maryland, Illinois) (Lynch and Whitcomb 1978, Robbins et al. 1989a, Robinson 1992, Temple and Cary 1988), it appears

to be relatively unimportant in others (West Virginia, eastern California, northwestern Pennsylvania, New Hampshire), yet declines have been recorded in these regions as well (Baird 1990, Hall 1984b, Marshall 1988, Sherry and Holmes 1992).

3. Why is it that in most studies migrants show declines whereas resident species do not? The following explanations for this observation have been given: (a) permanent residents are edge or open-country species that occur on forest sites because of artificial openings, (b) permanent residents have superior dispersal capabilities, (c) permanent residents have higher reproductive capacity (Lynch and Whitcomb 1978), and (d) many permanent residents are hole nesters, whereas all but two migrants are not. All of these ideas, however, are broad generalizations that do not apply to all permanent residents as a group when compared with migrants as a group.

It has also been argued that declines result because of the increasing number of small, low-quality breeding sites. If smaller patches are of poorer quality for migrants, however, they are also likely to be of poorer quality for residents, and for many of the same reasons (increased nest parasitism and nest predation, decreased microhabitat heterogeneity, etc.). We suggest that a major reason for resident persistence in patches (i.e., an evident lack of area sensitivity) is that overall, resident populations are not declining whereas those of migrants are. If resident populations are not declining, there is likely to be a sufficient number of excess individuals to continually recolonize poor patches (Pulliam 1988).

There is, in fact, only one factor that all Neotropical migrants share and that permanent residents do not—long-distance migration.

Successional Changes to
Breeding-ground Habitat

Several long-term, single-site studies have documented declines in Neotropical migrant populations in the absence of a change in the amount of fragmentation during the study period. In some of these studies, fragmentation had occurred but did not increase during the study (Ambuel and Temple 1983, Johnston and Winings 1987, Kendeigh 1982, Leck et al. 1988, Litwin 1986, Serrao 1985, Witham and Hunter 1992); in others, the studies were performed in extensive woodlands that had never been fragmented (Baird 1990, Hall 1984b, Holmes and Sherry 1988, Marshall 1988). Two studies have shown flat or increasing trends in migrants (Askins and Philbrick 1987, Wilcove 1988). Unfortunately, many of these investigations have included no data on the structure and composition of vegetation for the entire period over which the bird populations were surveyed, and the early surveys were often done by one person and later censuses by another, so that even a subjective assessment of change in the plant community was not available. There are, however, some important exceptions, and for these succession has been suggested as a significant factor in Neotropical migrant changes (Askins and Philbrick 1987, Hall 1984a, Litwin 1986, Litwin and Smith 1992).

Hall (1984a) studied a red-spruce forest on Shaver Mountain, West Virginia. Originally a mature red-spruce forest, the area was logged sometime between 1905 and 1915. Natural reseeding from neighboring spruce stands occurred shortly thereafter, as indicated by growth rings of representative individuals. Hall performed the surveys himself from 1959 to 1983 and so was able to correlate observed structural changes in the vegetation with commensurate changes in the bird community. He noted that maturation of the spruce was probably the main factor in the decline of the Magnolia Warbler and disappearance of the Chestnut-sided Warbler, Rufous-sided Towhee, and Purple Finch. However, he found that Magnolia Warblers on neighboring stands of mature red spruce also declined, despite a lack of obvious successional changes.

Litwin and Smith (1992) also suggested successional changes as an explanation for changes in the Neotropical bird community on their central New York site, though the successional changes on their site were quite mixed. They found that "a canopy that was broken and relatively open [in 1950] had closed [by 1980]. Canopy closure eliminated early successional tree species and reduced shrub density. Forest gaps created by selective logging and trails have since closed, but others have been created by loss of elm and beech" (Litwin and Smith 1992:490). Open areas on the site had also grown up to white pine and mixed second growth. Litwin and Smith concluded that these changes must have caused the decline or disappearance of the Red-eyed Vireo, Scarlet Tanager, Veery, Ovenbird, Northern Waterthrush, Canada Warbler, Black-and-white Warbler, Black-throated Green Warbler, Magnolia Warbler, Chestnut-sided Warbler, and American Redstart. They were really forced to this conclusion because there was no evidence of increased fragmentation or isolation of the site, and Breeding Bird Surveys for the region (1967–1981) generally showed an increasing or stable trend for nine of their species. Only for the Canada Warbler did the regional and local trends agree in showing a significant negative trend. However, it should be noted that, except possibly in the case of the Chestnut-sided Warbler, none of the reported vegetational changes could be positively associated with the disappearance of any of the listed species, and that the recent (1980–1988)

Table 4. Species in which declines or disappearances from eastern deciduous forest sites were first detected by Briggs & Criswell (1979)

Species	BBS 1966–1978[a]	BBS 1978–1988[b]
Yellow-billed Cuckoo	+[c] ***	−[e] ***
Eastern Wood-Pewee	−	−***
Acadian Flycatcher	+	−
Great Crested Flycatcher	+	−
Blue-gray Gnatcatcher	+	+
Veery	+	−***
Wood Thrush	+	−***
Red-eyed Vireo	+***	+
Yellow-throated Vireo	−	−
Northern Parula	+	−***
Black-and-white Warbler	+	+
American Redstart	+	−*
Prothonotary Warbler	+***	−
Ovenbird	+	−***
Louisiana Waterthrush	−	+
Kentucky Warbler	−	−**
Common Yellowthroat	+	−***
Hooded Warbler	+	+
Scarlet Tanager	+***	−**

Notes: + = a positive population trend; − = a negative population trend.
[a]Trend information based on U.S. Breeding Bird Survey data for 1966–1978 from Robbins et al. 1989b.
[b]Trend information based on U.S. Breeding Bird Survey data for 1978–1988 from Sauer & Droege 1992.
*$P < 0.10$
**$P < 0.05$
***$P < 0.01$

Breeding Bird Survey information showed significant rangewide declines for 7 of their 11 declining/disappearing species (Table 2).

These findings raise an interesting point. The first reports of migrant declines from single sites that had been studied for long periods were those published by Briggs and Criswell (1979) (Table 4). They found declines in 19 species over the 30 years they recorded bird populations, yet they could not point to any obvious changes in the sites to account for the population changes (though others have speculated that some change in the "matrix" of surrounding environments caused the declines). As in the study by Litwin and Smith (1992), rangewide Breeding Bird Survey results for 1966 to 1978 did not show significant, declining trends for most of the species (Table 4). However, the Breeding Bird Survey results for 1978 to 1988 showed significant declines for 10 of the 19 species.

Breeding Habitat Alteration
by White-tailed Deer

Recent studies have shown that white-tailed deer, as well as other herbivores, can have a major impact on the vegetation of forest communities (Alverson et al. 1988, Warren 1991). The reduction of understory vegetation may have an effect on birds using this habitat (Baird 1990, deCalesta 1994, DeGraaf et al. 1991, McShea and Rappole 1992). Ultimately, lack of regeneration of dominant tree species could affect midlevel and canopy species as well. We are currently evaluating these hypotheses using randomly located enclosures to test the response of migratory species to the exclusion of deer from patches of Appalachian oak forest (McShea and Rappole 1992). This study is still in its early stages.

Contaminant Poisoning

Contaminant poisoning was a major cause of ecosystem dysfunction in the 1950s and 1960s (Carson 1962), causing the decline or extirpation of many raptor and wetland species (Newton 1979, Risebrough and Peakall 1988). As of the late 1980s, populations of many of these birds had recovered, at least to a degree, as some of the most toxic substances have been banned. There continue to be local instances of contaminant poisoning, however, and there is evidence that some migrant species are picking up significant levels of contaminants during migration or on the wintering grounds (Henny et al. 1982, Lincer and Sherburne 1974).

Normal Population Fluctuations

All populations undergo normal fluctuations in their numbers. Some of these fluctuations are periodic, such as the responses to wet/dry cycles (Norwine and Bingham 1985) or prey density (Taylor 1984:114–124), whereas others are episodic in response to such events as disease outbreaks or severe wintering-ground weather (Wunderle et al. 1992). Some researchers have argued that not enough long-term data have been gathered to eliminate the possibility that observed declines are not the sort of variations to be expected of populations in changing environments (Holmes et al. 1986).

Variation in Food Resources on Breeding Sites. Important prey species for insectivorous birds show periodic cycles of population variation that presumably affect their predators. Such cycles have also been documented for species of warblers that exploit spruce budworms. These species show dramatic population increases during budworm outbreaks, followed by declines during budworm population lows (Morse 1989:118). As in the case of climatic variation, prey cycles should have relatively short-term and local or regional effects. Some researchers have argued that the contrast between the increases in many forest-dwelling migrants seen in the first decade of the U.S. Breeding Bird Survey followed by the decreases in the second decade are a result of this sort of predator-prey cycle. In fact, the Canadian Breeding Bird Survey has shown numerous short-term fluctuations but no consistent increasing or decreasing trend for Neotropical migrants over the past 25 years, though many species that migrate to the United States have shown declines (Erskine et al. 1992).

Climatic Events on the Breeding Ground. Downpours, hail, strong winds, and cold snaps, as well as longer-term cycles of temperature and precipitation, can have devastating local or even regional effects on bird populations. Blake et al. (1992) reported an apparent correlation between drought and declining populations of long-distance migrants in the Midwest from 1985 to 1989, although curiously they did not find comparable declines in short-distance migrants or residents. Knopf and Sedgwick (1987) recorded declines in Brown Thrashers and Rufous-sided Towhees in riparian habitats a year after floods along the South Platte River in Colorado. Similarly, Vega-Rivera and Rappole (1994) found that drought conditions in south Texas study sites in 1989 caused nearly complete abandonment of these sites by local breeding birds in 1990.

It should be possible to separate the local and regional effects to be expected from climatic and predator-prey cycles from rangewide trends by taking the following measures: (1) continuing to measure populations through the natural course of climatic variation, (2) contrasting trends from one part of the range with another, or (3) monitoring trends in numbers of transients drawn from broad areas of the range. Efforts to assess bird populations rangewide are still quite new in terms of long-term natural cycles. It may take some time to sort out the many complex factors that affect migrant populations unless some single factor overwhelms the system.

Procedural Biases

Assumptions. Nearly all the long-term studies and breeding bird counts performed to date have been based on singing birds. Such counts are subject to potential biases. Song-count accuracy rests on three assumptions: (1) the observer is able to detect and identify all territorial birds by song, (2) each singing bird represents the male of a pair of adult, breeding individuals, and (3) all males that are paired sing at the same rate throughout the course of the counting period. All three assumptions can be incorrect under some circumstances (Rappole et al. 1992, Rappole and Waggerman 1986).

Analytical Error. James et al. (1992) maintain that the statistical analyses used by the U.S. Fish and Wildlife Service Breeding Bird Survey (Sauer and Droege 1992) to analyze their data may be inappropriate for distinguishing regional from national population trends.

Sampling Error. For any statistical procedure involving sampling, the experimenter sets a level of probability for rejection of the null hypothesis (H_0), usually 5%. By definition, then, there is a 5% chance of making a Type I error, i.e., rejecting the null hypothesis when it is actually true. For all of the population measurement studies, the null hypothesis is that "observed changes in the population are the result of random variation." As Hutto (1988) has pointed out, there may be a greater than 5% chance that studies reporting declines are due to Type I errors because studies showing no declines may be less likely to be published. That is, out of 100 studies, we can expect 5 studies to show declines on the basis of chance alone, but these 5 studies are more likely to be published than the other 95.

Stopover/Wintering-ground Habitat Loss

Suggestions that migratory bird populations could decline as a result of stopover and winter habitat loss were made before such habitat losses had been documented, based on the following syllogism: if migratory birds are limited by availability of tropical habitats, and if tropical habitats are being destroyed, then migratory bird populations will decline (Morton 1980a, Rappole 1974, Vogt 1970). This conclusion stimulated considerable debate: first, regarding the degree of dependency of migrants on their stopover and wintering habitats; second, regarding the relative amounts of habitat loss in breeding versus nonbreeding habitats; and third, regarding the evidence for migratory bird declines. The logic of the syllogism assumes additional weight as information continues to be accumulated indicating migrant declines (Askins et al. 1990) and increasing rates of tropical habitat loss (Sader and Joyce 1988). None of these questions have been thoroughly resolved to the satisfaction of all investigators. At some point it will be possible to reach an agreement concerning the existence of declines, and it may also be possible to quantify the actual extent of remaining breeding and wintering habitat. However, it will continue to be difficult to assign causes for declines to one portion of the life cycle. As pointed out by Blake et al. (1992:428), "Although destruction of tropical forests and conversion to pasture and cropland has had and will continue to have (with increasingly detrimental effects) adverse impacts on many species . . . , direct evidence for such effects on long-distance migrants is limited." One could go further and say that direct evidence is absent, and likely to remain so. In fact, short of complete removal of winter habitat followed by disappearance of the dependent species, it is difficult to imagine what would constitute "direct evidence."

Conclusions

Establishment of a direct link between the reported declines detailed above and the various suggested breeding, stopover, or wintering-ground effects is problematic (Askins et al. 1990:39, Wilcove and Terborgh 1984). Obviously, measuring changes in populations of migrating, wintering, or breeding birds does not provide information on where the causes of the changes are located. However, population changes will have different characteristics depending on where the regulation is occurring, and documentation of the presence or absence of these characteristics may help to determine which portion of the life cycle is most critical in terms of population control. Such studies should be pursued aggressively. The answers are of considerably more than academic interest.

SPECIES ACCOUNTS

Pied-billed Grebe
Podilymbus podiceps

Canada (Quebec): Grèbe à Bec Bigarré

Guatemala: Zambullidor Piquipinto

Mexico: Zambullidor Piquigrueso

Puerto Rico: Zaramago

West Indies: Chien d'Eau, Grand
Plongeon, Plongeon, Zaramago,
Zaramagullon, Zaramagullon
Grande

RANGE. Breeds in southeastern Alaska and from central Canada south locally through temperate North America. In Mexico, local in summer, widespread in winter. Winters through most of breeding range from southern British Columbia and the central United States southward, casually farther north. Winters throughout West Indies, Central and South America from Colombia, northern Venezuela, and Guyana south to Chile and Argentina to Santa Cruz; also Aruba, Curaçao, Bonaire, Trinidad, Tobago. Northern birds winter south to Panama. Panamanian birds breed and winter in Panama.

STATUS. Common; most widespread grebe in North America. Declining in the western region of the United States and Canada.

37

HABITAT. Inhabits ponds with much shoreline and emergent vegetation, marshes with areas of open water 30 to 75 cm deep, and marshy inlets and bays. Found on freshwater ponds, sloughs, flooded areas, marshy parts of lakes and rivers, occasionally on brackish lagoons or estuarine waters with weak tidal influence. A solitary nester; generally only one pair nests per pothole. Constructs a floating nest, usually in shallow water but sometimes in water 1–2 m deep, and well concealed in emergent vegetation. Builds nest around or anchored to reeds, rushes, or bushes and usually within 15 m of open water.

SPECIAL HABITAT REQUIREMENTS. Fresh marshes, sluggish streams, ponds 7 ha or less, with some emergent vegetation.

Further Reading. United States and Canada: Faaborg 1976, Glover 1953, Godfrey 1966, Palmer 1962, Sealy 1978a, Terres 1980, Wetmore 1924. Mexico and Central America: Edwards 1972, Land 1970, Monroe 1968, Peterson and Chalif 1973, Ridgely and Gwynne 1989, Stiles and Skutch 1989. South America: Hilty and Brown 1986, Meyer de Schauensee 1966. Caribbean: Bond 1947, Herklots 1961, Raffaele 1989.

Eared Grebe
Podiceps nigricollis

Canada (Quebec): Grèbe à Cou Noir

Costa Rica and Mexico: Zambullidor Mediano

Guatemala: Zambullidor Cuellinegro, Zambullidor Orejón

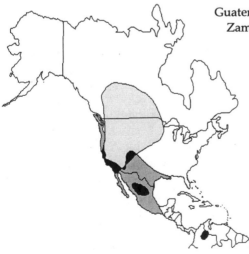

RANGE. Breeds from south-central British Columbia and Manitoba south to Baja California and south-central Texas. Winters inland from California, Nevada, Utah, New Mexico, and central Texas and on the Pacific Coast from southern British Columbia south to Guatemala. Casual in eastern United States and also in Old World. Occasionally winters in Costa Rica, rarely El Salvador and Guatemala. In Mexico, widespread in winter except on Yucatán Peninsula.

STATUS. Large breeding colonies and winter flocks common. Long-term population increase, especially in the western United States.

HABITAT. When breeding, inhabits marshy lakes and ponds and large pools in streams or rivers in the prairie region of North America. Less typically inhabits marshes with some open water. During winter found on salt lakes, bays, estuaries, and seacoasts. All records south of Mexico have been on freshwater lakes. Nests in compact colonies or sometimes singly, in sheltered areas or in shallow water away from emergent vegetation. Builds a floating nest that is a platform of marsh vegetation built up from the bottom or anchored to reeds.

SPECIAL HABITAT REQUIREMENTS. Freshwater lakes and ponds larger than 7 ha, with shallow margins and emergent vegetation.

Further Reading. United States and Canada: Faaborg 1976, Farrand 1983a, Grinnell and Miller 1944, Low and Mansell 1983, Palmer 1962, Terres 1980, Verner and Boss 1980, Wetmore 1924. Mexico and Central America: Edwards 1972, Land 1970, Peterson and Chalif 1973, Stiles and Skutch 1989.

Western Grebe

Aechmophorus occidentalis
(includes Clark's Grebe,
Aechmophorus clarkii)

Canada (Quebec): Grèbe de l'Ouest
Mexico: Zambullidor Achichilique

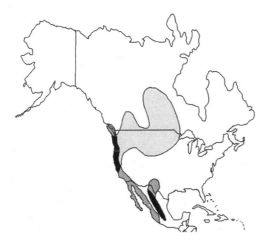

RANGE. Breeds from southeastern Alaska and south-central British Columbia to southwestern Manitoba south to southern California, New Mexico, northwestern Iowa, and western Minnesota. Winters along the Pacific Coast from southern British Columbia to Baja California, and from Utah, Colorado, New Mexico, and western and southern Texas south into Mexico. In Mexico, a migrant and winter visitor along western coasts south to Jalisco; locally in the interior from Chihuahua to Puebla.

STATUS. Locally abundant.

HABITAT. Inhabits fairly extensive areas of open water bordered by tall emergent plants. Found on marshes, lakes, and bays; in winter may be found on salt, brackish, and fresh water where small fishes are abundant. In Mexico on large lakes, brackish lagoons, and coastal marine waters. Nests in colonies of hundreds, even thousands, of pairs at some lakes in extensive areas of shallow (30 cm) open water bordered by tules or rushes.

SPECIAL HABITAT REQUIREMENTS. Open, fresh lakes bordered by rushes or tules.

Further Reading. United States and Canada: Godfrey 1966, Grinnell and Miller 1944, Lindvall and Low 1982, Palmer 1962, Terres 1980, Wetmore 1924. Mexico: Edwards 1972, Peterson and Chalif 1973.

American White Pelican

Pelecanus erythrorhynchos
(formerly White Pelican)

Canada (Quebec): Pélican Blanc

Costa Rica: Pelícano Blanco Americano

Guatemala, Mexico, and Puerto Rico: Pelícano Blanco

West Indies: Alcatraz Blanco

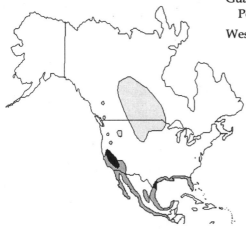

RANGE. Breeds from south-central British Columbia to central Manitoba and southwestern Ontario south locally to extreme northern California east to northern Colorado, northeastern South Dakota, and southwestern Minnesota. Sporadic on central coast of Texas and from central to southern California. Winters along the Pacific Coast from central California and southern Arizona south to Central America, and from Florida and the Gulf States south; casually throughout breeding range in western North America. In Mexico, winters mainly along both coasts. Rare winter visitor in Guatemala. Casual in the West Indies on the Bahamas, Cuba, Isle of Pines, Grand Cayman, Jamaica, and Puerto Rico.

STATUS. Locally common throughout breeding range.

HABITAT. Found primarily on lakes, also rivers, estuaries, and shallow coastal bays and inlets. Loafs on beaches, sandbars, and driftwood. Nests on

41

the ground in colonies of a few to several hundred pairs on small, relatively flat islands, without tall (>1 m) obstructions, with loose earth suitable for heaping into nest mounds. Occasionally nests on floating islands of marsh plants.

SPECIAL HABITAT REQUIREMENTS. Islands for nesting; protection from human disturbance.

Further Reading. United States and Canada: Godfrey 1966, Knopf 1979, Knopf and Kennedy 1981, Lingle and Sloan 1980, Palmer 1962, Terres 1980. Mexico and Central America: Edwards 1972, Land 1970, Peterson and Chalif 1973, Ridgely and Gwynne 1989, Stiles and Skutch 1989.

Brown Pelican
Pelecanus occidentalis

Canada (Quebec): Pélican Brun

Costa Rica and Guatemala: Pelícano Pardo

Mexico: Pelícano Pardo

Puerto Rico: Alcatraz, Pelícano Pardo

Venezuela: Alcatraz

West Indies, Trinidad, and Tobago: Grand-Gosier

RANGE. Atlantic Coast from North Carolina to Venezuela. Pacific Coast from British Columbia to Chile. Abundant along Pacific Panama coast, especially in Panama Bay and the Pearl Islands; less numerous but still fairly common along entire Caribbean coast. Migrant to northern Brazil (Rio Uraricuera), coast of Ecuador, Humbolt Current waters along the coasts of Peru and Chile (to Corral and occasionally Chiloe Island); accidental in Tierra del Fuego, Argentina. Virtually throughout the West Indies, Trinidad, and Tobago.

STATUS. Threatened.

HABITAT. Breeds on islands away from mammalian predators; nests in colonies on the ground or in low trees and shrubs, commonly among mangroves. Winters on coastal waters and inland lakes, sandbars, mudflats, or beaches of outer islands.

SPECIAL HABITAT REQUIREMENTS. Islands away from mammalian predators, and waters providing a good supply of fish.

Further Reading. United States and Canada: Bull and Farrand 1977, Godfrey 1966, Terres 1980. Mexico and Central America: Edwards 1972, Land 1970, Peterson and Chalif 1973, Ridgely and Gwynne 1989, Stiles and Skutch 1989. South America: Hilty and Brown 1986, Meyer de Schauensee 1966. Caribbean: Bond 1947, Herklots 1961, Raffaele 1989.

Double-crested Cormorant

Phalacrocorax auritus

Canada (Quebec): Cormoran à Aigrettes

Mexico: Cormorán Orejudo

Puerto Rico: Cormorán Crestado

West Indies: Cormoril, Cortúa, Cortúa de Mar

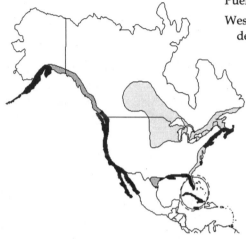

RANGE. Breeds in the southeastern Bering Sea, southern Alaska, and from southwestern British Columbia and northern Alberta to Newfoundland south along Atlantic and Pacific Coasts; very locally throughout interior of North America. Winters along the Pacific Coast from the Aleutians and southern Alaska south to Baja California and Guerrero, Mexico; common throughout most of Mexico. On the Atlantic Coast from New England south; in the Mississippi and Rio Grande Valleys; and along the Gulf Coast south to Central America. Sightings are increasing in Puerto Rico. Northernmost Bahamas, Cuba, and Isle of Pines; vagrant elsewhere in the West Indies east to Guadeloupe.

STATUS. Common; increasing in eastern and central United States and Canada.

HABITAT. Inhabits coastal areas, bays, estuaries, marine islands, freshwater lakes, ponds, rivers, sloughs, and swamps. Has a pronounced preference for perching in trees, on rocks, buoys, or other objects that overhang or project from the water. Nests in colonies of a few to 3,500 pairs on rocky islands, cliffs facing water, or in stands of live or dead trees in or near water. In the Northeast and along the Pacific Coast, nests on the ground, on rocky islands, or on cliffs. Inland and in Florida, usually nests in trees.

SPECIAL HABITAT REQUIREMENTS. Undisturbed nesting sites and conven-
ient, dependable food source within a foraging radius of 10 to 15 km from
roost or colony.

Further Reading. United States and Canada: Farrand 1983a, Godfrey 1966,
Palmer 1962, Terres 1980. Mexico: Edwards 1972, Peterson and Chalif 1973.
Caribbean: Bond 1947, Raffaele 1989.

Neotropic Cormorant

Phalacrocorax brasilianus
(formerly Olivaceous
Cormorant, *Phalacrocorax
olivaceus*)

Chile: Cormorán Negro

Costa Rica: Cormorán Neotropical,
Pato de Agua, Pato Chancho

Guatemala: Cormorán Bigua, Maleche

Mexico and Puerto Rico: Cormorán
Oliváceo

Venezuela: Cotúa Olivácea

West Indies: Cormoril, Cotúa, Cotúa
de Agua Dulce

RANGE. Coasts of Louisiana and Texas south to Brazil; also Cuba, Isle of
Pines, Bahamas, and Mexico, from Sonora in the west and Tamaulipas in the
east through central and southern Mexico including Yucatán Peninsula. The
only cormorant in Costa Rica and Honduras; probably resident in Guatemala. In Panama, common along both coasts and on larger bodies of fresh-

water. May occur in great numbers in Panama Bay just after the onset of the dry season when upwellings in inshore waters attract schools of small fish. Most numerous in the Caribbean lowlands, and may be found in almost any aquatic habitat. Breeds on Cuba, Isle of Pines, Great Inagua, San Salvador, and Cat Island; casual on Jamaica and Dominica.

STATUS. Common.

HABITAT. Freshwater lakes, reservoirs, ponds, and coastal islands. Prefers shallow clear water at low elevations. Nests in colonies in living or dead trees in or near water. In Colombia, common on freshwater lakes and rivers and in salt and estuarine habitats on both coasts; chiefly lowlands, occasionally (usually immatures) to temperate-zone lakes.

SPECIAL HABITAT REQUIREMENTS. Relatively undisturbed areas for breeding.

Further Reading. United States and Canada: Bull and Farrand 1977, Terres 1980. Mexico and Central America: Edwards 1972, Land 1970, Peterson and Chalif 1973, Ridgely and Gwynne 1989, Stiles and Skutch 1989. South America: Hilty and Brown 1986, Meyer de Schauensee 1979, Meyer de Schauensee and Phelps 1978. Caribbean: Bond 1947, Herklots 1961, Raffaele 1989.

Anhinga
Anhinga anhinga

Canada (Quebec): Anhinga
Costa Rica: Aninga, Pato Aguja
Guatemala: Pato Aguja
Mexico: Anhinga Americana
Venezuela: Cotúa Agujita
West Indies: Cotúa Real, Marbella

RANGE. Breeds from central and eastern Texas east to coastal North Carolina and south to southern Brazil and Ecuador. Winters in southeastern United States from central South Carolina, Georgia, Florida, and the Gulf Coast southward. Occasionally disperses north of breeding range. Resident in tropics generally south to Brazil and northern Argentina. West of the Andes only to western Ecuador. In Colombia, most numerous in the forest zones north and east of the Andes. Resident in Honduras and Guatemala. Probably resident in Trinidad and Tobago.

STATUS. Common throughout range, but recently declining in the eastern United States.

HABITAT. Inhabitats quiet or slow-moving, often rather murky waters. Usually found in wooded freshwater swamps, streams, or tree-fringed lakes with water lilies, lotus, and other aquatic vegetation. Found in cypress swamps, freshwater sloughs of sawgrass and reeds with scattered willow clumps, or mangrove-bordered salt and brackish bays, lagoons, and tidal streams. Primarily a freshwater bird, but will range to marine coasts. Nests in small groups with herons and egrets. May appropriate nests of Great and Snowy Egrets or Little Blue Herons. Nests are usually 1 to 3 m above water.

SPECIAL HABITAT REQUIREMENTS. Quiet, sheltered waters with some trees for perching; minimal human disturbance.

Further Reading. United States and Canada: Farrand 1983a, Godfrey 1966, Oberholser 1974a, Palmer 1962, Pough 1951, Terres 1980. Mexico and Central America: Edwards 1972, Land 1970, Monroe 1968, Ridgely and Gwynne 1989. South America: Hilty and Brown 1986, Meyer de Schauensee 1966, Meyer de Schauensee and Phelps 1978. Caribbean: Bond 1947, Herklots 1961.

American Bittern
Botaurus lentiginosus

Canada (Quebec): Butor Américain
Costa Rica: Avetoro Norteño, Puncus
Guatemala: Avetoro Norteamericana
Mexico: Garza Norteña de Tular
Puerto Rico: Yaboa Americana
West Indies: Ave Toro, Guanaba Rojo

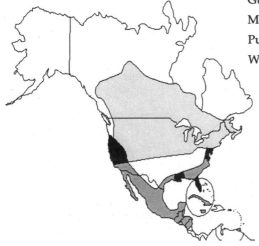

RANGE. Breeds from extreme southeastern Alaska, central British Columbia, and southwestern Northwest Territories to central Quebec and Newfoundland south to southern California, New Mexico, Texas, and Florida; breeds rarely south of northern California, Utah, Great Plains states, Ohio River valley, and Virginia. Winters from southern British Columbia, Utah, New Mexico, central parts of Gulf States, and southern New England south to southern California, Gulf of Mexico, and along Atlantic Coast. In Mexico winters locally mainly in the lowlands. A rare migrant and winter visitor in Guatemala. Winters in small numbers in the Bahamas and Greater Antilles, in particular Cuba; rare winter visitor to Puerto Rico; casual in Lesser Antilles.

STATUS. Rather common, but elusive; may be declining, especially in the western United States and Canada.

HABITAT. Inhabits freshwater or saltwater marshes, bogs, swamps, wet meadows, or wherever the ground is wet and where tall, emergent vegetation such as cattails, bulrushes, and reeds are present. Usually a solitary nester, but may form loose colonies in favorable habitat. Typically nests on flimsy platform of cattails, reeds, or sedges, just above water in emergent vegetation, occasionally on the ground among grasses or in shrubs. Usually perches on the ground, sometimes on a log or stump, or on cattails 1 m above water, rarely in trees.

SPECIAL HABITAT REQUIREMENTS. Wetlands with tall, dense emergent vegetation. This species is totally dependent on wetland habitats throughout its range.

Further Reading. United States and Canada: Farrand 1983a, Godfrey 1966, Grinnell and Miller 1944, Hunter 1990, Low and Mansell 1983, Palmer 1962, Robbins et al. 1983, Weller 1961. Mexico and Central America: Edwards 1972, Land 1970, Peterson and Chalif 1973, Ridgely and Gwynne 1989, Stiles and Skutch 1989. Caribbean: Bond 1947, Raffaele 1989.

Least Bittern
Ixobrychus exilis

Canada (Quebec): Petit Butor

Costa Rica: Avetorillo Pantanero, Mirasol

Guatemala: Avetorillo Pantanero

Mexico: Garcita de Tular

Puerto Rico: Betlén, Gaulín, Martinetito

Trinidad and Tobago: Quioc Jaune

Venezuela: Garza Enana

West Indies: Crabier, Garcita, Martinete, Martinetito, Martín García

RANGE. Breeds locally in western North America from southern Oregon south through California to Mexico; in eastern North America from southern Manitoba, Ontario, Quebec, and New Brunswick south to Texas and Florida. Resident and winter populations extend through Middle America and the Greater Antilles; also resident in South America east of the Andes from Colombia and Venezuela south to central Argentina; west of the Andes in coastal Peru.

STATUS. Locally common, but elusive. May be declining throughout the southeastern United States.

HABITAT. Inhabits freshwater marshes, bogs, and swamps with dense cattails, reeds, bulrushes, buttonbush, sawgrass, smartweeds, arrowheads, and other tall aquatic and semiaquatic vegetation. Prefers marshes with scattered bushes or other woody growth. Less commonly found in coastal brackish marshes and mangrove swamps. Nests singly in dense stands of emergent vegetation approximately 0.5 m above water that is up to 1 m deep, and close to open water.

SPECIAL HABITAT REQUIREMENTS. Freshwater wetlands surrounded by tall aquatic vegetation. This species is dependent on wetland habitats throughout its range.

Further Reading. United States and Canada: Godfrey 1966, Hunter 1990, Low and Mansell 1983, Palmer 1962, Terres 1980, Weller 1961. Mexico and Central America: Edwards 1972, Land 1970, Peterson and Chalif 1973, Ridgely and Gwynne 1989, Stiles and Skutch 1989. South America: Hilty and Brown 1986, Meyer de Schauensee 1966, Meyer de Schauensee and Phelps 1978. Caribbean: Bond 1947, Herklots 1961, Raffaele 1989.

Great Blue Heron
Ardea herodias

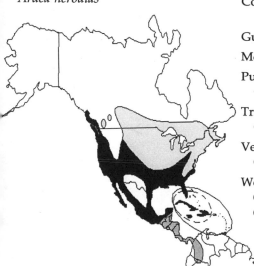

Canada (Quebec): Grand Héron Bleu

Costa Rica: Garzón Azulado, Garza Ceniza

Guatemala: Garzón Azulado

Mexico: Garzón Cenizo

Puerto Rico: Garzón Azulado, Garzón Ceniciento, Garzón Cenizo

Trinidad and Tobago: Aileronne à Calotte Blanche

Venezuela and Mexico: Garzón Cenizo

West Indies: Crabier Noir, Garcilote, Garzón Cienzo, Gironde, Guardacosta

RANGE. Breeds from southern Alaska, coastal and southern British Columbia, southeastern Northwest Territories, and central Manitoba east to Nova Scotia and south, locally, throughout the United States and much of Mexico to Guerrero, Veracruz, the Gulf Coast, and southern Florida, including the Keys, Cuba, the coast of the Yucatán Peninsula, and Los Roques of the northern Venezuelan coast. Winters from southern-coastal Alaska, coastal British Columbia, central United States, and southern New England south throughout Mexico with stragglers to Colombia and Venezuela. Resident on islands from Aruba to Trinidad and Tobago and common during northern winter in Puerto Rico. A rare migrant or winter visitor in Trinidad, Tobago, the Caribbean lowlands, Honduras, Panama, Costa Rica, and Guatemala. Winters throughout the West Indies.

STATUS. Common throughout range; has been increasing in the eastern and central United States.

HABITAT. Inhabits a wide variety of freshwater and saltwater habitats including ponds, lakes, streams, rivers, marshes, wet meadows, tidal flats, sandbars, and shallow bays, or wherever shallow water or marsh vegetation is present. Generally nests in colonies, preferably in an isolated patch of woodland or on an island. Builds nest in the tops of the tallest trees, live or dead, often above 15 m, but also in bushes, on rock ledges, sea cliffs, in tules, rushes, and on the ground. In colonies, may build dozens of nests, which are used repeatedly, in the crown of the same tree. In mixed heronries, typically nests in highest parts of trees while other heron species occupy lower parts of same trees. May travel as far as 16 km from nest sites to foraging areas.

SPECIAL HABITAT REQUIREMENTS. Open water or wetland habitats; protection from clearing of trees in forested wetlands and from human disturbance.

Further Reading. United States and Canada: Farrand 1983a, Godfrey 1966, Low and Mansell 1983, Palmer 1962, Pough 1951, Terres 1980, Verner and Boss 1980. Mexico and Central America: Edwards 1972, Land 1970, Monroe 1968, Peterson and Chalif 1973, Ridgely and Gwynne 1989, Stiles and Skutch 1989. South America: Hilty and Brown 1986, Meyer de Schauensee 1966, Meyer de Schauensee and Phelps 1978. Caribbean: Bond 1947, Herklots 1961, Raffaele 1989.

Great Egret
Casmerodius albus

Canada (Quebec): Grande Aigrette

Costa Rica and Guatemala: Garceta Grande, Garza Real

Mexico: Garzón Blanco

Puerto Rico: Garza Real, Garzón Blanco

Trinidad and Tobago: Grand Aigrette

Venezuela: Garza Blanca Real

West Indies: Crabier Blanc, Garzón, Garzón Blanco, Garzón Real

RANGE. Breeds from southern Oregon and Idaho south to Arizona, and from Minnesota, southern Ontario, and Maine south through Middle and South America to southern Chile and central Argentina. Resident in the West Indies. Winters from northern California across the southern United States and south along the Atlantic Coast from North Carolina south through the breeding range.

STATUS. Common throughout range; increasing in the western United States.

HABITAT. Inhabits streams, ponds, lakes, rice fields, freshwater and salt-water marshes and lagoons, and mudflats. Nests singly or in colonies, often with other herons, ibises, Wood Storks, cormorants, and Anhingas. Usually nests in woods or thickets near water as long as there is adequate support for the nest. Builds nest from 1 to 12 m above ground, depending on sub-strate. Sensitive to disturbance by people when nesting and may flush at the slightest provocation. After feeding during the day, flies singly or in small

53

groups to a communal roost in trees or shrubs. Gregarious during all seasons.

SPECIAL HABITAT REQUIREMENTS. Open water or wetland habitats near woodlands; protection from clearing of trees within forested wetlands and from human disturbance.

Further Reading. United States and Canada: Farrand 1983a, Godfrey 1966, Grinnell and Miller 1944, Low and Mansell 1983, Palmer 1962, Terres 1980. Mexico and Central America: Edwards 1972, Land 1970, Monroe 1968, Peterson and Chalif 1973, Ridgely and Gwynne 1989, Stiles and Skutch 1989. South America: Hilty and Brown 1986, Meyer de Schauensee 1966, Meyer de Schauensee and Phelps 1978. Caribbean: Bond 1947, Herklots 1961, Raffaele 1989.

Snowy Egret
Egretta thula

Canada (Quebec): Aigrette Neigeuse

Costa Rica: Garceta Nirosa

Guatemala: Garceta Nevada, Garcita Blanca

Mexico: Garza Dedos Dorados

Puerto Rico: Garza Blanca

Trinidad and Tobago: Aigrette à Panache

Venezuela: Chusmita, Garcita Blanca

West Indies: Aigrette Blanche, Crabier Blanc, Garza Blanca, Garza Real, Garza de Rizos

RANGE. Breeds from northern California and Montana south to central and eastern Texas, along the lower Mississippi Valley, and from Maine south along the Atlantic and Gulf Coasts to South America. After breeding, disperses north to Oregon, Nebraska, the Great Lakes, and Atlantic Canada. Winters from northern California, southwestern Arizona, the Gulf Coast, and coastal South Carolina south throughout the breeding range. Found virtually throughout the West Indies, although rare in the Lesser Antilles.

STATUS. Common; breeding range expanding northward; populations generally increasing, especially in the central and western United States.

HABITAT. Inhabits ponds; borders of lakes; freshwater, brackish, and saltwater marshes and swamps; mangroves; stream courses; tidal flats; rice fields; and sometimes dry fields. Nests in colonies (many are coastal), sometimes with thousands of pairs, or in smaller colonies with other herons, ibises, cormorants, and Anhingas, or even singly. In western United States, commonly nests on the ground in cattail marshes; in other areas, may nest up to 9 m high in trees and shrubs.

SPECIAL HABITAT REQUIREMENTS. Wetlands.

Further Reading. United States and Canada: Farrand 1983a, Godfrey 1966, Grinnell and Miller 1944, Loftin 1967, Low and Mansell 1983, Palmer 1962, Terres 1980. Mexico and Central America: Edwards 1972, Land 1970, Monroe 1968, Peterson and Chalif 1973, Ridgely and Gwynne 1989, Stiles and Skutch 1989. South America: Hilty and Brown 1986, Meyer de Schauensee 1966, Meyer de Schauensee and Phelps 1978. Caribbean: Bond 1947, Herklots 1961, Raffaele 1989.

Little Blue Heron
Egretta caerulea

Canada (Quebec): Petit Héron Bleu

Costa Rica: Garceta Azul

Guatemala: Garcita Azul, Garza Gris

Mexico and Venezuela: Garza Azul

Puerto Rico: Garza Azul, Garza Pinta

Trinidad and Tobago: Aigrette Bleu

West Indies: Aigrette Blanch (imm.), Aigrette Bleue, Crabier Blanc, Crabier Noir, Garza Azul, Garza Blanca (imm.), Garza Común, Garza Pinta

RANGE. In western North America, breeds in southern California and the Pacific slope of Middle America; in eastern North America from Maine south along the Atlantic coastal plain through Middle America; in the West Indies; in northern South America; and locally and sporadically in parts of the southern and central United States. Stragglers appear during the postbreeding period in North and South Dakota, Michigan, southern Ontario, southern Quebec, Nova Scotia, and other localities where the species is not known to breed. Winters from coastal Virginia south throughout most of breeding range.

STATUS. Common; declining in the eastern United States.

HABITAT. Prefers freshwater ponds, lakes, marshes, meadows, and marshy shores of streams, but also inhabits mangroves and other brackish and salt-water coastal habitats. Roosts in trees and shrubs at night. Nests in colonies of up to 100 in trees, usually hardwoods, over or near freshwater. Tends to nest on the edge of mixed colonies, often in company with the Tricolored Heron.

SPECIAL HABITAT REQUIREMENTS. Wooded wetland habitats; open wetlands near forest cover.

Further Reading. United States and Canada: Farrand 1983a, Godfrey 1966, Low and Mansell 1983, Palmer 1962, Terres 1980. Mexico and Central America: Edwards 1972, Land 1970, Monroe 1968, Peterson and Chalif 1973, Ridgely and Gwynne 1989, Stiles and Skutch 1989. South America: Hilty and Brown 1986, Meyer de Schauensee 1966, Meyer de Schauensee and Phelps 1978. Caribbean: Bond 1947, Herklots 1961, Raffaele 1989.

Tricolored Heron
Egretta tricolor
(also known as Louisiana Heron)

Canada (Quebec): Héron à Ventre Blanc

Costa Rica and Guatemala: Garceta Tricolor

Mexico: Garza Vientriblanca

Puerto Rico and Venezuela: Garza Pechiblanca

Trinidad and Tobago: Aigrette à Ventre Blanc

West Indies: Crabier, Garza Morada, Garza Pechiblanco, Garza de Vientre Blanco

RANGE. Breeds in central Baja California, Gulf Coast of Texas, Louisiana, and Alabama, north along the Atlantic Coast to southern Maine, south along both coasts of Middle America to Ecuador, Colombia, and Venezuela, on the Pacific Coast of South America to Peru, on the Caribbean-Atlantic Coast to northeastern Brazil, and through the Bahamas and Greater Antilles; rare or casual in the Lesser Antilles. Wanders north to southern California, Nevada, Arkansas, and Oklahoma. Winters in southern Baja California, southeastern Texas, the Gulf and Atlantic Coasts (north to New Jersey), and south through the remainder of the breeding range.

STATUS. Population stable.

HABITAT. Swamps, bayous, coastal ponds, salt marshes, mangroves, mudflats, and lagoons. Nests on the ground or up to 7 m high in a tree.

SPECIAL HABITAT REQUIREMENTS. Open water or wetland habitats; mangrove swamps (Puerto Rico).

57

Further Reading. United States and Canada: Bull and Farrand 1977, Godfrey 1966, Terres 1980. Mexico and Central America: Edwards 1972, Land 1970, Monroe 1968, Peterson and Chalif 1973, Ridgely and Gwynne 1989, Stiles and Skutch 1989. South America: Hilty and Brown 1986, Meyer de Schauensee 1966, Meyer de Schauensee and Phelps 1978. Caribbean: Bond 1947, Herklots 1961, Raffaele 1989.

Reddish Egret

Egretta rufescens

Costa Rica: Garceta Rojiza

Guatemala: Garceta Rojiza, Garza Rojiza

Mexico: Garza Piquirrosa

Puerto Rico: Garza Rojiza

Venezuela: Garza Rojiza

West Indies: Crabier, Garza

RANGE. Breeds in Baja California and along Gulf Coast to Florida, south to Guatemala, Cuba, Hispaniola, and the Bahamas, with stragglers to Colorado, Illinois, and South Carolina. Winters from southern Florida to Colombia and Venezuela with stragglers to Jamaica and Puerto Rico. Formerly more widespread but was exterminated in Florida in the early twentieth century by plume hunters.

STATUS. Uncommon to rare.

HABITAT. In Texas breeds on dry coastal islands in brushy thickets of yucca and pricklypear; in Florida found mostly in mangroves. Nests on ground or up to 5 m high in low bushes or trees. In winter, found in coastal wetlands.

SPECIAL HABITAT REQUIREMENTS. Open water or shallow, protected coastal wetlands; protection of nesting colonies from human disturbance is essential. Also, protection and enhancement of mangrove and coastal scrub nesting and estuarine foraging habitats are critical.

Further Reading. United States and Canada: Bull and Farrand 1977, Godfrey 1966, Hunter 1990, Terres 1980. Mexico and Central America: Edwards

1972, Land 1970, Peterson and Chalif 1973, Ridgely and Gwynne 1989, Stiles and Skutch 1989. South America: Hilty and Brown 1986, Meyer de Schauensee 1966, Meyer de Schauensee and Phelps 1978. Caribbean: Bond 1947, Raffaele 1989.

Cattle Egret
Bubulcus ibis

Canada (Quebec): Héron Garde-boent

Costa Rica: Garcilla Bueyera, Garza del Ganado

Guatemala: Garcilla Bueyera

Mexico: Garza Ganadera

Puerto Rico: Garza del Ganado, Garza Africana

Venezuela: Garcita Reznera

West Indies: Crabier Garde-boeuf, Garrapatosa, Garza Africana, Garza Ganadera

RANGE. Breeds across the eastern half of the United States and in the West from southern California, Arizona, and New Mexico south through Middle America, the West Indies, and much of South America. Many individuals disperse after breeding as far as southern British Columbia, south-central Canada, and the Maritime Provinces. Winters from the southern United States south through the breeding range. This Old World species was first recorded in South America in the late 1800s and has been steadily increasing its range over the Western Hemisphere since then.

STATUS. Common; recently declining in the central United States.

HABITAT. Frequents a great variety of habitats including pastures, freshwater and salt marshes, fallow and plowed fields, orchards, citrus groves, road shoulders and median strips, vacant lots, lawns, and other open grassy areas. Least shy and least aquatic of North American herons; usually found in close association with large hoofed mammals, particularly cattle. Nests colonially, often with other herons and ibises, in both freshwater and saltwater habitats, on islands, in willows and tamarisks along watercourses, occasionally in cypress swamps with an understory of buttonbush, or in scrub oaks in marshlands. Also nests in redcedar, red maple, and pines. Usually builds nests at heights of 2 to 4 m, but up to 10 m in heronries.

SPECIAL HABITAT REQUIREMENTS. Wetlands for nesting, and open areas for feeding. Roosts in mangroves (Puerto Rico).

Further Reading. United States and Canada: Farrand 1983a, Godfrey 1966, Palmer 1962, Terres 1980. Mexico and Central America: Edwards 1972, Land 1970, Peterson and Chalif 1973, Ridgely and Gwynne 1989, Stiles and Skutch 1989. South America: Hilty and Brown 1986, Meyer de Schauensee 1966, Meyer de Schauensee and Phelps 1978. Caribbean: Bond 1947, Raffaele 1989.

Green Heron
Butorides virescens

Canada (Quebec): Héron Vert

Costa Rica: Garcilla Verde

Guatemala: Garcita Verde

Mexico: Garcita Oscura

Puerto Rico: Martinete

Venezuela: Chicuaco Cuello Rojo

West Indies: Aquaita Caiman, Caali, Caga-leche, Cagon, Crabier, Cra-cra, Cuaco, Kio, Martinete, Rac-rac, Valet de Caiman

RANGE. Breeds from Washington and Oregon south through much of the southwestern United States; in eastern North America from the eastern edge of the Great Plains, southern Ontario, and New Brunswick south through Middle America, the West Indies, and a large portion of South America. Stragglers appear during the postbreeding period in eastern Washington, Idaho, southern Canada, and other localities where the species is not found as a breeder. Winters from the southern United States south through the breeding range.

STATUS. Common, locally abundant; long-term increase in the West.

HABITAT. Found in a wide variety of freshwater and saltwater habitats, primarily those in riparian deciduous zones. These include wet woodlands, lakeshores, ponds, rivers, streams, swamps, and marshes. Commonly alights on trees, stumps, or submerged debris, but roosts on or close to the ground. Generally a solitary nester, but sometimes nests in colonies of six or more pairs. Nest may be built away from water in dry woodlands and orchards, on a low tussock or muskrat house, or in trees near water; often it is a dense tangle in crowns of middle-aged trees, typically 3 to 5 m above ground, but up to 10 m.

SPECIAL HABITAT REQUIREMENTS. Wooded wetlands or open-water habitats.

Further Reading. United States and Canada: Godfrey 1966, Low and Mansell 1983, Palmer 1962, Terres 1980, Verner and Boss 1980. Mexico and Central America: Edwards 1972, Land 1970, Monroe 1968, Peterson and Chalif 1973, Ridgely and Gwynne 1989, Stiles and Skutch 1989. South America: Hilty and Brown 1986, Meyer de Schauensee 1966, Meyer de Schauensee and Phelps 1978. Caribbean: Bond 1947, Raffaele 1989.

Black-crowned Night-Heron

Nycticorax nycticorax

Canada (Quebec): Bihoreau à Couronne Noire

Costa Rica: Martinete Coroninegro

Guatemala: Garza de Noche Coroninegra, Martinete Común, Martinete Coroninegro

Mexico: Garza Nocturna Coroninegra

Puerto Rico: Yaboa Real

Trinidad and Tobago: Crabier Batali

Venezuela: Guaco

West Indies: Coq d'Eau, Coq de Nuit, Crabier, Gallinaza (imm.), Guanaba de la Florida, Guanaba Lominegro, Rey Congo, Yaboa Real

RANGE. Breeds in the Western Hemisphere from central Washington and east-central Alberta to southern Quebec, northeastern New Brunswick, and Nova Scotia south locally through the United States, Middle America, the Bahamas, Greater Antilles, and South America to Tierra del Fuego. Wanders widely. After breeding, disperses throughout much of the United States, not restricted to its breeding range. Winters in the southwestern United States, lower Ohio Valley, Gulf Coast, and southern New England south throughout the breeding range. An uncommon and local resident in Puerto Rico. Winters virtually throughout the West Indies, most individuals being visitors from North America.

STATUS. Common throughout most of its range. Has declined in the eastern United States and Canada.

HABITAT. Inhabits a wide variety of freshwater, brackish, and saltwater habitats, including lakes, ponds, marshes, wooded swamps, slow streams with pools, or rivers. Roosts by day, usually in a well-foliaged tree, not necessarily near feeding grounds. Nests in small to large colonies, usually with other heron species, in almost any habitat: wooded areas near coastal marshes, spruce groves on marine islands, hardwood forests on offshore islands, swamps, cattail marshes on prairies, clumps of tall grass on dry ground, apple orchards, and sometimes city parks.

SPECIAL HABITAT REQUIREMENTS. Open-water or wooded wetland habitats; freshwater swamps (Puerto Rico).

Further Reading. United States and Canada: Farrand 1983a, Godfrey 1966, Grinnel and Miller 1944, Low and Mansell 1983, Palmer 1962, Terres 1980, Verner and Boss 1980. Mexico and Central America: Edwards 1972, Land 1970, Monroe 1968, Peterson and Chalif 1973, Ridgely and Gwynne 1989, Stiles and Skutch 1989. South America: Hilty and Brown 1986, Meyer de Schauensee 1966, Meyer de Schauensee and Phelps 1978. Caribbean: Bond 1947, Herklots 1961, Raffaele 1989.

Yellow-crowned Night-Heron
Nyctanassa violacea

Canada (Quebec): Bihoreau Violacé

Costa Rica: Martinete Cabecipinto

Guatemala: Garza de Noche de Corona Amarilla, Martinete Azul, Martinete Coronigualdo

Mexico: Garza Nocturna Coroniclara

Puerto Rico: Yaboa Común

Trinidad and Tobago: Crabier à Croissant

Venezuela: Chicuaco Enmascarado

West Indies: Coq d'Eau, Coq de Nuit, Crabier de Bois, Crabier Gris, Crabier de Montagne, Guanaba, Guanaba Real, Rey Congo, Yaboa

RANGE. Breeds from Nebraska and southeastern Minnesota south and east to the lower Ohio Valley and along the eastern Atlantic coastal plain from Massachusetts south through Middle America and the West Indies; in South America south and west to Ecuador and along the northern and eastern coast of the continent from Colombia and Venezuela to southern Brazil. Winters from coastal South Carolina south throughout the breeding range.

STATUS. Much less common than Black-crowned Night-Heron.

HABITAT. Inhabits both freshwater and saltwater habitats, usually lush river swamps, but also tidal flats, stagnant backwaters or bayous of large cypress swamps, mangrove swamps, or dry, rocky, almost waterless thickets on certain islands. Nests in small to large colonies, sometimes with Black-crowned, Little Blue, Tricolored, and Great Blue Herons, or singly, in trees or bushes and sometimes on the ground. Often nests in willows close to water, in mangroves, or in bald cypresses, usually 3 to 4 m above ground.

SPECIAL HABITAT REQUIREMENTS. Wooded swamps such as mangroves and gallery forests.

Further Reading. United States and Canada: Farrand 1983a, Godfrey 1966, Low and Mansell 1983, Palmer 1962, Terres 1980. Mexico and Central America: Edwards 1972, Land 1970, Peterson and Chalif 1973, Ridgely and Gwynne 1989, Stiles and Skutch 1989. South America: Hilty and Brown 1986, Meyer de Schauensee 1966, Meyer de Schauensee and Phelps 1978. Caribbean: Bond 1947, Herklots 1961, Raffaele 1989.

White Ibis

Eudocimus albus

Canada (Quebec): Ibis Blanc

Costa Rica: Coco, Ibis Blanco

Guatemala and Mexico: Ibis Blanco

Puerto Rico: Coco Blanco

West Indies: Coco, Coco Blanco

RANGE. Breeds along the Atlantic coastal plain from North Carolina south through Middle America, the West Indies, and northern South America. Stragglers appear during the postbreeding period in California, Colorado, South Dakota, Missouri, Illinois, Virginia, New Jersey, New York, Vermont, Quebec, and other localities where the species is not known to breed. Winters in breeding range from coastal South Carolina southward.

STATUS. Common; long-term increase, especially in the central part of U.S. range.

HABITAT. Coastal saltwater, brackish, or freshwater marshes; mangroves. Nests in trees 1 to 5 m above water, also in sawgrass and bulrushes in Florida, Louisiana, and Mexico.

SPECIAL HABITAT REQUIREMENTS. Open water or coastal wetlands.

Further Reading. United States and Canada: Bull and Farrand 1977, Godfrey 1966, Terres 1980. Mexico and Central America: Edwards 1972, Land

1970, Monroe 1968, Peterson and Chalif 1973, Ridgely and Gwynne 1989, Stiles and Skutch 1989. South America: Hilty and Brown 1986, Meyer de Schauensee 1966. Caribbean: Bond 1947, Raffaele 1989.

Glossy Ibis
Plegadis falcinellus

Canada (Quebec): Ibis Luisant

Costa Rica: Coco Negro, Ibis Morito

Mexico: Ibis Negro

Puerto Rico: Cigüeño, Coco Prieto, Ibis Instro

Venezuela: Corocoro Castaño

West Indies: Cigüeña, Coco, Coco Oscuro, Coco Prieto, Pecheur

RANGE. Breeds in North America locally from Maine and Rhode Island south to Florida, and west on the Gulf Coast to Louisiana. Also inland in Arkansas, the upper Midwest, and southern Canada. Winters from northern Florida and the Gulf Coast of Louisiana south through the West Indies to Venezuela. Winters in the Americas from northern Florida and the Gulf Coast of Louisiana south through the Greater Antilles, northwestern Costa Rica, and northern Venezuela.

STATUS. Locally common; formerly an irregular breeding bird in North America in small colonies along the Atlantic Coast, but since the late 1970s has increased in numbers.

HABITAT. Freshwater, brackish, and saltwater habitats, primarily marshes and estuaries. Prefers shallow pools bordered by shrubs and emergent vegetation. Nests in small colonies, usually with herons or other waders, in a variety of habitats: willows, mangroves, tropical buttonwood and salt myrtle (Florida); willows, gum, swamp maple, bay, and buttonbush in cypress swamps (South Carolina); barrier beach forests (New Jersey coast); and cattail marshes. Nests on the ground in marshes, or up to 3 m high in shrubs and trees growing in water, in sites well covered with vegetation.

SPECIAL HABITAT REQUIREMENTS. Wetlands.

Further Reading. United States and Canada: Burger and Miller 1977, DeGraaf and Rudis 1986, Low and Mansell 1983, Palmer 1962, Terres 1980. Mexico and Central America: Hilty and Brown 1986, Ridgely and Gwynne 1989, Stiles and Skutch 1989. South America: Meyer de Schauensee 1966, Meyer de Schauensee and Phelps 1978. Caribbean: Bond 1947, Raffaele 1989.

White-faced Ibis
Plegadis chihi

Canada (Quebec): Ibis à Face Blanche

Costa Rica: Ibis Cariblanco

Mexico: Ibis Cara Blanca

RANGE. Breeds locally in central and western United States from California, Oregon, Idaho, Montana, North Dakota, and Minnesota south to coastal Texas; recorded in Middle America from Mexico, Guatemala, El Salvador, Honduras, and Nicaragua; also in South America from Ecuador, Peru, Colombia, Venezuela, southern Brazil, Bolivia, Uruguay, Paraguay, northern Argentina, and Chile. Winters from southern California and Texas south through breeding range.

STATUS. Uncommon; nesting populations have been generally increasing in several parts of western United States.

HABITAT. Inhabits wetland habitats, preferably marshes and sloughs or ponds surrounded by low bushes or willows and emergent vegetation. Also in tule or bulrush swamps, in centers of ponds, and in irrigated rice fields. Roosts in marshes in the evenings. Generally nests in mixed colonies in large beds of bulrushes or reeds up to 1 m above water; on floating mats of dead plants, in cattails and hardstem and alkali bulrush. Generally nests in areas well covered with vegetation.

SPECIAL HABITAT REQUIREMENTS. Freshwater marshes and sloughs.

Further Reading. United States and Canada: Burger and Miller 1977, Godfrey 1966, Oberholser 1974a, Palmer 1962, Ryder 1967, Terres 1980. Mexico and Central America: Edwards 1972, Peterson and Chalif 1973, Stiles and Skutch 1989. South America: Meyer de Schauensee 1966.

Roseate Spoonbill
Ajaia ajaja

Costa Rica and Guatemala: Espátula Rosada

Mexico: Ibis Espátula

West Indies: Cuchareta, Espátula, Sebiya, Spatule

RANGE. Breeds from southern Florida, Louisiana, Texas, and both coasts of northern Mexico south through Middle and South America to central Argentina and Chile; also Cuba, Isle of Pines, Hispaniola, and southern Bahamas. Winters throughout breeding range; partial migrant from extreme northern and southern portions of the breeding range.

STATUS. Common.

HABITAT. Mangroves in Florida, in low bushes along coastal islands, and even on treeless spoil banks along the Intracoastal Waterway in Texas. Coastal salt and brackish waters. Nests from 3 to 5 m above ground in a low tree or bush.

SPECIAL HABITAT REQUIREMENTS. Open water and wetlands; protection from human disturbance.

Further Reading. United States and Canada: Bull and Farrand 1977, Godfrey 1966, Terres 1980. Mexico and Central America: Edwards 1972, Land 1970, Peterson and Chalif 1973, Ridgely and Gwynne 1989, Stiles and Skutch 1989. South America: Hilty and Brown 1986, Meyer de Schauensee 1966. Caribbean: Bond 1947, Raffaele 1989.

Wood Stork
Mycteria americana

Canada (Quebec): Cigogne Américaine

Costa Rica: Cigüeñón, Garzón, Guairón

Guatemala: Cigüeñón, Garzón Pulido

Mexico: Cigüeño Americano

West Indies: Cayama (Cuba), Coco, Faisan

RANGE. Breeds from Florida, Georgia, and South Carolina south through Middle and South America to central Argentina; also the Bahamas and Greater Antilles. Winters throughout most of breeding range; partial migrant from extreme northern and southern portions of the breeding range; performs postbreeding migrations from Mexico into southern California, Texas, and Louisiana, moving northward in April–May and southward in August–September.

STATUS. Declining and endangered in the United States, common in the rest of its range.

HABITAT. Breeds in giant cypress, and loss of this habitat is a factor contributing to the decline of Wood Storks in North America. Nests above 20 m in trees. In mangroves nests are located just above water.

SPECIAL HABITAT REQUIREMENTS. Giant cypress and wetlands; protection from human disturbance.

Further Reading. United States and Canada: Godfrey 1966, Knopf 1979, Terres 1980. Mexico and Central America: Edwards 1972, Land 1970, Peterson and Chalif 1973, Ridgely and Gwynne 1989, Stiles and Skutch 1989. South America: Hilty and Brown 1986, Meyer de Schauensee 1966. Caribbean: Bond 1947.

Fulvous Whistling-Duck
Dendrocygna bicolor

Canada (Quebec): Dendrocygne Fauve

Chile: Pato Silbón

Costa Rica: Pijije Camelo

Guatemala: Pichichi Colorado, Pijije Colorado

Mexico: Pato Pijije Alioscuro

Trinidad and Tobago: Ailes Rouge, Ouikiki

Venezuela: Yugaso Colorado

West Indies: Vingeon Rouge, Yaguasín

RANGE. Breeds from southern California to southwestern Arizona, and from central and eastern Texas and the Gulf Coast of Louisiana south to Nayarit,

Jalisco, the Valley of Mexico, and northern Veracruz; locally in southern Florida. Wanders sporadically northward to Middle Atlantic states. Winters from southern California, southern Arizona, the Gulf Coast, and southern Florida south to Mexico. In Mexico mainly in lowlands; in the east from Tamaulipas to Campeche; in the west from southern Sonora to Oaxaca. Occasional in northern Baja California, central Mexico, Chiapas, and Yucatán. U.S. birds rarely make it farther south than mid-Mexico. In Guatemala seen only on Lake Retana in the winter months. Has occurred widely in the West Indies as a vagrant.

STATUS. Fairly common, but population levels fluctuate.

HABITAT. Inhabits marshlands, wet meadows, and in North America, primarily flooded agricultural land and rice fields. Does not ordinarily frequent woodlands. Loafs among dense bulrushes or far out on marshy ponds. Prefers to nest in rice fields on low, contour levees, as well as just above water among rice plants and weeds growing between levees. Also nests in bulrushes, knotgrass, and dense beds of cattails, on hummocks in marshes, at the edges of ponds and swamps, or in rank tall grasses of wet meadows; rarely in tree cavities.

SPECIAL HABITAT REQUIREMENTS. Broad, open marshlands.

Further Reading. United States and Canada: Baldwin et al. 1964, Bellrose 1976, Cottam and Glazener 1959, Godfrey 1966, Johnson 1965, Terres 1980. Mexico and Central America: Edwards 1972, Land 1970, Peterson and Chalif 1973, Ridgely and Gwynne 1989, Stiles and Skutch 1989. South America: Hilty and Brown 1986, Meyer de Schauensee 1966, Meyer de Schauensee and Phelps 1978. Caribbean: Bond 1947, Raffaele 1989.

Black-bellied Whistling-Duck

Dendrocygna autumnalis

Costa Rica: Pijije Común

Guatemala: Pichichi Común, Pichichil, Pijije Común

Mexico: Pato Pijije Aliblanco

Trinidad and Tobago: Ailes Blanches, Ouikiki

RANGE. Breeds from Arizona and Texas south through Middle America and northern South America. Winters from central Mexico south through the breeding range. In Mexico, mainly in low coastal regions; southern Sonora to Chiapas, and Tamaulipas to the Yucatán Peninsula; less common in the interior. Fairly common on the Pacific Coast of Guatemala with a possible increase in the population in the winter as birds move down from Mexico. A vagrant to the Lesser Antilles and Cuba. West Indian records occur for all seasons of the year.

STATUS. Rather common within its breeding range, elsewhere a straggler.

HABITAT. Prefers open woodlands, groves or thicket borders of ebony, mesquite, retama, huisache, and cacti near banks and shallows of rivers, ponds, or marshes. In semiarid south Texas, uses small reservoirs and stock tanks. May loaf on shores of small ponds and frequently perches in trees; usually does not alight or venture on deep water. Nests in cavities of elms, willows, live oaks, ebony, mesquite, hackberry, and other trees; also in nest boxes and on the ground. Nest trees may be standing in or located up to 1 km from water, whereas ground nests are usually in grazed brush pastures, well hidden under low shrubs, and usually near water. Uses natural cavities with entrance holes ranging from 10 × 12 cm to 18 × 30 cm, and located about 3 m above ground or water.

SPECIAL HABITAT REQUIREMENTS. Shallow waters or wetlands, and natural cavities in trees or depressions in ground near water for nesting.

Further Reading. United States and Canada: Baldwin et al. 1964, Bellrose 1976, Bolen 1967a, 1967b, Bolen and Forsyth 1967, Godfrey 1966, Johnsgard 1975b, Meanley and Meanley 1958, Terres 1980. Mexico and Central America: Edwards 1972, Land 1970, Peterson and Chalif 1973, Ridgely and Gwynne 1989, Stiles and Skutch 1989. South America: Hilty and Brown 1986, Meyer de Schauensee 1966. Caribbean: Bond 1947, Herklots 1961, Raffaele 1989.

Greater White-fronted Goose
Anser albifrons

Canada (Quebec): Oie à Front Blanc
Mexico: Ganso Manchado

RANGE. Breeds from northern Alaska south to Bristol Bay and Cook Inlet, east across northern Yukon, northwestern Northwest Territories, and southern Victoria Island to northeastern Northwest Territories. Winters from southern British Columbia south along the coastal states, on the Gulf Coast from Texas and Louisiana south to Mexico, and rarely in the lower Mississippi Valley from Missouri southward. A winter visitor mainly in northern and central Mexico, occasionally south to Tabasco in the east and Chiapas in the west. Casual to Surinam.

STATUS. Common throughout range.

HABITAT. Inhabits the borders of shallow marshes and lakes, riverbanks and islands, deltas, dry knolls, and hills near rivers and ponds in arctic tundra. Generally found in areas characterized by dwarf birch, willows, bilberries, crowberries, Labrador tea, cottongrasses, and sphagnum in moist depressions, and by reindeer moss and cetraria on drier sites. Rests on shallow ponds and sloughs in marshes. Typically nests in depressions on the ground in tall grass bordering tidal sloughs or in sedge marshes, usually within 100 m of water, or on hummocks along rivers, streams, and lakes. Generally does not nest in colonies but may be found in loose colonies of 15 to 20 pairs in favored locations. Winters in sheltered inland and coastal marshes and on open terrain and pasturelands with small bodies of water.

SPECIAL HABITAT REQUIREMENTS. Wetlands in arctic tundra.

Further Reading. United States and Canada: Bellrose 1976, Dzubin et al. 1964, Godfrey 1966, Johnsgard 1975b, Pough 1951, Terres 1980. Mexico: Edwards 1972, Peterson and Chalif 1973.

Snow Goose
Chen caerulescens

Canada (Quebec): Oie Blanche
Mexico: Ganso Cerúleo (Blanco)

RANGE. Breeds from northern Alaska east along the arctic coast and islands to Baffin Island, south to Southampton Island, and along both coasts of Hudson Bay to the head of James Bay. Winters from Puget Sound south to the interior valleys of California and Mexico, from Kansas and Missouri south to the Gulf Coast, and along the Atlantic Coast from New York to Florida. During migration, found in large numbers on staging areas in the Dakotas, Minnesota, Iowa, and Nebraska. The first record for the species south of Mexico was in Honduras in 1957. Winter visitor to southern Mexico and the eastern lowlands.

STATUS. Locally abundant.

HABITAT. Breeds on islands of the Canadian Arctic Archipelago or is found within 8 km of salt water on flat tundra of marsh grasses and sedges, in limestone basins, on islands of river deltas, or on plains usually drained by large rivers that open early in the season. Nests in a shallow depression on the ground in large, loose colonies, on dry sites, primarily in unspoiled, primitive areas. Nests, well concealed by tundra grasses and sedges, as close as 5 to 6 m from each other on flat land. During winter, uses both freshwater and saltwater marshes and wet prairies.

SPECIAL HABITAT REQUIREMENTS. Wetlands on arctic tundra.

75

Further Reading. United States and Canada: Cooch 1964, Farrand 1983a, Godfrey 1966, Lemieux 1959, Monroe 1968, Terres 1980, Verner and Boss 1980. Mexico: Edwards 1972, Peterson and Chalif 1973.

Wood Duck
Aix sponsa

Canada (Quebec): Canard Huppé
Mexico: Pato Arco Iris
West Indies: Huyuyo

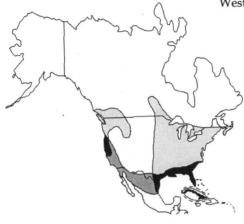

RANGE. Breeds in western North America from southern British Columbia and southwestern Alberta south to central California and western Montana; in eastern North America from east-central Saskatchewan east to Prince Edward Island and Nova Scotia south to central and southeastern Texas and the Gulf Coast. In the West, winters irregularly throughout the breeding range; in the East, winters primarily in the southern parts of the breeding range. In Mexico recorded in Sinaloa, Valley of Mexico, San Luis Potosí, and Tamaulipas. Winters in the West Indies south to the northern Bahamas and probably western Cuba; casual on Jamaica, Puerto Rico, and Saba.

STATUS. Common; has increased in recent years primarily because of maturing forests with tree cavities and the availability of artificial nest boxes.

HABITAT. Inhabits woodlands near shallow, quiet inland lakes, swamps, river bottoms, ponds, marshes, and streams where nest sites are available. Important forest types are central and southern floodplain forests, red-maple swamps, temporarily flooded oak forests, and northern bottomland hardwoods. Prefers to nest in natural cavities 5 to 15 m above ground with entrance holes 10 cm in diameter, cavity depths of 60 cm, and cavity bottoms measuring 25 × 25 cm. Uses nest trees in (or up to 1 km from) water 10 to 50 cm deep. Readily accepts nest boxes provided with nesting materials of wood shavings or sawdust.

SPECIAL HABITAT REQUIREMENTS. Nest holes in trees or nest boxes in or near still or slow-moving water.

Further Reading. United States and Canada: Bellrose 1976, Godfrey 1966, Grice and Rogers 1965, McGilvrey 1968, Palmer 1976b, Terres 1980. Mexico: Edwards 1972, Peterson and Chalif 1973. Caribbean: Bond 1947, Raffaele 1989.

Green-winged Teal
Anas crecca

Canada (Quebec): Sarcelle Européenne

Guatemala: Cerceta Alioscura, Cerceta Aliverde

Mexico: Cerceta Común

Puerto Rico: Pato Aliverde

West Indies: Pato de la Carolina, Pato Serrano, Sarcelle

RANGE. Breeds from Alaska and northwestern and southern Northwest Territories to north-central Labrador and Newfoundland south to central Oregon, Colorado, southern Ontario and Quebec, and Nova Scotia; breeds locally from southern California east to southern New Mexico, Iowa, and Pennsylvania, and on the Atlantic Coast to Delaware. Winters from southern Alaska and southern British Columbia to New Brunswick and Nova Scotia south to Central America; also in the Hawaiian Islands. In Honduras, a rare winter migrant from the United States. In Tobago, a common migrant and winter resident. In Guatemala, a rare migrant and winter resident found in the Pacific lowland and on the volcanic lakes; could occur in the Caribbean lowland. Migrates and winters throughout Mexico. A rare migrant and winter resident in the West Indies, apparently most numerous in the Caribbean in Cuba (Oct.–Apr.).

STATUS. Relatively common throughout most of range; recently has increased in the central region of the United States and Canada.

HABITAT. Inhabits inland waters with dense rushes or other emergent vegetation on mixed and shortgrass prairies; also northern boreal forests. May be found resting on mudbanks or stumps, or perching on low limbs of dead trees. Nests in a depression on dry ground at the base of shrubs, under a log, or in dense grass, usually 1 to 100 m (but up to 0.5 km) from water. Winters in both freshwater and brackish marshes, ponds, streams, and estuaries.

SPECIAL HABITAT REQUIREMENTS. Lakes, marshes, ponds, pools, and shallow streams.

Further Reading. United States and Canada: Bellrose 1976, Godfrey 1966, Harrison 1975, Palmer 1976a, Terres 1980. Mexico and Central America: Edwards 1972, Land 1970, Monroe 1968, Peterson and Chalif 1973, Ridgely and Gwynne 1989. Caribbean: Bond 1947, Herklots 1961, Raffaele 1989.

Mallard

Anas platyrhynchos
(includes Mexican Duck,
formerly *Anas diazi*, now
considered conspecific with
Anas platyrhynchos)

Canada (Quebec): Canard Malard

Costa Rica: Pato Cabeciverde

Cuba: Pato Inglés

Guatemala: Anade Real, Pato
   Cabeciverde, Pato Real

Mexico: Pato de Collar

Puerto Rico: Pato Oscuro

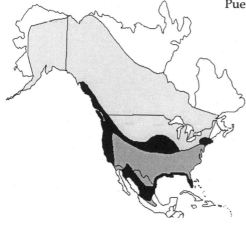

RANGE. Breeds from northern Alaska east to southern Northwest Territories and across to southern Maine south to California, the southern Great Basin, and New Mexico, and from Oklahoma east through the Ohio Valley to Virginia. Winters generally from southern Alaska and southern Canada south to central Mexico. Introduced and established in the Hawaiian Islands. In Costa Rica, once a rare migrant but no recent reports. In Puerto Rico, an extremely rare winter migrant and resident. In Trinidad and Tobago, an occasional winter resident and migrant. In Honduras, a rare migrant sometimes seen in the Bay of Fonesca area and the Sula Valley. In Guatemala, a migrant and winter resident only rarely; found in the Pacific lowland. A rare winter resident in Cuba and the Bahamas; casual to Jamaica, Puerto Rico, and St. Croix.

STATUS. The most common and widely distributed duck in North America.

HABITAT. Inhabits ponds, lakes, rivers, streams, marshes, wet meadows, and wooded swamps of primarily mixed and shortgrass prairie; also boreal forest regions and subarctic deltas. Typically nests on the ground in dry or slightly marshy areas within 100 m of water, sometimes as far as 2 km away in grasslands. Conceals nest well in snowberry clumps, among weeds and grasses, in pastures, stubble, cultivated fields, or marsh vegetation; rarely, in cavities, on hollowed tops of stubs, or in tree crotches. Winters on inland ponds and rivers with some open water; less commonly in coastal marshes.

Further Reading. United States and Canada: Bellrose 1976, DeGraaf and Rudis 1986, Godfrey 1966, Johnsgard 1975b, Palmer 1976a, Terres 1980. Mexico and Central America: Edwards 1972, Land 1970, Monroe 1968, Peterson and Chalif 1973, Ridgely and Gwynne 1989, Stiles and Skutch 1989. Caribbean: Bond 1947, Herklots 1961, Raffaele 1989.

Northern Pintail
Anas acuta
(formerly Pintail)

Canada (Quebec): Canard Pilet

Costa Rica: Pato Rabudo

Guatemala: Pato Chano, Pato Pechiblanco, Pato Rabudo

Mexico: Pato Golondrino

Puerto Rico: Pato Pescuecilargo

West Indies: Canard Pilet, Pato Guineo, Pato Pescuecilargo

RANGE. Breeds from northern Alaska across northern Canada to northern and eastern Quebec, New Brunswick, and Nova Scotia to California, across to the Great Lakes, St. Lawrence River, and Maine. Winters from southern Alaska south to northern New Mexico, and east to central Missouri and the Ohio Valley (uncommonly); along the Atlantic Coast from Massachusetts south throughout the southern United States to South America. Winters throughout Mexico. In Honduras a regular migrant and winter resident. In Panama a rare migrant and winter resident reported mostly from Bocas del Toro, Los Santos, and Cocle; no longer occurs in the Canal area with any regularity. An inconsistent but regular visitor to Puerto Rico. A fairly common winter resident in the West Indies east to Hispaniola; casual in the Lesser Antilles but rare in the Virgin Islands. In Colombia a very erratic and local migrant and winter resident found in the lower Magdalena Valley, Cauca Valley, and western and eastern Andes below 2,600 m.

STATUS. Has been declining for a long time, especially in the western and central United States.

HABITAT. Found in a wide variety of habitats, but typically inhabits open country with low vegetation and with many scattered small, shallow bodies of water. Frequents lakes, rivers, marshes, and ponds in grasslands, barrens, dry tundra, open boreal forest, and cultivated fields. Often builds a nest in a hollow on dry ground, sometimes concealed by grasses or shrubs, usually within 100 m (occasionally 0.5 km) of water. Nests in stubble fields, in a dry portion within a large marsh, or in lightly grazed pasture, but generally avoids timbered or extensively brushy areas. Winters on freshwater and brackish coastal marshes, shallow lagoons, mudflats along rivers, and sheltered marine waters. In Costa Rica in winter found on open freshwater lagoons, marshes, and sloughs. In Puerto Rico and the Virgin Islands prefers freshwater but sometimes found on salt ponds.

SPECIAL HABITAT REQUIREMENTS. Mudbanks or exposed water margins and shallow wetlands for feeding.

Further Reading. United States and Canada: Bellrose 1976, DeGraaf and Rudis 1986, Godfrey 1966, Johnsgard 1975b, Krapu 1974, Palmer 1976a. Mexico and Central America: Edwards 1972, Land 1970, Monroe 1968, Peterson and Chalif 1973, Ridgely and Gwynne 1989, Stiles and Skutch 1989. South America: Hilty and Brown 1986, Meyer de Schauensee 1966. Caribbean: Bond 1947, Raffaele 1989.

Blue-winged Teal
Anas discors

Canada (Quebec): Sarcelle à Ailes Bleues

Costa Rica and Mexico: Cerceta Aliazul Clara

Guatemala: Cerceta Aliazul, Pato Azulejo, Pato Cureto

Puerto Rico: Pato Zarcel

Trinidad and Tobago: Sarcelle à Croissants

Venezuela: Barraquete Aliazul

West Indies: Pato de la Florida, Pato Zarcel, Sarcelle

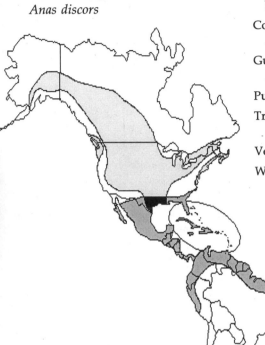

RANGE. Breeds from east-central Alaska and southwestern Northwest Territories to southern Quebec and southwestern Newfoundland, south to northeastern California, east across to central Louisiana, central Tennessee, and eastern North Carolina. Winters from southern California to western and southern Texas, the Gulf Coast, and North Carolina on the Atlantic Coast south to South America. In Mexico, a widespread migrant and winter visitor. In Guatemala, a common migrant and a fairly common winter resident in the Pacific lowland and on the volcanic lakes; uncommon in the Caribbean lowland and Peten. In Honduras, by far the most abundant migrant and winter resident duck—widespread in the lowlands of both coasts. In Panama and Costa Rica, even though it is a migrant, it is easily the most common duck. In Colombia, the most abundant and widespread migrant and winter resident among waterfowl. A migrant and winter resident in the Guyanas, Venezuela north of the Orinoco River, Ecuador and Peru south to Junín and rarely to northern Chile, Brazil in Maranhão and Pará, Uruguay, Argentina to Buenos Aires. Found throughout the West Indies, where it is the most common winter resident duck. In Trinidad and Tobago, a common winter migrant.

STATUS. Common throughout range.

HABITAT. Prefers wetlands on rolling tallgrass prairie but is also found in mixed shortgrass prairie and boreal and deciduous forests. More of a shoreline inhabitant than one of open water, preferring calm water or sluggish currents to fast water. Uses rocks protruding from water, muskrat houses, trunks or limbs of fallen trees, or bare stretches of shoreline or mudflats as resting sites. Nests on dry ground in dense grassy sites such as hay fields and sedge meadows where the vegetation ranges from 20 to 60 cm high at the onset of nesting, usually within several hundred meters of open water; occasionally on a sedge tussock or muskrat house, in slough grass, or in alfalfa fields. Nests communally in good habitat. Winters on shallow inland freshwater marshes and on brackish and saltwater marshes. In Costa Rica and Panama in winter prefers freshwater marshes, ponds, and sloughs, but occurs in virtually any shallow, still, or slowly flowing water in the lowlands or at middle elevations (i.e., river pools, salt ponds, estuaries, flooded fields). In Puerto Rico and the Virgin Islands found in all aquatic habitats from fresh to hypersaline salt ponds.

SPECIAL HABITAT REQUIREMENTS. Marshes, sloughs, ponds, lakes, and sluggish streams.

Further Reading. United States and Canada: Bellrose 1976, Bennett 1938, DeGraaf and Rudis 1986, Godfrey 1966, Johnsgard 1975b, Palmer 1976a. Mexico and Central America: Edwards 1972, Land 1970, Monroe 1968, Peterson and Chalif 1973, Ridgely and Gwynne 1989, Stiles and Skutch 1989. South America: Hilty and Brown 1986, Meyer de Schauensee 1966, Meyer de Schauensee and Phelps 1978. Caribbean: Bond 1947, Herklots 1961, Raffaele 1989.

Cinnamon Teal
Anas cyanoptera

Canada (Quebec): Sarcelle Cannelle

Costa Rica: Cerceta Castaña

Guatemala and Puerto Rico: Pato
  Canela

Mexico: Cerceta Aliazul Café

Venezuela: Barraquete Colorado

RANGE. Breeds from southern British Columbia east to southwestern Saskatchewan (probably) and south to northern Baja California, Jalisco, Chihuahua, Tamaulipas, and central Texas. Winters from central California, southern Nevada, central Utah, southeastern Arizona, southern New Mexico, and central Texas south through Middle America to Colombia (occasionally), northern Venezuela, and probably Ecuador; occasionally to Cuba. In Mexico a common migrant and winter visitor throughout. In Guatemala a rare migrant and winter resident in the Pacific lowland. In Costa Rica a casual migrant and winter resident in the Tempisque basin. In Honduras a rare migrant or spring visitor. In Puerto Rico and the Virgin Islands rare but apparently increasing. Resident in South America in Colombia (eastern Andes and the Cauca and Magdalena Valleys) and from central Peru, Bolivia, Paraguay, and southern Brazil south to the Straits of Magellan.

STATUS. Common in the western United States.

HABITAT. Inhabits small, shallow wetlands, including areas with alkaline waters, but may also be found around larger and deeper lakes. Nests on the ground in dense grasses less than 0.5 m high, in cattails or reeds near water, or in a hollow in the ground, often 30 m or more from water. Broods may be moved as far as 1 km from the nest site to good brood cover of lush emergent vegetation adjacent to water with abundant food. Winters primarily on freshwater, though occasionally found in marine habitats.

SPECIAL HABITAT REQUIREMENTS. Shallow lake margins, ponds bordered by tules and grasses, sloughs, marshes, sluggish streams, reservoirs, and irrigation ditches.

Further Reading. United States and Canada: Bellrose 1976, Godfrey 1966, Grinnell and Miller 1944, Johnsgard 1975b, Low and Mansell 1983, Palmer 1976a, Verner and Boss 1980. Mexico and Central America: Edwards 1972, Land 1970, Monroe 1968, Peterson and Chalif 1973, Ridgely and Gwynne 1989, Stiles and Skutch 1989. South America: Hilty and Brown 1986, Meyer de Schauensee 1966, Meyer de Schauensee and Phelps 1978. Caribbean: Bond 1947, Raffaele 1989.

Northern Shoveler
Anas clypeata

Canada (Quebec): Canard Souchet

Costa Rica and Guatemala: Pato Cuchara

Mexico: Pato Cucharón

Puerto Rico: Pato Cuchareta

West Indies: Canard Souchet, Pato Cuchareta

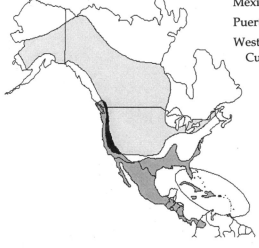

RANGE. Breeds from northern Alaska to northern Manitoba, south to northwestern and eastern Oregon, northern Utah, Colorado, Nebraska, Missouri, and central Wisconsin. Winters from the coast of southern British Columbia to central Arizona east to the Gulf Coast and to South Carolina on the Atlantic Coast south to South America. In Mexico a widespread winter visitor. In Honduras a fairly common migrant and winter resident on the Caribbean slope; in Panama a rare winter visitor. In Costa Rica a fairly common winter resident in Guanacaste, elsewhere local and in small numbers. In Colombia usually a regular migrant and winter resident found in the Magdalena and Cauca Valleys. Found virtually throughout the West Indies but uncommon in Puerto Rico and extremely rare in St. Croix, St. Thomas, and St. John. In Trinidad a rare winter migrant.

STATUS. Fairly common; declining in some areas west of the Mississippi River.

HABITAT. Prefers shallow prairie marshes, particularly with abundant plant and animal life floating on the surface, but also potholes, sloughs, and marshes in taiga, forests, and (less commonly) cultivated country. Tolerates a wide range of water conditions, from clean and clear to muddy, flowing

86

to stagnant, and considerably alkaline. Loafs on mudbanks or low sloping shorelines with short or flattened vegetation. Nests on dry ground in a slight hollow, preferably in short grasses within 100 m of water, but will nest in hay fields, meadows, and rarely bulrushes if grasses are not available. Seldom nests in weed patches, and avoids woody vegetation such as willows. Winters in freshwater and brackish habitats. In Panama in winter, found in freshwater ponds in the lowlands on both slopes. In Trinidad in winter, found in fresh and brackish marshes and on mudflats.

SPECIAL HABITAT REQUIREMENTS. Shallow waters with muddy bottoms, surrounded by dry grassy areas for nesting.

Further Reading. United States and Canada: Bellrose 1976, DeGraaf and Rudis 1986, Godfrey 1966, Johnsgard 1975b, Palmer 1976a, Poston 1974. Mexico and Central America: Edwards 1972, Land 1970, Monroe 1968, Peterson and Chalif 1973, Ridgely and Gwynne 1989, Stiles and Skutch 1989. South America: Hilty and Brown 1986, Meyer de Schauensee 1966. Caribbean: Bond 1947, Herklots 1961, Raffaele 1989.

Gadwall
Anas strepera

Canada (Quebec): Canard Chipeau
Mexico: Pato Friso, Pato Pinto

Gadwall

RANGE. Breeds from southern Alaska and southern Yukon to the New Brunswick–Nova Scotia border, south locally to southern California, northern Texas, central Minnesota, and northern Pennsylvania and on the Atlantic Coast to North Carolina. Winters from southern Alaska, southern British Columbia, and Colorado to southern South Dakota, Iowa, the southern Great Lakes, and Chesapeake Bay on the Atlantic Coast south to Mexico and the Gulf Coast. In Mexico winters to Guerrero and Tabasco. Winters in Cuba and Jamaica; rare elsewhere in the West Indies.

STATUS. Uncommon, but numbers have increased substantially since the mid-1970s, and the range is expanding eastward.

HABITAT. Inhabits prairie marshes, sloughs, ponds, or small lakes in grasslands in both freshwater and brackish habitats. Generally avoids wetlands bordered by woodlands or thick brush, preferring those bordered by dense, low herbaceous vegetation or shrubby willows and with grassy islands. Nests on the ground on a well-drained site on islands in lakes, in upland meadows or pastures, in alfalfa fields, or on prairies, usually within 50 m of water. Prefers to nest in uplands rather than over water, especially in dense, coarse herbaceous vegetation and under shrubby willows. Prefers to winter in freshwater marshy habitats but can be found on open water of any kind.

SPECIAL HABITAT REQUIREMENTS. Shallow water for feeding; marshes or grassy areas near water for nesting.

Further Reading. United States and Canada: Bellrose 1976, DeGraaf and Rudis 1986, Godfrey 1966, Johnsgard 1975b, Palmer 1976a, Terres 1980. Mexico: Edwards 1972, Peterson and Chalif 1973. Caribbean: Bond 1947.

American Wigeon
Anas americana

Canada (Quebec): Canard Siffleur d'Amérique

Costa Rica and Venezuela: Pato Calvo

Guatemala: Chano, Pato Calvo, Pato Frontino

Mexico: Pato Chalcuán, Pato Panadero

Puerto Rico: Pato Cabeciblanco

West Indies: Moni-blanco, Pato Cabeciblanco, Pato Lavanco

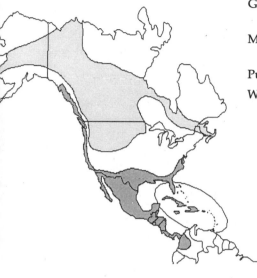

RANGE. Breeds from central Alaska and central Yukon to New Brunswick and southern Nova Scotia, south to northeastern California, central Colorado, South Dakota, southern Ontario, and northern New York, sporadically to the Atlantic Coast. Winters throughout the Hawaiian Islands and from southern Alaska to southern Nevada, sporadically across the central United States to the southern Great Lakes and Ohio Valley, and on the Atlantic Coast from Nova Scotia south throughout the southern United States to Central America. Winters throughout Mexico. In Guatemala a fairly common migrant and winter resident found along the Pacific Coast and on the interior lakes up to 1,500 m. In Honduras a fairly common migrant and winter visitor on the Caribbean slope. In Costa Rica a regular winter resident usually in fairly small numbers in Tempisque basin. In Panama a locally fairly common but somewhat irregular winter migrant. In Colombia an uncommon and erratic migrant and winter visitor. Occurs virtually throughout the West Indies. In Trinidad and Tobago seen occasionally on the Caroni Swamp and the Point-à-Pierre reservoir.

STATUS. Common.

89

HABITAT. Inhabits freshwater wetlands and lakes from tundra to shortgrass and mixed prairie, preferring permanent to temporary waters. Nests in a hollow on dry ground on an island or on shore, in tall grasses or weeds, or at the base of a tree or bush, as far as 400 m from water. Commonly associates with diving ducks, and in winter frequents coastal marshes and bays, wet meadows, and shallow freshwater and brackish ponds. In Panama in winter found on ponds, lakes, and freshwater marshes. In Costa Rica in winter prefers marshes, pond margins, and sloughs with open shorelines. In Guatemala in winter often found on the larger lakes.

SPECIAL HABITAT REQUIREMENTS. Large lakes, ponds, marshes, sluggish streams and rivers, with open water and exposed shoreline.

Further Reading. United States and Canada: Baldwin et al. 1964, Bellrose 1976, DeGraaf and Rudis 1986, Godfrey 1966, Johnsgard 1975b, Palmer 1976a, Terres 1980, Verner and Boss 1980. Mexico and Central America: Edwards 1972, Land 1970, Monroe 1968, Peterson and Chalif 1973, Ridgely and Gwynne 1989, Stiles and Skutch 1989. South America: Hilty and Brown 1986, Meyer de Schauensee 1966, Meyer de Schauensee and Phelps 1978. Caribbean: Bond 1947, Herklots 1961, Raffaele 1989.

Canvasback
Aythya valisineria

Canada (Quebec): Morillon à Dos Blanc

Guatemala: Porrón Dorsiblanco

Mexico: Pato Coacoxtle

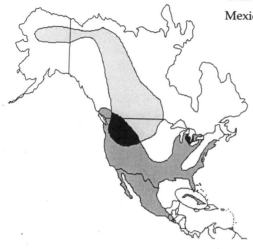

RANGE. Breeds from central Alaska and northern Yukon to eastern Ontario and south to south-coastal Alaska; locally in inland areas to northeastern California across to northern Utah, central New Mexico, northwestern Iowa, and southern Ontario. Winters along the Pacific Coast from the central Aleutians and south-coastal Alaska south to Baja California, from Arizona and New Mexico to the Great Lakes, and on the Atlantic Coast from New England south to the Gulf Coast and Mexico. In Mexico in winter south to central Mexico (Michoacán, Federal District, and Veracruz). In Guatemala and Honduras a rare winter migrant from the United States. A rare fall and winter visitor to Cuba and Puerto Rico.

STATUS. Locally common; numbers are declining, especially in the western parts of its range, because of loss of breeding habitat through drainage and drought.

HABITAT. In breeding season, prefers shallow prairie marshes, 4 ha or less, or other permanent wetlands with stable water levels, bordered by cattails and bulrushes, with little, if any, wooded vegetation around the shoreline. Usually nests over water 15 to 60 cm deep in bulrushes, reeds, or cattails, sometimes on a muskrat house, rarely on dry ground. Attaches nest to surrounding plants or builds on a mat of floating dead plants, 1 to 20 m from edge of open water. Large lakes of 60 ha or more, marshes, and rivers with

91

submerged beds of sago pondweed are favored during migration. Winters primarily on estuaries and sheltered bays, sometimes on deep, freshwater lakes, where wild celery and pondweeds thrive.

SPECIAL HABITAT REQUIREMENTS. Marshes, ponds, lakes, and rivers bordered by emergent vegetation and with enough open water for taking off and landing.

Further Reading. United States and Canada: Bellrose 1976, DeGraaf and Rudis 1986, Evans and Bartels 1981, Godfrey 1966, Johnsgard 1975b, Palmer 1976a, Stoudt 1982, Terres 1980. Mexico and Central America: Edwards 1972, Land 1970, Peterson and Chalif 1973, Ridgely and Gwynne 1989. Caribbean: Bond 1947.

Redhead
Aythya americana

Canada (Quebec): Morillon à Tête Rouge

Guatemala: Pato Cabecicolorado, Porrón Americano

Mexico: Pato Cabecirrojo, Pato Cabeza Roja

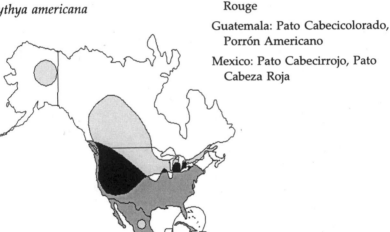

RANGE. Breeds from central British Columbia and southwestern Northwest Territories to northwestern and central Minnesota, south to southern California, the Texas Panhandle, and northern Iowa; locally in south-central and southeastern Alaska, and sporadically in eastern North America. Winters from British Columbia on the Pacific Coast, in the interior from Nevada to the middle Mississippi and Ohio Valleys and the Great Lakes, and from New England on the Atlantic Coast south throughout the southern United States to Mexico. Winters throughout Mexico. In Guatemala a rare winter visitor to be looked for in the Pacific lowlands and on the volcanic lakes up to 1,500 m. Winters south to the Bahamas, Cuba, and Jamaica.

STATUS. Common; generally increasing, especially in the western United States.

HABITAT. Inhabits freshwater marshes, sloughs, ponds, and shallow lakes bordered by hardstem bulrush, cattails, reeds, or sedges in prairies. Usually nests in emergent vegetation or on a floating mat of dead plant material over shallow water 15 to 60 cm deep, fairly close to shore, but occasionally on dry ground or over water 1.3 m deep. Tends to be semiparasitic, sometimes laying eggs in nests of other waterfowl, especially the Canvasback. Prefers to rear broods in potholes at least 0.5 ha in size, with deeper water than those used for nesting. Winters primarily on freshwater and brackish lakes, rivers, and estuaries, in areas well protected from heavy wave action. In Guatemala in winter found especially on open freshwater.

SPECIAL HABITAT REQUIREMENTS. Wetlands at least 0.25 ha in size bordered by permanent, dense emergent vegetation, with stable water levels during the nesting season.

Further Reading. United States and Canada: Bellrose 1976, DeGraaf and Rudis 1986, Godfrey 1966, Johnsgard 1975b, Lokemoen 1966, Palmer 1966, Tate and Tate 1982, Weller 1964. Mexico and Central America: Edwards 1972, Land 1970, Peterson and Chalif 1973. Caribbean: Bond 1947.

Ring-necked Duck
Aythya collaris

Canada (Quebec): Morillon à Collier

Costa Rica: Porrón Collarejo

Guatemala: Pato de Collar, Porrón Acollarado

Mexico: Pato Piquianillado

Puerto Rico: Pato del Medio

Venezuela: Pato Zambullidor de Collar

West Indies: Cabezón, Pato del Medio, Pato Negro

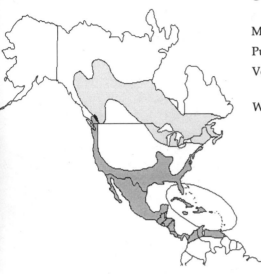

RANGE. Breeds in east-central and southeastern Alaska, and from central British Columbia and northwestern and southern Northwest Territories to Newfoundland and Nova Scotia, south to northeastern California, southeastern Arizona, northern Illinois, and Massachusetts. Winters on the Pacific Coast from southeastern Alaska, in the interior from southern Nevada to the lower Mississippi and Ohio Valleys, and on the Atlantic Coast from New England south through the southern United States to Panama. Winters throughout Mexico. A rare migrant and winter visitor throughout Guatemala. In Honduras a regular but uncommon migrant and winter visitor in the Caribbean lowlands. In Costa Rica an uncommon to locally common winter resident especially in Guanacaste (mainly Tempisque basin); rare and local on the Caribbean slope and in Valle Central. In Panama an occasional winter visitor in very small numbers; reported from Chiriquí, the Canal area (Miraflores Lake and Pedro Miguel), and eastern Panama province. An uncommon and local visitor to Puerto Rico; rare on St. Croix, and extremely rare on St. John and St. Thomas. Winters as far south as Barbados, St. Lucia, and St. Andrew; also Trinidad and Margarita Island.

STATUS. Common; long-term increase in the western part of its range; recently in eastern part also.

HABITAT. Inhabits shallow, dense bogs, swamps, and marshes, especially those with sweetgale or leatherleaf cover, typically having a pH range of 5.5 to 6.8, and preferably near or in woodlands. Also uses small potholes, sloughs, and beaver flowages near larger wooded lakes or rivers with submerged and emergent vegetation. Nests on floating mats of vegetation, among hummocks, in clumps of marsh vegetation or on islands, on relatively dry sites usually within a few feet of water; seldom in emergent vegetation over water. Winters on fresh or brackish marshes, lakes, and estuaries, rarely on strictly saline waters. In Panama, Costa Rica, and Guatemala found on ponds, lakes, and freshwater marshes.

SPECIAL HABITAT REQUIREMENTS. Wetlands with an expanse of open water.

Further Reading. United States and Canada: Bellrose 1976, DeGraaf and Rudis 1986, Godfrey 1966, Johnsgard 1975b, Mendall 1958, Palmer 1976a. Mexico and Central America: Edwards 1972, Land 1970, Peterson and Chalif 1973, Ridgely and Gwynne 1989, Stiles and Skutch 1989. South America: Meyer de Schauensee 1966, Meyer de Schauensee and Phelps 1978. Caribbean: Bond 1947, Herklots 1961, Raffaele 1989.

Lesser Scaup
Aythya affinis

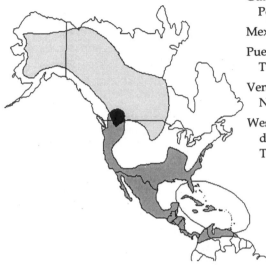

Canada (Quebec): Petit Morillon

Costa Rica: Porrón Menor

Guatemala: Pato Bola, Pato Chaparra, Porrón Menudo

Mexico: Pato Boludo Menor

Puerto Rico: Pato Pechiblanco, Pato Turco

Venezuela: Pato Zambullidor del Norte

West Indies: Canard Tête-noire, Pato del Medio, Pato Morisco, Pato Turco

RANGE. Breeds from central Alaska to northern Manitoba and western Ontario, south to southern interior British Columbia, northern Wyoming, and northwestern and central Minnesota; casually or irregularly east to southern Ontario and west-central Quebec, and south to northeastern California and Colorado, northern Illinois, and northern Ohio. Winters from southern Alaska and southern British Columbia and Utah to the southern Great Lakes region and New England, south throughout the southern United States to South America. Winters throughout Mexico. In Guatemala a fairly common migrant and winter resident found on the large volcanic lakes. In Honduras a common winter migrant outnumbered only by Blue-winged Teal. In Costa Rica an uncommon to rare winter resident, regular only in Guanacaste. Has declined in the Caribbean. Now uncommon in Puerto Rico and uncommon to rare in the Virgin Islands. In Panama a locally fairly common winter visitor (Nov.–Mar.) in less disturbed portions of larger lakes. In Colombia probably a rare and sporadic winter visitor to 1,000 m in the Cauca Valley and to 2,600 m in the eastern Andes from southern Boyaca to Sabana de Bogota. A common winter duck in the Bahamas and Greater Antilles, but rare in the Lesser Antilles. In Trinidad and Tobago a frequent to infrequent winter visitor.

STATUS. Abundant.

HABITAT. Inhabits grass-margined wetlands in prairie and forested habitats, with the largest breeding concentrations found in marshes of hardstem bulrush bordering lakes. Usually nests in upland areas adjacent to water but also on islands, in wet meadows, in shallows at edges of bays and sloughs among bulrushes, or on tussocks in marshes. Conceals nest well in hollows usually on dry ground, in grasses, nettles, low brush, even under driftwood. Winters on sheltered bays, estuaries, coastal marshes, and freshwater lakes, preferring a more sheltered habitat than does Greater Scaup. In Costa Rica prefers freshwater ponds, lakes, and sloughs where water is reasonably clear and at least 1 m deep. In Puerto Rico and the Virgin Islands almost always found in flocks on bodies of open water. In Guatemala generally on large bodies of open water.

SPECIAL HABITAT REQUIREMENTS. Lakes, ponds, potholes, marshes, and sloughs bordered by grasses.

Further Reading. United States and Canada: Bellrose 1976, Godfrey 1966, Johnsgard 1975b, Palmer 1976a, Terres 1980. Mexico and Central America: Edwards 1972, Land 1970, Peterson and Chalif 1973, Ridgely and Gwynne 1989, Stiles and Skutch 1989. South America: Hilty and Brown 1986, Meyer de Schauensee 1966, Meyer de Schauensee and Phelps 1978. Caribbean: Bond 1947, Herklots 1961, Raffaele 1989.

Hooded Merganser
Lophodytes cucullatus

Canada (Quebec): Bec-scie Couronné

Mexico: Pato Mergo Copetón

Puerto Rico: Merganse de Caperuza

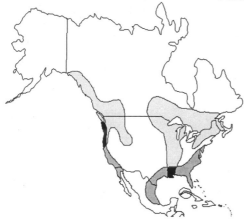

RANGE. Breeds from southern Alaska to Nova Scotia, south to Oregon and Idaho, east to Maine and Massachusetts, and locally in the Mississippi Valley and southeastern United States. Winters on freshwater from British Columbia and New England south to California, Texas, and Florida. Rare migrant and winter visitor to northern Mexico, recorded to Baja California, Tamaulipas, Veracruz, Michoacán, and the Valley of Mexico. A rare winter visitor in the West Indies, where it has been recorded from the Bahamas, Cuba, Puerto Rico, St. Croix, and Martinique.

STATUS. Locally common or rare, though has recently increased in the eastern part of its range.

HABITAT. Inhabits wooded, clear freshwater habitats, preferably water with sandy, gravelly, or cobbled bottoms. Prefers fast-flowing water, but also uses standing water as long as it is clear, contains abundant small fish and invertebrates, and has available nest sites. Easily disturbed, tends to avoid areas of human activity. In Wisconsin, brood habitat was described as rivers with high levels of food resources, fast currents (0.2–0.3 m/sec), and wide (15–20 m), moderately deep (0.5 m) channels with cobbled bottoms and heavy surrounding cover of mixed hardwoods. A cavity nester that uses almost any hole or hollow tree, at any height, as long as it is large enough for the female and her nest. The nest tree is usually within a few meters of, or standing in, water. Prefers flooded shoreline with standing trees, and with snags or stumps interspersed, but will nest in other locations, including nest boxes.

SPECIAL HABITAT REQUIREMENTS. Wooded, clear streams, rivers, swamps, ponds, and lakes with cavity trees.

Further Reading. United States and Canada: Bellrose 1976, Farrand 1983a, Godfrey 1966, Johnsgard 1975b, Kitchen and Hunt 1969, Morse et al. 1969, Palmer 1976a, Terres 1980. Mexico and Central America: Edwards 1972, Peterson and Chalif 1973. Caribbean: Bond 1947, Raffaele 1989.

Red-breasted Merganser
Mergus serrator

Canada (Quebec): Bec-scie à Poitrine Rousse

Mexico: Pato Mergo Pechicastaño

Puerto Rico: Merganse Pechirrojo

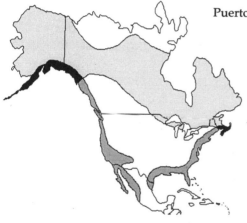

RANGE. Breeds from northern Alaska to east-central Northwest Territories, northern Baffin Island, Labrador, and Newfoundland south to the Aleutian Islands, northern British Columbia, central Minnesota, northern New York, and Nova Scotia, casually south along the Atlantic Coast to Long Island, New York. Winters primarily along coasts and on large inland bodies of water from southern Alaska, the Great Lakes, and Nova Scotia south to Baja California, southern Texas, and the Gulf Coast. In Mexico winters mainly on both coasts of Baja California and in Sonora, Sinaloa, Chihuahua, and Tamaulipas. Casual in the West Indies, where it has been recorded from the Bahamas, Cuba, and Puerto Rico.

Red-breasted Merganser

STATUS. Common.

HABITAT. Prefers to breed on small islands or islets with low, prostrate woody vegetation or other natural features to cover the nest, and with open shores, gravel bars, or rocks to provide roosting and preening areas for drakes and young. Nests on the ground under low cover, generally within 10 m of water, preferably on islands, but also on riverbanks and lakeshores, in marshes, on rocky islets, or in bank recesses. Nests may be under low conifer boughs, under or between boulders in shallow cavities, in tall grass, heather, or bracken, or under driftwood. Although this species prefers inland waters, it may be found along the coasts on shores and on marine islets. Winters mainly in estuaries and sheltered bays, less frequently on inland freshwater.

SPECIAL HABITAT REQUIREMENTS. Rivers, ponds, and lakes with some overhead cover nearby for nesting.

Further Reading. United States and Canada: Bellrose 1976, Clapp et al. 1982, Godfrey 1966, Johnsgard 1975b, Palmer 1976b. Mexico: Edwards 1972, Peterson and Chalif 1973. Caribbean: Bond 1947.

Ruddy Duck
Oxyura jamaicensis

Canada (Quebec): Canard Roux

Guatemala: Pato Café, Pato Rojizo

Mexico and Puerto Rico: Pato Chorizo, Pato Rojizo Alioscuro

West Indies: Canard Plongeon, Coucouraime, Pato Chorizo, Pato Espinoso, Pato Rojo

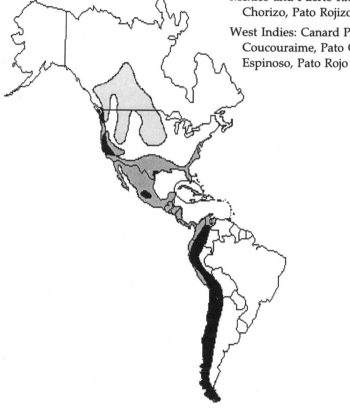

RANGE. Breeds in east-central Alaska and from central and northeastern British Columbia to western Ontario and south to southern California, western and southern Texas, and southwestern Louisiana, with some scattered breeding east to Nova Scotia and south to Florida. Birds in the Valley of Mexico, Greater Antilles, Lesser Antilles south to Grenada, and South America are resident. Resident populations are augmented by migrants. Winters from southern British Columbia, Idaho, Colorado, Kansas, the Great Lakes, and on the Atlantic Coast from Massachusetts south throughout the southern United States to Mexico. In Guatemala a rare winter visitor on large lakes

101

in the volcanic highlands. In Honduras a rare winter migrant from the United States.

STATUS. Common.

HABITAT. Inhabits permanent freshwater and alkaline prairie marshes having dense stands of cattails, bulrushes, whitetop, and reeds, and relatively stable water levels. Usually nests over shallow water in emergent vegetation, on a floating mat of vegetation, or on a platform built up from the floor of the marsh. Occasionally nests on a muskrat house, floating log, or old coot nest. "Dump" nests are common in marshes with fluctuating water levels. Nests in both large and small marshes, from potholes less than 1 ha in size to 500-ha sloughs. Commonly loafs and rests on water well out from shore. Prefers large bodies of shallow freshwater and brackish water, especially with areas of aquatic plant growth, during migration. Winters on ice-free inland waters or on sheltered shallow brackish or saltwater coastal waters.

SPECIAL HABITAT REQUIREMENTS. Open water close to dense emergent vegetation with muskrat channels or natural passageways to allow movement between the nest site and open water. Freshwater ponds (Puerto Rico).

Further Reading. United States and Canada: Bellrose 1976, Clapp et al. 1982, Cottam 1939, DeGraaf and Rudis 1986, Godfrey 1966, Johnsgard 1975b, Joyner 1969, Palmer 1976b, Siegfried 1976. Mexico and Central America: Edwards 1972, Land 1970, Monroe 1968, Peterson and Chalif 1973, Ridgely and Gwynne 1989. Caribbean: Bond 1947, Raffaele 1989.

Black Vulture
Coragyps atratus

Canada (Quebec): Vautour Noir
Chile: Gallinazo, Jote, Queluy
Costa Rica: Zopilote Negro
Guatemala: Zopilote, Zopilote Negro
Mexico: Carroñero Común
Peru: Gallinazo Cabeza Negra
Trinidad and Tobago: Corbeau
Venezuela: Zamuro

RANGE. Resident from southern Arizona and western Texas to southern Illinois, southern Indiana, and New Jersey south to the Gulf Coast, southern Florida, and Central and South America to Argentina. May retreat from northern range limits in winter, with North American individuals migrating as far as Panama.

STATUS. Common, generally increasing; range is extending slightly northward.

HABITAT. Nearly ubiquitous except in heavily forested regions. Found in the southern Great Plains, southeastern pine forests, oak-hickory forests, and intermediate oak-pine forests. Does not construct a nest. Frequently lays eggs in hollow bases of trees or stumps, rarely more than 3 to 5 m above ground, but also on the ground, under dense or thorny vegetation, in cavities of rocks, on the floor of caves, on cliff ledges, or in abandoned buildings. Winter roosts in the northeastern United States (Pennsylvania) preferentially contain large or dense conifers or large deciduous trees (Thompson et al. 1990).

103

SPECIAL HABITAT REQUIREMENTS. Savanna, thorn forest, second growth.

Further Reading. United States and Canada: Brown and Amadon 1968, Coleman and Fraser 1989, Farrand 1983a, Godfrey 1966, Harrison 1979, Heintzelman 1979, Johnson 1965, Scott et al. 1977, Tate and Tate 1982, Terres 1980. Mexico and Central America: Edwards 1972, Land 1970, Monroe 1968, Peterson and Chalif 1973, Ridgely and Gwynne 1989, Stiles and Skutch 1989. South America: Hilty and Brown 1986, Meyer de Schauensee 1966. Caribbean: Herklots 1961.

Turkey Vulture
Cathartes aura

Canada (Quebec): Vautor à Tête Rouge

Chile: Gallinazo

Costa Rica: Zopilote Cabecirrojo

Guatemala: Vinda

Mexico: Aura Cabecirrojo, Aura Común

Trinidad and Tobago: Corbeau à Tête Rouge

Venezuela: Oripopo, Zamuro

West Indies: Aura, Aura Tinosa, Carrion Crow (Bahamas), John Crow (Jamaica)

RANGE. Breeds from southern British Columbia, western Ontario, extreme southern Ontario, and Massachusetts south throughout the remaining continental United States to South America. North American individuals winter from northern California, Arizona, Texas, Nebraska, the Ohio Valley, and Pennsylvania south to the Gulf Coast, Florida, and northwestern South America. In Guatemala more birds are probably present during migration as the northern populations move through. In Colombia, where numbers are augmented by migrants from the United States, this species is most abundant in the northwest. Found in Cuba (including coastal cays), Isle of Pines, Cayman Islands, Jamaica, Hispaniola, Puerto Rico, and northwestern Bahamas; casual on Bimini, New Providence, and St. Croix.

STATUS. Common; long-term increase in the eastern United States.

HABITAT. Uses a wide variety of habitats, from tropical and temperate forests to open plains and deserts, lowlands, and mountains. Does not build a nest. Lays eggs on the floor of caves (preferably with two entrances), on the ground in dense shrubs, in hollow logs or stumps, on rocky outcrops or ledges (especially in the western United States), in swamps, in hollow snags, in old hawk nests, or on the floor in abandoned buildings. The eggs are usually well hidden from view and inaccessible to predators. Preens and roosts in tall snags or trees with open branches. May gather in groups of up to 70 birds to roost at night. Communal winter roosts in eastern United States characterized by large or dense conifers or large deciduous trees (Thompson et al. 1990).

Further Reading. United States and Canada: Brown and Amadon 1968, Coleman and Fraser 1989, DeGraaf and Rudis 1986, Godfrey 1966, Grinnell and Miller 1944, Heintzelman 1979, Johnson 1965, Prior 1990, Sprunt 1955, Terres 1980, Verner and Boss 1980, Wilbur and Jackson 1983. Mexico and Central America: Edwards 1972, Land 1970, Monroe 1968, Peterson and Chalif 1973, Ridgely and Gwynne 1989, Stiles and Skutch 1989. South America: Hilty and Brown 1986, Meyer de Schauensee 1966, Meyer de Schauensee and Phelps 1978. Caribbean: Bond 1947, Herklots 1961, Raffaele 1989.

Osprey

Pandion haliaetus

Canada (Quebec): Aigle Pecheur

Canada (Quebec): Aigle Pecheur

Chile, Costa Rica, Mexico, and
Venezuela: Aguila Pescadora

Puerto Rico: Aguila de Mar, Aguila
Pescadora

Trinidad and Tobago: Gavilán
Pecheur

West Indies: Aguila de Mar, Aiglon,
Guaraguao de Mar, Guincho,
Halcón Pescador, Malfini de la Mer

RANGE. Breeds from northwestern Alaska and northern Yukon to central Labrador and Newfoundland south locally to Baja California, central Arizona, southern Texas, the Gulf Coast, and southern Florida. Winters from central California, southern Texas, the Gulf Coast, and Florida south to Argentina. In Mexico a widespread winter visitor. In Guatemala, Costa Rica, and Honduras a common migrant and winter resident, with the local population being augmented by northern birds. In Panama fairly common during the northern winter around larger bodies of water in the lowlands of both slopes. In Colombia a fairly common winter resident throughout; some nonbreeding birds may be present year-round. Mainly a winter resident south to northern Chile, northern Argentina, and Uruguay. A fairly common winter resident in Puerto Rico and the Virgin Islands. Breeds in the Bahamas and on cays of Cuba. North American individuals winter throughout the West Indies. A winter visitor to both Trinidad and Tobago.

STATUS. Locally common to uncommon; had declined precipitously, primarily because of pesticides but also because of habitat destruction and human disturbance, but has been increasing throughout its range in North America. Reduced hatching success continues in eastern United States because of exposure to pesticide contamination on wintering grounds outside the United States (Steidl et al. 1991).

HABITAT. Nearly cosmopolitan distribution, occurring on every continent except Antarctica. Occupies a wide range of habitats in close association with water, primarily lakes and rivers, and along coastal waters with adequate supplies of fish. Nests in loose colonies or singly, and uses a wide variety of structures to support large stick nests, which may be 20 m or more above ground. Prefers a snag in or near water, with a broken top or side limbs able to support the nest. Prefers tall snags that provide good visibility and security. Also nests on pilings, utility poles, duck blinds, buildings, steel towers for transmission lines, windmills, channel markers, a wide variety of living or dead trees, wooden platforms in marshes, on cliffs, and sometimes on the ground. Nest site may be used by the same pair year after year.

SPECIAL HABITAT REQUIREMENTS. Elevated nest sites near water with rich fish resources; will use nest platforms where natural sites are absent. Freedom from human disturbance during breeding season.

Further Reading. United States and Canada: Bird 1983, DeGraaf and Rudis 1986, Godfrey 1966, Hagan and Walters 1990, Heintzelman 1979, Hunter 1990, Johnson 1965, Levenson and Koplin 1984, Sprunt 1955, Steidl and Griffin 1991, Vahle et al. 1988, Van Daele and Van Daele 1982, Zarn 1974a. Mexico and Central America: Edwards 1972, Land 1970, Monroe 1968, Peterson and Chalif 1973, Ridgely and Gwynne 1989, Stiles and Skutch 1989. South America: Hilty and Brown 1986, Meyer de Schauensee 1966, Meyer de Schauensee and Phelps 1978. Caribbean: Bond 1947, Herklots 1961, Raffaele 1989.

American Swallow-tailed Kite

Elanoides forficatus
(formerly Swallow-tailed Kite)

Canada (Quebec): Milan à Queue Fourchue

Costa Rica: Elanio Tijereta

Guatemala: Elanio Tijereta, Gavilán Tijereta

Mexico: Milano Tijereta

Trinidad: Queue en Ciseaux

Venezuela: Gavilán Tijereta

RANGE. Breeds locally from South Carolina south to Florida, west to Louisiana, and from southeastern Mexico (Campeche and Quintana Roo) south through most of Middle America and Trinidad to Bolivia, northern Argentina, and southern Brazil. North and Central American birds winter in South America. In Mexico a rare winter migrant. In Guatemala a fairly common migrant. In Honduras a fairly common migrant confined to the Caribbean lowlands. In Costa Rica and Panama a common migrant. A migrant to Colombia and northern Ecuador and probably to southern Brazil (Paraná, Mato Grosso). An uncommon migrant in Trinidad. Occurs rarely in the Lesser Antilles. North American individuals migrate to South America via Cuba, Grand Cayman, and Jamaica.

STATUS. Locally common.

HABITAT. Inhabits open river-bottom forests with adjacent semiprairie land, freshwater marshes bordering large lakes, lowland cypress swamps, and pine glades. Nests in the very tops of tall, slender living trees, usually 20 to 30 m, but up to 60 m, above ground. In Florida, usually nests in pines or in black mangroves. Selects trees in open, thinly wooded areas, or along the edge of trails or openings so the birds can approach the nest unimpeded.

SPECIAL HABITAT REQUIREMENTS. Very tall living trees for nesting.

Further Reading. United States and Canada: Brown and Amadon 1968, Godfrey 1966, Heintzelman 1979, Oberholser 1974a, Terres 1980. Mexico and Central America: Edwards 1972, Land 1970, Monroe 1968, Peterson and Chalif 1973, Ridgely and Gwynne 1989, Stiles and Skutch 1989. South America: Hilty and Brown 1986, Meyer de Schauensee 1966, Meyer de Schauensee and Phelps 1978. Caribbean: Bond 1947, Herklots 1961.

Mississippi Kite
Ictinia mississippiensis

Costa Rica: Elanio Colinegro
Guatemala: Gavilán de Mississippi
Mexico: Milano Migratorio

RANGE. Breeds from central Arizona to north-central Kansas, southern Illinois, western Kentucky, the northern portions of the Gulf States, and South Carolina south to central and southeastern New Mexico across the Gulf Coast and north-central Florida. Expanding breeding range along its northern border. Winters in South America. Migrates in flocks through eastern and southern Mexico (Tamaulipas, Veracruz, Oaxaca, Tabasco, Chiapas). In Guatemala a rare migrant. In Honduras a regular and fairly common migrant. In Costa Rica a common migrant in both spring and fall in the lowlands of both slopes and Valle Central. In Panama an uncommon to briefly common migrant with records from Bocas del Toro, both slopes of the Canal area in eastern Panama province, San Blas, and Darién. Passes through Panama in mid-March to late-April in spring and in October in fall. In Colombia apparently a rare migrant. Winters in Paraguay and northern Argentina. Migration routes and winter range are not well known.

STATUS. Common in the southern Great Plains States, uncommon in the southeastern states. Expanding its breeding range in North America.

HABITAT. Inhabits forests, open woodlands, and prairies. Found on the prairies of Kansas and in bald-cypress swamps and pinelands in the Gulf States, scrub-oak country in Oklahoma, and mesquite–sand sagebrush rangeland in Texas. In the East, nests in riparian habitats or in large pines, oaks, and sweetgums of large woods. In the Great Plains, nests in shelterbelts, farm woodlots, lawn trees in towns, or any small grove of trees. Also nests in scrub oaks and mesquite. Depending on the tree, places nest from 3 to 40 m above ground. In Arizona, nests in cottonwoods taller than 16 m in open groves or in scattered clumps surrounded by dense riparian scrubland of salt cedar and velvet mesquite 2 to 10 m tall.

SPECIAL HABITAT REQUIREMENTS. Riparian nesting habitats.

Further Reading. United States and Canada: Brown and Amadon 1968, Farrand 1983a, Glinski and Gennaro 1988, Glinski and Ohmart 1983, Heintzelman 1979, Oberholser 1974a, Parker and Ogden 1979, Sprunt 1955, Terres 1980. Mexico and Central America: Edwards 1972, Land 1970, Peterson and Chalif 1973, Ridgely and Gwynne 1989. South America: Hilty and Brown 1986, Meyer de Schauensee 1966, Meyer de Schauensee and Phelps 1978.

Northern Harrier
Circus cyaneus
(formerly Marsh Hawk)

Canada (Quebec): Busard des Marais

Costa Rica: Aguilucho Norteño

Guatemala: Aguilucho de Ciénaga,
Aguilucho Norteño

Mexico: Aguililla Rastrera

Puerto Rico: Aquilucho Pálido,
Gavilán de Ciénaga

Venezuela: Aguilucho Pálido

West Indies: Gavilán de Ciénaga,
Gavilán Sabanero, Malfini Savane

RANGE. Breeds from northern Alaska to southern Quebec and Newfoundland south to northern Mexico, Illinois, and Virginia. Winters from southern British Columbia, southern Ontario, and Massachusetts south through the United States, Middle America, and the West Indies to northern South America. A resident population of the species (*Circus cyaneus hudsonius*) occurs throughout much of western and southern South America. In Guatemala, Costa Rica, and Panama a rare migrant and winter resident most numerous on the Pacific slope. In Colombia a rare migrant and winter resident; numbers are decreasing. A casual migrant to northwestern Venezuela. In Puerto Rico and the Virgin Islands an uncommon to rare winter visitor.

STATUS. Common, but declining in the Midwest.

HABITAT. Typically inhabits sloughs, wet meadows, fresh or salt marshes, swamps, prairies, and plains. Nests singly or sometimes semicolonially, on the ground in a variety of sites, but usually near or above water. Nest is in tall grass in open fields, in swamps with low shrubs and clearings, sometimes built over water on a stick foundation, sedge tussock, or willow clump, or on a knoll of dry ground. Generally roosts on the ground or perches on very low objects such as fence posts or tree stumps. During the nonbreeding season, inhabits areas far removed from nesting habitat. Roosts in undisturbed fields or marshes in winter.

SPECIAL HABITAT REQUIREMENTS. Open country with herbaceous or low woody vegetation for concealing nests.

Further Reading. United States and Canada: DeGraaf and Rudis 1986, Evans 1982, Godfrey 1966, Heintzelman 1979, Johnsgard 1990, Low and Mansell 1983, McAtee 1935, Serrentino 1992, Sprunt 1955, Tate and Tate 1982, Terres 1980. Mexico and Central America: Edwards 1972, Land 1970, Peterson and Chalif 1973, Ridgely and Gwynne 1989, Stiles and Skutch 1989. South America: Hilty and Brown 1986, Meyer de Schauensee 1966, Meyer de Schauensee and Phelps 1978. Caribbean: Bond 1947, Raffaele 1989.

Sharp-shinned Hawk
Accipiter striatus

Canada (Quebec): Epervier Brun

Costa Rica, Guatemala, and Mexico: Gavilán Pajarero; Gavilán Pechirrufo Menor

Cuba: Garrapina

Dominican Republic: Guaraguaito de Sierra

Haiti: Emouchet

Puerto Rico: Falcón de Sierra, Gavilán de Sierra, Halcón de Sierra

Venezuela: Gavilán Arrastrador

West Indies: Gavilán Colilargo, Halcón, Halconcito

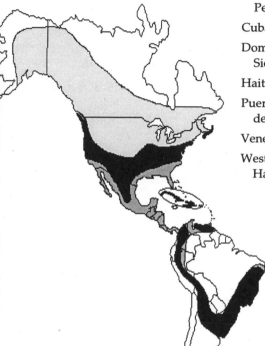

RANGE. Breeds from western and central Alaska and northern Yukon to southern Labrador and Newfoundland, south to central California, southern Texas, the northern parts of the Gulf States, and South Carolina. Winters from southern Alaska and the southernmost portions of the Canadian provinces south through the United States to Panama. In Mexico, U.S. birds migrate and winter through most of the country. In Guatemala and Costa Rica, an uncommon migrant and winter resident on both the Caribbean and Pacific slopes. In Panama an uncommon winter visitor. Birds south of Panama are residents there. A few North American individuals winter in the Bahamas and in the Greater Antilles.

STATUS. Fairly common; long-term increase in the eastern and western part of its range.

HABITAT. Primarily inhabits dense, even-aged coniferous forests 25 to 60 years old in the Pacific Northwest (Reynolds et al. 1982) and also mixed conifer-birch-aspen forests. Less commonly inhabits other woodland types except in mountainous areas. Usually nests in trees with dense foliage, primarily conifers, typically 12 to 11 m above ground and below a well-developed canopy. Nests may be in small groves of conifers surrounded by deciduous trees. Generally constructs a new nest each year in the immediate area of the previous year's nest. During migration and in winter may occur in almost any type of habitat containing trees or shrubs.

SPECIAL HABITAT REQUIREMENTS. Dense coniferous-deciduous forest.

Further Reading. United States and Canada: DeGraaf and Rudis 1986, Evans 1982, Godfrey 1966, Heintzelman 1979, Johnsgard 1990, Jones 1979, Platt 1976, Reynolds and Meslow 1984, Tate and Tate 1982.

Cooper's Hawk
Accipiter cooperii

Canada (Quebec): Epervier de Cooper

Costa Rica and Guatemala: Gavilán de Cooper

Mexico: Gavilán Palomero, Gavilán Pechirrufo Mayor

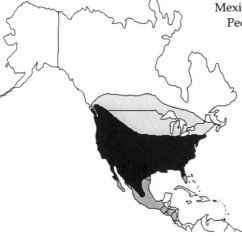

RANGE. Breeds from southern British Columbia and central Alberta east to southern Quebec and Maine (rare in New Brunswick, Prince Edward Island, and Nova Scotia) south to Baja California, Sinaloa, Chihuahua, Nuevo Léon, southern Texas, Louisiana, central Mississippi, central Alabama, and Florida. Winters from Washington, Colorado, and southern Minnesota to New England south through the southern United States to Costa Rica. Winters through much of Mexico. In Guatemala a rare migrant and winter resident recorded only in the highlands west of Guatemala City and at Cobán. In Costa Rica a very rare migrant and winter resident in the highlands; winters from Cordillera de Tilarán to Cordillera de Talamanca (Cerro de la Muerte) from 1,500 to 3,000 m. Accidental in Colombia.

STATUS. Uncommon; has been increasing in the eastern United States.

HABITAT. Inhabits older stands of various types of coniferous, mixed, and deciduous forests and open woodlands including small woodlots, riparian woodlands in dry country, open arid pinyon woodlands, and forested mountainous regions. Usually nests in deciduous or coniferous trees near the edge of a wooded area, with large open fields and water nearby. Places nest from 6 to 20 m above ground (usually 10–15 m). Occasionally uses old crow nests. In winter and on migration, may use almost any habitat containing trees or shrubs.

SPECIAL HABITAT REQUIREMENTS. Mature coniferous or deciduous woodlands.

Further Reading. United States and Canada: DeGraaf and Rudis 1986, Evans 1982, Godfrey 1966, Heintzelman 1979, Johnsgard 1990, Jones 1979, Reynolds et al. 1982, Reynolds and Meslow 1984. Mexico and Central America: Edwards 1972, Land 1970, Peterson and Chalif 1973, Ridgely and Gwynne 1989, Stiles and Skutch 1989.

Common Black-Hawk
Buteogallus anthracinus
(formerly Black Hawk)

Costa Rica, Guatemala, and
  Venezuela: Gavilán Cangrejero
Mexico: Aguililla Negra Menor
West Indies: Gavilán Batista, Halcón
  Cangrejero

RANGE. Resident in central Arizona, southwestern Utah, southern New Mexico, and western Texas, south through Central America to Colombia (northernmost populations move southward during winter). In Mexico found on lowlands and foothills of Sonora, Chihuahua, and Tamaulipas south and east through Chiapas, and the Yucatán Peninsula. Found in Cuba, including coastal cays, Isle of Pines, and St. Vincent. A vagrant to Grenada, the Grenadines, and St. Lucia. In Costa Rica locally common along both coasts near water. In Panama fairly common along the Caribbean coast and ranging inland to some extent along larger rivers. In Colombia fairly common to 500 m; locally along the eastern base of the Andes from Norte de Santander to southern Meta. In Trinidad found in Caroni and Oiopuche swamps, the valleys of the northern range, and Cedros Peninsula.

STATUS. Rare; threatened in Arizona and New Mexico.

HABITAT. An obligate of riparian areas. Optimum habitat consists of a flowing stream bordered by mature riparian forests. Also inhabits broad alluvial valleys, narrow rocky canyons, and marshes near the coast. In Costa Rica found in coastal mangroves, beaches, mudflats, rivers and streams, swamps, and marshes. Nests in trees from 5 to 30 m above ground, preferably within a grove of trees rather than in a lone tree. Builds nests in cottonwood, sycamore, alder, mesquite, willow, velvet ash, ponderosa pine, and Douglas-fir. May use same nest in successive years.

SPECIAL HABITAT REQUIREMENTS. Mature, relatively undisturbed riparian habitat with permanent streams and tall (25–30 m) trees for nesting.

Further Reading. United States and Canada: Heintzelman 1979, Oberholser 1974a, Schnell 1979. Mexico and Central America: Edwards 1972, Land 1970, Peterson and Chalif 1973, Ridgely and Gwynne 1989, Stiles and Skutch 1989. South America: Hilty and Brown 1986, Meyer de Schauensee 1966, Meyer de Schauensee and Phelps 1978. Caribbean: Bond 1947, Herklots 1961.

Gray Hawk
Buteo nitidus

Costa Rica and Venezuela: Gavilán Gris

Guatemala: Gavilán Gris, Ratonero Gris

Mexico: Aguililla Gris

RANGE. Breeds from U.S.-Mexican border to northern Argentina. Only northernmost part of the population usually migrates south for the non-breeding season.

STATUS. Uncommon to locally fairly common. Historically probably resided in greater numbers in relatively pristine subtropical thorn-scrub environs of southern Texas and southeastern Arizona.

HABITAT. Favors the vicinity of water and seems to require substantial stands of cottonwoods and willows for nesting and hunting (U.S.). Found in wood edges, riverine forests, and semiarid groves elsewhere in its range. Nest is about 0.5 m across, often hidden in leaves at top of a hackberry, cottonwood, or mesquite, 5 to 15 m up along a large stream; built of sticks and lined with evergreen-leaved twigs. Nesting habitat is a complex blend of wetland habitats (tropical-subtropical riparian deciduous woodlands of mesquite and hackberry) bordering warm-temperate riparian deciduous forests of cottonwoods, willows, sycamore, and ash. Winter habitat is probably upland tropical-subtropical deciduous forests, woodlands, and scrublands; riparian forests and woodlands; and native habitats altered for agriculture and livestock.

SPECIAL HABITAT REQUIREMENTS. Stands of cottonwoods and willows near open water for nesting.

Further Reading. United States and Canada: Glinski 1988, Robbins et al. 1983, Terres 1980. Mexico and Central America: Edwards 1972, Land 1970, Peterson and Chalif 1973, Ridgely and Gwynne 1989, Stiles and Skutch 1989. South America: Meyer de Schauensee and Phelps 1978.

Red-shouldered Hawk
Buteo lineatus

Canada (Quebec): Buse à Épaulettes Rousses

Mexico: Aguililla Pechirrojiza

RANGE. Breeds from northern California south, west of the Sierra Nevada divide, to Baja California, and from eastern Nebraska, central Minnesota, southern Ontario, and southern New Brunswick south to Mexico to Veracruz in the east and Sinaloa in the west. Winters primarily from eastern Kansas and central Missouri to southern New England south to northern Mexico, but also sporadically throughout breeding range.

STATUS. Common, but has been increasing in the western United States.

HABITAT. Inhabits forested wetlands, wooded river swamps, bottomlands, and wooded margins of marshes, often close to cultivated fields. Nests 5 to

20 m above ground in tall trees. Usually builds nest 7 to 15 m above ground in a main fork and close to the tree trunk. Has built nests in oak, pine, bald cypress, mangrove, cottonwood, birch, beech, sycamore, yellow-poplar, ash, sweetgum, and maple. Occasionally uses an abandoned hawk, crow, or squirrel nest as a foundation for a new nest; often uses the same nest site year after year. Seems to prefer mature forests and is usually more common in lowland areas than in mountainous regions.

SPECIAL HABITAT REQUIREMENTS. Riparian deciduous woodlands with tall trees for nesting.

Further Reading. United States and Canada: Bednarz and Dinsmore 1982, Forbush and May 1955, Godfrey 1966, Heintzelman 1979, McAtee 1935, Portnoy and Dodge 1979, Sprunt 1955, Stewart 1949, Tate and Tate 1982. Mexico: Edwards 1972, Peterson and Chalif 1973.

Broad-winged Hawk
Buteo platypterus

Canada (Quebec): Petite Buse

Costa Rica and Mexico: Gavilán Aludo

Guatemala: Gavilán Aludo, Ratonero de Paso

Mexico: Aguililla Migratoria Menor

Venezuela: Gavilán Bebehumo

West Indies: Gavilán Bobo, Guaraguao de Bosque, Malfini, Manger-Poulet

RANGE. Breeds in central Alberta and central Saskatchewan, and from central Manitoba to New Brunswick and Nova Scotia south to eastern Texas, the Gulf Coast, and Florida. Winters in southern Florida and from Mexico to South America. In Mexico a common migrant throughout the northeast and south; a few birds may stay for the winter. In Guatemala a common migrant and rare winter resident. In Panama and Costa Rica abundant as a migrant but also a common winter resident in forests. It is perhaps the most abundant hawk in Panamanian and Costa Rican woodlands during the northern winter months. In Colombia a common migrant and winter resident west of the Andes and on the eastern slope of the eastern Andes. A winter resident in French Guiana, Venezuela, Colombia, and western South America south to western Peru, east of the Andes to Bolivia, northwestern Brazil, and southwestern Mato Grosso. Resident in Tobago, migrant to Trinidad. Continental North American individuals winter for the most part in Central and northern South America and Trinidad.

STATUS. Common throughout its range.

HABITAT. Inhabits a variety of forests: continuous dry woodlands of oaks, beeches, maples, and mixed conifer-hardwoods around lakes, streams, and swamps. Normally nests near water in a variety of tree species, from 6 to 30 m, but as low as 1 to 3 m, above ground. Nest site preference is probably related to life form of the tree species and characteristics of the site rather than to prevalence of a particular tree species. Black and yellow birch are commonly selected for nesting in New England. Sometimes uses old crow, hawk, or squirrel nests. In migration, when conditions are favorable, forms large flocks, or "kettles," soars to the top of thermals, and then glides to another, thus saving energy during the long flight to the wintering area. Sensitive to tropical deforestation (Morton 1992).

Further Reading. United States and Canada: DeGraaf and Rudis 1986, Farrand 1983a, Fitch 1974, Forbush and May 1955, Godfrey 1966, Heintzelman 1979, Johnsgard 1990, Matray 1974, Rosenfield 1984, Rusch and Doerr 1972, Sprunt 1955, Terres 1980. Mexico and Central America: Edwards 1972, Land 1970, Monroe 1968, Peterson and Chalif 1973, Ridgely and Gwynne 1989, Stiles and Skutch 1989. South America: Hilty and Brown 1986, Meyer de Schauensee 1966, Meyer de Schauensee and Phelps 1978. Caribbean: Bond 1947, Raffaele 1989.

Swainson's Hawk
Buteo swainsoni

Canada (Quebec): Buse de Swainson
Costa Rica: Gavilán de Swainson
Guatemala: Azacuán, Ratonero de Swainson
Mexico: Aguililla Migratoria Mayor
Venezuela: Gavilán Langostero

RANGE. Breeds locally in east-central Alaska, Yukon, and Northwest Territories, and from central Alberta and central Saskatchewan to western Illinois south to southern California, central and southern Texas, and western Missouri. Winters primarily on the pampas of southern South America, casually north to the southwestern United States and southeastern Florida. Migrates in loose flocks through Mexico (except the Yucatán Peninsula) and along the Pacific slope of Central America. In Guatemala and Honduras a common migrant. In Argentina a winter resident mainly south to Río Negro. In Panama an abundant winter migrant, but only a few birds remain as winter residents in open country on the Pacific slope. Stragglers recorded from central and southeastern Colombia, northwestern Venezuela, western Peru, eastern Brazil, Paraguay, and eastern Bolivia.

121

STATUS. Common; has increased in western parts of the breeding range.

HABITAT. Inhabits prairies, plains, shrubsteppes, deserts, large mountain valleys, savannas, open pine-oak woodlands, and cultivated lands with scattered trees. Nests up to 30 m above the ground in isolated trees, in willows and cottonwoods along wetlands and drainages, in windbreaks in fields and around farmsteads, in giant cacti, or on the crossbars of telephone poles. Occasionally nests on the ground, on low cliffs, on rocky pinnacles, or on cutbanks. May repair and use same nest year after year; sometimes builds on old Black-billed Magpie nests.

Further Reading. United States and Canada: Bechard et al. 1990, Bednarz 1988, Dunkle 1977, Farrand 1983a, Godfrey 1966, Heintzelman 1979, Johnsgard 1990, Restani 1991, Risebrough et al. 1989, Sprunt 1955, Tate and Tate 1982, Terres 1980. Mexico and Central America: Edwards 1972, Land 1970, Monroe 1968, Peterson and Chalif 1973, Ridgely and Gwynne 1989, Stiles and Skutch 1989. South America: Hilty and Brown 1986, Meyer de Schauensee 1966, Meyer de Schauensee and Phelps 1978.

Zone-tailed Hawk
Buteo albonotatus

Costa Rica: Gavilán Colifajeado

Guatemala: Gavilán Ratonero,
    Ratonero Califajeado

Mexico: Aguililla Aura

Venezuela: Gavilán Negro

RANGE. Reaches the northern limit of its range in the southwestern United States, breeding from central Arizona to western Texas and south to Bolivia,

Paraguay, southern Brazil, and northern Argentina. Northernmost populations are partly migratory, but populations in Central and South America appear to be resident. The migratory behavior of this species is poorly understood.

STATUS. Locally common to uncommon.

HABITAT. Open country with scattered trees or thickets, especially near marshes or streams, oak and sycamore woodlands in breeding season (U.S.), riparian woodlands, montane conifers; in Mexico, montane pine-oak association; in South America, riverbanks, open woodlands, and dry deciduous woodlands. The nest is a platform of sticks often lined with leafy twigs, high in a tree, often in gallery woodland and rarely within 16 km of another Zone-tailed Hawk nest. In Mexico breeds in the pine-oak belt of the mountains and winters in the lowlands.

SPECIAL HABITAT REQUIREMENTS. Open country with marshes and streams with little human disturbance. Protection of known nest trees, especially those in vulnerable riparian areas, probably a useful management practice.

Further Reading. United States and Canada: Snyder and Glinski 1988, Terres 1980. Mexico and Central America: Edwards 1972, Land 1970, Peterson and Chalif 1973, Ridgely and Gwynne 1989, Stiles and Skutch 1989. South America: Meyer de Schauensee and Phelps 1978.

Red-tailed Hawk
Buteo jamaicensis

Canada (Quebec): Buse à Queue Rousse

Costa Rica: Gavilán Colirrojo

Guatemala: Gavilán de Cola Colorada, Ratonero Colirrojo

Mexico: Aguililla Colirrufa

West Indies: Gavilán de Monte, Guaraguao, Malfini

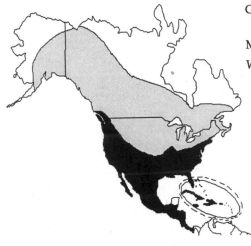

RANGE. Breeds from western and central Alaska and central Yukon to New Brunswick and Nova Scotia south to Central America. Winters from southern Canada throughout the remainder of the breeding range. Northern migrants occur through much of Mexico except the Yucatán Peninsula. In Guatemala and Honduras the resident population is augmented by U.S. migrants and winter residents. In Costa Rica and Panama a few individuals are sometimes seen migrating in spring and fall. A casual and possible winter visitor from North America on New Providence, Eleuthera, and Great Inagua. Also occurs on Grand Bahama, Abaco, Andros, the Greater Antilles, and some of the northern Lesser Antilles.

STATUS. Common; increasing throughout its breeding range.

HABITAT. Inhabits a wide variety of habitats throughout its range, preferring mixed country of open pasture, fields, meadows, or swampy areas interspersed with coniferous or deciduous woods. Among buteos, the most tolerant of human disturbance (Dobkin 1992). Inhabits deserts and plains with scattered trees and open mountain forests, generally avoiding dense, unbroken woodlands and tundra. Usually nests in a tall tree in or at the edge of a woodland, or in an isolated tree in an open area. Frequently selects the largest and tallest tree available; constructs nest in a crotch from 10 to 30 m above ground. In treeless areas, nests on rocky cliffs, shrubs, or cacti.

Further Reading. United States and Canada: Austin 1964, Bednarz and Dinsmore 1982, DeGraaf and Rudis 1986, Farrand 1983a, Fitch et al. 1946, Forbush and May 1955, Gilmer et al. 1983, Godfrey 1966, Heintzelman 1979, Johnsgard 1990, Speiser and Bosakowski 1988, Terres 1980. Mexico and Central America: Edwards 1972, Land 1970, Monroe 1968, Peterson and Chalif 1973, Ridgely and Gwynne 1989, Stiles and Skutch 1989. Caribbean: Bond 1947, Raffaele 1989.

Ferruginous Hawk
Buteo regalis

Canada (Quebec): Buse Rouilleuse
Mexico: Aguililla Real

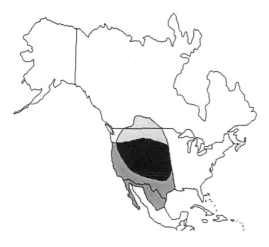

RANGE. Breeds from eastern Washington, southern Alberta, and southern Saskatchewan south to eastern Oregon, Nevada, northern and southeastern Arizona, northern New Mexico, north-central Texas, western Oklahoma, and Kansas. Winters primarily from the central and southern parts of breeding range south to Mexico. Much of the total population may winter in the southwestern United States. Winters occasionally to the arid highlands of Central Mexico.

STATUS. Common; increasing across the breeding range.

125

HABITAT. Inhabits the semiarid western plains and arid intermountain regions including shrubsteppes and badlands; prefers relatively unbroken ungrazed prairie grasslands, with scattered trees, rock outcrops, or tall trees along creek bottoms for nesting sites. Cultivated landscapes cannot sustain populations (Gilmer and Stewart 1983). Prefers tall trees for nesting; will use a wide variety of sites, including ground nests on riverbed mounds, cutbanks, low hills, buttes, and small cliffs; in open country uses short trees, powerline structures, and haystacks. Tree nests are usually in the upper canopy, from 2 to 15 m above ground. Nests are often used year after year, but cultivation and other human disturbance adversely affect nesting density. Generally winters on the southern plains.

SPECIAL HABITAT REQUIREMENTS. Open grasslands with elevated nesting sites and abundant small mammals, especially jackrabbits, ground squirrels, or prairie dogs (Smith et al. 1981).

Further Reading. United States and Canada: Blair and Schitoskey 1982, Evans 1982, Godfrey 1966, Hall et al. 1988, Heintzelman 1979, Johnsgard 1990, Schmutz 1984, Snow 1974a, Sprunt 1955, Tate and Tate 1982, Weston 1969, Woffinden and Murphy 1983. Mexico: Edwards 1972, Peterson and Chalif 1973,

American Kestrel
Falco sparverius

Canada (Quebec): Crécerelle Américaine

Costa Rica: Cernícalo Americano

Guatemala: Cernícalo Americano, Clis-Clis

Mexico: Halcón Cernícalo

Venezuela: Halcón Primito

West Indies: Cernícalo, Cuyaya, Gli-gli, Gri-gri, Halcón, Pri-pri

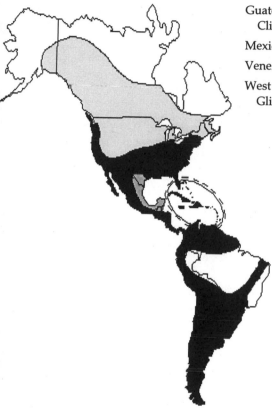

RANGE. Breeds from western and central Alaska and southern Yukon to northern Ontario, southern Quebec, and southern Newfoundland south through Middle America, the West Indies, and South America, excluding the Amazon basin. Winters from southeastern Alaska, southern British Columbia, and the central United States south throughout the breeding range.

STATUS. Common; resident subspecies declining in extreme southern United States (Florida and southern Alabama, Georgia, and South Carolina) and in the West but increasing in the central United States.

127

HABITAT. Widely distributed in habitats that include deserts, forest openings, marshes, grasslands, agricultural and suburban areas, towns, and cities. Prefers to nest in natural tree cavities with small entrances, or in cavities excavated by flickers. If these are unavailable, nests in a variety of sites including niches in rocky cliffs, under eaves of buildings, in old magpie nests, cavities in cacti, unused chimneys, or nest boxes. Nest sites are usually along roadways, streams, ponds, or forest edges, from 1 to 20 m above ground, though typically from 3 to 10 m. Frequently perches on fence posts, utility poles and wires, and in trees. Occupies the same types of habitats during winter as during the breeding season.

SPECIAL HABITAT REQUIREMENTS. Open country with low vegetation, cavities in trees with dbh greater than 30 cm, and elevated perches from which to sight prey.

Further Reading. United States and Canada: Balgooyen 1976, Bond 1947, Cade 1982, DeGraaf and Rudis 1986, Farrand 1983a, Godfrey 1966, Heintzelman 1979, Hunter 1990, Johnsgard 1990, Smith et al. 1972, Thomas et al. 1979. Mexico and Central America: Edwards 1972, Land 1970, Monroe 1968, Peterson and Chalif 1973, Ridgely and Gwynne 1989, Stiles and Skutch 1989. South America: Hilty and Brown 1986, Meyer de Schauensee 1966, Meyer de Schauensee and Phelps 1978. Caribbean: Bond 1947, Raffaele 1989.

Merlin
Falco columbarius

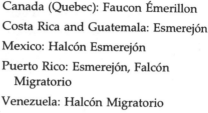

Canada (Quebec): Faucon Émerillon

Costa Rica and Guatemala: Esmerejón

Mexico: Halcón Esmerejón

Puerto Rico: Esmerejón, Falcón Migratorio

Venezuela: Halcón Migratorio

West Indies: Emouchet, Gri-gri de Montagne, Halcón, Halconito

RANGE. Breeds from northwestern Alaska and northern Yukon to Labrador and Newfoundland, south to southern Alaska, eastern Oregon, northern Minnesota, southern Quebec, New Brunswick, and Nova Scotia. Winters west of the Rockies from south-central Alaska, southern British Columbia, Wyoming, and Colorado southward, locally across southern Canada, and in the eastern United States from Maryland, the Gulf Coast, and southern Texas south through Middle America and the West Indies to northern South America from northwestern Peru to northern Venezuela and Trinidad. Accidental in Bahia, Brazil.

STATUS. Uncommon, but showing long-term increase.

HABITAT. Inhabits open areas such as forest edges, bogs, and lakes in boreal and moist Pacific Coastal forests, grassland and shrubsteppe, and prairie-parkland of the northern Great Plains. Generally nests in trees from 2 to 20 m above ground, often in old stick nests of crows, ravens, magpies, or other raptors, in or near open areas, and generally near water. Occasionally nests on the ground, on bare ledge of a cliff, or in cavities in trees. Prairie birds prefer to nest in isolated groves of trees near water, and in wooded areas

129

along rivers, generally in coniferous trees. Some remain in prairie habitat even in winter; others will use almost any habitat type encountered in the winter range. Conversion of sagebrush/grassland to cropland greatly lowers quality of foraging habitat (Becker and Sieg 1987). In Puerto Rico and the Virgin Islands in winter, frequents coastal lakes and lagoons where shorebirds abound.

Further Reading. United States and Canada: Cade 1982, Evans 1982, Fox 1964, Godfrey 1966, Heintzelman 1979, Hunter 1990, Johnsgard 1990, McAtee 1935, Sprunt 1955, Trimble 1975. Mexico and Central America: Edwards 1972, Land 1970, Monroe 1968, Peterson and Chalif 1973, Ridgely and Gwynne 1989, Stiles and Skutch 1989. South America: Hilty and Brown 1986, Meyer de Schauensee 1966. Caribbean: Bond 1947, Herklots 1961, Raffaele 1989.

Peregrine Falcon
Falco peregrinus

Canada (Quebec): Faucon Pélerin

Chile: Gavilán, Halcón

Costa Rica, Guatemala, Mexico, and
   Venezuela: Halcón Peregrino

Peru: Halcón Real

Puerto Rico: Falcón Peregrino,
   Gavilán Raye

West Indies: Halcón de Patos

RANGE. Breeds from northern Alaska, Banks, Victoria, southern Melville, Somerset, and northern Baffin Islands and Labrador south to Baja California, southern Arizona, New Mexico, western and central Texas, and Colorado; recently reintroduced and reestablished as a breeding bird in parts of the northeastern United States. Winters from southern Alaska, the Queen Charlotte Islands, coastal British Columbia, the central and southern United States, and New Brunswick south to South America. In Guatemala, Costa Rica, Honduras, and Panama a rare migrant and winter resident. In Colombia an uncommon migrant and winter resident west of the Andes to at least 2,800 m. A winter visitor throughout the West Indies. In Trinidad and To-

131

bago a fairly common winter visitor seen in Caroni Swamp and St. Giles Islet.

STATUS. Subspecies *anatum* rare and endangered; numbers appear to be stable or slightly increasing since the 1970s except in parts of the Yukon and northern Rocky Mountains. Castastrophic decline after 1950s because of reproductive impairment (eggshell thinning) and excess adult mortality because of organochlorine pesticide contamination meant that by 1960 there were no known active aeries in the northeastern United States. Reestablished as a breeding bird through introductions in the northeastern United States. Subspecies *tundrius* threatened.

HABITAT. Usually inhabits open country, from tundra and seacoasts to high mountains and more open forested regions, preferably where there are rocky cliffs with ledges overlooking rivers, lakes, or other water and an abundance of birds. Sometimes breeds in cities. Prefers to nest in a shallow depression scraped in gravel and debris on a high cliff ledge, pothole, or small cave that provides sanctuary from disturbance. Bluffs, slopes, pinnacles, cutbanks, and seastacks are also used as nest sites in the far north. Other nest sites include old stick nests of ravens and hawks, ledges of tall buildings, and historically, holes and stubs of large trees. Tends to return to the same nesting site in subsequent years.

SPECIAL HABITAT REQUIREMENTS. Cliffs or other nesting habitat near water, and an abundance of prey.

Further Reading. United States and Canada: Cade 1960, 1982, Cade et al. 1988, Craig 1986, DeGraaf and Rudis 1986, Evans 1982, Godfrey 1966, Hunter 1990, Johnsgard 1990, Johnson 1965, Skaggs et al. 1988. Mexico and Central America: Edwards 1972, Land 1970, Monroe 1968, Peterson and Chalif 1973, Ridgely and Gwynne 1989, Stiles and Skutch 1989. South America: Hilty and Brown 1986, Meyer de Schauensee 1966, Meyer de Schauensee and Phelps 1978. Caribbean: Bond 1947, Herklots 1961, Raffaele 1989.

Prairie Falcon
Falco mexicanus

Canada (Quebec): Faucon des Prairies
Mexico: Halcón Pálido

RANGE. Breeds from southeastern British Columbia, southern Alberta, southern Saskatchewan, and northern North Dakota south to Baja California, New Mexico, and northern Texas. Winters from the breeding range in southern Canada south to Mexico. Winters to central Mexico, rarely to Oaxaca.

STATUS. Locally common; long-term decline in western part of range.

HABITAT. Inhabits prairies, deserts, riverine escarpments, canyons, foothills, and mountains in relatively arid western regions. Occupies open, treeless terrain that accommodates its low-level style of hunting. Wintering birds are found away from the breeding areas in intermountain valleys and on the Great Plains. Nests on cliffs, from low rock outcrops of 10 m to vertical cliffs more than 100 m high. Prefers cliffs with a sheltered ledge with loose debris or gravel for a nest scrape, overlooking treeless country for hunting. Also nests in larger caves in cliffs and in vertical or columnar cracks with lodged material. Sometimes uses old nests of ravens, hawks, or eagles.

SPECIAL HABITAT REQUIREMENTS. Suitable nesting sites on cliffs in open country free of human disturbance.

133

Further Reading. United States and Canada: Cade 1982, Enderson 1964, Evans 1982, Godfrey 1966, Heintzelman 1979, Johnsgard 1990, McAtee 1935, Snow 1974b, Sprunt 1955, Terres 1980. Mexico: Edwards 1972.

Yellow Rail
Coturnicops noveboracensis

Canada (Quebec): Râle Jaune
Mexico: Ralito Pálido Norteño

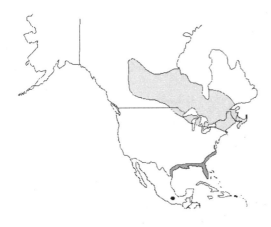

RANGE. Breeds locally from northwestern Alberta and southwestern Northwest Territories to southern Quebec and New Brunswick south to southern Alberta, North Dakota, southern Wisconsin, southern Ontario, Massachusetts, and Connecticut. Winters from coastal North Carolina south to southern Florida, west along the Gulf Coast to central and southeastern Texas; in Mexico; and locally from Oregon south to southern California. In Mexico it occurs as a resident in Río Lerma marshes in the state of México.

STATUS. Depending on locality, common, rare, or casual.

HABITAT. Highly secretive; spends most of its time beneath dense, rank vegetation. Inhabits shallow, freshwater, grassy, and sedge marshes and wet meadows. Prefers to nest in drier portions of marsh, usually where ground is damp but there is no standing water. Conceals nest in dense clumps of marsh grasses, with surrounding vegetation forming canopy over the nest. In fall and winter, lives in high margins of fresh and saltwater marshes, savannas, grain fields, hay fields, and among garden crops.

SPECIAL HABITAT REQUIREMENTS. Marshes or wet meadows.

Further Reading. United States and Canada: Anderson 1977, Bull and Farrand 1977, Devitt 1939, Godfrey 1966, Terres 1980, Walkinshaw 1939. Mexico: Edwards 1972, Peterson and Chalif 1973.

Black Rail
Laterallus jamaicensis

Costa Rica: Polluela Negra

Guatemala: Gallineta Negra, Polluela Negra

Mexico: Ralito Negruzco

Puerto Rico: Gallito Negro

RANGE. Breeds locally along the Atlantic Coast from New York south to central Florida; on the Gulf Coast in eastern Texas and western Florida; in Kansas, and from California to South America. Winters along the coast of California from the breeding range north to Tomales Bay; in the Imperial and lower Colorado River Valleys of southeastern California; along the Gulf Coast from southeastern Texas east to Florida; and in South America. In Mexico resident only in northwestern Baja California. In Guatemala recorded only in Duenas in winter in 1903; since this species is so easily overlooked, it is possibly present in greater numbers than this single record would indicate. In Costa Rica most likely a migrant and winter resident, known only from scattered sightings in Bebedero basin (Taboga, repeatedly), Peninsula de Osa (Rancho Quemado), Río Frío district (Medio Queso); possibly widespread but overlooked in the lowlands of both slopes. It is unclear whether this bird occurs in Panama. In Puerto Rico and Jamaica it was once apparently a breeding resident but was likely extirpated by the introduced mongoose; now an extremely rare winter visitor. A winter resident in Cuba and a transient in the Bahamas.

STATUS. Locally common.

HABITAT. Inhabits coastal salt marshes, and occasionally inland freshwater marshes and wet meadows. Uses grain fields and hay meadows to some extent. Prefers higher portions of a marsh, where vegetation is rank and dense and the ground is damp. Completely hides nest in clumps of marsh grass or at the base of glasswort, in or along the edge of a marsh. Usually places nest on matted dead grass, but sometimes directly on damp ground. Life history poorly known; spends much of its time under matted grasses in high salt marshes.

SPECIAL HABITAT REQUIREMENTS. Brackish and coastal salt marshes with areas of dense, but not necessarily tall, cover and damp soil. Wet grassy areas, fresh or salt (Puerto Rico).

Further Reading. United States and Canada: DeGraaf and Rudis 1986, Hunter 1990, Pough 1951, Terres 1980, Todd 1977. Mexico and Central America: Edwards 1972, Land 1970, Ridgely and Gwynne 1989, Stiles and Skutch 1989. Caribbean: Bond 1947, Raffaele 1989.

King Rail
Rallus elegans

Canada (Quebec): Râle Élégant

Mexico: Ralón Barrado Rojizo

West Indies: Gallinuela de Agua Dulce, Martillera

RANGE. Breeds locally from eastern Nebraska and central Minnesota to Connecticut south through northwestern and central Kansas, central

Oklahoma, and most of the eastern United States to western and southern Texas, central Mississippi and Alabama, and southern Florida. Winters primarily from southern Georgia, Florida, the southern portions of the Gulf States, and southern Texas south to Mexico. In Mexico a casual visitor reported in Tamaulipus, Guanajuato, and Veracruz. Found on Cuba and the Isle of Pines.

STATUS. Uncommon; populations declining in Midwest prairies, low elsewhere.

HABITAT. Inhabits coastal and inland brackish and freshwater marshes with abundant vegetation (especially sedges, bulrushes, and cattails), roadside ditches, tidal rivers, rice fields, and upland fields near marshes. Forages and nests along waterways made by muskrat (distribution coincides closely with that of muskrat). Conceals nests with a cone-shaped or round canopy of vegetation overhead; usually locates nests less than 0.5 m above shallow water on grass or sedge tussocks or on hummocks among cattails. Not known to breed in salt marshes, but wintering birds inhabit coastal brackish, saltwater (rarely), and freshwater marshes.

SPECIAL HABITAT REQUIREMENTS. Wetlands with abundant vegetation and fairly stable water levels during the breeding season.

Further Reading. United States and Canada: Bateman 1977, DeGraaf and Rudis 1986, Godfrey 1966, Johnsgard 1975a, Meanley 1969, Tate and Tate 1982. Mexico: Edwards 1972, Peterson and Chalif 1973. Caribbean: Bond 1947.

Virginia Rail
Rallus limicola

Canada (Quebec): Râle de Virginie
Mexico: Ralo Barrado Rojizo

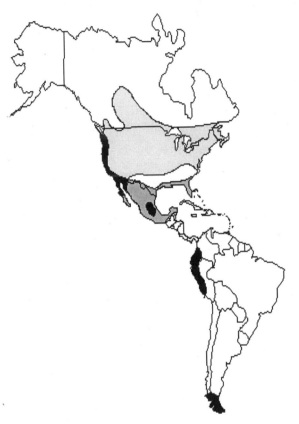

RANGE. Breeds locally from southern British Columbia and northwestern Alberta to southern Quebec, New Brunswick, and southwestern Newfoundland south to Baja California, northern Texas, Kansas, Ohio, Virginia, and along the Atlantic Coast to North Carolina. Winters from southern British Columbia and Washington, northern Mexico, the Gulf states, and Georgia south to Guatemala. Resident in central Mexico, Ecuador, Peru, southern Chile, and Argentina.

STATUS. Common.

HABITAT. Inhabits freshwater marshes, marshy borders of lakes and streams, and occasionally brackish and salt marshes. Prefers areas with shal-

138

low water and abundant emergent vegetation. Usually constructs nests in cattails or sedges just above shallow water, often covered with a loose canopy of vegetation and well attached to surrounding vegetation or on a clump of grass or tussock.

SPECIAL HABITAT REQUIREMENTS. Wetlands with sedge and cattail edge.

Further Reading. United States and Canada: Berger 1951, DeGraaf and Rudis 1986, Godfrey 1966, Horak 1970, Johnsgard 1975a, Low and Mansell 1983, Walkinshaw 1937, Zimmerman 1977. Mexico and Central America: Edwards 1972, Land 1970, Peterson and Chalif 1973. South America: Hilty and Brown 1986, Meyer de Schauensee 1966. Caribbean: Bond 1947.

Sora
Porzana carolina

Canada (Quebec): Râle de Carolina

Costa Rica: Polluela Sora

Guatemala: Gallineta Careta, Polluela Sora

Mexico: Ralo Barrado Grisáceo

Puerto Rico: Gallito

Trinidad and Tobago: Poule-savanne à Gorge Noir

Venezuela: Turura Migratorio

RANGE. Breeds from southern Yukon and west-central and southwestern Northwest Territories to west-central and southern Quebec and southwestern Newfoundland, south locally to Baja California, central Arizona, New

Mexico, central Illinois, and Maryland. Winters from central California to the Gulf Coast and southern South Carolina south to South America; occasionally north to southern Canada. In Mexico winter residents can be found (probably widely) in marshes and mangroves. In Guatemala an uncommon and local migrant and winter visitor on the Pacific slope and in the Petén; occasionally recorded elsewhere. In Costa Rica a widespread but local migrant and winter resident found in the lowlands and casually to 1,500 m on both slopes including Valle Central; most numerous in Tempisque basin and the Río Frío area. Winter resident in the tropical to temperate zone in Guyana, Venezuela, and Columbia south to Ecuador and central Peru. Generally an uncommon visitor to Puerto Rico and the Virgin Islands from October to April. Occurs as a migrant from North and Central America in both Trinidad and Tobago.

STATUS. Recently increasing in the central United States and Canada.

HABITAT. Prefers freshwater marshes with rank vegetation, but also inhabits brackish and salt marshes, ponds, swamps, bogs, wet grassy meadows, and sloughs, especially those with sedges and cattails, sometimes in mangroves (Puerto Rico). Constructs a well-concealed nest which may be fastened to or supported by reed stems on a raised platform of vegetation. Locates nest 15 cm to more than 30 cm above water, or occasionally on the ground. Generally nests over water 15 to 20 cm deep, preferably among sedges. In late summer, concentrates in areas where food is plentiful, such as rice fields or other seed-producing areas.

SPECIAL HABITAT REQUIREMENTS. Wetlands with abundant, dense vegetation.

Further Reading. United States and Canada: DeGraaf and Rudis 1986, Godfrey 1966, Horak 1970, Low and Mansell 1983, Meanley 1965, Odom 1977, Walkinshaw 1940, Webster 1964. Mexico and Central America: Edwards 1972, Land 1970, Peterson and Chalif 1973, Ridgely and Gwynne 1989, Stiles and Skutch 1989. South America: Hilty and Brown 1986, Meyer de Schauensee 1966. Caribbean: Bond 1947, Raffaele 1989.

Purple Gallinule
Porphyrula martinica

Canada (Quebec): Gallinule Pourprée

Costa Rica: Calamón Morado,
Gallareta Morada, Gallina de Agua

Guatemala: Calamón Morado,
Gallareta Morada

Mexico: Gallareta Morada

Puerto Rico: Gallareta Azul, Gallareta
Inglesa

Trinidad and Tobago: Poule d'Eau à
Cachet Bleu

West Indies: Cascamiol, Gallareta
Azul, Gallareta Inglesa, Gallareta
Platanera, Poule d'Eau à Cachet
Bleu, Poule Sultana, Sultan

RANGE. Breeds along the Atlantic Coast from Maryland and Delaware south through Central America to South America, and in eastern and southern Texas, the Gulf States, and Florida; locally in southern Illinois, western Tennessee, and central Ohio. Winters from southern Texas, Louisiana, and Florida south throughout remainder of breeding range. In Mexico found locally from Sonora and Tamaulipas, south and east to Chiapas and Quintana Roo. In Guatemala a rare and local resident in the lowlands. In Costa Rica a locally common resident countrywide. Resident in Colombia and perhaps Panama, with migrants possibly augmenting local residents. Resident in Trinidad and Tobago. In Puerto Rico an uncommon to rare permanent resident over most of the coast, but seen regularly at Cartagena Lagoon. Casual in the Bahamas, the Greater Antilles east to Puerto Rico, and the Lesser Antilles from Barbuda south. Wanders freely among the islands.

STATUS. Uncommon throughout breeding range.

HABITAT. Characteristically inhabits marshy wetlands with a variety of emergent marsh vegetation, especially where pickerelweed and plants with floating leaves, such as water lilies, are abundant. Walks freely on lily pads and other floating vegetation, alights readily on bushes, and climbs about in

141

branches over water. Usually nests on an island of floating water plants. Builds a well-concealed nest, sometimes suspended and woven into marsh vegetation or willow thickets up to 2 m above shallow water.

SPECIAL HABITAT REQUIREMENTS. Freshwater marshes and swamps, ponds, and channels of slow-moving water with well-vegetated edges.

Further Reading. United States and Canada: Cramp and Simmons 1980, Godfrey 1966, Holliman 1977, Johnsgard 1975a, Low and Mansell 1983, Terres 1980. Mexico and Central America: Edwards 1972, Land 1970, Peterson and Chalif 1973, Ridgely and Gwynne 1989, Stiles and Skutch 1989. South America: Hilty and Brown 1986, Meyer de Schauensee 1966. Caribbean: Bond 1947, Herklots 1961, Raffaele 1989.

Common Moorhen
Gallinula chloropus
(formerly Common Gallinule)

Canada (Quebec): Gallinule Commune

Costa Rica and Mexico: Gallareta Frentirroja

Guatemala: Gallareta de Frente Colorada, Gallareta Frentirroja, Polla de Agua

Puerto Rico: Gallareta Común

Trinidad and Tobago: Coq-lagon, Poule d'Eau à Cachet Rouge

West Indies: Dagareta, Gallareta Pico, Gallareta Pico Colorado, Gallareta Pico Rojo, Poule d'Eau, Poule d'Eau à Cachet Rouge, Yagareta

RANGE. Breeds locally in California, central Arizona, and northern New Mexico, and from central Minnesota and southern Wisconsin to Vermont and Massachusetts, south to South America. Winters in eastern North America from South Carolina and the Gulf Coast south, elsewhere throughout the breeding range, occasionally north to Utah, Minnesota, southern Ontario, and New England. In Mexico a local resident or migrant found in marshes nearly throughout the country. In Guatemala an uncommon and local migrant and winter visitor on the Pacific slope. In Costa Rica a resident, and possibly migrant, during the northern winter. In Panama locally common in the lowlands on both slopes; widespread and increasing in the Canal area. A fairly common resident of Caroni Marsh in Trinidad. In Colombia a local resident and possible northern migrant found up to 3,100 m in the Caribbean region. Found virtually throughout the West Indies; a common permanent resident in Puerto Rico, less so in the Virgin Islands.

STATUS. Locally common.

HABITAT. Inhabits freshwater marshes, lakes, ponds, slow-flowing streams and rivers, and nearly any body of water with emergent vegetation such as cattails, bulrushes, reeds, sedges, and burreeds growing in water more than 30 cm deep. Also inhabits rice fields in the southern United States. Builds nest in emergent vegetation 1 to 2 m high, on a hummock or other clump of vegetation over water that is about 1 m deep. Locates nest at or just above water level, concealing it with a canopy of surrounding plants. Occasionally nests in shrubs such as willow or alder. Builds brood platforms or uses muskrat houses or platforms built by coots.

SPECIAL HABITAT REQUIREMENTS. Emergent vegetation growing in water 30 cm to 1 m deep, and areas of open water.

Further Reading. United States and Canada: DeGraaf and Rudis 1986, Fredrickson 1971, Godfrey 1966, Johnsgard 1975a, Krauth 1972, Low and Mansell 1983, Strohmeyer 1977. Mexico and Central America: Edwards 1972, Peterson and Chalif 1973, Stiles and Skutch 1989. South America: Hilty and Brown 1986, Meyer de Schauensee 1966. Caribbean: Bond 1947, Raffaele 1989.

American Coot
Fulica americana

Canada (Quebec): Foulgue Américaine

Costa Rica: Focha Americana

Guatemala: Focha Americana, Gallareta Común

Mexico: Gallareta Americana

Puerto Rico: Gallinazo Americana

West Indies: Dagareta Negra, Gallareta, Gallareta Pico Blanco, Gallinazo, Judelle, Yagareta

RANGE. Breeds from the southern Yukon, southwestern Quebec, southern New Brunswick, and Nova Scotia south locally through Middle America, the West Indies, and Andean South America. Winters from southeastern Alaska and British Columbia south through the Pacific states and from Colorado and Arizona to the lower Mississippi and Ohio Valleys and Maryland south through the southeastern United States through the breeding range. In Mexico, winters along both coasts and locally in the interior throughout the country. In Guatemala, a common migrant and winter visitor on the volcanic lakes. In Costa Rica, primarily a migrant and winter resident, October to April, widespread in the lowlands on both Atlantic and Pacific slopes; large concentrations only in Tempisque basin. In Honduras, a common migrant and winter visitor, most abundant on the Caribbean slope but also found on the Pacific slope and in the Swan Islands. Formerly seen in rafts of up to 1,000 birds on Lake Yojoa and the flood-fallow lakes in the San Pedro Sula–La Lima region in the 1950s. In Panama, an uncommon to locally common winter resident on lakes and ponds in the lowlands of both Atlantic and

Pacific slopes; numbers seem to have declined somewhat in recent years in the Canal area, though some birds still winter regularly on Gatun Lake. Locally common resident in Colombia. A fairly common visitor to Puerto Rico but uncommon in the Virgin Islands during most months except summer. Found in the Bahamas and Greater Antilles, a vagrant to the Lesser Antilles.

STATUS. Common; long-term continuing decline in the eastern United States, recent decline in the western United States and Canada.

HABITAT. Inhabits marshes, lakes, ponds, sloughs, potholes, and marshy borders of creeks and rivers, or ephemeral habitats when conditions are suitable. Prefers wetlands with interspersed emergent vegetation, especially cattails and bulrushes. Constructs floating display platforms, egg nests, and brood nests located in and anchored to emergent vegetation, usually within 1 m of open water. May also nest on top of a muskrat house. During migration, found on rivers, lakes, ponds, reservoirs, or sewage lagoons; in winter, on freshwater or brackish waters.

SPECIAL HABITAT REQUIREMENTS. Freshwater wetlands, with shallow water about 1 m deep and emergent vegetation interspersed with areas of open water.

Further Reading. United States and Canada: Cramp and Simmons 1980, DeGraaf and Rudis 1986, Fjeldsa 1977, Fredrickson 1970, Godfrey 1966, Gullion 1954, Johnsgard 1975a, Jones 1940, Kiel 1955, Thomas et al. 1979. Mexico and Central America: Edwards 1972, Land 1970, Monroe 1968, Peterson and Chalif 1973, Ridgely and Gwynne 1989, Stiles and Skutch 1989. South America: Hilty and Brown 1986, Meyer de Schauensee 1966. Caribbean: Bond 1947, Raffaele 1989.

Sandhill Crane
Grus canadensis

Canada (Quebec): Grue Canadienne
Mexico: Grulla Gris
West Indies: Grulla

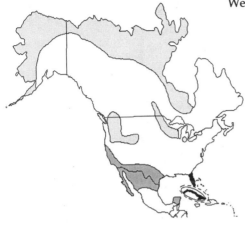

RANGE. Breeds from western and central Alaska and northern Yukon to Baffin Island, south locally to southern Alaska, northeastern California, Colorado, southern Minnesota, southern Michigan, and western Quebec. Resident from southern Mississippi, southern Alabama, and southern Georgia south to Florida. Winters from central California and southeastern Arizona to the Gulf Coast and southern Georgia south to Mexico. Winters in Mexico to the northern states and south in the interior to Jalisco and Puebla (occasionally to Yucatán Peninsula). Resident from southern Mississippi, southern Alabama and southern Georgia south through Florida to Cuba and the Isle of Pines.

STATUS. Northern subspecies (Lesser, Canadian, and Greater) are migratory and locally common; nonmigratory subspecies (Cuban, Florida, and Mississippi) are threatened or endangered. Long-term, continued population increases in northern subspecies.

HABITAT. Inhabits prairies, tundra, open pinewood flats, and other open areas. Breeds in or near shallow marshes, bogs, sloughs, margins of lakes, ponds, and river deltas. In mountainous regions, inhabits isolated, well-watered river valleys, marshes, and meadows. Occasionally inhabits relatively small marshes and patches of prairie in forested country. Usually nests in or near shallow wetlands adjacent to feeding grounds. Nest is located on a mound of emergent vegetation, sticks, grass, moss, or mud among rushes, sedges, grasses, or other tall, dense vegetation. Pairs mate for life. During winter, roosts in flocks at night on low damp ground or in shallow water, and flies to feeding grounds at dawn.

SPECIAL HABITAT REQUIREMENTS. Shallow wetlands adjacent to a meadow, cultivated fields, or open woodlands, and free of human disturbance.

Further Reading. United States and Canada: Cramp and Simmons 1980, Farrand 1983a, Godfrey 1966, Johnsgard 1975a, Lewis 1977, Terres 1980, Walkinshaw 1949. Mexico and Central America: Edwards 1972, Peterson and Chalif 1973. Caribbean: Bond 1947.

Whooping Crane
Grus americana

Canada (Quebec): Grue Blanche d'Amérique

Mexico: Grulla Blanca

RANGE. Breeds in the vicinity of Wood Buffalo National Park in southwestern Northwest Territories and in northern Alberta. Introduced at Grays Lake National Wildlife Refuge, Idaho. Winters in the vicinity of Arkansas National Wildlife Refuge on the Gulf Coast of Texas, occasionally northeast to southern Louisiana. Former visitor to Mexico; vagrants should be looked for in Tamaulipas.

STATUS. Endangered. Numbers fewer than 150 individuals in the traditional wild flock, about 13 in the introduced Rocky Mountain flock, and about 55 in captivity.

HABITAT. Inhabits marshy areas interspersed with shallow potholes having soft marly bottoms and a pH range of 7.6 to 8.3. Primarily inhabits aspen parkland, but also in northern coniferous forest, shortgrass plains, northern mixed forest, river deltas, and tundra. Nests on a mound of bulrushes in shallow water, on islands, or along shores of large wetlands where there is a heavy cover of bulrushes. Pairs mate for life and return to the same general area each year but not to the same nest site. Winters on tallgrass prairies, salt flats, coastal marshes, lagoons, and brackish waters.

SPECIAL HABITAT REQUIREMENTS. Large, shallow wetlands that provide visibility over a wide area and are free of human disturbance.

Further Reading. United States and Canada: Allen 1952, Farrand 1983a, Godfrey 1966, Mackenzie 1977, McNulty 1966, Novakowski 1966, Terres 1980. Mexico: Edwards 1972, Peterson and Chalif 1973. Caribbean: Bond 1947.

Black-bellied Plover
Pluvialis squatarola

Canada (Quebec): Pluvier à Ventre Noir

Chile: Chorlo Arctico

Costa Rica: Chorlito Gris

Guatemala: Chorlito Gris, Chorlito Pechinegro

Mexico: Chorlo Axilinegro

Puerto Rico and Venezuela: Playero Cabezón

Trinidad and Tobago: Gros Pluvier Doré

RANGE. Breeds from northern Alaska south to western Alaska, and from northwestern Northwest Territories and Banks Island, southern Melville, Devon, and western and southern Baffin Islands, south to the Yukon River, north-central and northeastern Northwest Territories, and Southampton and Coats Islands. Winters primarily in coastal areas from southern British Columbia and New Jersey south along both coasts of the United States and South America to Argentina. In Mexico a migrant and winter resident on both coasts, occasionally inland. In Guatemala and Costa Rica on the Pacific Coast during migration and winter. In Panama a common migrant and winter visitor on both slopes. A common visitor to Puerto Rico, the Virgin Islands, and islands from Aruba to Trinidad and Tobago. In Colombia a fairly common migrant and winter resident on both the Caribbean and Pacific Coasts. Winters chiefly along the Pacific Coast of South America to Concepción, Chile. Rarer on Atlantic Coast where recorded south to Uruguay and occasionally Buenos Aires, Argentina.

STATUS. Common.

HABITAT. Breeds on moist to dry upland rolling tundra. Nests in a depression on the ground in relatively dry sites on or near a ridge, often in a prominent area affording a wide view. Usually locates nest on gravelly ground, sometimes with large boulders or with sparse vegetation of lichens, dryad, saxifrage, willows, sedges, or grasses. In other seasons frequents mudflats, beaches, shores of ponds and lakes, flooded fields, and salt marshes. Commonly associates with other shorebirds, especially Willets, American Golden-Plovers, Red Knots, and curlews.

SPECIAL HABITAT REQUIREMENTS. Tundra during breeding season.

Further Reading. United States and Canada: Bent 1929, Godfrey 1966, Hussell and Page 1976, Palmer 1967, Terres 1980. Mexico and Central America: Edwards 1972, Land 1970, Monroe 1968, Peterson and Chalif 1973, Rappole et al. 1983, Ridgely and Gwynne 1989, Stiles and Skutch 1989. South America: Hilty and Brown 1986, Johnson 1965, Meyer de Schauensee 1966, Meyer de Schauensee and Phelps 1978, Rappole et al. 1983. Caribbean: Bond 1947, Herklots 1961, Raffaele 1989, Rappole et al. 1983.

American Golden-Plover

Pluvialis dominica
(formerly Lesser Golden
Plover)

Canada (Quebec): Pluvier Doré
d'Amérique

Chile: Chorlo Dorado

Costa Rica: Chorlito Dorado Menor

Guatemala: Chorlito Dorado
Americano

Mexico: Chorlo Axiliclaro

Puerto Rico and Venezuela: Playero
Dorado

RANGE. Breeds from northern Alaska and northern Yukon to Banks, Devon, and northern Baffin Islands south to central Alaska, northwestern British Columbia, southwestern Northwest Territories, northern Ontario, and Southampton and southern Baffin Islands. Winters in southern South America. In Mexico a rare migrant on the coasts and in the interior; in Guatemala primarily a spring migrant. In Costa Rica a rare fall migrant, rare and local winter resident, and uncommon to sporadically common spring migrant. In Panama a rare to locally uncommon migrant on or near both coasts; most

151

reports are from the Canal area. Transient in the West Indies; in Puerto Rico and the Virgin Islands a rare fall migrant and extremely rare spring migrant. A regular migrant to Trinidad and Tobago in small numbers. In Colombia an uncommon fall migrant and possible rare winter resident; migrates chiefly east of the Andes. Winters mostly in southern South America to Argentina. Very rare in spring in Chile.

STATUS. Common, once abundant; declined because of hunting at the end of the nineteenth century but is recovering.

HABITAT. Breeds on arctic and subarctic tundra beyond tree limit, usually where the ground cover is lichens and mosses. Nests in a depression on dry ground, preferably on higher sites such as banks of gullies or streams but not necessarily near water. Young birds quickly move to wetter areas such as sphagnum swamps. In migration, occupies shortgrass pastures, plowed fields, burned-over meadows, and, in coastal areas, beaches and mudflats.

SPECIAL HABITAT REQUIREMENTS. Dry, grassy tundra in breeding season; upland fields, even golf courses, in migration (Caribbean). Winters on the Pampas of Argentina.

Further Reading. United States and Canada: Bent 1929, Cramp and Simmons 1983, Godfrey 1966, Hussell and Page 1976, Palmer 1967, Terres 1980. Mexico and Central America: Edwards 1972, Land 1970, Monroe 1968, Peterson and Chalif 1973, Rappole et al. 1983, Ridgely and Gwynne 1989, Stiles and Skutch 1989. South America: Hilty and Brown 1986, Johnson 1965, Meyer de Schauensee 1966, Meyer de Schauensee and Phelps 1978, Rappole et al. 1983. Caribbean: Bond 1947, Herklots 1961, Raffaele 1989, Rappole et al. 1983.

Snowy Plover
Charadrius alexandrinus

Canada (Quebec): Pluvier Neigeux

Chile: Chorlo Nevado

Costa Rica: Chorlitejo Patinegro, Chorlito, Turillo

Mexico: Chorlito Alejandrino

Puerto Rico: Playero Blanco

Venezuela: Frailecito

West Indies: Becassine, Cabezón, Corredor, Frailecito, Marítimo, Patilla, Playante, Playero, Titire de Playa

RANGE. Breeds along the Pacific Coast from southern Washington to Baja California, and locally from southern Oregon, western Nevada, southwestern Montana, central Kansas, and north-central Oklahoma south to southeastern California and north-central Texas; also along the Gulf Coast from Florida west to Texas and northeastern Tamaulipas; in the southern Bahamas, Greater and Lesser Antilles, Curaçao, and east to Margarita Island; on the Pacific Coast of Oaxaca; and along the Pacific coast of South America in Peru and Chile. Winters on islands and in coastal areas from northern Oregon and the Gulf Coast south to Costa Rica. In Costa Rica a very rare migrant in both spring and fall. In Honduras a fall accidental. In Panama apparently a casual winter visitor, but perhaps somewhat overlooked. In Colombia hypothetical with no specimens. American migrants winter casually to the western Caribbean. Any birds in Chile are resident there, with no migrants augmenting the population. In the region of the West Indies breeds and winters in the Bahamas (Andros, Exuma, and Watling's Island southward), Greater Antilles, and northern Lesser Antilles; also islands in the south Caribbean. An uncommon and extremely local resident in Puerto Rico and Ane-gada.

STATUS. Locally common; suffering a serious decline in the southern and middle Pacific Coast regions of the United States. Western subspecies *nivosus* threatened.

HABITAT. Inhabits dry sandy coastal beaches above the wash of the tides, sand spits or bars separating the ocean from coastal wetlands, estuarine mar-gins, alkali flats, dry lake beds, or the shores of salt ponds and alkali lakes. Prefers open habitats; avoids thick vegetation and narrow beaches littered with driftwood or backed by bluffs. Nests singly or sometimes in loose col-onies on extensive, open sandy beaches devoid of, or sparsely covered with, vegetation or driftwood. Generally nests near water, but occasionally farther away if no formidable barrier is between the nest and water, in scrape on the ground, usually among small rocks, kelp, or other objects.

SPECIAL HABITAT REQUIREMENTS. Open sandy nesting habitat, preferably near water. Inlet stabilization may cause direct loss of foraging habitat. In winter, flats where crystallized salt lines the water's edge (Raffaele 1989:72).

Further Reading. United States and Canada: Bent 1929, Farrand 1983a, God-frey 1966, Harrison 1979, Hunter 1990, Page and Stenzel 1981, Palmer 1967, Tate and Tate 1982, Terres 1980. Mexico and Central America: Edwards 1972, Monroe 1968, Peterson and Chalif 1973, Rappole et al. 1983, Ridgely and Gwynne 1989, Stiles and Skutch 1989. South America: Hilty and Brown 1986, Johnson 1965, Meyer de Schauensee 1966, Meyer de Schauensee and Phelps 1978, Rappole et al. 1983. Caribbean: Bond 1947, Raffaele 1989, Rappole et al. 1983.

Wilson's Plover
Charadrius wilsonia

Canada (Quebec): Pluvier de Wilson

Costa Rica: Chorlitejo Picudo,
Chorlito Gritón, Turillo

Guatemala and Mexico: Chorlito
Piquigrueso

Puerto Rico: Playero Marítimo

Venezuela: Playero Picagrueso

RANGE. Breeds on the Atlantic Coast from New Jersey to Florida, on the Gulf Coast to Texas, and south in the Caribbean area to Venezuela, Guyana, and through the West Indies; on the Pacific Coast from Baja California and Sonora to Peru. In Mexico resident on the coasts of Baja California, Sonora, Sinaloa, and Nayarit; winters locally along the east coast (Gulf of Mexico). In Guatemala a rare migrant; possibly a resident, to be looked for on the Pacific Coast. In Costa Rica migrants and winter visitors augment resident populations on both coasts and are locally abundant on Pacific side (Golfo de Nicoya, Gulfo Dulce), uncommon on Caribbean side. Along Caribbean coast of Honduras and in the Caribbean islands apparently rare but probably regular on migration and in winter. In Panama a locally fairly common migrant and winter resident along both coasts, but much more numerous and widespread on the Pacific Coast; particularly numerous at the Pacific entrance to the canal. A common permanent resident in Puerto Rico and the Virgin Islands. A winter visitor to Trinidad; no definite records for Tobago. In Colombia local populations may be augmented by U.S. migrants. In South America most birds are resident.

STATUS. Uncommon, rather local.

HABITAT. Breeds on shingle beaches, dunes, dry sandy edges of coastal pools. Sometimes nests in loose colonies, more often single and scattered. Nest is a scraped hollow in the sand.

SPECIAL HABITAT REQUIREMENTS. Coastal beach, dune nesting sites near water. Borders of fresh ponds (Puerto Rico).

Further Reading. United States and Canada: Bull and Farrand 1977, Godfrey 1966, Terres 1980. Mexico and Central America: Edwards 1972, Land 1970, Monroe 1968, Peterson and Chalif 1973, Rappole et al. 1983, Ridgely and Gwynne 1989, Stiles and Skutch 1989. South America: Hilty and Brown 1986, Meyer de Schauensee 1966, Meyer de Schauensee and Phelps 1978, Rappole et al. 1983. Caribbean: Bond 1947, Herklots 1961, Raffaele 1989, Rappole et al. 1983.

Semipalmated Plover

Charadrius semipalmatus

Canada (Quebec): Pluvier à Collier

Chile: Chorlo Semipalmeado

Costa Rica: Chorlitejo Semipalmeado, Chorlito, Turillo

Guatemala and Mexico: Chorlito Semipalmeado

Puerto Rico and Venezuela: Playero Acollarado

RANGE. Breeds from northern Alaska, northern Yukon, and Banks, Victoria, and central Baffin Islands to the northern Labrador coast, south to western Alaska, southwestern and central British Columbia, northern Manitoba, central Quebec, and southern Nova Scotia. Nonbreeding birds often summer in the wintering areas south to Panama. Winters primarily in coastal areas from central California, the Gulf Coast, and South Carolina south to Argentina. Migrates and winters along both coasts of Mexico. In Panama a very common migrant and common winter resident. A common migrant and winter visitor on both coasts of Colombia. A winter resident recorded from all South American countries with seacoasts to Llanguihue, Chile, and Santa Cruz, Argentina. Occurs in the West Indies throughout the year but most numerous as a fall migrant. A common winter visitor to Puerto Rico and the Virgin Islands.

STATUS. Common.

HABITAT. Found during spring and fall migration on beaches, mudflats, lakeshores, riverbanks, freshly plowed fields, shallow marshes, gravel pits, and peat banks. Breeds in dry arctic tundra, sometimes quite far from water, and prefers lichen-grown gravelly tundra or areas of rubble and patches of stranded debris. Nests in unsheltered scrapes on level ground in gravelly or sandy soil, or in moss or lichens and above the high-water line in loose colonies. Avoids grassy areas. In winter occupies vast undisturbed coastal wetlands such as those on the Caribbean coast of Venezuela.

SPECIAL HABITAT REQUIREMENTS. In breeding season, level, well-drained gravelly tundra; in winter, undisturbed coastal wetlands and tidal flats.

Further Reading. United States and Canada: Cottam and Hanson 1938, Godfrey 1966, Palmer 1967, Sutton and Parmelee 1955, Terres 1980. Mexico and Central America: Edwards 1972, Land 1970, Monroe 1968, Peterson and Chalif 1973, Rappole et al. 1983, Ridgely and Gwynne 1989, Stiles and Skutch 1989. South America: Hilty and Brown 1986, Johnson 1965, Meyer de Schauensee 1966, Meyer de Schauensee and Phelps 1978, Rappole et al. 1983. Caribbean: Bond 1947, Herklots 1961, Raffaele 1989, Rappole et al. 1983.

Piping Plover

Charadrius melodus

Canada (Quebec): Pluvier Siffleur
Mexico: Chorlito Melódico
Puerto Rico: Playero Melódico

RANGE. Breeds locally from south-central Alberta to south-central Manitoba, south to eastern Montana and central and eastern Nebraska; in the Great Lakes region from northern Michigan and southern Ontario south to the southern shores of Lake Michigan and Lake Ontario; and in coastal areas from Newfoundland south to Virginia. Winters on the coast from South Carolina south to Florida and west to eastern Texas, sparsely in Bahamas and Greater Antilles. In Mexico winters south to the coast of Tamaulipas. In Puerto Rico a rare winter visitor. There are also several records from St. Croix and one from Anegada. Accidental on Barbados.

STATUS. Great Lakes populations endangered. Endangered in Canada. Great Plains population threatened in United States. Atlantic Coast population increasing in New England, declining from Delaware to North Carolina.

HABITAT. Inhabits exposed, sparsely vegetated sandy shores and islands of shallow lakes and ponds, dry sandy ocean beaches, higher portions of beach near dunes, recent dredge spoils, and large open sandy areas, especially where scattered grass tufts are present. Nests in a hollow in sand, well beyond high tide on ocean beaches, on raised sand spits, or on the lower slopes of dunes. Generally nests on narrow beaches as little as 2 m wide. Sometimes nests under tufts of grass. Adults tend to return to the same breeding area year after year. In winter found on beaches, margins of lagoons, and areas of rubble.

SPECIAL HABITAT REQUIREMENTS. Unspoiled, undeveloped beaches with little vegetation.

Further Reading. United States and Canada: Bent 1929, Cairns 1982, De-Graaf and Rudis 1986, Godfrey 1966, Harrison 1979, Palmer 1967, Tate and Tate 1982, Terres 1980, Wershler 1986a, Wilcox 1959. Mexico and Central America: Edwards 1972, Peterson and Chalif 1973, Rappole et al. 1983, Stiles and Skutch 1989. Caribbean: Bond 1947, Raffaele 1989, Rappole et al. 1983.

Killdeer
Charadrius vociferus

Canada (Quebec): Pluvier Kildir

Costa Rica: Chorlitejo Tildío, Chorlitejo de Dos Collares, Pijije, Tildío

Guatemala: Chorlitejo, Chorlito Tildío, Collarejo

Mexico: Chorlito Tildío

Puerto Rico: Playero Sabanero

Venezuela: Playero Gritón

West Indies: Chevalier de Terre, Collier, Frailecillo, Gritón, Playero Sabanero, Titire Sabanero

RANGE. Breeds from east-central and southeastern Alaska and southern Yukon to central Quebec, western Nova Scotia, Prince Edward Islands, and western Newfoundland, south to Southern Baja California, central Mexico, Tamaulipas, the Gulf Coast, and southern Florida; also in the southern Bahamas and Greater Antilles, and in western South America along the coast of Peru. Winters from southern British Columbia across the central United States to New England and south throughout the remainder of North America to Colombia and Ecuador. Winters widely throughout Mexico. In Panama an uncommon and somewhat local migrant and winter resident throughout. Recorded as a migrant to Tobago; no records for Trinidad. A very common and local winter resident in Colombia. A rare winter resident in Venezuela. Found throughout the West Indies as a transient or winter visitor; resident in the Bahamas and Greater Antilles. In Puerto Rico a common permanent resident, uncommon in the Virgin Islands; local populations augmented in the fall, winter, and spring by northern migrants.

STATUS. Common throughout most of range but long-term decline in the western United States, recent (since early 1980s) decline in the central United States.

HABITAT. Indiscriminately occupies open areas, but favors open dry uplands, meadows, pastures, and disturbed or heavily grazed areas where grass is short, sparse, or absent. Nests in a scrape on gravelly or bare ground, occasionally on a flat or gently sloping roof of a building, often with a few pebbles, grasses, or weeds in the scrape. During migration and in winter more associated with wetland habitats—beaches, watercourses, and mudflats—as well as open fields.

SPECIAL HABITAT REQUIREMENTS. Open areas with closely cropped or sparse vegetation in breeding season. Wet fields, edges of freshwater ponds (Puerto Rico).

Further Reading. United States and Canada: Cramp and Simmons 1983, DeGraaf and Rudis 1986, Godfrey 1966, Palmer 1967, Terres 1980. Mexico and Central America: Edwards 1972, Land 1970, Monroe 1968, Peterson and Chalif 1973, Rappole et al. 1983, Ridgely and Gwynne 1989, Stiles and Skutch 1989. South America: Hilty and Brown 1986, Meyer de Schauensee 1966, Meyer de Schauensee and Phelps 1978, Rappole et al. 1983. Caribbean: Bond 1947, Herklots 1961, Raffaele 1989, Rappole et al. 1983.

Mountain Plover
Charadrius montanus

Canada (Quebec): Pluvier
Montagnard

Mexico: Chorlito Llanero

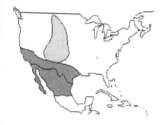

RANGE. Breeds from extreme southern Alberta and northern Montana south to central and southeastern New Mexico, western Texas, western Oklahoma, and western Missouri. Winters from central California, southern Arizona, and central and coastal Texas south to the grassy plains of northern Mexico.

STATUS. Common but declining.

HABITAT. A species of the high plains and arid regions of western valleys and hills, usually found far from water. Generally avoids mountainous areas and prefers areas dominated by blue grama and buffalo grass. Decline of the Mountain Plover has followed the decline of the bison and the increase in agriculture. Plowed fields are not good habitat, and ungrazed fields have vegetation that is too tall. Prefers shortgrass prairie that is moderately grazed. Nests in scrapes on flat ground, preferably in areas of blue grama–buffalo grass with scattered clumps of cacti and western wheatgrass or in prairie-dog towns (Knowles et al. 1982). Avoids tall vegetation. In winter, congregates in flocks of 15 to several hundred on alkali flats, agricultural land, grazed pastures, or other open arid habitats.

SPECIAL HABITAT REQUIREMENTS. Shortgrass prairie and arid plains.

Further Reading. United States and Canada: Bent 1929, Farrand 1983a, Godfrey 1966, Graul 1975, Palmer 1967, Terres 1980, Wershler 1986b. Mexico: Edwards 1972, Peterson and Chalif 1973, Rappole et al. 1983.

American Oystercatcher
Hematopus palliatus

Canada (Quebec): Huîtrier
  Américaine
Coasta Rica: Ostrero Americano
Guatemala: Ostrero
Mexico: Ostrero Blanquinegro
Puerto Rico: Caracolero, Ostrero
Venezuela: Caracolero

RANGE. Breeds on Atlantic and Pacific Coasts from Massachusetts and Long Island, New York, and Baja California south to Argentina. Breeds on both coasts of Mexico, including islands in the Gulf of California and off western Mexico; along the Gulf Coast west to central Texas and south to the Yucatán Peninsula and Cozumel Island; in the Bahamas and Greater and Lesser Antilles; and along the coasts of Middle and South America south to south-central Argentina and central Chile. Very local on the Atlantic Coast of the United States; migrants augment local populations. In Guatemala recorded locally along the Pacific Coast; possibly a rare resident. In Costa Rica a migrant on both coasts in fall and spring; an uncommon winter and nonbreeding summer resident on the Pacific Coast mainly in Golfo de Nicoya. In Honduras apparently a straggler or very rare migrant along the Caribbean coast. Panamanian birds are residents, as are birds in Puerto Rico, the Virgin Islands, and West Indies.

STATUS. Uncommon to rare. Declined along the Atlantic Coast throughout the latter part of the nineteenth century, becoming extirpated from Massachusetts and possibly southern Maine to Virginia. After 1900 gradually expanded range northward (Kramer 1948), and since 1969 has increased steadily (Humphrey 1988).

HABITAT. Inhabits large expanses of relatively undisturbed shore and flats. Nest is a hollow in the sand on dry flat beaches, well above the reach of high tides, without lining and on a little mound from which the incubating bird has a lookout. This is a mainly sedentary species that never moves far from the sea.

SPECIAL HABITAT REQUIREMENTS. Large expanses of undisturbed shore and flats for breeding, and a source of shellfish. Stony beaches, rocky headlands (Puerto Rico).

Further Reading. United States and Canada: Brewster 1885, Bull and Farrand 1977, Godfrey 1966, Terres 1980. Mexico and Central America: Edwards 1972, Land 1970, Monroe 1968, Peterson and Chalif 1973, Rappole et al. 1983, Ridgely and Gwynne 1989, Stiles and Skutch 1989. South America: Hilty and Brown 1986, Meyer de Schauensee 1966, Meyer de Schauensee and Phelps 1978, Rappole et al. 1983. Caribbean: Bond 1947, Raffaele 1989, Rappole et al. 1983.

Black-necked Stilt
Himantopus mexicanus

Canada (Quebec): Echasse Américaine

Costa Rica: Cigüeñuela Cuellinegro

Guatemala: Candelero, Cigüeñuela Cuellinegra, Soldadito

Mexico: Avoceta Piquirrecta

Puerto Rico: Viuda

Venezuela: Viuda Patilarga

West Indies: Arcaguet, Cachiporra, Echasse, Pete-pete, Pigeon d'Etang, Viuda, Zancudo

RANGE. Breeds locally on the Atlantic Coast from southern New Jersey south to southern Florida, and from southern Oregon, southern Colorado, central Kansas, the Gulf Coast of Texas, and southern Louisiana south to South America. Winters from central California, the Gulf Coast of Texas and Louisiana, and southern Florida south to South America. In Mexico found in the lowlands of both slopes; local in interior. In Guatemala resident populations are augmented by U.S. migrants and winter residents; uncommon in the lowlands. In Costa Rica and Honduras a migrant and winter visitor. In Panama uncommon to locally common and seasonably common; much more numerous on the Pacific Coast. In Colombia usually fairly common in saline coastal lagoons and freshwater marshes. A migrant south to Bolivia and Argentina. In the West Indies breeds in the Bahamas, Greater Antilles, and northern Lesser Antilles (no records of any U.S. migrants wintering in Puerto Rico or the Virgin Islands, where this species is a resident); a vagrant in the southern Lesser Antilles. A migrant to both Trinidad and Tobago.

STATUS. Common.

HABITAT. Inhabits shallow freshwater and brackish ponds, alkaline lakes, wet meadows, open marshes, and flooded fields and pastures. Nests in slight depressions on the ground, on sandy or gravelly shores, along drier margins of ponds and lakes, or on deep, well-built, floating platforms over shallow water in loose colonies. Also nests on hummocks, on small islands, or under clumps of vegetation; nest may be well concealed or in the open. Commonly associates with other shorebirds, especially avocets, godwits, and curlews.

SPECIAL HABITAT REQUIREMENTS. Shallow wetlands. Borders of salt ponds, mangrove swamps (Puerto Rico).

Further Reading. United States and Canada: Bent 1927, Farrand 1983a, Godfrey 1966, Palmer 1967, Terres 1980. Mexico and Central America: Edwards 1972, Land 1970, Monroe 1968, Peterson and Chalif 1973, Rappole et al. 1983, Ridgely and Gwynne 1989, Stiles and Skutch 1989. South America: Hilty and Brown 1986, Rappole et al. 1983. Caribbean: Bond 1947, Herklots 1961, Raffaele 1989, Rappole et al. 1983.

American Avocet
Recurvirostra americana

Canada (Quebec): Avocette Américaine

Costa Rica and Guatemala: Avoceta Americana

Mexico and Puerto Rico: Avoceta Piquicurva

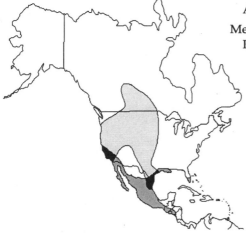

RANGE. Breeds from southeastern British Columbia and central Alberta to Minnesota, south locally to southern California, northern Utah, and southern New Mexico, and east to central Kansas and coastal Texas. Winters mostly in coastal lowlands from northern California and southern Texas south to

Mexico, and locally in southern Florida. Winters locally through much of Mexico, but only casually and increasingly rarely south of Mexico. In Guatemala an uncommon winter resident found on the Pacific Coast at sea level. In Honduras a rare winter visitor. In Costa Rica a casual winter visitor to the northwestern lowlands (Tempisque basin, vicinity of Golfo de Nicoya). Accidental in Puerto Rico (2 records), St. Croix (1 record), and the British Virgin Islands (1 record). A migrant to Tobago only.

STATUS. Common.

HABITAT. Inhabits the borders of muddy saline, alkaline, and freshwater ponds, lakes, and marshes, particularly favoring shallow alkaline lakes, wet meadows, and pastures with scattered open pools. Eggs are laid in a scrape on the ground. Nests in colonies on dry, sun-baked mudflats near water, on low gravelly or sandy islands with scant vegetation, or in marshes bordering shallow lakes. If the water level rises to flood the nest, adds vegetation to raise the nest above the water level. Commonly associates with Black-necked Stilts, godwits, and Lesser Yellowlegs.

SPECIAL HABITAT REQUIREMENTS. Wetlands bordered by open flats or areas with scattered tufts of grass.

Further Reading. United States and Canada: Bent 1927, Farrand 1983a, Gibson 1971, Godfrey 1966, Harrison 1979, Low and Mansell 1983, Palmer 1967, Terres 1980. Mexico and Central America: Edwards 1972, Land 1970, Monroe 1968, Peterson and Chalif 1973, Rappole et al. 1983, Ridgely and Gwynne 1989, Stiles and Skutch 1989. South America: Meyer de Schauensee 1966, Rappole et al. 1983. Caribbean: Bond 1947, Herklots 1961, Raffaele 1989, Rappole et al. 1983.

Greater Yellowlegs
Tringa melanoleuca

Canada (Quebec): Grand Chevalier à Pattes Jaunes

Chile: Pitoitoi Grande

Costa Rica: Patiamarillo Mayor, Pijije, Zarceta

Guatemala: Archibebe, Becasina, Patigualdo Grande

Mexico: Patiamarillo Mayor

Puerto Rico: Playero Guineillo Grande

Trinidad and Tobago: Chin-chin

Venezuela: Tigüi-Tigüi Grande

RANGE. Breeds from southern Alaska, southwestern Northwest Territories, and south-central British Columbia east across the northern and central portions of the Canadian provinces to central and southern Labrador, Newfoundland, and northeastern Nova Scotia. Winters from Oregon and southern Nevada to southern Texas, the Gulf Coast, and coastal South Carolina south to South America. Nonbreeding birds sometimes remain on the wintering grounds, especially along both U.S. coasts. Migrates to and winters on the coastal lowlands of Mexico, also in the interior. In Central America an uncommon or locally common migrant and winter visitor. In Colombia a common migrant and winter visitor on both coasts. A regular and relatively common visitor to Chile during the northern winter; winters both inland and

167

coastally occasionally as far south as the Straits of Magellan. Present in the West Indies year-round but most numerous in fall and spring; common in the Virgin Islands and Puerto Rico in the fall, fairly common in the winter and spring. A fairly common winter visitor to Trinidad and Tobago, more numerous in the fall.

STATUS. Common.

HABITAT. The Nearctic boreal region to the edge of subarctic coniferous forest zone, where it inhabits swampy muskegs or bogs with scattered trees, wet clearings and pools, or tundra. Perches freely when breeding, often alighting on tops of trees, bushes, or dead stubs. Nest is a depression in the ground, usually near trees, logs, or stumps, on a dry wooded ridge or on recently burned ground, and normally near water. During migration and in winter, frequents shallow fresh, brackish, and salt waters, mudflats, river bars, tidal flats, marshes and pools, rain pools in fields, and damp grassy meadows.

SPECIAL HABITAT REQUIREMENTS. Muskeg and tundra.

Further Reading. United States and Canada: Cramp and Simmons 1983, Godfrey 1966, Low and Mansell 1983, Palmer 1967. Mexico and Central America: Edwards 1972, Land 1970, Monroe 1968, Peterson and Chalif 1973, Rappole et al. 1983, Ridgely and Gwynne 1989, Stiles and Skutch 1989. South America: Hilty and Brown 1986, Johnson 1965, Meyer de Schauensee 1966, Rappole et al. 1983. Caribbean: Bond 1947, Herklots 1961, Raffaele 1989, Rappole et al. 1983.

Lesser Yellowlegs
Tringa flavipes

Canada (Quebec): Petit Chevalier à Pattes Jaunes

Chile: Pitoitoi Chico

Costa Rica: Patiamarillo Menor, Pijije, Zarceta

Guatemala: Archibebe, Becasina, Patigualdo Chico

Mexico: Patiamarillo Menor

Puerto Rico: Playero Guineillo Pequeño

Trinidad and Tobago: Pieds Jaunes

Venezuela: Tigüi-Tigüi Grande

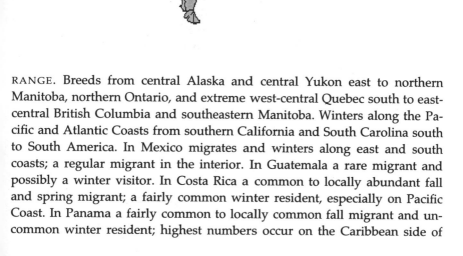

RANGE. Breeds from central Alaska and central Yukon east to northern Manitoba, northern Ontario, and extreme west-central Quebec south to east-central British Columbia and southeastern Manitoba. Winters along the Pacific and Atlantic Coasts from southern California and South Carolina south to South America. In Mexico migrates and winters along east and south coasts; a regular migrant in the interior. In Guatemala a rare migrant and possibly a winter visitor. In Costa Rica a common to locally abundant fall and spring migrant; a fairly common winter resident, especially on Pacific Coast. In Panama a fairly common to locally common fall migrant and uncommon winter resident; highest numbers occur on the Caribbean side of the Canal area. In Colombia a fairly common migrant and winter visitor. Winters in South America both inland and coastally, occasionally as far south as the Straits of Magellan. A common winter visitor to both Trinidad and Tobago. Found throughout the year in the West Indies, with North American individuals augmenting local populations as migrants and winter residents. A common visitor to Puerto Rico and the Virgin Islands in all months except June and July, when uncommon. Much more common than Greater Yellowlegs in the Virgin Islands.

STATUS. Common; declining, especially in the western United States.

HABITAT. The Nearctic coniferous forest zone, from boreal and subarctic regions into the low Arctic; occurs mainly inland, and to some extent upland. Prefers grassy meadows and bogs, natural clearings, or burned areas in forest with scattered stumps and fallen logs, often far from open water. Nests singly or in loose colonies. Nest is a depression on the ground, located on a dry sloping bank, ridge, or level plateau; in open high woodland with sparse, fairly low undergrowth; in swampy muskeg; or on undrained land surrounded by farmland. Outside the breeding season, inhabits shallow prairie sloughs in open country, muddy shores of lakes and marshy ponds, sewage beds, river margins, and inland and coastal marshes.

SPECIAL HABITAT REQUIREMENTS. Tundra and muskeg.

Further Reading. United States and Canada: Bent 1927, Cramp and Simmons 1983, Godfrey 1966, Low and Mansell 1983, Palmer 1967. Mexico and Central America: Edwards 1972, Land 1970, Monroe 1968, Peterson and Chalif 1973, Rappole et al. 1983, Ridgely and Gwynne 1989, Stiles and Skutch 1989. South America: Hilty and Brown 1986, Johnson 1965, Meyer de Schauensee 1966, Rappole et al. 1983. Caribbean: Bond 1947, Herklots 1961, Raffaele 1989, Rappole et al. 1983.

Solitary Sandpiper
Tringa solitaria

Canada (Quebec): Chevalier Solitare

Costa Rica: Andarríos Solitario, Tiguiza

Guatemala: Andarríos Solitario, Becasineta Solitaria

Mexico: Playero Charquero

Puerto Rico and Venezuela: Playero Solitario

RANGE. Breeds from central and south-coastal Alaska and northern Yukon to northern and central Ontario, east through central Quebec to central and southern Labrador, and south to northwestern and central British Columbia, southern Manitoba, and northern Minnesota. Winters from the Gulf Coast, southeastern Georgia, and Florida south to Argentina. Migrates and winters throughout the Caribbean. A moderately common migrant and winter visitor locally throughout Mexico. In Central America a fairly common migrant and uncommon winter visitor. In Panama a fairly common migrant and less numerous winter resident; this bird and the Spotted Sandpiper are by far the two most widespread shorebirds in Panama. In Colombia a migrant and winter resident in small numbers. Winters along coastal and inland waters south to western Peru, Bolivia, and Río Negro, Argentina. In Puerto Rico a common fall migrant, uncommon in winter and spring. A migrant and winter visitor to both Trinidad and Tobago. A fairly common transient but rare winter resident in the West Indies; in the Virgin Islands uncommon in fall, rare in winter and spring.

STATUS. Common.

HABITAT. Inland muskeg with scattered mature trees or clumps of trees near freshwater lakes and ponds in the coniferous forest belt of boreal and subarctic regions. On the breeding grounds perches freely on treetops, twigs, limbs, and stumps. Nests above ground in old nests of American Robins, Cedar Waxwings, Rusty Blackbirds, and Eastern Kingbirds. Usually uses nests in coniferous trees that border muskeg, open bogs, or a lake. Outside of the breeding season occurs inland along shallow freshwater woodland streams, ponds, bogs, flooded marshes, stagnant pools, mudflats, and barnyard puddles.

SPECIAL HABITAT REQUIREMENTS. Muskegs.

Further Reading. United States and Canada: Bent 1927, Cramp and Simmons 1983, Godfrey 1966, Palmer 1967, Pough 1951. Mexico and Central America: Edwards 1972, Land 1970, Monroe 1968, Peterson and Chalif 1973, Rappole et al. 1983, Ridgely and Gwynne 1989, Stiles and Skutch 1989. South America: Hilty and Brown 1986, Johnson 1965, Meyer de Schauensee 1966, Meyer de Schauensee and Phelps 1978, Rappole et al. 1983. Caribbean: Bond 1947, Herklots 1961, Raffaele 1989, Rappole et al. 1983.

Willet
Catoptrophorus semipalmatus

Canada (Quebec): Chevalier Semi-palmé

Chile: Playero Grande de Alas Blancas

Costa Rica and Guatemala: Piguilo

Mexico: Playero Pihuihui

Puerto Rico and Venezuela: Playero Aliblanco

West Indies: Becassine Aile-blanche, Chorlo, Playero Aliblanco, Zarapico Real

RANGE. Breeds locally from eastern Oregon, central Alberta, and south-western Manitoba, south to northeastern and east-central California, western and northern Nebraska, and eastern South Dakota; locally along the Atlantic and Gulf Coasts from Prince Edward Island south to southern Florida and west to southern Texas. Occurs sporadically (nonbreeding birds) in summer as far south as northern South America. Winters along the Pacific and Atlantic Coasts from northern California and Virginia south through Central America and northern South America to northeast Brazil on the Atlantic Coast and to Peru (occasionally northern Chile) on the Pacific Coast. Migrants stay away from the plains of Bolivia, Paraguay, and Argentina. In Mexico, breeds in Tamaulipas and winters along the coasts, mainly on the Pacific side. A common migrant and fairly common winter resident throughout Central America. In Colombia a common migrant and winter resident on both coasts and occasionally inland. Fairly common in Puerto Rico and the Virgin Islands. A winter visitor to both Trinidad and Tobago.

Willet

STATUS. Common; may be declining.

HABITAT. Inhabits tidal and coastal marshes, beaches, sandy islands with tall and thick grasses, open pastures, dry uplands near water, and dunes in the eastern United States. Prefers inland prairies and plains, alkali flats, and grassy dikes, usually near water in the western United States. Nests semicolonially in a depression on the ground or in a thick clump of vegetation, sometimes far from water in the West. Locates nest in open areas on a sandy beach or in well-hidden areas in low grasses. Associates freely with godwits, curlews, large plovers, and some shorebirds. Often perches on bushes, trees, fences, posts, and buildings.

SPECIAL HABITAT REQUIREMENTS. Moist plains and prairies in western North America, coastal marshes and nearby grassy areas in the eastern United States.

Further Reading. United States and Canada: Bent 1927, Farrand 1983a, Godfrey 1966, Low and Mansell 1983, Palmer 1967, Ryan and Renken 1987, Stenzel et al. 1976, Terres 1980, Wilcox 1980. Mexico and Central America: Edwards 1972, Land 1970, Monroe 1968, Peterson and Chalif 1973, Rappole et al. 1983, Ridgely and Gwynne 1989, Stiles and Skutch 1989. South America: Hilty and Brown 1986, Johnson 1965, Meyer de Schauensee 1966, Meyer de Schauensee and Phelps 1978, Rappole et al. 1983. Caribbean: Bond 1947, Herklots 1961, Raffaele 1989, Rappole et al. 1983.

Wandering Tattler
Heteroscelus incanus

Canada (Quebec): Chevalier Errant
Costa Rica: Correlimos Vagamundo
Mexico: Playero Sencillo

RANGE. Breeds in North America in northwestern Canada and Alaska. Winters along the Pacific Coast and on islands from southern California (rarely Oregon and Washington) south to Galápagos Islands, Colombia, and Ecuador; also Hawaiian Islands, Marianas, and Philippines south to Fiji, Samoa, Society Islands, and the Tuamotu Archipelago. In Mexico winters mainly along the Pacific Coast of Baja California and the offshore islands; recorded also on the west coast from Sonora to Guerrero. In Costa Rica an uncommon to occasionally numerous migrant along the outer Pacific Coast in the fall and spring; occasional individuals, mainly on offshore islands, in both winter and summer. Fairly common year-round on Cocos Island; summering birds are nonbreeding yearlings. Recorded on Malpelo Island off the Pacific Coast of Colombia; as of yet unrecorded on Colombia's coast, but probably a casual winter visitor or winter resident to rocky shores and offshore islands. Casual winter visitor to coastal Ecuador and Peru.

STATUS. Rather rare.

HABITAT. Breeding territories are centered on valleys with a suitable stream and are located in treeless tundra or in mountains above treeline. Nest is a depression in dead, flat grass near the edge of a lowland brackish pond or suitable stream. Winters on rocky coasts, shell beaches, and rocky coves. Migrates mostly over water, but sometimes seen on the shores of ponds well away from the sea.

SPECIAL HABITAT REQUIREMENTS. Mountain streams above timberline for breeding.

Further Reading. United States and Canada: Bull and Farrand 1977, Godfrey 1966, Richards 1988, Terres 1980. Mexico and Central America: Edwards 1972, Monroe 1968, Peterson and Chalif 1973, Rappole et al. 1983, Ridgely and Gwynne 1989, Stiles and Skutch 1989. South America: Hilty and Brown 1986, Meyer de Schauensee 1966, Rappole et al. 1983.

Spotted Sandpiper
Actitis macularia

Canada (Quebec): Maubèche Branle-
queue

Chile: Playero Manchado

Costa Rica: Alzacolita, Andarríos
Maculado, Piririza, Tiguiza

Guatemala: Alzacolito, Andarríos
Maculado

Mexico: Playerito Alzacolita

Puerto Rico and Venezuela: Playero
Coleador

Trinidad and Tobago: Ricuit

RANGE. Breeds from central Alaska and central Yukon to Labrador and Newfoundland, south to southern Alaska, southern California, and central Arizona, and east to the northern portions of the Gulf States, North Carolina, Virginia, and eastern Maryland. Winters from southwestern British Columbia, southern Arizona, southern New Mexico, southern Texas, the southern portions of the Gulf States, and coastal South Carolina to Chile and Argentina. Occasionally nonbreeding birds remain on the wintering ground in summer. A migrant throughout Mexico, wintering on both coasts and at low altitudes inland. The most widespread shorebird in Costa Rica, and probably in the rest of Central America as well. A common winter visitor and perhaps also a migrant in both Trinidad and Tobago. A very common migrant and winter resident in virtually all coastal and inland waters in Colombia. In South America winters chiefly inland, regularly south to Bolivia and southern Brazil, occasionally to Chile, Uruguay, and Argentina. Reliable sight records from the Tarapacá region of Chile, more rarely from points farther south. A winter resident or transient in the West Indies throughout the year; a common winter visitor to Puerto Rico and the Virgin Islands.

STATUS. Common; declining since early 1980s in the eastern United States and Canada.

HABITAT. Inhabits the edges of ponds, lakes, rivers, and streams and open terrain with temporary pools. Nests alone or in loose colonies, building nest on the ground among thick, tall grasses, occasionally under a bush or log, and usually near water. Sometimes found far from water in dry fields, pastures, and weedy shoulders of roads, occasionally on coastal beaches and dunes. Roosts on stumps, stranded logs, or rocks affording a clear view. In winter, frequents watercourses shaded by trees, and prefers shallow, muddy lagoons, creeks, canals, and higher mudflats.

SPECIAL HABITAT REQUIREMENTS. Margins of freshwater bodies. Mangrove edges, borders of streams (Caribbean).

Further Reading. United States and Canada: Cramp and Simmons 1983, DeGraaf and Rudis 1986, Godfrey 1966, Knowles 1942, Palmer 1967. Mexico and Central America: Edwards 1972, Land 1970, Monroe 1968, Peterson and Chalif 1973, Rappole et al. 1983, Ridgely and Gwynne 1989, Stiles and Skutch 1989. South America: Hilty and Brown 1986, Johnson 1965, Meyer de Schauensee 1966, Meyer de Schauensee and Phelps 1978, Rappole et al. 1983. Caribbean: Bond 1947, Herklots 1961, Raffaele 1989, Rappole et al. 1983.

Upland Sandpiper
Bartramia longicauda

Canada (Quebec): Maubèche des
    Champs
Chile: Batita
Costa Rica: Gansa, Pradero
Guatemala: Chorlito Pradero
Mexico: Zarapito Ganga
Puerto Rico: Ganga
Venezuela: Tibi-Tibe

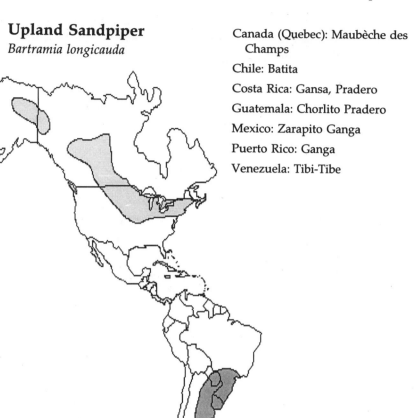

RANGE. Breeds locally from north-central Alaska, northern Yukon, and northern Alberta to southern Quebec, central Maine, and southern New Brunswick south to northeastern Oregon, central Colorado, north-central Texas, central Missouri, West Virginia, and Maryland. Migrates through Central America, the Caribbean, and northern South America. Winters in Brazil, Paraguay, Uruguay, and Argentina. In Mexico a spring and fall migrant through the Valley of Mexico. In Guatemala a migrant reported as rare in published accounts, but may be overlooked as it passes through at night. An uncommon and local migrant in much of Central America. A transient in the West Indies. In Puerto Rico a rare spring and fall migrant, accidental in winter. There are two records each from Anegada, St. Thomas, and St. Croix. A very rare migrant in Trinidad and Tobago. In Colombia an uncom-

179

mon to fairly common spring and fall migrant. Winters inland chiefly in southern Brazil from São Paulo south to Uruguay and on the Pampas of Paraguay and Argentina south to Río Negro.

STATUS. Uncommon; once abundant, but numbers have declined in the northeastern United States because of agricultural practices and loss of grassland habitats. Loss of habitat to agriculture on South American wintering grounds has also been implicated in decline (Dobkin 1992). Population showing long-term increase in the central United States and Canada.

HABITAT. Inhabits grassy open areas, ranging from sandy, sparsely vegetated flats to open, grassy bogs and muskeg. Most often found in rich pastureland, hay fields, and alfalfa fields. During the breeding season alights freely on fence posts, telephone poles, and other elevated sites. During migration, frequents alfalfa fields, pastures, prairie-dog towns, and rarely shores and mudflats. Nest is a depression on the ground among rank grasses, along sloughs in prairies, or in clearings of spruce muskeg, in loosely spaced colonies. Conceals nest by covering with nearby vegetation. Absent from tallgrass prairie less than 10 ha in size.

SPECIAL HABITAT REQUIREMENTS. Open grasslands.

Further Reading. United States and Canada: Andrle and Carroll 1988, Carter 1992, Cramp and Simmons 1983, Godfrey 1966, Higgins and Kirsh 1975, Palmer 1967, Samson 1980, Tate and Tate 1982. Mexico and Central America: Edwards 1972, Land 1970, Monroe 1968, Osborne and Peterson 1984, Peterson and Chalif 1973, Rappole et al. 1983, Ridgely and Gwynne 1989, Stiles and Skutch 1989. South America: Hilty and Brown 1986, Johnson 1965, Meyer de Schauensee 1966, Meyer de Schauensee and Phelps 1978, Rappole et al. 1983. Caribbean: Bond 1947, Herklots 1961, Raffaele 1989, Rappole et al. 1983.

Eskimo Curlew
Numenius borealis

Canada (Quebec): Courlis Esquimau
Chile and Mexico: Zarapito Boreal
Guatemala: Zarapito Esquimal
Puerto Rico: Playero Artico

RANGE. Nearly extinct. Formerly bred in northwestern Northwest Territories, possibly to western Alaska. Wintered in south-central Brazil south through Paraguay and Uruguay to southern Argentina and Chile. Some migrants formerly passed nonstop over Mexico. Once a rare migrant in Puerto Rico and the Lesser Antilles. As of 1961, in both Trinidad and Tobago, it had not been seen for more than 100 years. Up to the close of the nineteenth century, the Eskimo Curlew was one of the most common birds on the North American continent, but it is now on the verge of extinction. The rigidity of the species' migratory routes and the fact that part of these routes passed through densely populated areas proved to be the species' undoing. After nesting safely in the far north, most of the birds passed through Labrador

181

down the Atlantic Coast in the fall, then south nonstop over the Caribbean to the north coast of South America and from there by easy stages to the wintering grounds on the Argentine Pampas. In spring after reaching Colombia or Venezuela, the birds turned west, crossing the Gulf of Mexico and following the Mississippi River up through the Midwestern states to the breeding grounds in Canada. Because of their table quality, and their very great numbers, the birds were shot in Argentina and on spring migration in the Mississippi Valley. The take was enormous, facilitated by the fact that the birds flew in compact flocks, as many as 20 to 30 birds falling at a shot. Hunters sometimes bagged 2,000 birds in a single day. Although the species was once thought extinct, and may possibly be, sightings from Texas (Galveston to Rockport, 1959–1963, with photographs from Galveston in spring 1962 and Padre Island 1972) and Manitoba (May 1980) among others, indicate that a few pairs may still exist (Am. Ornithol. Union 1983:184).

STATUS. Endangered.

HABITAT. Breeds on arctic tundra. Winters on grasslands in Argentina.

SPECIAL HABITAT REQUIREMENTS. Arctic tundra for breeding.

Further Reading. United States and Canada: Bull and Farrand 1977, Godfrey 1966, Richards 1988, Robbins 1980, Terres 1980. Mexico and Central America: Land 1970, Peterson and Chalif 1973, Rappole et al. 1983. South America: Johnson 1965, Meyer de Schauensee 1966, Rappole et al. 1983. Caribbean: Bond 1947, Herklots 1961, Rappole et al. 1983.

Whimbrel
Numenius phaeopus

Canada (Quebec): Courlis Corlieu

Chile: Zarapito Común

Costa Rica: Cherela, Zarapito
  Trinador, Zarceta

Guatemala: Chorlo, Zarapito Trinador

Mexico: Zarapito Cabecirrayado

Puerto Rico: Playero Pico Corvo

Trinidad and Tobago: Bec Corchu

Venezuela: Chorlo Real

RANGE. Breeds in North America in western and northern Alaska east along the arctic coast to northwestern Northwest Territories south to Denali National Park, southwestern Yukon, and along the western side of Hudson Bay to northwestern James Bay. Nonbreeding birds summer on the coasts of California and New Jersey, south to South Carolina. Winters locally from California to Galapagos Islands and southern Chile, rarely on the coasts of Texas, Louisiana, and South Carolina. Also winters on the Caribbean coast of Colombia and Venezuela to Brazil and southern Argentina. In Mexico a common migrant and winter visitor on the Pacific Coast, seldom reported on the Atlantic Coast. In Central America a fairly common migrant and winter visitor, especially on the Pacific Coast. A rare visitor to Puerto Rico. Rare but regular on St. Croix; extremely rare elsewhere in Virgin Islands, where it has been recorded from St. Thomas, Anegada, Beef Island, and Necker Island. Recorded in the West Indies in all seasons; North American individuals occur as migrants and apparently as rare winter residents. A regular but never abundant winter visitor in Trinidad and Tobago. In Colombia a common migrant and winter resident on both coasts. Winters along all coasts south to Bahia, Brazil, and to Tierra del Fuego, Chile; casually to Tierra del Fuego, Argentina. This bird breeds the farthest north and winters the farthest south of any bird in the Western Hemisphere.

STATUS. Common.

HABITAT. Breeds in bare open areas of tundra or comparatively dry moorland that has perhaps the occasional stunted tree. Winters in areas with tidal flats exposed at low water, salt marshes at high water, beach grass near sandy shores, estuaries, and tidal creeks. Nest is a saucer-shaped depression on top of a low hummock of mosses or grasses on arctic tundra. In migration found in flooded fields, shallow lake edges, and river bars.

SPECIAL HABITAT REQUIREMENTS. Arctic tundra for breeding.

Further Reading. United States and Canada: Bull and Farrand 1977, Godfrey 1966, Richards 1988, Terres 1980. Mexico and Central America: Edwards 1972, Land 1970, Monroe 1968, Peterson and Chalif 1973, Rappole et al. 1983, Ridgely and Gwynne 1989, Stiles and Skutch 1989. South America: Hilty and Brown 1986, Johnson 1965, Meyer de Schauensee 1966, Meyer de Schauensee and Phelps 1978, Rappole et al. 1983. Caribbean: Bond 1947, Herklots 1961, Raffaele 1989, Rappole et al. 1983.

Long-billed Curlew
Numenius americanus

Canada (Quebec): Courlis à Long Bec

Costa Rica, Guatemala, and Mexico:
Zarapito Piquilargo

RANGE. Breeds from south-central British Columbia to southern Manitoba, south to northeastern California, central Utah, central New Mexico, and northern Texas, and east to southwestern Kansas. Winters from central California, southern Texas, southern Louisiana, and coastal South Carolina south to Mexico. In Mexico a rather rare migrant and winter visitor on both coasts; locally inland to Guatemala. A rare winter resident in Central America. Formerly occurred on Cuba, Jamaica, and possibly the Lesser Antilles, but not recorded from anywhere in the West Indies in the twentieth century.

STATUS. Fairly common, once fairly abundant; decline is due to past hunting pressure and loss of habitat caused by grazing and agriculture. Population has been increasing in the western United States but declining in the central United States and Canada. Listed as a candidate for federal threatened and endangered status.

HABITAT. Inhabits grasslands ranging from moist meadowland to very dry prairie. When at or near water, often loosely associates with godwits, Willets, and yellowlegs. During the breeding season, commonly perches on bushes, low trees, dirt mounds, rocks, stumps, fence posts, utility poles, or other elevated sites. Nests in a slight hollow on the ground, usually in flat areas among short grasses such as cheatgrass and bluegrass. Locates nest in moist

185

or arid areas far from water. In other seasons, frequents wet habitats such as the shallow margins of inland and coastal waters, open areas of marshes, intertidal zones, or sandbars.

SPECIAL HABITAT REQUIREMENTS. Prairies or grassy meadows.

Further Reading. United States and Canada: Allen 1980, Godfrey 1966, Palmer 1967, Stenzel et al. 1976, Tate and Tate 1982. Mexico and Central America: Edwards 1972, Land 1970, Monroe 1968, Peterson and Chalif 1973, Rappole et al. 1983, Ridgely and Gwynne 1989, Stiles and Skutch 1989. Caribbean: Bond 1947, Herklots 1961, Raffaele 1989, Rappole et al. 1983.

Hudsonian Godwit
Limosa haemastica

Canada (Quebec): Barge Hudsonienne
Chile: Zarapito de Pico Recto
Costa Rica: Aguja Lomiblanca
Mexico: Limosa Ornamentada
Puerto Rico: Barga Aliblanca
Venezuela: Becasa de Mar

RANGE. Breeds locally in south-coastal and western Alaska, western Northwest Territories, northwestern British Columbia, and around Hudson Bay. Migrates through the Caribbean and northern South America. Winters coastally and inland in Paraguay, Uruguay, southern Brazil, Chile, and Argentina to Tierra del Fuego. In Mexico a very rare migrant normally migrating from the United States to South America via the Caribbean (Cuba, Hispaniola, Puerto Rico, Trinidad, and coastal Venezuela). In Costa Rica a rare migrant. In Panama known from one sighting on the Caribbean side of the Canal area. An extremely rare fall migrant to Puerto Rico and St. Croix. A transient in the West Indies, where it is recorded from the Bahamas, Cuba, Hispaniola, Guadeloupe, and Barbados, questionably from Dominica and Martinique. In

Colombia a hypothetical species with no specimens. In Chile a more or less regular visitor to the Magallanes region, a straggler elsewhere. Migrates east of the Andes.

STATUS. Locally common.

HABITAT. Inhabits wet bogs and marshes in open expanses along the northern edge of boreal forests and moist tundra, near tidal or fluvial shorelines. Nests in a depression on top of a hummock in a sedge marsh or meadow with water up to 5 to 10 cm deep. Frequently locates nest under a dwarf birch, occasionally on a sedge or grass tussock, under or among small willows, or near a fallen spruce tree, but usually where the vegetation is at least 10 to 15 cm high and thick enough to conceal the nest from the sides. In other seasons, frequents fresh, brackish, and salt waters on beaches, mudflats, marshes, flooded fields, and shallow ponds.

SPECIAL HABITAT REQUIREMENTS. Extensive sedge marshes and meadows near tidal flats. In migration through Caribbean, grassy freshwater pond edges, mudflats.

Further Reading. United States and Canada: Bent 1927, Farrand 1983a, Godfrey 1966, Hagar 1966, Mackenzie 1977. Mexico and Central America: Edwards 1972, Peterson and Chalif 1973, Rappole et al. 1983, Ridgely and Gwynne 1989, Stiles and Skutch 1989. South America: Hilty and Brown 1986, Johnson 1965, Meyer de Schauensee and Phelps 1978, Rappole et al. 1983. Caribbean: Bond 1947, Herklots 1961, Raffaele 1989, Rappole et al. 1983.

Marbled Godwit
Limosa fedoa

Canada (Quebec): Barge Marbrée
Chile: Zarapito Monteado
Costa Rica: Aguja Canela
Guatemala: Aguja Jaspeada
Mexico: Limosa Canela
Puerto Rico: Barga Jaspeada

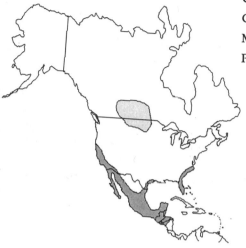

RANGE. Breeds from central Alberta to southern Manitoba and northern Ontario south to central Montana, northeastern South Dakota, and northwestern Minnesota. Winters from central California, western Nevada, the Gulf Coast, and coastal South Carolina south to South America. Some nonbreeding birds occur in the winter range during summer. In Mexico a common but irregular winter visitor on both coasts, casual inland. In Central America an uncommon migrant and winter visitor. An extremely rare migrant and possible winter resident in Puerto Rico, St. Croix, and Anegada; rare elsewhere in the West Indies, where it is recorded from Cuba, Jamaica, Hispaniola, Grenada, and Carriacou, questionably from Guadeloupe and Martinique. No records in the twentieth century for Trinidad or Tobago, but it may occasionally occur as a fall migrant. In Colombia a hypothetical species with no specimens and only one sight record. Winters casually on Pacific Coast to northern Chile.

STATUS. Common.

HABITAT. Inhabits grassy plains, broad flat wet meadows, and prairie sloughs, usually near lakes, rivers, or streams; preferentially where grass is short and sparse. Moderate grazing maintains habitat. Nests singly or pos-

189

sibly semicolonially. Nest is in a slight hollow in short grasses, often in plain sight and usually near water. During migration, frequents open coastal beaches and lake shores; in winter, tidal flats in sheltered bays, next to inlets, and on open beaches. Gregarious and often in the company of Willets and American Avocets.

SPECIAL HABITAT REQUIREMENTS. Short, sparse grass cover along wetland shorelines. Mudflats, marshes (Puerto Rico).

Further Reading. United States and Canada: Bent 1927, Farrand 1983a, Godfrey 1966, Palmer 1967. Mexico and Central America: Edwards 1972, Land 1970, Monroe 1968, Peterson and Chalif 1973, Rappole et al. 1983, Ridgely and Gwynne 1989, Ryan et al. 1984, Stiles and Skutch 1989. South America: Hilty and Brown 1986, Johnson 1965, Meyer de Schauensee 1966, Rappole et al. 1983. Caribbean: Bond 1947, Herklots 1961, Raffaele 1989, Rappole et al. 1983.

Ruddy Turnstone
Arenaria interpres

Canada (Quebec): Tourne-pierre Roux
Chile: Chorlo Vuelvepiedras
Costa Rica: Vuelvepiedras Rojizo
Guatemala: Vuelvepiedras
Mexico: Vuelvepiedras Común
Puerto Rico and Venezuela: Playero
Turco

RANGE. Breeds in North America from northern Alaska and the Canadian arctic islands south to western Alaska, and Southampton, Coats, and Mansel Islands; probably also to the northern portions of the Northwest Territories. Nonbreeding birds may be found throughout the winter range in summer. Winters along the Pacific and Atlantic Coasts from central California and New York south to Chile and Argentina. In Mexico a moderately common migrant and winter visitor on both coasts. In Guatemala a rare migrant and winter visitor on both the Caribbean and Pacific Coasts at sea level. In Costa Rica, Honduras, and Panama a common migrant in fall and spring, a locally common winter resident; more common on Pacific Coast. A common visitor to, and winter resident in, the West Indies in all months except June and July when uncommon. A migrant to Trinidad and Tobago; some birds may spend the winter. In Colombia a common migrant and winter resident on both the Caribbean and Pacific Coasts. Winters along coasts south to Buenos Aires, Argentina, and to the Llanquihue region of Chile.

STATUS. Common.

HABITAT. Inhabits flat, lichen-covered, mossy, or gravelly tundra near the seacoast in a variety of boreal habitats. Often perches on boulders, stakes, pilings, piers, and boats during breeding season. Nests in a depression in tundra in exposed or sheltered sites such as beside a rock or clump of vegetation. Uses a wet area near the nest site for brood rearing. In other seasons, frequents rocks, reefs, and mussel beds of the intertidal zone, sandy beaches, and solid mudflats.

SPECIAL HABITAT REQUIREMENTS. Dry, dwarf-shrub tundra near the coast.

Further Reading. United States and Canada: Bent 1929, Farrand 1983a, Godfrey 1966, Palmer 1967, Terres 1980. Mexico and Central America: Edwards 1972, Land 1970, Monroe 1968, Peterson and Chalif 1973, Rappole et al. 1983, Ridgely and Gwynne 1989, Stiles and Skutch 1989. South America: Hilty and Brown 1986, Johnson 1965, Meyer de Schauensee 1966, Rappole et al. 1983. Caribbean: Bond 1947, Herklots 1961, Raffaele 1989, Rappole et al. 1983.

Surfbird
Aphriza virgata

Canada (Quebec): Echassier du Ressac
Chile: Chorlo de los Rompientes
Costa Rica: Chorlito de Rompientes
Guatemala: Playero de los
    Rompientes
Mexico: Playero Roquero

RANGE. Breeds in south-central Alaska. Winters in southeastern Alaska down the Pacific Coast of Mexico, especially outer Baja California, to the Straits of Magellan and western Tierra del Fuego, Chile. Although a regular visitor to Chile during the northern winter, not common on the country's coast and is most likely to be found on the wave-washed, seaweed-covered outlying rocks of promontories and headlands.

STATUS. Common.

HABITAT. Breeds high in the mountains of south-central Alaska. Nest is a small depression of bare, dry rock lined with a few lichens and mosses or dried bits of leaves. Winters on rocky coasts of the Pacific.

SPECIAL HABITAT REQUIREMENTS. High mountain habitat in south-central Alaska.

Further Reading. United States and Canada: Bull and Farrand 1977, Godfrey 1966, Palmer 1967, Richards 1988, Terres 1980. Mexico and Central America: Edwards 1972, Land 1970, Monroe 1968, Peterson and Chalif 1973, Rappole et al. 1983, Ridgely and Gwynne 1989, Stiles and Skutch 1989. South America: Johnson 1965, Meyer de Schauensee 1966, Rappole et al. 1983.

Red Knot
Calidris canutus

Canada (Quebec): Bécasseau à Poitrine Rousse

Chile: Playero Artico

Costa Rica: Correlimos Grande

Guatemala: Correlimos Gordo, Playero Pechrrufo

Mexico: Playero Piquicorto

Puerto Rico: Playero Gordo

Trinidad and Tobago: Poule Couchant

Venezuela: Playero Pecho Rufo

RANGE. Breeds in North America from northwestern and northern Alaska and the Canadian arctic islands east to Ellesmere Island and south to southern Victoria and Southampton Islands. Winters along the Pacific and Atlantic Coasts from southern California and Massachusetts south to the Straits of Magellan. Nonbreeding birds occasionally summer in the wintering range. In Mexico and Central America a rather rare and irregular migrant, especially on the Pacific Coast. In Colombia probably a rare migrant, but only a few sightings and no specimens. An uncommon migrant to the Bahamas and Greater Antilles in the fall. Rare but regular on St. Croix, and extremely rare elsewhere in the Virgin Islands, where it is known from St. John, Anegada, and Necker Island. A migrant to Trinidad only.

STATUS. Locally common; once abundant, but numbers were reduced by shooting in the late nineteenth century.

HABITAT. Inhabits high inland plains, plateaus, and elevated slopes covered with glacial gravel and frost-riven rocks and shales, sometimes several miles from the coast. Nests in a shallow depression among dryads, lichens, and other tundra vegetation, rubble, and gravel, and usually on high, dry hills and plateaus. In migration and in winter, occurs mainly along the coast on exposed mudflats, sand spits, beaches, matted salt marshes, and river deltas.

SPECIAL HABITAT REQUIREMENTS. Barren or stony tundra and dry nesting sites.

Further Reading. United States and Canada: Bent 1927, Farrand 1983a, Godfrey 1966, Palmer 1967, Sperry 1940, Terres 1980. Mexico and Central America: Edwards 1972, Land 1970, Monroe 1968, Peterson and Chalif 1973, Rappole et al. 1983, Ridgely and Gwynne 1989, Stiles and Skutch 1989. South America: Hilty and Brown 1986, Johnson 1965, Meyer de Schauensee 1966, Meyer de Schauensee and Phelps 1978, Rappole et al. 1983. Caribbean: Bond 1947, Herklots 1961, Raffaele 1989, Rappole et al. 1983.

Sanderling
Calidris alba

Canada (Quebec): Sanderling

Chile: Playero Común

Costa Rica, Puerto Rico, and
 Venezuela: Playero Arenero

Guatemala: Playero Arenero

Mexico: Playerito Correlón

Trinidad and Tobago: Becasse Blande

RANGE. Breeds in North America from northern Alaska and Prince Patrick Island to northern Ellesmere Island, south to northern Northwest Territories, and Southampton and northern Baffin Islands. Migrates and winters throughout Central America, the Caribbean, and South America. Nonbreeding birds occur on the wintering grounds during the summer. Winters locally in the Aleutian Islands, and from southern Alaska, the Gulf Coast, and Massachusetts south along the Atlantic and Pacific Coasts to Chile and Argentina. In Mexico a common winter visitor on both coasts. In Central America a migrant and possible winter visitor on both coasts. A fairly common winter resident and transient in the West Indies throughout the year, although rare in the Virgin Islands in all seasons. A winter visitor to Trinidad and Tobago.

In Colombia a fairly common transient and winter resident on both coasts. Winters on all coasts south to the Aysén region of Chile and Tierra del Fuego, Argentina. Unquestionably the most abundant visitor to Chile during the northern winter months.

STATUS. Common.

HABITAT. Inhabits high arctic tundra, particularly dry clay–mixed stony plains sparsely covered with willows, dryad, and saxifrage. Nests in a slight hollow or depression on the ground on stony, well-drained ridge tops, gentle slopes, or level alluvial plains. Usually locates nest at the edge of, or in, clumps of low plants and within several hundred meters of a marshy pond, but occasionally nests up to 1 km or more from wet tundra. Outside the breeding season, frequents sandy coastal beaches and tidal flats. Inland migrants inhabit sandbars along rivers and on lake beaches.

SPECIAL HABITAT REQUIREMENTS. Dry tundra for breeding; sandy beaches during migration and winter.

Further Reading. United States and Canada: Bent 1927, Cramp and Simmons 1983, Farrand 1983a, Godfrey 1966, Palmer 1967, Parmelee 1970. Mexico and Central America: Edwards 1972, Land 1970, Monroe 1968, Peterson and Chalif 1973, Rappole et al. 1983, Ridgely and Gwynne 1989, Stiles and Skutch 1989. South America: Hilty and Brown 1986, Johnson 1965, Meyer de Schauensee 1966, Meyer de Schauensee and Phelps 1978. Caribbean: Bond 1947, Herklots 1961, Raffaele 1989, Rappole et al. 1983.

Semipalmated Sandpiper
Calidris pusilla

Canada (Quebec): Bécausseau Semi-palmé

Chile: Playero Semipalmeado

Costa Rica: Correlimos Semipalmeado

Guatemala: Correlimos Semipalmeado, Playero Semipalmeado

Mexico: Playerito Semipalmeado

Puerto Rico: Playero Gracioso

Venezuela: Playero Semipalmeado

RANGE. Breeds from the arctic coast of western and northern Alaska north to Victoria and central Baffin Islands, east to northern Labrador, and south to western Alaska and central Northwest Territories across to northern Ontario, northern Quebec, and coastal Labrador. Winters from southern Florida south along Caribbean and Atlantic Coasts to Argentina and along Pacific Coast to northern Chile. Nonbreeding birds may summer in coastal North America south to the Gulf Coast and Panama. In Mexico a rather rare migrant and winter visitor along the east coast. In Central America a fairly common migrant and winter visitor. The population wintering in Costa Rica consists of short-billed birds from the westernmost part of the breeding range. A common visitor to the West Indies throughout the year, especially in the fall, but generally uncommon in the Virgin Islands. A common winter visitor in large and small flocks to both Trinidad and Tobago. The distribution has been difficult to determine because of identification difficulties among similar shorebirds in winter plumage. This bird's yearly migration from the Arctic to as far south as Argentine Patagonia and back is a round-trip of more than 32,000 km.

STATUS. Abundant.

HABITAT. Inhabits subarctic and low- to high-arctic tundra from coasts, dunes, borders of tidal inlets, and deltas to damp grassy flats in interior and wet riverside tundra. Often occurs near lakes or pools, shifting from first areas uncovered by melting snow and surface ice to others becoming clear shortly afterward, including upland tundra. Nests in a slight depression on the ground amid short sedges, sometimes in sand on grassy dunes or in low wet tundra near small lakes. In other seasons, frequents mudflats, sandy beaches, and wet meadows, favoring the vicinity of water on tidal flats, lagoons, and ponds.

SPECIAL HABITAT REQUIREMENTS. Grassy or hummocky tundra in breeding season. All wetland edges in winter.

Further Reading. United States and Canada: Bent 1927, Cramp and Simmons 1983, Farrand 1983a, Godfrey 1966, Palmer 1967, Terres 1980. Mexico and Central America: Edwards 1972, Land 1970, Monroe 1968, Peterson and Chalif 1973, Rappole et al. 1983, Ridgely and Gwynne 1989, Stiles and Skutch 1989. South America: Hilty and Brown 1986, Johnson 1965, Meyer de Schauensee 1966, Meyer de Schauensee and Phelps 1978, Rappole et al. 1983. Caribbean: Bond 1947, Herklots 1961, Raffaele 1989, Rappole et al. 1983.

Western Sandpiper
Calidris mauri

Canada (Quebec): Bécausseau du Nord-Ouest

Costa Rica: Becacina, Correlimos Occidental, Patudo

Guatemala: Correlimos Occidental, Playero Occidental

Mexico, Puerto Rico, and Venezuela: Playerito Occidental

RANGE. Breeds on islands in the Bering Sea and along the coasts of western and northern Alaska. Winters from California and North Carolina south along both the Pacific and Atlantic Coasts to Peru and Surinam. In Mexico a common migrant and winter visitor on both coasts. In Costa Rica the most abundant small sandpiper on both coasts but especially on the Pacific Coast. A fairly common visitor to Puerto Rico, but rare in the Virgin Islands; a transient and winter resident elsewhere in the West Indies and Trinidad. In Colombia a very common migrant and winter resident on tidal mudflats and coastal lagoons on both coasts.

STATUS. Common.

HABITAT. Inhabits a complex mosaic of wet low-lying grass and sedge marshes dotted with small pools and lakes, and relatively well-drained heath-covered tundra such as on hummocks, ridges, and better-drained slopes of hills that are vegetated with mosses, lichens, dwarf shrub heath, dwarf birch, willows, and some herbs, grasses, and sedges. Nests on the ground in loose colonies or singly. Nest is usually well camouflaged under low vegetation and located in dry or moist areas, from the upper slopes of hills down to the marsh edge, or within a marsh on a patch of heath tundra. After fledging, moves into marshes, frequenting margins of lakes and rivers. Outside of the breeding season, frequents mudflats, beaches, shores of lakes and ponds, and flooded fields.

SPECIAL HABITAT REQUIREMENTS. In breeding season patches of dwarf shrub–heath tundra (at least 0.25 ha) interspersed among wet marshes. Wetland edges in winter.

Further Reading. United States and Canada: Bent 1927, Cramp and Simmons 1983, Godfrey 1966, Holmes 1971, Palmer 1967. Mexico and Central America: Edwards 1972, Land 1970, Monroe 1968, Peterson and Chalif 1973, Rappole et al. 1983, Ridgely and Gwynne 1989, Stiles and Skutch 1989. South America: Hilty and Brown 1986, Meyer de Schauensee 1966, Meyer de Schauensee and Phelps 1978, Rappole et al. 1983. Caribbean: Bond 1947, Herklots 1961, Raffaele 1989, Rappole et al. 1983.

Least Sandpiper
Calidris minutilla

Canada (Quebec): Bécausseau Minuscule

Chile: Playero Enano

Costa Rica: Becacina, Correlimos Menudo, Menudillo, Patudo

Guatemala: Correlimos Enano, Playerito Menudo

Mexico: Playerito Mínimo

Puerto Rico and Venezuela: Playerito Menudo

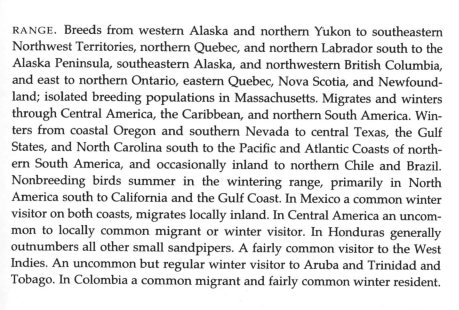

RANGE. Breeds from western Alaska and northern Yukon to southeastern Northwest Territories, northern Quebec, and northern Labrador south to the Alaska Peninsula, southeastern Alaska, and northwestern British Columbia, and east to northern Ontario, eastern Quebec, Nova Scotia, and Newfoundland; isolated breeding populations in Massachusetts. Migrates and winters through Central America, the Caribbean, and northern South America. Winters from coastal Oregon and southern Nevada to central Texas, the Gulf States, and North Carolina south to the Pacific and Atlantic Coasts of northern South America, and occasionally inland to northern Chile and Brazil. Nonbreeding birds summer in the wintering range, primarily in North America south to California and the Gulf Coast. In Mexico a common winter visitor on both coasts, migrates locally inland. In Central America an uncommon to locally common migrant or winter visitor. In Honduras generally outnumbers all other small sandpipers. A fairly common visitor to the West Indies. An uncommon but regular winter visitor to Aruba and Trinidad and Tobago. In Colombia a common migrant and fairly common winter resident.

Least Sandpiper

STATUS. Very common.

HABITAT. Inhabits open grass or sedge bogs and marshes in the northern spruce forest just south of treeless tundra, or among complexes of pools and water channels with scattered knolls and hummocks. Nest is a depression in a mossy hummock, plant tuft, clump of grass, or sometimes on the ground, usually in marshy cover but sometimes in drier upland near water. Outside of the breeding season, prefers wet, muddy, or grassy areas such as muddy shores of grass-fringed marshes or estuaries, grassy wet meadows, and grass-bordered mudflats of lakes, ponds, or rivers; found less frequently on sandy beaches.

SPECIAL HABITAT REQUIREMENTS. Wetlands of subarctic boreal forests and tundra. Mudflats in winter.

Further Reading. United States and Canada: Cottam and Hanson 1938, Cramp and Simmons 1983, Godfrey 1966, Low and Mansell 1983, Palmer 1967, Pough 1951, Terres 1980. Mexico and Central America: Edwards 1972, Land 1970, Monroe 1968, Peterson and Chalif 1973, Rappole et al. 1983, Ridgely and Gwynne 1989, Stiles and Skutch 1989. South America: Hilty and Brown 1986, Johnson 1965, Meyer de Schauensee 1966, Meyer de Schauensee and Phelps 1978, Rappole et al. 1983. Caribbean: Bond 1947, Herklots 1961, Raffaele 1989, Rappole et al. 1983.

White-rumped Sandpiper
Calidris fuscicollis

Canada (Quebec): Bécausseau à Croupion Blanc

Chile: Playero de Lomo Blanco

Costa Rica: Becasina, Correlimos Lomiblanco, Patudo

Guatemala: Correlimos Lomiblanco, Playero de Rabadilla Blanca

Mexico: Playerito de Rabadilla Blanca

Puerto Rico: Playero Rabadilla Blanca

Venezuela: Playero de Rabadilla Blanca

RANGE. Breeds from northern Alaska to northern Bylot Island, south to the mainland coasts of Northwest Territories, northwestern Hudson Bay, Southampton Island, and southern Baffin Island. Migrates through Central America, the Caribbean, and northern South America. Winters throughout South America (status unknown in Uruguay). In Mexico a rare migrant recorded on the Yucatán Peninsula and Cozumel Island. In Central America an uncommon to rare migrant. An uncommon migrant in the West Indies in the fall and rare in the spring. Winters chiefly east of the Andes from southern Brazil and Paraguay to Tierra del Fuego; also on Aruba, Trinidad, and Tobago. Because this species migrates east of the Andes, it is very rare in Chile except in the very southern portion, where it is the most abundant visiting shorebird—even more common than Baird's Sandpiper.

STATUS. Common.

HABITAT. Inhabits both lowland and upland tundra, frequently around bog pools, on dry ridges, or among grassy tussocks near rivers or lakes. Nests in mossy depressions in clumps of grasses and sedges in the uplands, or in mossy hummocks on well-vegetated tundra that is persistently wet, often near marshy ponds and lake shores. Conceals nest, usually among grasslike plants, including narrow-leaved cottongrass, grass rush, water sedges, and mosses. Usually closely associated with moist, open terrain, but tolerates occasional freezing and snow cover and a wide range of temperatures. In migration, prefers shallow grassy pools, wet meadows, and marshes, but also occurs on sandbars, mudflats, and beaches.

SPECIAL HABITAT REQUIREMENTS. Mossy or grassy tundra.

Further Reading. United States and Canada: Bent 1927, Cramp and Simmons 1983, Drury 1961, Godfrey 1966, Palmer 1967, Parmelee et al. 1968, Terres 1980. Mexico and Central America: Edwards 1972, Land 1970, Monroe 1968, Peterson and Chalif 1973, Rappole et al. 1983, Ridgely and Gwynne 1989, Stiles and Skutch 1989. South America: Hilty and Brown 1986, Johnson 1965, Meyer de Schauensee 1966, Meyer de Schauensee and Phelps 1978, Rappole et al. 1983. Caribbean: Bond 1947, Herklots 1961, Raffaele 1989, Rappole et al. 1983.

Baird's Sandpiper
Calidris bairdii

Canada (Quebec): Bécausseau de Baird

Costa Rica: Correlimos de Baird

Mexico: Playerito de Baird

Puerto Rico and Venezuela: Playero de Baird

RANGE. Breeds from western and northern Alaska to Ellesmere Island, south to central Alaska, northern Northwest Territories, Southampton Island, and south-central Baffin Island. Winters in South America. In Mexico a rather rare migrant chiefly through the interior, locally on high mountain lakes. In Panama rare to locally uncommon as a fall migrant, very rare in the spring; recorded only from the Canal area, where mostly seen on wet, shortgrass fields on the Caribbean side. Known from only 10 sightings on St. Croix since 1982. Not yet recorded from elsewhere in the region of the Virgin Islands or Puerto Rico. In Colombia a fall migrant in small numbers and apparently a rare to very rare spring migrant. Winters chiefly in southern

South America, locally in the Andes of Ecuador and central Peru; apparently not recorded from Brazil. Except for the Sanderling, this is the most abundant visiting shorebird in Chile. Unlike the Sandering, it is found in the interior, even along the higher lakes of the Andes, as well as along the coast.

STATUS. Common.

HABITAT. Inhabits dry coastal and alpine tundra, particularly barren, exposed ridges, terrace banks, and raised beaches that are sparsely covered with low matted vegetation. Nests in a shallow depression on the ground, usually in windblown and lichen-strewn areas with large patches of bare soil. Sometimes nests in grassy areas in a tuft of vegetation or among lichen-covered rocks. Prefers sheltered places and frequents muddy, sandy, and grassy areas near water, including irrigated fields, shores of lakes and ponds, alpine tundra, and marshes. During migration, prefers inland more than coastal habitats.

SPECIAL HABITAT REQUIREMENTS. Dry tundra.

Further Reading. United States and Canada: Cramp and Simmons 1983, Drury 1961, Farrand 1983a, Godfrey 1966, Palmer 1967, Terres 1980. Mexico and Central America: Edwards 1972, Land 1970, Monroe 1968, Peterson and Chalif 1973, Rappole et al. 1983, Ridgely and Gwynne 1989, Stiles and Skutch 1989. South America: Hilty and Brown 1986, Johnson 1965, Meyer de Schauensee 1966, Meyer de Schauensee and Phelps 1978, Rappole et al. 1983. Caribbean: Bond 1947, Herklots 1961, Raffaele 1989, Rappole et al. 1983.

Pectoral Sandpiper
Calidris melanotos

Canada (Quebec): Bécausseau à
   Poitrine Cendrée

Chile: Playero Pectoral

Costa Rica: Becacina, Correlimos
   Pechirrayado, Correlimos Pectoral,
   Patudo

Guatemala: Correlimos Patiamarillo,
   Playero Pinto

Mexico: Playero Pechirrayado

Puerto Rico: Playero Manchado

Trinidad and Tobago: Grosse Becasse

Venezuela: Tin-Guin

RANGE. Breeds from western and northern Alaska to Bathurst and Devon Islands, south to western Alaska, central and southeastern Northwest Territories, and the south coast of Hudson Bay. Migrates through Central America, the Caribbean, and northern South America. Winters in wet meadows and grassy mudflats in Peru, Bolivia, Paraguay, Uruguay, Chile (irregularly, from Africa to Osorno), southeastern Brazil, and Argentina. In Mexico a rare migrant in the coastal lowlands and the interior; absent on the Yucatán Peninsula. In Central America a locally common migrant in both fall and spring, but no winter records. A fairly common fall but rare spring migrant to the West Indies; sometimes flocks of as many as 5,000 occur. A fairly common migrant and winter resident in Trinidad, Tobago, and Aruba in fall. In Co-

209

lombia a fairly common fall migrant and a very uncommon or rare spring migrant.

STATUS. Common.

HABITAT. Inhabits a variety of tundra habitats on flat terrain that is poorly drained, usually wet, and characterized by low grasses and sedges, dwarf shrubs, and cottongrass tussocks. Nests in a depression on dry ground in areas with a continuous cover of grasses and sedges. Hides nest well, usually under a tree or bush. Outside of the breeding season, prefers grassy terrain bordering moving or still waters; only rarely found on open mudflats.

SPECIAL HABITAT REQUIREMENTS. Dry nesting sites on arctic tundra.

Further Reading. United States and Canada: Bent 1927, Cramp and Simmons 1983, Godfrey 1966, Low and Mansell 1983, Palmer 1967, Pitelka 1959. Mexico and Central America: Edwards 1972, Land 1970, Monroe 1968, Peterson and Chalif 1973, Rappole et al. 1983, Ridgely and Gwynne 1989, Stiles and Skutch 1989. South America: Hilty and Brown 1986, Johnson 1965, Meyer de Schauensee 1966, Meyer de Schauensee and Phelps 1978, Rappole et al. 1983. Caribbean: Bond 1947, Herklots 1961, Raffaele 1989, Rappole et al. 1983.

Dunlin
Calidris alpina

Canada (Quebec): Bécausseau à Dos Roux

Mexico: Playero Dorsirrojo

RANGE. Breeds in North America on tundra and along the arctic coast of Greenland, Southampton Island, northwestern and southern coasts of Hudson Bay, Alaska, and northwestern Northwest Territories. Winters along coasts from southeastern Alaska to Baja California and from Massachusetts to southern Florida and along the Gulf Coast to Texas into Mexico.

STATUS. Common.

HABITAT. Breeds in a dry spot on wet grass or sedge tundra. The nest is made of grasses and leaves and is located on a grassy hummock or other dry site in moist tundra, or in a coastal salt marsh. In migration and winter favors coastal habitats with mud or sand flats, beaches, tidal inlets and lagoons, and lake or river shores; also flooded grassland.

SPECIAL HABITAT REQUIREMENTS. Mudflats, open beaches in winter.

Further Reading. United States and Canada: Bull and Farrand 1977, Godfrey 1966, Terres 1980. Mexico: Edwards 1972, Peterson and Chalif 1973.

Stilt Sandpiper

Calidris himantopus

Canada (Quebec): Bécausseau à Échasses

Chile: Playero de Patas Largas

Costa Rica: Becasina, Correlimos Patilargo, Patudo

Guatemala: Correlimos Patilargo, Playero Patilargo

Mexico: Playero Zancón

Puerto Rico and Venezuela: Playero Patilargo

Trinidad and Tobago: Chevalier

RANGE. Breeds from northern Alaska, northern Yukon, northwestern Northwest Territories, and southern Victoria Island southeast to southeastern Northwest Territories, northeastern Manitoba, and northern Ontario; probably also south locally in Canada to borders of the taiga. Winters primarily in South America from northern Bolivia to southern Brazil and northern Argentina, casually northward to southeastern California, the Gulf Coast, and Florida. A rather rare and irregular migrant in eastern and southern Mexico. In Central America a fairly common but local migrant in fall and in spring on both coasts. A fairly common visitor to the West Indies in fall and winter, but rare in spring. A winter visitor only to Trinidad and Aruba. In Colombia a very uncommon migrant and a rare winter resident on the coasts and inland. With a migratory route similar to that of the Baird's Sandpiper, this bird may be a fairly common visitor in southern Chile during the northern winter.

212

STATUS. Uncommon.

HABITAT. Breeds in sedge meadows interrupted by old beach ridges, eskers, or other elevated areas dominated by dwarf birch, heaths, willows, crowberries, and dryads. Sometimes occurs in wet tundra areas with fairly high willows, or on much drier slopes with moderate vegetative cover. Avoids truly barren ridgetops. Nests in a depression on the ground in relatively open areas of dry tundra, usually atop a hummock or on a low, well-drained gravel ridge; occasionally nests next to a shrub. Locates nest site independent of standing water. May reuse old nests. Moves young from drying sedge meadows to wet areas.

SPECIAL HABITAT REQUIREMENTS. Well-drained sedge meadows in arctic tundra with elevated sites for nesting.

Further Reading. United States and Canada: Cramp and Simmons 1983, Godfrey 1966, Jehl 1973, Palmer 1967. Mexico and Central America: Edwards 1972, Land 1970, Monroe 1968, Peterson and Chalif 1973, Ridgely and Gwynne 1989, Stiles and Skutch 1989. South America: Hilty and Brown 1986, Meyer de Schauensee 1966, Meyer de Schauensee and Phelps 1978. Caribbean: Herklots 1961, Raffaele 1989.

Buff-breasted Sandpiper

Tryngites subruficollis

Canada (Quebec): Bécausseau Roussâtre

Costa Rica: Praderito Pechianteado, Zarceta

Mexico: Playerito Pradero

Trinidad and Tobago: Petit Pieds Jaune

Venezuela: Playero Dorado

RANGE. Breeds from northern Alaska to Banks, Melville, Bathurst, and Devon Islands, south to northwestern Northwest Territories and to Jenny Lind and King William Islands. Migrates through Central America, the Caribbean, and South America excluding Guyana, Surinam, and French Guiana. Winters inland (mostly) in Brazil, Bolivia, Paraguay, Uruguay, and Argentina. No records from Chile. In Mexico and Central America a rare migrant to be looked for in prairies, fields, rice-growing areas. A very rare migrant in Trinidad, doubtful in Tobago. In Colombia an uncommon fall migrant. Rare but widespread on migration through the West Indies, mostly in early fall.

STATUS. Uncommon; once abundant, but numbers were reduced by several decades of hunting pressure.

HABITAT. Prefers raised and grassy terrain, sometimes by streams, but avoids marshy areas. Nests in a shallow cavity in dry, mossy, or grassy tundra, sometimes near water or on high and dry banks. Only occasionally occurs on beaches and along shores in migration, favoring shortgrass prairies, burned-over grasslands, cotton fields, recently plowed fields, sun-baked stubble, and barren, recently inundated lands.

SPECIAL HABITAT REQUIREMENTS. Dry, grassy tundra.

Further Reading. United States and Canada: Bent 1927, Cramp and Simmons 1983, Godfrey 1966, Palmer 1967. Mexico and Central America: Edwards 1972, Monroe 1968, Peterson and Chalif 1973, Rappole et al. 1983, Ridgely and Gwynne 1989, Stiles and Skutch 1989. South America: Hilty and Brown 1986, Meyer de Schauensee 1966, Meyer de Schauensee and Phelps 1978, Rappole et al. 1983. Caribbean: Bond 1947, Herklots 1961, Rappole et al. 1983.

Short-billed Dowitcher
Limnodromus griseus

Canada (Quebec): Bécausseau Roux

Costa Rica and Guatemala: Agujeta Común

Mexico: Costurero Marino

Puerto Rico: Chorlo Pico Corto

Venezuela: Becasina Migratoria

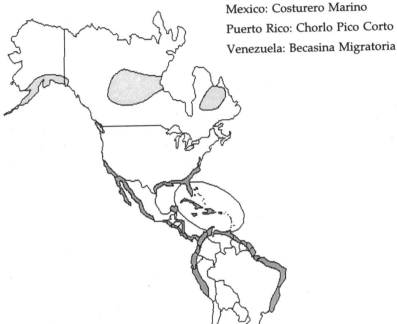

RANGE. Breeds from southern Yukon to north-central Northwest Territories, south to east-central British Columbia, and across to central Saskatchewan; from the interior of the Ungava Peninsula south to northern Ontario; and in coastal regions of southern Alaska. Winters in central California, southern Arizona, the Gulf Coast, and coastal South Carolina. Migrates through Central America, the Caribbean, and northern South America. Winters throughout the Caribbean and in South America as far south as southern Peru and Brazil, also Aruba and Trinidad and Tobago. Nonbreeding birds often occur south to wintering grounds in summer. In Mexico and Guatemala a rather rare winter visitor and migrant on both coasts. In Costa Rica and Honduras a locally abundant migrant in fall and spring and a common to abundant winter resident. A locally common migrant and fairly common winter resident in Puerto Rico, but rather uncommon in much of the West Indies. In Colombia a fairly common migrant and winter resident, with nonbreeders present year-round on both coasts.

STATUS. Common.

HABITAT. Primarily inhabits coniferous forest and muskeg with thin floating mats of moss and long grass, but also inhabits swampy coastal tundra with low scrub of willow, alder, birch, and a few taller larch and spruce. Nests in a hollow in mosses, in a clump of grasses, or on dry ground in wet areas. Sometimes nests in a small clearing in coniferous forest near muskeg, but not under trees or in broken terrain. After breeding, moves to open prairie lakes and sloughs. During migration and winter, occurs on mud and sand flats in sheltered bays and estuaries, on borders of shallow pools in salt marshes, on sandy beaches, and in flooded fields.

SPECIAL HABITAT REQUIREMENTS. In breeding season, swampy coastal tundra or muskeg. In winter, tidal mudflats.

Further Reading. United States and Canada: Cramp and Simmons 1983, Godfrey 1966, Palmer 1967, Sperry 1940. Mexico and Central America: Edwards 1972, Land 1970, Monroe 1968, Peterson and Chalif 1973, Rappole et al. 1983, Ridgely and Gwynne 1989, Stiles and Skutch 1989. South America: Hilty and Brown 1986, Meyer de Schauensee 1966, Meyer de Schauensee and Phelps 1978, Rappole et al. 1983. Caribbean: Bond 1947, Herklots 1961, Raffaele 1989, Rappole et al. 1983.

Long-billed Dowitcher
Limnodromus scolopaceus

Canada (Quebec): Bécausseau à Long-bec

Costa Rica: Agujeta Piquilarga, Agujeta Silbona

Guatemala: Agujeta Piquilargo

Mexico: Costurero de Agua Dulce

ꟼuerto Rico: Chorlo Pico Largo

RANGE. Breeds in North America in coastal western and northern Alaska, northern Yukon, and northwestern Northwest Territories. Winters from central California, southern Arizona, southern New Mexico, central Texas, the Gulf Coast, and southern Florida south to Panama. In Mexico a moderately common winter visitor on the Pacific Coast, locally inland. In Central America a rare migrant and winter visitor (Guatemala), possible winter resident (Costa Rica), and probably rare winter resident (Panama). The distribution in the region of Puerto Rico and the Virgin Islands is uncertain. Probably occurs in Trinidad, Tobago, and Colombia, but difficulties in identification and separation of dowitchers render range uncertain.

STATUS. Common.

HABITAT. Found on arctic continental coastal belts and marginally within the subarctic just beyond treeline. Inhabits grassy and sedge tundra, with or without scattered low woody vegetation and usually near shallow freshwater. Nests in a shallow depression in a tuft of grass or in moss, on dry or moist ground, usually near freshwater. In migration and winter, prefers grassy margins of shallow, muddy freshwater pools and occasionally saltwater habitats. Associates freely with other shorebirds, including the larger plovers.

SPECIAL HABITAT REQUIREMENTS. Grassy tundra and wet meadows.

Further Reading. United States and Canada: Bent 1927, Cramp and Simmons 1983, Godfrey 1966, Palmer 1967, Sperry 1940. Mexico and Central America: Edwards 1972, Land 1970, Monroe 1968, Peterson and Chalif 1973, Rappole et al. 1983, Ridgely and Gwynne 1989, Stiles and Skutch 1989.

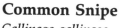

Common Snipe
Gallinago gallinago

Canada (Quebec): Bécaussine Ordinaire

Costa Rica: Becasina Común, Becada

Guatemala: Becardón, Becasina

Mexico: Agachona Común

Puerto Rico: Becasina

Venezuela: Becasina Chillona

RANGE. Breeds from northern Alaska and Yukon Territory to southern Northwest Territories, northern Quebec, and central Labrador, south to central California, Arizona, and Colorado across to West Virginia, New England, and the Maritime Provinces. Winters from southeastern Alaska, southern British Columbia, the central United States, and Virginia south through Middle America and the West Indies to central South America. Populations in northern and central South America are resident, whereas those in southern South America are austral migrants.

STATUS. Common, but declining across North America.

HABITAT. Inhabits wetlands, especially fens, bogs, swamps, and marshes, primarily in peatlands scattered within spruce, fir, and larch boreal forest. Occupies areas with fairly dense, low woody growth such as willows and alders, and with a ground cover of sphagnum, sedges, and grasses, preferably near open pastures or other clearings. Nests in a scrape on fairly dry ground or in a tussock of grass or sedge, usually in wet habitats but occasionally at the edge of wetlands. Conceals nest, sometimes covering it with an arch of dry vegetation. Also inhabits areas of decomposed wet plant litter along ponds, meandering rivers and brooks, and other marshy sites. In winter occupies wet, marshy habitats, wet meadows, flooded fields, and stream edges.

SPECIAL HABITAT REQUIREMENTS. In breeding season, bogs, fens, and swamps with moist organic soils near open areas free of obstacles or high vegetation. In winter, grassy freshwater edges.

Further Reading. United States and Canada: DeGraaf and Rudis 1986, Fogarty and Arnold 1977, Godfrey 1966, Palmer 1967, Sperry 1940, Tuck 1972. Mexico and Central America: Edwards 1972, Land 1970, Monroe 1968, Peterson and Chalif 1973, Rappole et al. 1983, Ridgely and Gwynne 1989, Stiles and Skutch 1989. South America: Hilty and Brown 1986, Meyer de Schauensee 1966, Meyer de Schauensee and Phelps 1978, Rappole et al. 1983. Caribbean: Bond 1947, Herklots 1961, Raffaele 1989, Rappole et al. 1983.

Wilson's Phalarope
Phalaropus tricolor

Canada (Quebec): Phalarope de Wilson

Chile: Pollito de Mar Boreal

Costa Rica: Falaropo de Wilson, Falaropo Tricolor

Guatemala and Puerto Rico: Falaropo de Wilson

Mexico: Falaropo Piquilargo

RANGE. Breeds from southern Yukon and northern Alberta to southern Michigan and southwestern Quebec south to south-central California, east-central Arizona, west-central New Mexico, northern Texas, eastern South Dakota, northern Illinois, northern Indiana, and northern Ohio; a rare breeder in Massachusetts since the 1970s. Winters primarily in the southern half of South America, casually as far north as southern California and southern Texas. In Mexico a rather rare migrant mainly through western Mexico on islands and the coast. In Central America a rare to locally common migrant, especially in fall. A very rare, but increasingly frequent, fall-winter visitor to Puerto Rico and the Virgin Islands. In Colombia a rare fall migrant on the Pacific Coast and inland west of the Andes. Winters in highlands of

221

Peru; also inland and coastal waters of Chile, Bolivia, Paraguay, Uruguay, and Argentina south to Chubut. Accidental in Brazil. Unlike the other phalaropes, this species migrates over land and spends the winter months on or near the shore and around inland waters, rather than on the open ocean.

STATUS. Uncommon.

HABITAT. Once inhabited natural prairies, but now found mainly on highly disturbed mixed-grass prairies dotted with small glacial potholes. Also found in taiga broken by moist, grassy muskeg and many small lakes and pools, and in farming country of aspen-grove parklands. Inhabits rolling uplands as high as 2,000 m in elevation. Nests semicolonially in a scrape on the ground around damp meadows with marsh grasses and sedges or rushes. Also nests by shallow sloughs fringed with short grasses or sedges, by lake shores, and in hay meadows or pastures, often up to 100 m from water. Outside the breeding season, mainly found on inland wetlands but sometimes on saline or alkaline depressions.

SPECIAL HABITAT REQUIREMENTS. Shallow water bordered by low grasses or sedges.

Further Reading. United States and Canada: Bent 1927, Cramp and Simmons 1983, Godfrey 1966, Hohn 1967, Low and Mansell 1983, Palmer 1967. Mexico and Central America: Edwards 1972, Land 1970, Monroe 1968, Peterson and Chalif 1973, Rappole et al. 1983, Ridgely and Gwynne 1989, Stiles and Skutch 1989. South America: Hilty and Brown 1986, Johnson 1965, Meyer de Schauensee 1966, Rappole et al. 1983.

Red-necked Phalarope
Phalaropus lobatus

Canada (Quebec): Phalarope
  Hyperboré

Chile: Pollito de Mar Boreal

Costa Rica and Guatemala: Falaropo
  Picofino

Mexico: Falaropo Piquifino

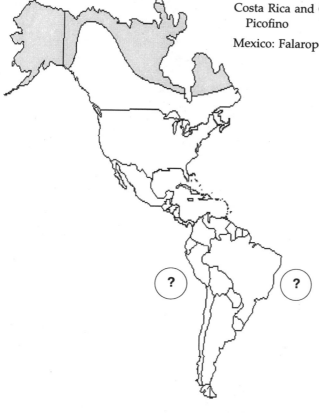

RANGE. Breeds in North America from northern Alaska and southern Victoria Island to east-central Northwest Territories and southern Baffin Island, south to northwestern British Columbia and northern Alberta across to northern Quebec, and locally along coast of Labrador. Winters mainly at sea in the Southern Hemisphere, largely in tropical and subtropical oceans. Rarely seen near land. A common migrant off the Pacific coasts of Mexico and Central and South America; rare and irregular inland. Not reported from Caribbean coast. A rare straggler to Chile; the species' main wintering grounds are farther north on the fringes of the Humbolt Current off the coasts of Peru and Ecuador. Occurs as a vagrant in the West Indies.

223

Red-necked Phalarope

STATUS. Uncommon.

HABITAT. Inhabits the wetter portions of flat alluvial plains, sedge-grass marshlands, clearings in alder and willow scrub, and heath-covered slopes above alder and willow scrub. Nests in a small hollow in moss or among sedges, usually atop a small hummock surrounded by water or near a marshy pond or small stream. In winter, occurs near upwellings or where other local conditions produce a high biomass of accessible food organisms.

SPECIAL HABITAT REQUIREMENTS. Wet grassy or sedgy terrain interspersed with pools, boreal clearings, or tundra.

Further Reading. United States and Canada: Bent 1927, Clapp et al. 1983, Cramp and Simmons 1983, Godfrey 1966, Hohn 1967, Palmer 1967. Mexico and Central America: Edwards 1972, Land 1970, Monroe 1968, Peterson and Chalif 1973, Rappole et al. 1983, Ridgely and Gwynne 1989, Stiles and Skutch 1989. South America: Hilty and Brown 1986, Johnson 1965, Meyer de Schauensee 1966, Rappole et al. 1983.

Red Phalarope
Phalaropus fulicaria

Canada (Quebec): Phalarope Roux
Chile: Pollito de Mar Rojizo
Costa Rica: Falaropo Rojo
Mexico: Falaropo Piquigrueso

RANGE. Breeds in North America on parts of mainland and islands within the Arctic Circle, south to the upper west coast of Hudson Bay and west to western and northern Alaska. A common migrant in the Pacific off Mexico and Central America; rare near the coast. Uncommon but regular off the Pacific Coast of Colombia, mostly more than 8 km from shore. Winters chiefly at sea from Colombia to Peru, but regularly to Chile. Although all species of phalaropes reach Chile during the northern winter, this species is by far the most abundant and can be found from 80 to 160 km from the coast. Occurs as a vagrant in the West Indies.

STATUS. Common.

HABITAT. Breeds on dry or moist tundra not far from sea, also on coastal islands. Nest is in a depression in tundra, well concealed and domed with grasses, with a runway through the grass leading to the nest. Favors pools with muddy shorelines. Winters in open ocean off Pacific Coast of South America where upwellings make the ocean rich in plankton. Migrates over water.

SPECIAL HABITAT REQUIREMENTS. Tundra for breeding.

Further Reading. United States and Canada: Bull and Farrand 1977, Godfrey 1966, Richards 1988, Terres 1980. Mexico and Central America: Edwards 1972, Peterson and Chalif 1973, Rappole et al. 1983, Stiles and Skutch 1989. South America: Hilty and Brown 1986, Johnson 1965, Rappole et al. 1983.

Laughing Gull
Larus atricilla

Canada (Quebec): Mouette Rieuse d'Amérique

Costa Rica: Gaviota Reidora

Guatemala and Mexico: Gaviota Gritona

Mexico: Gaviota Atricila

Puerto Rico: Gaviota Gallega

West Indies: Gallego Común, Galleguito, Gaviota Boba, Gaviota Cabecinegra, Gaviota Gallega, Goeland, Mangui, Mauve à Tête Noire, Pigeon de la Mer

RANGE. Breeds on the Pacific Coast of western Mexico in Sonora and Sinaloa (formerly to southern Salton Sea, southern California), on the Atlantic Coast from northern Nova Scotia south to Florida, along the Gulf Coast to Texas, and through the West Indies to islands off the Coast of Venezuela and the state of Yucatán. Nonbreeding birds occur in summer on the Salton Sea, Great Lakes, Gulf-Caribbean Coast of Middle America, and western coast of Mexico. Winters from North Carolina south to Brazil and Peru. Accidental in Greenland, New Mexico, Colorado, Nebraska, South Dakota, Wisconsin, and Tennessee.

STATUS. Common; showing long-term increase.

HABITAT. Breeds on coastal islands or on tufts of grass or reeds in saltwater marshes and along beaches. The nest is usually large and constructed of weeds, sedges, and grasses, built in a colony where eggs are usually above tidal waters. Winters in coastal areas, occasionally inland waters.

SPECIAL HABITAT REQUIREMENTS. Nesting areas away from excessive disturbance and mammalian predators.

Further Reading. United States and Canada: Godfrey 1966, Terres 1980. Mexico and Central America: Edwards 1972, Land 1970, Monroe 1968, Peterson and Chalif 1973, Rappole et al. 1983, Ridgely and Gwynne 1989, Stiles and Skutch 1989. South America: Hilty and Brown 1986, Meyer de Schauensee 1966, Rappole et al. 1983. Caribbean: Bond 1947, Herklots 1961, Raffaele 1989, Rappole et al. 1983.

Franklin's Gull
Larus pipixcan

Canada (Quebec): Mouette de
Franklin

Costa Rica and Guatemala: Gaviota
de Franklin

Mexico: Gaviota Apipizca

RANGE. Breeds from eastern Alberta and central Saskatchewan to western Minnesota, south locally to east-central Oregon, northwestern Wyoming, and northwestern Iowa. Migrates through Central America and winters from Honduras south to Colombia, Ecuador, Peru, and southern Chile. Winters rarely in southern coastal California, casually along the Gulf Coast of Texas and Louisiana.

STATUS. Common; showing long-term decline.

HABITAT. Breeds exclusively in shallow freshwater marshes and sloughs in the temperate prairie belt. Favors shallow wetlands up to 15 cm deep with bulrushes, cattails, whitetop, and common reeds, preferably near cultivated lands. Nests in colonies numbering from a few hundred up to 50,000 pairs. Nest, usually near open water, is built on masses of marsh vegetation, often floating on water and anchored to surrounding vegetation. In migration and in winter, inhabits sandy beaches, sandbars, fields, and pastures.

SPECIAL HABITAT REQUIREMENTS. Marshes and sloughs with sparse emergent vegetation (fewer than 10 plants <1 m tall/sq m).

Further Reading. United States and Canada: Clapp et al. 1983, Cramp and Simmons 1983, Godfrey 1966, Guay 1968, Low and Mansell 1983, McAtee and Beal 1912. Mexico and Central America: Edwards 1972, Land 1970, Monroe 1968, Peterson and Chalif 1973, Rappole et al. 1983, Ridgely and Gwynne 1989, Stiles and Skutch 1989. South America: Hilty and Brown 1986, Meyer de Schauensee 1966, Rappole et al. 1983.

Bonaparte's Gull
Larus philadelphia

Canada (Quebec): Mouette de Bonaparte

Costa Rica: Gaviota de Bonaparte

Mexico: Gaviota Menor

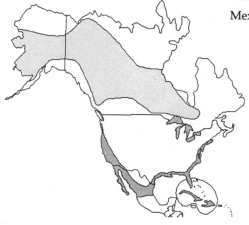

RANGE. Breeds from western and central Alaska and central Yukon to northern Manitoba, south to southern British Columbia, central Saskatchewan, southern Manitoba, and southern James Bay. Occurs in summer (nonbreeding birds) south to coastal areas in California and New England, and in the interior to the Great Lakes. Winters from Washington south along the

Pacific Coast into Mexico and from the Great Lakes south through the Ohio and Mississippi Valleys to the Gulf Coast. Also winters in Cuba, Hispaniola, and the Lesser Antilles.

STATUS. Common.

HABITAT. Inhabits coastal and interior lowlands, primarily black-fly-infested muskeg swamps in taiga up to treeline. Nests in dispersed colonies or singly, from 1 to 6 m above ground in branches or stumps of spruce, fir, and tamarack, and near water. May also nest in reeds, on mudflats of temporary potholes, and in clumps of bulrushes. Outside the breeding season it occurs on freshwater lakes, rivers, and sloughs, wet meadows, flooded fields, estuaries, shallow coastal waters, bays, and inlets.

SPECIAL HABITAT REQUIREMENTS. Ponds or lakes in swampy muskeg flanked by short to medium conifers.

Further Reading. United States and Canada: Bent 1921, Clapp et al. 1983, Cramp and Simmons 1983, Godfrey 1966. Mexico and Central America: Edwards 1972, Peterson and Chalif 1973, Rappole et al. 1983, Ridgely and Gwynne 1989, Stiles and Skutch 1989. Caribbean: Bond 1947, Rappole et al. 1983.

Ring-billed Gull
Larus delawarensis

Canada (Quebec): Goéland à Bec Cerclé

Costa Rica and Puerto Rico: Gaviota Piquianillada

Guatemala: Gaviota Piquipinta

Mexico: Gaviota de Delaware

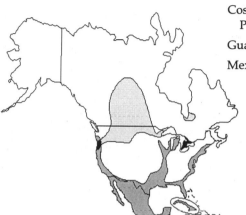

RANGE. Breeds in western North America from southern interior British Columbia and northeastern Alberta to north-central Manitoba south to northeastern California, south-central Colorado, and northeastern South Dakota; in eastern North America from north-central Ontario to southern Labrador south to eastern Wisconsin and northern Illinois across to central New Hampshire and New Brunswick. Winters along the Pacific Coast from southern British Columbia south to Mexico, Guatemala, El Salvador, and Panama; in the interior from the Great Lakes to Mexico and the Gulf Coast; and along the Atlantic Coast from the Gulf of St. Lawrence to Cuba, Jamaica, the Lesser Antilles, Trinidad, and Tobago.

STATUS. Common; increasing across the breeding range.

HABITAT. From boreal regions to temperate prairies, inhabits small to moderately sized rocky islands and occasionally peninsulas in large freshwater lakes, rivers, or ponds (a few colonies are on oceanic islands or coasts). Usually avoids densely settled areas. Frequently nests in mixed colonies with other larids, including Herring and California Gulls. Usually nests on the ground in flat, elevated, sparsely vegetated areas, but sometimes in low trees. Outside the breeding season, frequents harbors, refuse dumps, sewage out-

231

lets, reservoirs, lakes, ponds, streams, coastal bays, estuaries, beaches, and mudflats. Roosts on exposed sandbars and islands.

SPECIAL HABITAT REQUIREMENTS. Islands and peninsulas covered with low vegetation.

Further Reading. United States and Canada: Clapp et al. 1983, Cramp and Simmons 1983, Godfrey 1966, Jarvis and Southern 1976, Vermeer 1970. Mexico and Central America: Edwards 1972, Land 1970, Monroe 1968, Peterson and Chalif 1973, Rappole et al. 1983, Ridgely and Gwynne 1989, Stiles and Skutch 1989. South America: Hilty and Brown 1986, Meyer de Schauensee 1966, Rappole et al. 1983. Caribbean: Bond 1947, Raffaele 1989, Rappole et al. 1983.

California Gull
Larus californicus

Canada (Quebec): Goéland de California

Mexico: Gaviota Californiana

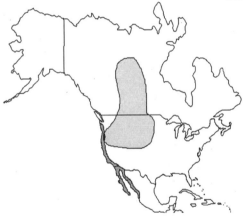

RANGE. Breeds from southwestern Northwest Territories south to northern Utah and north-central Colorado, and west to southern interior British Columbia and northeastern California. Winters from southern Washington and eastern Idaho south, along the Pacific Coast to Mexico; casual in inland Mexico and on the east coast to Veracruz.

STATUS. Common.

HABITAT. Inhabits barren islands on fresh, brackish, or alkaline lakes, shores of lakes or ponds, and marshes. Favors sites with low, sparse vegetation. A colonial nester, often in mixed colonies with the Ring-billed Gull. Avoids dense herbaceous cover, and constructs its nest in a scrape on the ground in elevated, boulder-strewn areas. Outside the breeding season occurs on seacoasts, bays, lagoons, estuaries, mudflats, and irrigated fields and other agricultural lands.

SPECIAL HABITAT REQUIREMENTS. Open sandy or gravelly lakeshores or islands.

Further Reading. United States and Canada: Godfrey 1966, Greenhalgh 1952, Johnsgard 1979, Vermeer 1970. Mexico: Edwards 1972, Peterson and Chalif 1973, Rappole et al. 1983.

Herring Gull
Larus argentatus

Canada (Quebec): Goéland Argenté

Costa Rica, Guatemala, and Puerto Rico: Gaviota Argenta

Mexico: Gaviota Plateada

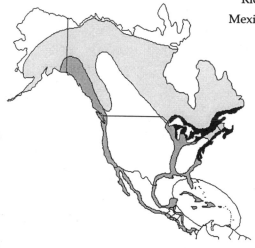

RANGE. Breeds in North America from northern Alaska and northern Yukon to east-central Northwest Territories, western Baffin Island, and northern Labrador south to south-central British Columbia, central Alberta, northern Minnesota, northeastern Illinois, northern Ohio, northern New York, and along the Atlantic Coast to northeastern South Carolina. Winters from south-

233

ern Alaska, the Great Lakes region, and Newfoundland south, mostly at sea and along coasts, large rivers, and lakes south to Panama and throughout the Caribbean.

STATUS. Abundant, but long-term decline in eastern North America.

HABITAT. Uses a wide variety of habitats, including sandy, rocky, or wooded islands, stabilized sand dunes, margins of tundra lakes, cordgrass marshes, cliffs, meadows, and buildings. Usually nests in exposed sites on the ground in small to large colonies, but occasionally in trees. Prefers to nest in low sites; depending on habitat may nest at the base of boulders, stumps, or bushes on grassy slopes, near large, tall clumps of vegetation, on drift adjacent to salt marshes, or on rocky or grassy substrates. In winter, occurs primarily along the shore of the ocean or other bodies of water, concentrating on beaches and in areas where food is likely to be abundant.

SPECIAL HABITAT REQUIREMENTS. Nesting sites must be free of terrestrial predators and within 40 km of a dependable source of food.

Further Reading. United States and Canada: Burger and Shisler 1978, Clapp et al. 1983, Cramp and Simmons 1983, Forbush and May 1955, Godfrey 1966. Mexico and Central America: Edwards 1972, Land 1970, Monroe 1968, Peterson and Chalif 1973, Rappole et al. 1983, Ridgely and Gwynne 1989, Stiles and Skutch 1989. Caribbean: Bond 1947, Raffaele 1989, Rappole et al. 1983.

Western Gull
Larus occidentalis

Canada (Quebec): Goéland de l'Ouest
Mexico: Gaviota Occidental

RANGE. Breeds on the Great Salt Lake in Utah and coastally from northern Washington south along the Pacific Coast to Magdalena Bay on the west coast of Baja California. Winters along the Pacific Coast of the United States south to southern Baja California and Nayarit, Mexico.

STATUS. Common.

HABITAT. Breeds on rocky edges of offshore islands or grassy slopes near beaches. Rarely ventures far inland. In winter found along sandy shores or nearby lagoons along coasts of the mainland and off-shore islands.

SPECIAL HABITAT REQUIREMENTS. Offshore islands with adequate breeding habitat.

Further Reading. United States and Canada: Bent 1921, Farrand 1983b, Godfrey 1966, Terres 1980. Mexico: Edwards 1972, Peterson and Chalif 1973, Rappole et al. 1983.

Glaucous-winged Gull
Larus glaucescens

Canada (Quebec): Goéland à Ailes Glauques

Mexico: Gaviota Aliglauca

RANGE. Breeds in North America from southern Alaska south to Oregon. Winters from southeastern Alaska to southern Baja California.

STATUS. Abundant.

HABITAT. Breeds on rocky cliffs among seabird colonies. The nest is a mound of dried straw, seaweeds, kelp, sometimes including feathers and fish bones. Winters along ocean bays and nearby lagoons and estuaries, or rocky or sandy seashores. May follow broad river valleys inland, but otherwise stays near the coast.

SPECIAL HABITAT REQUIREMENTS. Rocky cliffs for breeding.

Further Reading. United States and Canada: Bent 1921, Farrand 1983b, Godfrey 1966, Terres 1980. Mexico: Edwards 1972, Rappole et al. 1983.

Sabine's Gull
Xema sabini

Canada (Quebec): Mouette de Sabine
Costa Rica: Gaviota de Sabine
Mexico: Gaviota Colihendida

RANGE. Breeds in North America along the arctic coasts of North America from western Alaska east across the Canadian arctic. Apparently winters in the south Atlantic Ocean and off the Pacific Coasts of Panama, Colombia, Ecuador, Peru, and northern Chile, but winter distribution still not well understood.

STATUS. Common on breeding grounds.

HABITAT. Breeds on the arctic tundra. Nests singly or colonially in a hollow in tundra grasses; nest usually lined with grass. Winters on offshore and pelagic waters; rarely in coastal waters, where seen far from shore.

SPECIAL HABITAT REQUIREMENTS. Arctic tundra for breeding.

Further Reading. United States and Canada: Bent 1921, Farrand 1983b, Godfrey 1966, Terres 1980. Mexico and Central America: Edwards 1972, Rappole et al. 1983, Ridgely and Gwynne 1989, Stiles and Skutch 1989. South America: Hilty and Brown 1986, Meyer de Schauensee 1966, Rappole et al. 1983.

Gull-billed Tern
Sterna nilotica

Canada (Quebec): Sterne à Gros Bec

Costa Rica: Charrán Piquinegro

Guatemala: Golondrina de Mar, Pagaza Piconegra

Mexico: Golondrina Marina Piquigruesa

Puerto Rico: Gaviota Piquigorda

Venezuela: Gaviota Pico Gordo

West Indies: Gaviota de Pico Corto, Oiseau Fou, Pigeon de la Mer

RANGE. Breeds locally in western North America from the Salton Sea in California and Sonora, Mexico, east to southeast Texas, Louisiana, southern Mississippi, western Florida, and casually in interior Florida, north along the Atlantic Coast to southern Maryland, rarely to Delaware, southern New Jersey, and Long Island, New York. Also nests throughout the Caribbean. Winters along Pacific Coast from Oaxaca, Mexico, to Ecuador and Peru and from southern Texas, southern Louisiana, and central Florida to the Caribbean coast of Central America and north and northeast coasts of South America to Argentina; also throughout the Caribbean.

STATUS. Once very common in the United States, now uncommon. The species was almost wiped out by egg collectors and the feather trade and has never recovered.

HABITAT. Breeds on sandy shell-strewn upper beaches above the reach of high tides, sparsely vegetated estuarine islands, dredge spoil, and shell mounds on the coast, and inland along muddy edges of lakes and rivers. The nest may be a depression sparsely lined with bits of shells, straw, or grasses; sometimes a more substantial nest of weeds and grasses near water or on low marshy islands where eggs may be laid on damp ground or on matted grasses; sometimes on floating reeds in marsh of lake or on an old muskrat house. Migration and winter habitats are little known. In Puerto Rico this tern is typically found over freshwater lakes and brackish lagoons.

SPECIAL HABITAT REQUIREMENTS. Undisturbed nesting areas.

Further Reading. United States and Canada: Farrand 1983b, Godfrey 1966, Hunter 1990, Terres 1980. Mexico and Central America: Edwards 1972, Land 1970, Monroe 1968, Peterson and Chalif 1973, Rappole et al. 1983, Ridgely and Gwynne 1989, Stiles and Skutch 1989. South America: Hilty and Brown 1986, Meyer de Schauensee 1966, Meyer de Schauensee and Phelps 1978, Rappole et al. 1983. Caribbean: Bond 1947, Raffaele 1989, Rappole et al. 1983.

Caspian Tern
Sterna caspia

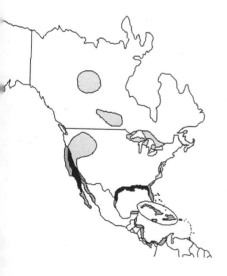

Canada (Quebec): Sterne Caspienne

Costa Rica: Pagaza Mayor, Pagaza Piquirrojo

Guatemala: Pagaza Piquirroja

Mexico: Golondrina Marina Grande Piquirroja

Puerto Rico: Gaviota de Caspia

Venezuela: Tirra Caspia

RANGE. Breeds locally in western North America from coastal and eastern Washington, eastern Oregon, northern Utah, and northwestern Wyoming south to southern California and western Nevada; in the interior from southern Northwest Territories to southern James Bay south to North Dakota, northeastern Illinois, and southern Ontario; at scattered localities along the Atlantic Coast from Newfoundland to South Carolina; and along the Gulf Coast from Texas east to Florida. Nonbreeding birds often summer in the James Bay and Great Lakes region and along both U.S. coasts. Winters primarily in coastal areas from California and North Carolina south to Mexico, Honduras, Panama, Colombia, and throughout the Greater Antilles.

STATUS. Common.

HABITAT. Usually found near the coast on sandy, stony, or shell beaches, barrier or spoil islands, islands with sand-gravel substrate with little or no vegetation, or on a shell berm in a salt marsh. Tends to occupy less developed and less polluted segments of the coast, but is also found inland along shorelines of large lakes. Usually nests in compact colonies, but occasionally singly in the vicinity of other tern species, in shallow depressions in the ground on bare sandy or rocky soil. In winter generally found along beaches and on isolated spits, often roosting with other larids. In migration, occurs along water courses or in large marshes.

SPECIAL HABITAT REQUIREMENTS. Sparsely vegetated islets or shorelines, protection of nesting colonies from excessive human disturbance.

Further Reading. United States and Canada: Clapp et al. 1983, Godfrey 1966, Hunter 1990, Johnsgard 1979, Ludwig 1965. Mexico and Central America: Edwards 1972, Land 1970, Monroe 1968, Peterson and Chalif 1973, Rappole et al. 1983, Ridgely and Gwynne 1989, Stiles and Skutch 1989. South America: Hilty and Brown 1986, Meyer de Schauensee 1966, Rappole et al. 1983. Caribbean: Bond 1947, Raffaele 1989, Rappole et al. 1983.

Royal Tern

Sterna maxima

Canada (Quebec): Sterne Royale

Costa Rica: Pagaza Real

Guatemala: Charrán Real, Golondrina de Mar

Mexico: Golondrina Marina Grande Piquinaranja

Puerto Rico: Gaviota Real

Venezuela: Tirra Canalera

West Indies: Foquette, Gaviota Real, Mauve, Oiseau Fou, Pigeon de la Mer

RANGE. Breeds along the coast of Virginia south to Texas and Mexico, wandering farther north in the summer. Winters from southern California east to the northern Gulf Coast and throughout Florida and north to Virginia; winters throughout Central America and coastal South America from Colombia to Peru and Argentina.

STATUS. Common.

HABITAT. Low sandy islands. Nest is a scraped depression in the sand. Nests colonially. Winters along coasts.

SPECIAL HABITAT REQUIREMENTS. Nesting sites near a good source of food. Remote cays in the Caribbean (Raffaele 1989:96).

Further Reading. United States and Canada: Bent 1921, Farrand 1983b, Godfrey 1966, Hunter 1990, Terres 1980. Mexico and Central America: Edwards 1972, Land 1970, Monroe 1968, Peterson and Chalif 1973, Rappole et al. 1983, Ridgely and Gwynne 1989, Stiles and Skutch 1989. South America: Hilty and Brown 1986, Meyer de Schauensee 1966, Meyer de Schauensee and Phelps 1978, Rappole et al. 1983. Caribbean: Bond 1947, Herklots 1961, Rappole et al. 1983.

Elegant Tern
Sterna elegans

Costa Rica: Pagaza Elegante

Guatemala: Charrán Elegante, Golondrina de Mar

Mexico: Golondrina Marina Elegante

RANGE. Breeds along both coasts of Baja California and intermittently in central California and Mexico. Winters along the Pacific Coast from Guatemala south to central Chile, most commonly from Ecuador south (rare north of Panama).

STATUS. Uncommon.

HABITAT. Undisturbed upper beaches for breeding. Migratory and winter habits are little known.

SPECIAL HABITAT REQUIREMENTS. Undisturbed nesting areas.

242

Further Reading. United States and Canada: Bent 1921, Godfrey 1966, Hunter 1990, Terres 1980. Mexico and Central America: Edwards 1972, Land 1970, Monroe 1968, Peterson and Chalif 1973, Rappole et al. 1983, Ridgely and Gwynne 1989, Stiles and Skutch 1989.

Sandwich Tern
Sterna sandvicensis

Canada (Quebec): Sterne Caugek

Costa Rica: Pagaza Puntiamarilla

Guatemala: Charrán, Golondrina de Mar Patinegra, Patinegro

Mexico: Golondrina Marina de Sandwich

Puerto Rico: Gaviota Piquiaguda

Venezuela: Gaviota Patinegra

West Indies: Gaviota de Pico Agudo

RANGE. Breeds locally from Virginia south to southern Florida and the Bahamas, off southern Cuba, off the state of Yucatán, Mexico, west across the Gulf Coast to southern Texas, and south into Mexico. Winters along the Pacific Coast from Oaxaca, Mexico, to Peru, and on the Atlantic Coast from Florida and the Caribbean along the eastern coast of South America to southern Argentina.

STATUS. Uncommon. Appears to be generally increasing in the Greater Antilles (Raffaele 1989:97).

HABITAT. Sandy islands for nesting. Nest is a scrape on sandy beaches above the high-water mark.

SPECIAL HABITAT REQUIREMENTS. Nesting areas near a reliable source of small fish; protection of nesting colonies from excessive human disturbance.

Further Reading. United States and Canada: Bent 1921, Farrand 1983b, Godfrey 1966, Hunter 1990, Terres 1980. Mexico and Central America: Edwards 1972, Land 1970, Monroe 1968, Peterson and Chalif 1973, Rappole et al. 1983, Ridgely and Gwynne 1989, Stiles and Skutch 1989. South America: Hilty and Brown 1986, Meyer de Schauensee 1966, Meyer de Schauensee and Phelps 1978, Rappole et al. 1983. Caribbean: Bond 1947, Raffaele 1989, Rappole et al. 1983.

Roseate Tern
Sterna dougallii

Canada (Quebec): Sterne Rosée

Mexico: Golondrina Marina de Dougall

Puerto Rico: Palometa

Venezuela: Tirra Rosada

West Indies: Carite, Gaviota, Mauve Blanche, Oiseau Fou, Palometa, Petite Mauve, Pigeon de la Mer

RANGE. Breeds along the Atlantic Coast locally from Nova Scotia south to Virginia, southern Florida, the Yucatán Peninsula, and the West Indies. Winters throughout the West Indies to Brazil and casually north to North Carolina.

STATUS. Locally common. The Northwest Atlantic population is listed as endangered (in the United States) because of human disturbance and loss of colony sites.

HABITAT. Nests in sandy spots on rocky or pebbly beaches on islands or along the shore. The nest is usually a scrape on the ground concealed in beach grass or in weedy places, sometimes lined with a few pieces of grass. Migrates and winters in coastal areas; almost never seen inland.

SPECIAL HABITAT REQUIREMENTS. Undisturbed nesting sites. Inshore areas, bays (Puerto Rico).

Further Reading. United States and Canada: Bent 1921, Godfrey 1966, Hunter 1990, Terres 1980. Mexico: Peterson and Chalif 1973, Rappole et al. 1983. South America: Meyer de Schauensee 1966, Rappole et al. 1983. Caribbean: Bond 1947, Raffaele 1989, Rappole et al. 1983.

Common Tern
Sterna hirundo

Canada (Quebec): Sterne Commune

Costa Rica: Charrán Común

Guatemala: Charrán Común, Golondrina de Mar

Mexico: Golondrina Marina Común

Puerto Rico: Gaviota Común

Venezuela: Tirra Medio Cuchillo

West Indies: Gaviota, Palometa, Petite Mauve

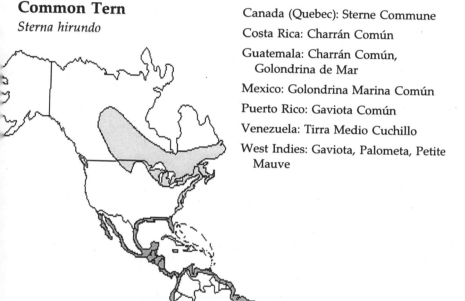

RANGE. Breeds from south-central Northwest Territories to southern Quebec and southern Labrador, Newfoundland, Nova Scotia, and Maine south to eastern Washington and southeastern Alberta, across to central Minnesota, northern Ohio, and northwestern Vermont, and locally along the Atlantic Coast to North Carolina; locally on the Gulf Coast in Texas, Mississippi, and western Florida. Winters in South America from Colombia and Venezuela south to Brazil, Peru, and Argentina. Migrates through the Caribbean, rarely along the coasts of southern California, South Carolina, Florida, and the Gulf Coast.

STATUS. Common; declining in eastern North America.

HABITAT. Inhabits a variety of habitats, mainly near water, often on islets, and usually in areas with little or no vegetation: sparsely vegetated sandy islands, barrier beaches, marshy islands, small islands in salt marshes, or low, small, rocky islands in lakes and rivers. Nests in colonies. Nest may vary from a slight hollow in sand or among pebbles to a well-built hollowed mound of grasses and seaweeds; may be in the open or near weeds, grasses, or bushes. Generally prefers sparse cover around the nest. After nesting, typically found along shorelines, on exposed rocks and old pilings, and inshore over shallow coastal waters.

SPECIAL HABITAT REQUIREMENTS. Nesting areas with scant vegetation, isolated from disturbance and predation, and in close proximity to a source of food. Shallow coastal waters (Caribbean).

Further Reading. United States and Canada: Burger and Lesser 1978, Forbush and May 1955, Godfrey 1966, Johnsgard 1979, Palmer 1941, Tate and Tate 1982. Mexico and Central America: Edwards 1972, Land 1970, Monroe 1968, Peterson and Chalif 1973, Rappole et al. 1983, Ridgely and Gwynne 1989, Stiles and Skutch 1989. South America: Hilty and Brown 1986, Meyer de Schauensee 1966, Meyer de Schauensee and Phelps 1978, Rappole et al. 1983. Caribbean: Bond 1947, Herklots 1961, Raffaele 1989, Rappole et al. 1983.

Forster's Tern
Sterna forsteri

Canada (Quebec): Sterne de Forster

Costa Rica: Charrán de Forster

Guatemala: Charrán de Forster, Golondrina de Mar

Mexico: Golondrina Marina de Forster

Puerto Rico: Gaviota de Forster

RANGE. Breeds from southeastern British Columbia and central Alberta to central Manitoba, south to southern California and south-central Idaho, across to central Kansas, northern Iowa, and northwestern Indiana; along the Atlantic Coast from southern New York south locally to North Carolina; and along the Gulf Coast from Texas east to Louisiana. Winters along the Pacific and Atlantic Coasts from central California and Virginia south to Honduras, possibly as far as Costa Rica and Panama.

STATUS. Common.

HABITAT. Primarily inhabits large saltwater and freshwater marshes; also found on marshy bays, marshy parts of islands, marshy edges of streams and lakes, sloughs, dikes in evaporation ponds, estuarine islands, marshes adjacent to barrier beaches, and dredge-spoil islets. Usually nests in small colonies, on mats of floating dead vegetation, flattened reeds and cattails, large muskrat houses near the edges of open pools of water, floating rootstalks of cattails, or sometimes in a shallow depression in sand or mud. May also locate nests on sand or gravel bars, beaches, or grassy islands. Sometimes uses old or abandoned nests of Western and Pied-billed Grebes. In winter, occurs in harbors, marshy bays, estuaries, lagoons, and inlets along coastal areas, occasionally occurring inland on lakes and ponds.

SPECIAL HABITAT REQUIREMENTS. Extensive marshy areas with vegetated nest sites partly open to water.

Further Reading. United States and Canada: Bergman et al. 1970, Clapp et al. 1983, Forbush and May 1955, Godfrey 1966, Johnsgard 1979, Low and Mansell 1983, McNicholl 1971. Mexico and Central America: Edwards 1972, Land 1970, Monroe 1968, Peterson and Chalif 1973, Rappole et al. 1983, Ridgely and Gwynne 1989, Stiles and Skutch 1989.

Least Tern

Sterna antillarum

Canada (Quebec): Petit Sterne

Costa Rica: Charrán de Chico, Charrán Menudo

Guatemala: Charrancito, Golondrina de Mar Chica

Mexico: Golondrina Marina Menor

Puerto Rico: Gaviota Pequeña

Venezuela: Gaviota Filico

West Indies: Golondrina de Mar, Gaviota Chica, Gaviotica, Oiseau Fou, Petite Mauve, Pigeon de la Mar

RANGE. Breeds along the Pacific Coast from central California south to southern Baja California and Guerrero, along the Atlantic Coast from Maine to Florida, and west along the Gulf Coast to Texas south along coastal Middle America to Honduras; also in the West Indies. Formerly bred inland in the United States along the Colorado, Red, Missouri, and Mississippi River systems. Winters mainly along the northern coast of South America from eastern Colombia to eastern Brazil.

STATUS. Endangered in Great Plains (Mississippi drainage) states and along California coast; recently declining along Atlantic coast.

HABITAT. Inhabits river sandbars, inland islands, broad areas of sand or gravel beaches, and newly cleared land along the coast. Nests singly or in scattered colonies. Nests in scrapes (with little or no lining) in sand or gravel (gravel or pebble substrates are preferred) above ordinary tides. In the Great Plains, uses same habitat as Piping Plover. In Oklahoma, frequents salt plains.

SPECIAL HABITAT REQUIREMENTS. Open, sandy coastal beaches, and river sandbars for nesting. Protection of nesting colonies from human disturbance is critical.

Further Reading. United States and Canada: Bent 1921, Godfrey 1966, Hunter 1990, Johnsgard 1979, Terres 1980, Tompkins 1959. Mexico and Central America: Edwards 1972, Land 1970, Monroe 1968, Peterson and Chalif 1973, Rappole et al. 1983, Ridgely and Gwynne 1989, Stiles and Skutch 1989. South America: Hilty and Brown 1986, Meyer de Schauensee 1966, Rappole et al. 1983. Caribbean: Bond 1947, Herklots 1961, Raffaele 1989, Rappole et al. 1983.

Sooty Tern
Sterna fuscata

Canada (Quebec): Sterne Fuligineuse

Costa Rica: Charrán Sombrío

Mexico: Golondrina Marina Dorsinegra

Puerto Rico: Gaviota Oscura

Venezuela: Gaviota de Veras

West Indies: Gaviota Monja, Gaviota Oscura, Oiseau Fou, Touaou

RANGE. Breeds from the Dry Tortugas (Florida) and coastal islands of Louisiana and Texas south throughout the West Indies, the Caribbean coast of the Yucatán Peninsula, and tropical islands of the Atlantic. Resident in western Mexico (Nayarit to Revilla Gigedo Islands). Winters in the Gulf of Mexico, the Caribbean, and along the northern coast of South America.

STATUS. Common.

HABITAT. Breeds on sandy or pebbly beaches or, in the tropics, on ledges of seaside cliffs. Often the nest is a scrape, but sometimes no nest is made.

249

Spends much of the nonbreeding season over the ocean, ranging widely in the Pacific.

SPECIAL HABITAT REQUIREMENTS. Nesting areas free of disturbance and a dependable source of small fish. Offshore nesting areas near cays (Caribbean).

Further Reading. United States and Canada: Godfrey 1966, Hunter 1990, Terres 1980. Mexico and Central America: Edwards 1972, Land 1970, Monroe 1968, Peterson and Chalif 1973, Rappole et al. 1983, Ridgely and Gwynne 1989, Stiles and Skutch 1989. South America: Hilty and Brown 1986, Meyer de Schauensee 1966, Meyer de Schauensee and Phelps 1978, Rappole et al. 1983. Caribbean: Bond 1947, Herklots 1961, Raffaele 1989, Rappole et al. 1983.

Black Tern
Chlidonias niger

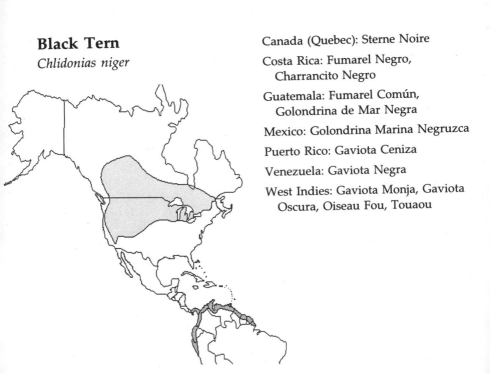

Canada (Quebec): Sterne Noire

Costa Rica: Fumarel Negro, Charrancito Negro

Guatemala: Fumarel Común, Golondrina de Mar Negra

Mexico: Golondrina Marina Negruzca

Puerto Rico: Gaviota Ceniza

Venezuela: Gaviota Negra

West Indies: Gaviota Monja, Gaviota Oscura, Oiseau Fou, Touaou

RANGE. Breeds from southwestern and east-central British Columbia and southwestern Northwest Territories to southern Quebec and New Brunswick, south locally to central California and Utah, across to Nebraska, Illinois, Pennsylvania, and Maine. Nonbreeding birds occur in summer on the Pacific Coast south to Panama, and in eastern North America to the Gulf Coast. Winters from Panama and northern coastal South America to Peru in the west and Surinam in the east. Occurs as a migrant in the Caribbean region.

STATUS. Common.

HABITAT. Found in taiga and on plains and prairies, where it inhabits shallow marshes, open areas of deeper marshes, reed-bordered sloughs, natural ponds, lakes, fish and stock ponds, shallow river impoundments, wet meadows, river oxbows, ditches, edges of streams, and swampy grasslands. Often nests in small colonies, but occasionally nests singly. Usually nests on a floating mass of vegetation, such as cattails and bulrushes, that is anchored to surrounding vegetation, on floating pieces of wood, or in a slight hollow atop a muskrat house. Prefers areas of emergent vegetation over water up to 1 m deep or near open water. Sometimes uses abandoned nests of other birds including grebes, Forster's Terns, and American Coots. In migration, frequents freshwater and saltwater, occurring along the coast and along marshes, rivers, lakes, and nearby cultivated fields.

SPECIAL HABITAT REQUIREMENTS. Aquatic habitats with extensive stands of emergent vegetation and large areas of shallow open water. In Caribbean, fresh and brackish ponds.

Further Reading. United States and Canada: Clapp et al. 1983, Forbush and May 1955, Godfrey 1966, Hunter 1990, Johnsgard 1979, Low and Mansell 1983, McNicoll 1971, Novak 1992, Tate and Tate 1982. Mexico and Central America: Edwards 1972, Land 1970, Monroe 1968, Peterson and Chalif 1973, Rappole et al. 1983, Ridgely and Gwynne 1989, Stiles and Skutch 1989. South America: Hilty and Brown 1986, Meyer de Schauensee 1966, Meyer de Schauensee and Phelps 1978, Rappole et al. 1983. Caribbean: Bond 1947, Herklots 1961, Raffaele 1989, Rappole et al. 1983.

Black Skimmer

Rynchops niger

Canada (Quebec): Bec-en-ciseaux

Costa Rica: Rayador Negro

Cuba, Puerto Rico, and Venezuela:
Pico de Tijera

Guatemala: Rayadora

Mexico: Rayador Americano

RANGE. Breeds locally from Massachusetts south to Florida and along the Gulf Coast to the Yucatán Peninsula. Also on the Pacific Coast from northwestern Mexico to northern Chile, on the Caribbean and Atlantic Coasts to Argentina, and along the larger rivers of northern and eastern South America to northern Argentina. Winters from North Carolina south to Argentina and from western Mexico to southern Chile and Argentina. Northern individuals may migrate as far south as Panama.

STATUS. Fairly common.

HABITAT. In breeding season, sandy beaches above the high-water mark. Nest is a scrape in the sand. In winter, shallow bays, estuaries, and creeks along the coast.

SPECIAL HABITAT REQUIREMENTS. Undisturbed nesting areas and a reliable source of food.

Further Reading. United States and Canada: Bent 1921, Farrand 1983b, Godfrey 1966, Hunter 1990, Terres 1980. Mexico and Central America: Edwards 1972, Land 1970, Monroe 1968, Peterson and Chalif 1973, Rappole et al. 1983, Ridgely and Gwynne 1989, Stiles and Skutch 1989. South America: Hilty and Brown 1986, Meyer de Schauensee 1966, Meyer de Schauensee and Phelps 1978, Rappole et al. 1983. Caribbean: Bond 1947, Raffaele 1989, Rappole et al. 1983.

White-crowned Pigeon
Columba leucocephala

Costa Rica: Paloma Coroniblanca

Mexico and Puerto Rico: Paloma Cabeciblanca

West Indies: Paloma Cabeciblanca, Paloma Casco Blanco, Paloma Coronita, Pigeon à Couronne Blanche, Ramier Tête-blanche, Torcaz Cabeciblanca

RANGE. Breeds in southern Florida on islets in the Florida Keys, throughout Florida Bay, the Lesser Antilles, and islets of Central America. Nonbreeding birds may occur in summer in southern peninsular Florida. Winters from southern peninsular Florida and the Florida Keys throughout most of the breeding range to Mexico, Belize, and Panama.

STATUS. Common.

HABITAT. Generally gregarious; breeds and roosts in large concentrations on brushy, small, low islands and keys, among coastal mangroves and pines. Nests colonially, but not with or near other colonial species. Builds nests 1 to 5 m or more above ground, on top of cacti or bushes, or high in mangroves; occasionally low and over water.

SPECIAL HABITAT REQUIREMENTS. Mangroves on undisturbed Caribbean islands (Raffaele 1989:100).

Further Reading. United States and Canada: Cottam and Knappen 1939, Farrand 1983b, Godfrey 1966, Harrison 1975, Terres 1980, Wiley and Wiley

1979. Mexico and Central America: Edwards 1972, Peterson and Chalif 1973, Rappole et al. 1983, Ridgely and Gwynne 1989, Stiles and Skutch 1989. Caribbean: Bond 1947, Raffaele 1989, Rappole et al. 1983.

Red-billed Pigeon
Columba flavirostris

Costa Rica and Guatemala: Paloma Piquirroja

Mexico: Paloma Morada Vientrioscura

RANGE. Breeds in the lower and middle Rio Grande Valley in southern Texas south through eastern and western Mexico to Costa Rica; mostly absent from northern parts of breeding range in winter, with northern birds withdrawing to Mexico.

STATUS. Local and uncommon.

HABITAT. Inhabits semiarid woodlands near water. In Texas, found in river thickets containing tall timber and a thick undergrowth of thorny shrubs including ebony blackbead, huisache, mesquite, bald cypress, great leucaena, Mexican ash, elm, black willow, and hackberry. Nests 2 to 10 m above ground on a horizontal tree branch, in a clump of small branches, or in a tree concealed by a tangle of vines or brush. Often perches on exposed bare branches high in trees.

SPECIAL HABITAT REQUIREMENTS. Tall, dense brush with small patches of tall trees.

Further Reading. United States and Canada: Bent 1932, Oberholser 1974a, Terres 1980. Mexico and Central America: Edwards 1972, Land 1970, Monroe 1968, Peterson and Chalif 1973, Rappole et al. 1983, Ridgely and Gwynne 1989, Stiles and Skutch 1989.

Band-tailed Pigeon

Columba fasciata

Canada (Quebec): Pigeon du Pacifique

Costa Rica and Mexico: Paloma
  Collareja

Guatemala: Paloma de Pico Amarillo,
  Paloma Piquigulada

Venezuela: Paloma Gargantilla

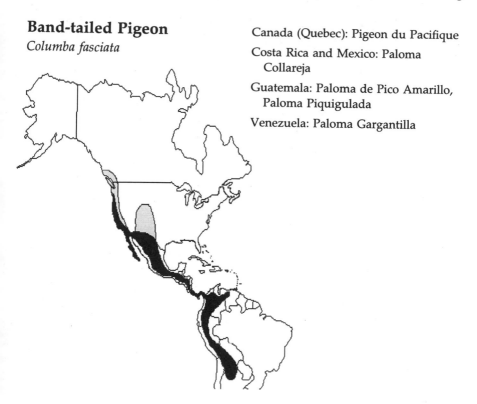

RANGE. Breeds from southwestern British Columbia south through the mountains of Washington, Oregon, California, and extreme western Nevada to Baja California; and from southern Nevada, Arizona, central Utah, north-central Colorado, New Mexico, and western Texas south through Middle America to northern Argentina. Winters from central California, central Arizona, and western Texas south through the breeding range, occurring widely in Mexico in foothills at low elevations.

STATUS. Locally common; declining, especially in the western United States.

HABITAT. Inhabits a variety of forests along the Pacific Coast: western hemlock, western redcedar, Douglas-fir, ponderosa pine, white fir, or incense cedar. In Oregon and Washington, prefers forests with a good interspersion of serial stages and openings; in California, prefers forests, woodlands, or chaparral with an abundance of oak. In the interior, occupies habitats ranging from montane oak woodlands to arid woodlands of pinyon pine and

255

oaks, and from agricultural areas near forests to berry-producing areas at 3,300 m elevation. Nests in coniferous or deciduous trees usually located near a clearing and on a moderate to steep slope or precipice. Conceals nest on horizontal branches typically 5 to 15 m high, rarely on the ground. Occasionally found in spruce-fir associations characterized by Engelmann spruce, subalpine fir, lodgepole pine, limber pine, and aspen, but prefers sites dominated by ponderosa pine and Gambel oak.

SPECIAL HABITAT REQUIREMENTS. Mature conifers or broad-leaved trees at least 2 m tall for nesting, and a source of mineral water or salt deposits in early fall and winter.

Further Reading. United States and Canada: Godfrey 1966, Jeffrey 1977, Johnsgard 1975a, Neff 1947, Peeters 1962. Mexico and Central America: Edwards 1972, Land 1970, Monroe 1968, Peterson and Chalif 1973, Rappole et al. 1983, Ridgely and Gwynne 1989, Stiles and Skutch 1989.

White-winged Dove
Zenaida asiatica

Canada (Quebec): Tourterelle à Ailes Blanches

Costa Rica, Guatemala, and Mexico: Paloma Aliblanca

Puerto Rico: Tortola Aliblanca

West Indies: Aliblanca, Tourterelle Aileblanche

RANGE. Breeds from southeastern California and southern Nevada to southwestern Texas south through Central America to western South America. Introduced and established in southern Florida. Winters generally within the breeding range, but northern birds are mostly migratory, casually ranging north to northern California and Colorado and occurring regularly along the Gulf Coast east to Florida.

STATUS. Locally abundant.

HABITAT. Occupies a variety of habitats in semiarid woodlands. Prefers tall, dense, or brushy riparian woodlands, with trees from 5 to 10 m tall and an understory of thorny shrubs. Also occupies desert scrub, desert grassland, oak woodland, chaparral, valleys of desert mountains, and shade and fruit trees of agricultural areas, country roadsides, and suburban residential areas. Builds nest on relatively horizontal branch of a variety of trees and large shrubs in shaded sites, preferably in the interior of woodlands and thickets.

SPECIAL HABITAT REQUIREMENTS. Nest sites with trees of low to medium height having dense foliage and fairly open ground cover. In Caribbean, uses coastal areas and scrub, also mangroves.

Further Reading. United States and Canada: Brown 1977, Cottam and Trefethen 1968, Godfrey 1966, Neff 1940, Oberholser 1974a. Mexico and Central America: Edwards 1972, Land 1970, Monroe 1968, Peterson and Chalif 1973, Rappole et al. 1983, Ridgely and Gwynne 1989, Stiles and Skutch 1989. South America: Hilty and Brown 1986, Meyer de Schauensee 1966, Rappole et al. 1983. Caribbean: Bond 1947, Rappole et al. 1983.

Mourning Dove

Zenaida macroura

Canada (Quebec): Tourterelle Triste

Costa Rica: Paloma Rabuda

Guatemala: Paloma Guirguira

Mexico: Paloma Común

Puerto Rico: Rabiche, Tortola Rabilarga

West Indies: Fifi, Rabiche, Tortola, Tortola Rabilarga, Tourterelle Queue-fine

RANGE. Breeds from southern and central Alberta to southern New Brunswick and Nova Scotia south to Mexico and the Greater Antilles. Winters primarily from northern California east across the central United States to Iowa, southern Michigan, southern Ontario, New York, and New England south throughout the breeding range to central Panama.

STATUS. Abundant; long-term decline in western United States, increase in the East.

HABITAT. Occupies a broad range of habitats from desert riparian areas to a variety of wooded habitats, but avoids densely forested regions. Inhabits open country, especially fields, orchards, or generally weedy areas having an abundance of grains and seeds. Also inhabits open mixed woodlands and wood edges, shelterbelts, church and cemetery sites, evergreen plantations, suburbs, and cities. Loosely colonial or solitary. Generally nests on horizontal branches in shrubs and trees, especially conifers up to 30 m, but typically 3 to 10 m, above ground, and occasionally on the ground, especially in shrub-steppe and grassland habitats.

SPECIAL HABITAT REQUIREMENTS. Open country with some bare ground.

Further Reading. United States and Canada: Davison and Sullivan 1963, DeGraaf and Rudis 1986, Forbush and May 1955, Godfrey 1966, Hanson and Kosack 1963, Howe and Flake 1989, Johnsgard 1975a, Keeler 1977. Mexico and Central America: Edwards 1972, Land 1970, Monroe 1968, Peterson and Chalif 1973, Rappole et al. 1983, Ridgely and Gwynne 1989, Stiles and Skutch 1989. Caribbean: Bond 1947, Raffaele 1989, Rappole et al. 1983.

Black-billed Cuckoo
Coccyzus erythropthalmus

Canada (Quebec): Coulicou à Bec Noir

Costa Rica and Mexico: Cuclillo Piquinegro

Guatemala: Cuco Piquinegro

Venezuela: Cuclillo Pico Negro

RANGE. Breeds from east-central and southeastern Alberta and southern Saskatchewan to New Brunswick and Nova Scotia south, at least locally, to eastern Colorado, north-central Texas, northern Arkansas, northern Alabama, and the Carolinas. Migrates through Central America and the Caribbean. Winters in Colombia, Venezuela, Ecuador, and Peru.

STATUS. Common; has recently declined in the eastern United States.

HABITAT. Prefers extensive areas of upland woods that provide a variety of trees, bushes, and vines. Also occurs in brushy pastures, hedgerows, shelterbelts, open woodlands, orchards, thickets, and along wooded roadsides. Selects nest sites that are well concealed by overhanging branches and leaf clusters. Usually nests in shrubs or on a low tree branch, typically 1 to 2 m above ground. Occasionally lays eggs in the nests of other birds. Nesting productivity often varies with outbreaks of caterpillars and other large insects.

SPECIAL HABITAT REQUIREMENTS. Low, dense, shrubby vegetation.

Further Reading. United States and Canada: Beal 1904, Bent 1940a, DeGraaf and Rudis 1986, Godfrey 1966, Herrick 1910, Johnsgard 1979, Sealy 1978b, Spencer 1943. Mexico and Central America: Edwards 1972, Land 1970, Monroe 1968, Peterson and Chalif 1973, Rappole et al. 1983, Ridgely and Gwynne 1989, Stiles and Skutch 1989. South America: Hilty and Brown 1986, Meyer de Schauensee 1966, Meyer de Schauensee and Phelps 1978, Rappole et al. 1983. Caribbean: Bond 1947, Herklots 1961, Rappole et al. 1983.

Yellow-billed Cuckoo
Coccyzus americanus

Canada (Quebec): Coulicou à Bec Jaune

Costa Rica: Cuclillo Piquigualdo

Guatemala: Cuco Piquigualdo, Cuclillo de Pico Amarillo

Mexico: Cuclillo Alirrojizo

Puerto Rico: Pájaro Bobo Piquiamarillo

Venezuela: Cuclillo Pico Amarillo

RANGE. Breeds from interior California and Utah east to southwestern Quebec and southern New Brunswick, south to Arizona and Mexico. Migrates through Middle America and the Caribbean. Winters in the Andes, northern South America, and eastern Brazil.

STATUS. Long-term and recent declines in the eastern and central regions of the United States, with precipitous declines in parts of the western United States.

HABITAT. Favors moderately dense thickets near watercourses, second-growth woodlands, deserted farmlands overgrown with shrubs and brush, and brushy orchards. Also inhabits thickets or understory in open woods, avoiding extremely dense woods and high elevations. Prefers to conceal nest in thick shrubs or brush overgrown with vines or in deciduous trees on horizontal limbs, typically 1 to 2 m above ground. Highly vulnerable to tropical deforestation (Morton 1992).

SPECIAL HABITAT REQUIREMENTS. Low, dense, shrubby vegetation. In western United States, contiguous riparian and other woodlands 40 to 80 ha or more (Laymon and Halterman 1989).

Further Reading. United States and Canada: Anderson and Laymon 1989, Beal 1904, Bent 1940a, DeGraaf and Rudis 1986, Godfrey 1966, Johnsgard 1979, Laymon and Halterman 1987, 1989, Preble 1957, Tate and Tate 1982. Mexico and Central America: Edwards 1972, Land 1970, Monroe 1968, Peterson and Chalif 1973, Rappole et al. 1983, Ridgely and Gwynne 1989, Stiles and Skutch 1989. South America: Hilty and Brown 1986, Meyer de Schauensee 1966, Meyer de Schauensee and Phelps 1978, Rappole et al. 1983. Caribbean: Bond 1947, Herklots 1961, Raffaele 1989, Rappole et al. 1983.

Mangrove Cuckoo
Coccyzus minor

Costa Rica: Cuclillo de Antifaz, Cuclillo Orejinegro

Guatemala: Cuclillo de Orejas Negras, Cuco Orejinegro

Mexico: Cuclillo Ventrisucio

Puerto Rico: Pájaro Bobo Menor

Trinidad and Tobago: Coucou Manioc Gris

Venezuela: Cuclillo de Manglar

West Indies: Arrierito, Arriero Chico, Boba, Carga-Agua, Coucou Manioc, Guacaira, Guagán, Pájaro Bobo, Petit Tacot, Primavera

RANGE. Breeds in southern Florida from Tampa Bay and Miami southward in coastal areas, including the Florida Keys, through Mexico and along the Gulf Coast to northeastern Brazil. Winters throughout the breeding range, but mostly south of Florida.

STATUS. Rare and local.

HABITAT. Inhabits red and black mangrove thickets and swamps near saltwater, and upland hardwood hummocks. Typically builds nest on a horizontal mangrove branch; nest is indistinguishable from the nest of the Yellow-billed Cuckoo.

SPECIAL HABITAT REQUIREMENTS. Mangrove swamps and hardwood hummocks.

Further Reading. United States and Canada: Bent 1940a, Farrand 1983b, Harrison 1975, Terres 1980. Mexico and Central America: Edwards 1972, Land 1970, Monroe 1968, Peterson and Chalif 1973, Rappole et al. 1983, Ridgely and Gwynne 1989, Stiles and Skutch 1989. South America: Hilty and Brown 1986, Meyer de Schauensee 1966, Meyer de Schauensee and Phelps 1978, Rappole et al. 1983. Caribbean: Bond 1947, Herklots 1961, Raffaele 1989, Rappole et al. 1983.

Flammulated Owl
Otus flammeolus

Canada (Quebec): Duc Nain
Guatemala: Autillo Gris, Lechucita
Mexico: Tecolote Ojioscuro Serrano

RANGE. Breeds locally from southern British Columbia, southern Idaho, and northern Colorado south to southern California, southern Arizona, southern New Mexico, western Texas, and Mexico. Winters in Mexico, casually north to southern California.

STATUS. Rare to locally common.

HABITAT. Inhabits forests of the western mountains, mostly from 1,400 to 2,300 m but as high as 3,300 m elevation. Prefers woods with dense, thicketlike cover close to relatively open areas. Favors mature ponderosa-pine forests but also occurs in forests of spruce-fir, Douglas-fir, lodgepole pine, aspen, and pinyon-juniper. Avoids cutover areas and forests less than 100 years old (Reynolds and Linkhart 1987). Usually nests in abandoned flicker or other woodpecker nest cavities from 2 to 10 m above ground in aspen, oaks, or pines. Will forcibly evict a flicker if an abandoned cavity is not available; rarely, nests in holes constructed by Bank Swallows.

SPECIAL HABITAT REQUIREMENTS. Closely associated with extensive stands of old-growth ponderosa pine and mixed old-growth ponderosa pine–Douglas-fir (Mannan and Meslow 1984). Some undergrowth or intermixture of oaks in the forest.

263

Further Reading. United States and Canada: Balda et al. 1975, Farrand 1983b, Heintzelman 1979, Howie and Ritcey 1987, Karalus and Eckert 1974, McCallum and Gehlbach 1988, Oberholser 1974a, Phillips et al. 1964. Mexico and Guatemala: Edwards 1972, Land 1970, Peterson and Chalif 1973.

Elf Owl
Micrathene whitneyi

Mexico: Tecolotito Colicorto

RANGE. Breeds from southeastern California, extreme southern Nevada, central Arizona, southwestern New Mexico, and western and southern Texas south to Mexico. Winters in Mexico.

STATUS. Common.

HABITAT. Prefers arid, low-elevation desert areas overgrown with cacti, mesquite, and creosote bush, or with agave, ocotillo, and cactus desert scrub. Also inhabits riparian cottonwood and willow groves; mesquite floodplains; walnut, sycamore, and oak woodlands; and juniper, pinyon pine, and oak woodlands up to 2,100 m in elevation. Avoids pure stands of pine, but will inhabit almost every other type of dry, woody vegetation. Nests in abandoned woodpecker holes, especially in saguaro cactus but also in agave bloom stalks, tree stumps, cottonwoods, mesquite, sycamore, pines, walnut oaks, or willows growing on mesas and desert slopes and in canyons. Nests generally 3 to 10 m above the ground.

SPECIAL HABITAT REQUIREMENTS. Trees, snags, cacti with abandoned woodpecker holes for nesting.

Further Reading. United States and Canada: Heintzelman 1979, Karalus and Eckert 1974, Ligon 1968, Oberholser 1974a. Mexico: Edwards 1972, Peterson and Chalif 1973, Rappole et al. 1983.

Burrowing Owl
Speotyto cunicularia

Costa Rica and Guatemala: Lechuza Llanera, Lechuza Terrestre

Mexico: Tecolote Zancón

Venezuela: Mochuelo de Hoyo

West Indies: Coucou, Coucouterre, Cucu

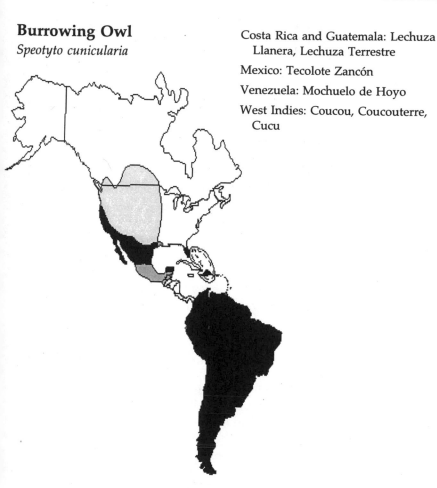

RANGE. Breeds from southern interior British Columbia to southern Manitoba south through eastern Washington, central Oregon, and California to Baja California, east to western Minnesota, western Missouri, Oklahoma, eastern Louisiana, and Florida, and south to Mexico and Central America. Populations in the Caribbean and South America are resident. Winters throughout breeding range except in northern portions in the Great Basin and Great Plains regions.

STATUS. Locally common; increasing in the western United States. A threatened species in Canada.

HABITAT. Prairie, plains, grasslands, deserts, open shrubsteppe, and sometimes open areas such as golf courses or airports. Depends greatly on mammals that dig burrows in open areas with short vegetation for nesting, roosting, and escaping. Commonly perches on fence posts, bushes, utility wires, roadside billboards, and burrow mounds. In the western United States, often nests in colonies in abandoned burrows of prairie dogs and ground squirrels; also nests in burrows of woodchucks, foxes, badgers, coyotes, and armadillos. In Florida, nests in gopher tortoise burrows. Can excavate its own burrow but usually enlarges burrows started by mammals and uses the same burrow for years if not disturbed. Control of colonial rodents and conversion of grassland to cropland have reduced populations. Management recommendations emphasize maintenance of pesticide-free zone of 600-m radius around nest burrows and provision of dense herbaceous small mammal habitat within owl home ranges (Haug and Oliphant 1990).

SPECIAL HABITAT REQUIREMENTS. Burrows of colonial burrowing mammals, especially prairie dogs and ground squirrels, in well-grazed grasslands.

Further Reading. United States and Canada: Butts 1973, Errington and Bennett 1935, Evans 1982, Green and Anthony 1989, Heintzelman 1979, Karalus and Eckert 1974, Konrad and Gilmer 1984, Marti 1974, Rich 1986, Tate and Tate 1982, Terres 1980, Zarn 1974b. Mexico and Central America: Edwards 1972, Land 1970, Monroe 1968, Peterson and Chalif 1973, Rappole et al. 1983, Ridgely and Gwynne 1989, Stiles and Skutch 1989. South America: Hilty and Brown 1986, Meyer de Schauensee 1966, Meyer de Schauensee and Phelps 1978, Rappole et al. 1983. Caribbean: Bond 1947, Rappole et al. 1983.

Long-eared Owl
Asio otus

Mexico: Buho Cornado Caricafé

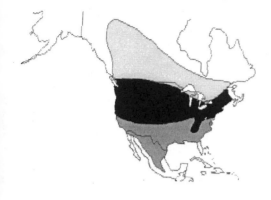

RANGE. Breeds from northern Yukon, southwestern Northwest Territories, northern Saskatchewan, and Nova Scotia, south to northern Baja California, southern Arizona, western and central Texas, Arkansas, northern Ohio, western Virginia, and New England. Winters from southern Canada south to Baja California, central Mexico, southern Texas, the Gulf Coast, and Georgia; casually to Florida.

STATUS. Locally common. Very nomadic.

HABITAT. Coniferous or mixed woodlands; sometimes deciduous forests, usually near open country. Also will inhabit open or dense thickets, parks, orchards, isolated woodlots, wooded swamps, riparian woodlands, and reservoir shorelines, even low-growing scrub if it is in the form of dense, tangled thickets. Most often uses old nests of large birds such as crows, hawks, ravens, herons, or magpies, but will use squirrel nests and natural tree cavities; dependent on dwarf-mistletoe brooms in Douglas-fir (Bull et al. 1989). Usually locates nest 5 to 10 m above the ground, but may nest on the ground or on ledges. Rarely, will construct own nest. Occurs up to 3,300 m. In winter, roosts communally in conifers or dense hardwoods.

SPECIAL HABITAT REQUIREMENTS. Dense (usually coniferous) vegetation for nesting and roosting cover.

Further Reading. United States and Canada: Armstrong 1958, Craig et al. 1988, DeGraaf and Rudis 1986, Heintzelman 1979, Johnsgard 1979, Karalus and Eckert 1974, Marks 1986, Marti 1976. Mexico: Edwards 1972.

Short-eared Owl

Asio flammeus

Canada (Quebec): Hibou des Marais

Costa Rica and Guatemala: Lachuza Campestre

Mexico: Buho Cornicorto Llanero

Venezuela: Lechuza Orejicorta

West Indies: Carabo, Chat-huant, Lechuza de Sabana, Mucaro Real, Mucaro de Sabana

RANGE. Breeds from northern Alaska and northern Yukon to northern Quebec and Labrador, south to central California, northern Nevada, Utah, Kansas, Missouri, northern Ohio, northern Virginia, and New Jersey. Winters generally in the breeding range from southern Canada south to Mexico. Individuals in Puerto Rico, Hispaniola, and South America are resident.

STATUS. Locally common; nomadic; declining across southern portions of its range. In western North America, loss of marshes and overgrazing or conversion of native prairie to crops have resulted in dramatic declines since the 1930s.

HABITAT. Primarily inhabits marshland and open grasslands, but also tundra, open fields, forest clearings, sagelands, deserts, sparse shrubsteppe, pastures, prairies, lower mountain slopes, canyons, arroyos, dunes, meadows, and other open habitats. Nests are sometimes in small loose colonies, placed in slight depressions on the ground, either in grassy cover among clumps of weeds or grasses or in exposed situations. Rarely, will nest in an excavated burrow. In winter, prefers open areas with little or no snow.

SPECIAL HABITAT REQUIREMENTS. Extensive open marshes or grasslands with an abundance of rodents.

Further Reading. United States and Canada: Bosakowski 1986, Clark 1975, DeGraaf and Rudis 1986, Godfrey 1966, Heintzelman 1979, Johnsgard 1979, 1988, Karalus and Eckert 1974, Low and Mansell 1983, Tate 1992, Tate and Tate 1982. Mexico and Central America: Edwards 1972, Land 1970, Peterson and Chalif 1973, Rappole et al. 1983, Stiles and Skutch 1989. South America: Hilty and Brown 1986, Meyer de Schauensee 1966, Meyer de Schauensee and Phelps 1978, Rappole et al. 1983. Caribbean: Bond 1947, Raffaele 1989, Rappole et al. 1983.

Lesser Nighthawk
Chordeiles acutipennis

Costa Rica: Añapero Menor

Guatemala: Tapacamino Halcón,
Tapacamino Menor

Mexico: Chotacabra Halcón

Venezuela: Aguaitacamino Chiquito

RANGE. Breeds from central interior California, southern Nevada, extreme southwestern Utah, central Arizona, central New Mexico, and central and southeastern Texas south to South America. Winters from Mexico south to South America. Withdraws in winter from northern Mexico. A fairly common migrant in Guatemala. A common migrant and winter visitor in Honduras, occurring in the lowlands of both coasts and the interior highlands. In Costa Rica a fairly common breeding, and perhaps permanent, resident, a common to abundant fall migrant, and a local winter resident. In Panama local breeders are augmented by northern migrants from late July to April. Northern breeders winter south to northwestern Colombia.

STATUS. Common, generally increasing.

HABITAT. Inhabits barren or partly brushy country in low deserts of the Southwest. Occurs around dry fields, dry washes and riverbeds, sandy flats, and broad, rocky, sparsely vegetated valleys. Generally lays eggs on bare ground in open sandy or gravelly areas but also in brushy areas, lowlands, hills, canyons, and dry rocky slopes and mesas, or on flat gravel and asphalt roofs.

Further Reading. United States and Canada: Bent 1940b, Harrison 1979, Oberholser 1974a, Terres 1980. Mexico and Central America: Edwards 1972,

Land 1970, Monroe 1968, Peterson and Chalif 1973, Ridgely and Gwynne 1989, Stiles and Skutch 1989. South America: Hilty and Brown 1986, Meyer de Schauensee 1966, Meyer de Schauensee and Phelps 1978.

Common Nighthawk
Chordeiles minor

Canada (Quebec): Engoulevant Commun

Costa Rica: Añapero Zumbón, Tapacaimes Común

Guatemala: Tapacaimes Zumbón

Mexico: Chotacabra Zumbón

Venezuela: Aguaitacamino Migratorio

West Indies: Capacho, Gaspayo, Querequeque, Querequete

RANGE. Breeds from southern Yukon and southern Northwest Territories to central Quebec and southern Labrador, south to southern California, southern Nevada, southern Arizona, Texas, the Gulf Coast, Florida, Mexico, and Central America. Winters in South America south to Argentina. Probably a migrant throughout except Baja California; rare on Yucatán Peninsula. A regular though uncommonly recorded migrant through Central America. The small numbers of breeding individuals in Costa Rica move south with the abundant northern migrants in the fall; a sporadically common spring migrant. An uncommon to fairly common migrant in Colombia in early fall.

Migrant and in part resident over most of South America south to Buenos Aires, Argentina. North American individuals occur as transients throughout the West Indies.

STATUS. Common; declining in the eastern United States and Canada.

HABITAT. Varied open habitats throughout most of North America. Prefers open habitats such as grasslands, sparse woods, or towns and cities. Also inhabits plowed fields, gravel beaches, railroad rights-of-way, and barren areas with rocky soils. Lays eggs on flat substrates such as gravelly ground, burned-over areas, gravel and asphalt rooftops, dry barren plains, bare rock, and partially vegetated soil, but always in the open.

Further Reading. United States and Canada: DeGraaf and Rudis 1986, Forbush and May 1955, Godfrey 1966, Johnsgard 1979, Tate and Tate 1982, Terres 1980. Mexico and Central America: Edwards 1972, Land 1970, Monroe 1968, Peterson and Chalif 1973, Ridgely and Gwynne 1989, Stiles and Skutch 1989. South America: Hilty and Brown 1986, Meyer de Schauensee and Phelps 1978. Caribbean: Bond 1947, Raffaele 1989.

Antillean Nighthawk
Chordeiles gundlachii

Puerto Rico: Capacho, Gaspayo, Piramidig, Querequeque Antillano

RANGE. Breeds in the Florida Keys, the Greater Antilles, and the Bahamas. Winters in South America.

STATUS. Uncommon.

HABITAT. Hunts for flying insects high over pastures and fields. Generally lays a single egg directly on the ground in an open area.

SPECIAL HABITAT REQUIREMENTS. Open and semi-open situations.

Further Reading. United States: Robbins et al. 1983. Caribbean: Raffaele 1989.

Common Poorwill
Phalaenoptilus nuttallii
(formerly Poor-will)

Canada (Quebec): Engoulevent de Nuttall

Mexico: Tapacamino Tevíi

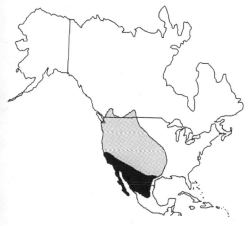

RANGE. Breeds from southern interior British Columbia, Montana, southeastern Alberta, and southwestern South Dakota south to Baja California, central Mexico, and central Texas and east to eastern Kansas. Winters in the southern part of the breeding range. In Mexico, U.S. migrants augment the resident population.

STATUS. Common; showing a long-term increase in the central regions of the United States and Canada.

HABITAT. Inhabits a variety of arid and semiarid open, usually rocky habitats in the West, from lowlands up to 3,600 m on mountain slopes. Inhabits sparse desert brushland, open prairies, open pinyon-juniper woodlands, mixed chaparral-grassland, brushy rocky canyons, mountain scrub, and pine-oak woodlands. Seems to prefer rocky habitats with scrubby cover or xeric woodlands. Lays eggs on gravelly ground, on a flat rock, or in a slight hollow scraped in the ground. Often lays eggs so that they are partially shaded by a bush, weeds, or a tuft of grass.

Further Reading. United States and Canada: Bent 1940a, Brigham and Barclay 1992, Farrand 1983b, Godfrey 1966, Oberholser 1974a, Terres 1980. Mexico: Edwards 1972, Peterson and Chalif 1973.

Chuck-will's-widow
Caprimulgus carolinensis

Canada (Quebec): Engoulevent de Caroline

Costa Rica: Chotacabra de Paso

Guatemala: Puercorrín de Paso, Tapacamino de Paso

Mexico: Tapacamino de Paso

Puerto Rico: Gabario Mayor

Venezuela: Aguaitacamino Americano

RANGE. Breeds from eastern Kansas, southern Iowa, and central Illinois to New Jersey and southern New York, south to south-central and southeastern Texas, the Gulf Coast, and southern Florida. Winters from southeastern Texas, Louisiana, and northern Florida to South America. A migrant and winter visitor through eastern sections of Mexico. An uncommon migrant and winter visitor to the lowlands of both coasts of Honduras, occurring in a wide variety of forested and semi-open situations. In Costa Rica a widespread but generally uncommon to rare migrant and winter resident. A rare to uncommon winter resident almost throughout Panama. A rare but regular migrant and winter visitor to Puerto Rico and the Virgin Islands. Winters south to Colombia, Venezuela, the Greater Antilles, and Bahamas; a vagrant to St. Martin, Saba, and Barbuda.

STATUS. Locally uncommon; declining in the eastern United States.

HABITAT. Prefers mixed oak and pine forests, but also inhabits live-oak groves, forest edges, and woodlands along river courses. Lays eggs on the ground on dead leaves, usually at the edges of forests, near roads or other clearings, usually with little or no undergrowth around the eggs.

SPECIAL HABITAT REQUIREMENTS. Pine-oak woods; in migration and winter, also open woodlands and palmetto thickets.

Further Reading. United States and Canada: Bent 1940a, Farrand 1983b, Forbush and May 1955, Godfrey 1966, Johnsgard 1979, Terres 1980. Mexico and Central America: Edwards 1972, Land 1970, Monroe 1968, Peterson and Chalif 1973, Ridgely and Gwynne 1989, Stiles and Skutch 1989. South America: Hilty and Brown 1986, Meyer de Schauensee 1966, Meyer de Schauensee and Phelps 1978. Caribbean: Bond 1947, Raffaele 1989.

Whip-poor-will
Caprimulgus vociferus

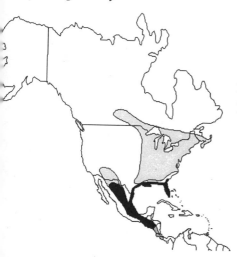

Canada (Quebec): Engoulevent Bois-pouri

Costa Rica: Chotacabra Gritón Ruidoso

Cuba: Guabario Chico

Guatemala: Cuerporruín Gritón

Mexico: Tapacamino Cuerporruín

Puerto Rico: Guabario

RANGE. Breeds from north-central Saskatchewan and southern Manitoba to southern Quebec and Nova Scotia, south to eastern Kansas, northeastern Texas, and northern Louisiana across to central Georgia; and from southern California, southern Nevada, central Arizona, and extreme western Texas south to Mexico. Winters from southern Texas, the Gulf Coast, and east-central South Carolina south to Central America. A migrant and permanent resident in much of Mexico. Wintering birds are found in the highlands of Guatemala. An uncommon resident and a fairly common winter visitor in Honduras. The eastern North American breeding population winters in much of Honduras. A casual to very rare winter resident along the Pacific Coast of Costa Rica. Winters casually to western Panama. Rare or casual as a winter resident in Cuba.

STATUS. Common; long-term and recent precipitous declines in eastern North America.

HABITAT. In the East, prefers open hardwood or mixed woodlands of pine, oak, and beech, particularly younger stands in fairly dry habitats; also favors stands with scattered clearings. In the Southwest, frequents densely wooded slopes of oak and pine in canyons and mountains. Lays eggs on dead leaves on well-drained ground; usually in areas of partial shade where there is no undergrowth. Often nests among trees at the edge of a clearing or path, sometimes laying eggs in the shade of a small bush.

SPECIAL HABITAT REQUIREMENTS. Open woodland, early successional forest or brushy field edges.

Further Reading. United States and Canada: DeGraaf and Rudis 1986, Forbush and May 1955, Godfrey 1966, Johnsgard 1979, Raynor 1941, Tate and Tate 1982, Terres 1980. Mexico and Central America: Edwards 1972, Land 1970, Monroe 1968, Peterson and Chalif 1973, Ridgely and Gwynne 1989, Stiles and Skutch 1989. South America: Hilty and Brown 1986. Caribbean: Bond 1947.

Black Swift
Cypseloides niger

Canada (Quebec): Martinet Noir

Costa Rica, Guatemala, Mexico, and Puerto Rico: Vencejo Negro

West Indies: Gros Martinet Noir, Hirondelle de Montague, Oiseau de la Pluie, Vencejo Negro

RANGE. Breeds locally in southeastern Alaska, northwestern and central British Columbia, southwestern Alberta, northwestern Montana, Colorado, and central Utah, south through Washington, Oregon, California, and Mexico to (locally) Central America and the Antilles. Winters (presumably) in western Mexico and Guyana. Individuals may wander or migrate casually to the vicinity of Guatemala, but the status of this species is uncertain. Any Black Swifts in Honduras are presumed to be resident. A breeding resident and migrant in Costa Rica, believed to migrate mostly or entirely on the Pacific slope. May be found as far south as northern South America. Winter range is still poorly known. Individuals may withdraw from the Greater Antilles to the south during the winter months. Accidental in the Virgin Islands. Found on St. Kitts, Guadeloupe, Dominica, Martinique, St. Lucia, St. Vincent, Grenada (transient), Barbados (transient), and doubtful in Isle of Pines. Individuals from these islands apparently winter in Guyana. Possible migrant to Trinidad.

STATUS. Rare or uncommon; long-term decline.

HABITAT. Roosts and nests in areas with rocky cliffs and canyons, especially near water, varying from ocean cliffs to mountain ledges, at elevations from sea level to 3,300 m. Nests in small colonies, from 5 to 15 pairs. Builds a bulky cup or disk of moss on a sea cliff, ledge, or cave, or in a crevice or ledge on a sheer, high, moist cliff face near or behind a waterfall, or over a pool.

SPECIAL HABITAT REQUIREMENTS. Crevices or ledges on rocky cliffs for nesting, preferably near or behind a waterfall.

Further Reading. United States and Canada: Bailey and Niedrach 1965, Godfrey 1966, Hunter and Baldwin 1962, Knorr 1961, Kondla 1973, Lack 1956, Terres 1980. Mexico and Central America: Edwards 1972, Land 1970, Monroe 1968, Peterson and Chalif 1973, Ridgely and Gwynne 1989, Stiles and Skutch 1989. South America: Meyer de Schauensee 1966. Caribbean: Bond 1947, Herklots 1961, Raffaele 1989.

Chimney Swift
Chaetura pelagica

Canada (Quebec): Martinet Ramoneur

Costa Rica: Vencejo de Paso

Guatemala: Vencejo de Chimenea, Vencejo Migratorio

Mexico: Vencejito de Paso

Venezuela and Virgin Islands: Vencejo de Chimenea

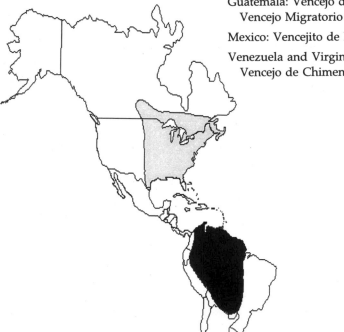

RANGE. Breeds east of the Rocky Mountains from east-central Saskatchewan and southern Manitoba to southern Quebec and New Brunswick, south to eastern New Mexico, south-central and southern Texas, the Gulf Coast, and south-central Florida. Winters in South America. A migrant through the east slope of Mexico. A migrant throughout much of Guatemala. An uncommon to common fall migrant and a common spring migrant in the lowlands and the islands off the north coast of Honduras. A sporadically common to abundant fall migrant on Costa Rica's Caribbean slope. Spring migration is along the Caribbean coast. Winters mostly in western Amazonia, also apparently in western Peru and northern Chile. Extremely rare in the Virgin Islands. No records from Puerto Rico. A fall and spring migrant recorded from the Bahamas, Cuba, Hispaniola, Jamaica, and the Cayman Islands.

STATUS. Common; but declining throughout the breeding range.

HABITAT. Not confined to any single habitat, as breeding range is largely dependent on suitable nesting sites. Presumably adapted to nesting in caves and probably hollow trees, now mostly nests in chimneys, silos, air shafts, and other structures. Prefers dark sheltered sites high above the ground. Constructs nest of twigs glued together with saliva; attaches nest to a vertical wall anywhere from near the top of a structure to more than 6 m below the top.

SPECIAL HABITAT REQUIREMENTS. Structures such as chimneys, silos, cisterns, wells, rafters, air shafts, and perhaps hollow trees for nest sites.

Further Reading. United States and Canada: Bailey and Niedrach 1965, DeGraaf and Rudis 1986, Fischer 1958, Forbush and May 1955, Godfrey 1966, Johnsgard 1979, Terres 1980. Mexico and Central America: Edwards 1972, Land 1970, Monroe 1968, Peterson and Chalif 1973, Ridgely and Gwynne 1989, Stiles and Skutch 1989. South America: Hilty and Brown 1986, Meyer de Schauensee 1966, Meyer de Schauensee and Phelps 1978. Caribbean: Bond 1947, Raffaele 1989.

Vaux's Swift

Chaetura vauxi

Canada (Quebec): Martinet de l'Ouest

Costa Rica: Vencejo Común, Vencejo Grisáceo

Guatemala: Vencejo Común, Vencejo de Vaux

Mexico: Vencejito Alirrápido

Venezuela: Vencejo de Vaux

RANGE. Breeds from southeastern Alaska, northwestern and southern British Columbia, northern Idaho, and western Montana south, chiefly from the Cascades and Sierra Nevada westward, to central California south to northern Mexico. Winters in Central America. A local breeder in eastern Mexico and a migrant in the west; winters in southern Mexico. Permanent residents in Middle America and Venezuela are augmented by North American migrants in the highlands of Guatemala and probably Honduras and Costa Rica. Southern limit of winter range still uncertain. A possible vagrant to Colombia.

STATUS. Uncommon.

HABITAT. Inhabits forested regions with large trees. In Montana, occurs in mixed forests of Douglas-fir, ponderosa pine, lowland fir, western larch, yellow birch, aspen, and cottonwoods; in California, inhabits ponderosa pine, mixed conifer, Jeffrey pine forests, and possibly black oak woodlands. Also inhabits river valleys in dense Douglas-fir and redwood forests. More abundant in old (>250 yr old) than younger (<165 yr) stands of Douglas-fir in Washington (Manuwal and Huff 1987). Nest is usually in a tall, hollow dead tree, or inside a burned-out stump, generally within 0.5 m but up to 2 m up from the bottom of the cavity. May depend on Pileated Woodpecker holes in old-growth conifer forests (Bull and Cooper 1991). Also nests in chimneys, but is not dependent on them.

SPECIAL HABITAT REQUIREMENTS. Hollow stubs or snags, preferentially in old stands, for nesting.

Further Reading. United States and Canada: Baldwin and Hunter 1963, Baldwin and Zaczkowski 1963, Bent 1940b, Godfrey 1966, Scott et al. 1977, Terres 1980, Verner and Boss 1980. Mexico and Central America: Edwards 1972, Land 1970, Monroe 1968, Peterson and Chalif 1973, Ridgely and Gwynne 1989, Stiles and Skutch 1989. South America: Meyer de Schauensee 1966.

White-throated Swift
Aeronautes saxatalis

Canada (Quebec): Martinet à Gorge Blanche

Mexico: Vencejito Pechiblanco

RANGE. Breeds from southern British Columbia, Idaho, Montana, and southwestern South Dakota south through the Pacific and southwestern states to Baja California, central Mexico, and Central America, and east to western Nebraska and western Texas. Winters from central California, central Arizona and, rarely, southern New Mexico south to Central America. Migrates or winters in the highlands nearly throughout Mexico. North American migrants pass through Guatemala to Honduras, which is the southern limit of this species' range.

STATUS. Common.

HABITAT. Inhabits areas with steep cliffs and deep canyons at elevations from near sea level to about 4,000 m. Inhabits primarily mountainous country but also coastal cliffs, rugged foothills, and desert canyons; ranges over adjacent valleys while foraging. Places nest in deep cracks and crevices in steep, rocky, often inaccessible cliff faces or canyons, from 3 to 60 m or more above the base. Sometimes nests in cracks in high walls of buildings such as bell towers or grain elevators.

SPECIAL HABITAT REQUIREMENTS. Crevices in cliffs for nesting.

Further Reading. United States and Canada: Bailey and Niedrach 1965, Bent 1940b, Godfrey 1966, Terres 1980, Verner and Boss 1980. Mexico and Central America: Edwards 1972, Land 1970, Monroe 1968, Peterson and Chalif 1973, Ridgely and Gwynne 1989.

Broad-billed Hummingbird
Cynanthus latirostris

Mexico: Colibrí Latirrostro

RANGE. Breeds from western Sonora, southeastern Arizona, southwestern New Mexico (Guadalupe Canyon), and very locally in western Texas, northern Chihuahua, and Tamaulipas south through Mexico to Oaxaca and Chiapas and east to Veracruz, Hidalgo, and Puebla. Winters in Mexico, casually north to southern Arizona.

STATUS. Common.

282

HABITAT. Prefers desert mountain canyons, riparian woodlands, and higher desert washes, especially where sycamores, cottonwoods, willows, and mesquite are present. Places nest on a branch of a small tree or on a stalk of a vine or shrub, usually 1 to 2 m above the ground.

SPECIAL HABITAT REQUIREMENTS. Red or red-and-yellow flowers for feeding.

Further Reading. United States and Canada: Cottam and Knappen 1939, Farrand 1983b. Johnsgard 1983, Moore 1939, Terres 1980. Mexico: Edwards 1972, Peterson and Chalif 1973.

White-eared Hummingbird
Hylocharis leucotis

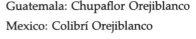

Guatemala: Chupaflor Orejiblanco
Mexico: Colibrí Orejiblanco

RANGE. Resident from Sonora, Chihuahua, Coahuila, Nuevo León, and Tamaulipas south through the highlands of Mexico, Guatemala, El Salvador, and Honduras to north-central Nicaragua. Irregular in summer in the mountains of southern Arizona, southwestern New Mexico, and western Texas; northernmost populations are partially migratory.

STATUS. Rare and irregular in United States.

HABITAT. Prefers the undergrowth of oak forests but also occurs in pine and dense pine-oak forests, high-mountain fir forests, partially open mountain country with scattered trees and shrubs, suburban gardens, and vacant lots with scattered shrubs and flowers. Nest sites are nearly always in shrubs or fairly low trees. Seems to prefer habitat with streams. Little is known about its habits in the United States, and there is no good evidence that it has ever nested in Arizona.

SPECIAL HABITAT REQUIREMENTS. Flowers, especially blue flowers such as salvia, for nectar.

Further Reading. United States and Canada: Cottam and Knappen 1939, Farrand 1983b, Johnsgard 1983, Phillips et al. 1964, Terres 1980. Mexico and Central America: Edwards 1972, Land 1970, Peterson and Chalif 1973.

Buff-bellied Hummingbird
Amazilia yucatanensis

Guatemala: Chupaflor Leonado, Chupaflor de Vientre Castaño

Mexico: Amazilia del Golfo

RANGE. Breeds from southern Texas in the lower Rio Grande Valley south to northern Guatemala, Belize, and casually to Honduras. A partial migrant in northern (subtropical) parts of breeding range. Casual north to central and eastern Texas, and southern Louisiana. Apparently also a rare visitor to Honduras during nonbreeding season.

STATUS. Uncommon (United States); formerly more common but has declined, perhaps as a result of habitat destruction and insecticide spraying north of the U.S.-Mexico border.

HABITAT. Inhabits semiarid lowlands dominated by woods or scrubby growth, preferring dense thickets, flowering bushes, and tangled vines along the banks of streams and ponds, and gullies. Also occurs in semiarid coastal scrub in open woods, chaparral thickets, farms and gardens, and in citrus groves. Places nest 1 to 2 m above the ground on a small drooping limb or horizontal fork of a twig in a small tree or bush such as ebony, blackbead, or hackberry, but sometimes in willow. Often chooses a nest site that is near a road, trail, or other clearing.

SPECIAL HABITAT REQUIREMENTS. Flowers for nectar.

Further Reading. United States and Canada: Johnsgard 1983, Oberholser 1974a, Terres 1980. Mexico and Central America: Edwards 1972, Land 1970, Monroe 1968, Peterson and Chalif 1973.

Violet-crowned Hummingbird
Amazilia violiceps

Mexico: Amazilia Occidental

RANGE. Breeds in southern Arizona in the Huachuca and Chiricahua Mountains, and in southwestern New Mexico in Guadalupe Canyon, south to central Mexico. Casual in southern California. With a few exceptions, withdraws into Mexico in winter.

STATUS. Rare and local; first discovered breeding in the United States in 1959.

HABITAT. Prefers riparian sycamore groves in desert mountain canyons. In the United States, generally associated with streamside plant life in the deserts and foothills of mountains. Builds a nest that is saddled on a horizontal limb in sycamores, 8 to 12 m above the ground.

Further Reading. United States and Canada: Farrand 1983b, Johnsgard 1983, Phillips et al. 1964, Terres 1980, Zimmerman and Levy 1960. Mexico: Edwards 1972, Peterson and Chalif 1973.

Blue-throated Hummingbird

Lampornis clemenciae

Mexico: Chupaflor Gorjiazul

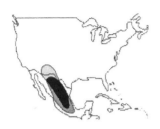

RANGE. Breeds in southeastern Arizona in the Huachuca and Chiricahua Mountains, and from western Texas south to Oaxaca, Mexico. Casual north to Colorado. Winters in Mexico, but occurs casually at Arizona feeders. A moderately common but irregular summer resident in the north-central highlands of Mexico, with wintering birds occurring farther south to the north-central Pacific Coast, the central highlands, and the north-central Atlantic Coast.

STATUS. Fairly common to uncommon.

HABITAT. Inhabits lush vegetation along wooded streamsides in mountain canyons. In Texas, found among bald cypress, pines, oaks, and bigtooth maples from 1,500 to 2,300 m in elevation. Builds nest in sites that are completely covered from above, such as vertical-walled canyons, rock overhangs, plant stalks, eaves of buildings, bridges, water towers, and inside buildings. May use the same nest site for several years, adding to the nest each time it is used.

SPECIAL HABITAT REQUIREMENTS. Nest sites that are sheltered from rain and sunlight, near an abundance of flowers for feeding, and within 1 to 2 m of a stream.

Further Reading. United States and Canada: Cottam and Knappen 1939, Farrand 1983b, Johnsgard 1983, Oberholser 1974a, Terres 1980. Mexico: Edwards 1972.

286

Magnificent Hummingbird

Eugenes fulgens
(formerly Rivoli's
Hummingbird)

Costa Rica: Colibrí Magnífico
Guatemala: Chupaflor Magnífico,
Chupaflor de Rivoli
Mexico: Chupaflor Coronimorado

RANGE. Breeds in western Colorado, and from southeastern Arizona, southwestern New Mexico, and western Texas south to Panama. Winters in Mexico and Central America, though a few remain at Arizona feeders. Southern limit of winter range is western Panama.

STATUS. Rare in Colorado, elsewhere uncommon to common.

HABITAT. Found above 1,500 m in deciduous woods along streams, and in pine or oak woods on mountain slopes and ridges. In Mexico, widespread at high altitudes, chiefly above 1,600 m (Blake 1953). Builds nest on a horizontal branch from 6 to 16 m above the ground. Uses a variety of trees for nesting, including cottonwoods, mountain maples, sycamores, alders, walnuts, pines, and Douglas-fir.

SPECIAL HABITAT REQUIREMENTS. Flowers for nectar.

Further Reading. United States and Canada: Cottam and Knappen 1939, Farrand 1983b, Johnsgard 1983, Terres 1980. Mexico and Central America: Edwards 1972, Land 1970, Ridgely and Gwynne 1989, Stiles and Skutch 1989.

Lucifer Hummingbird
Calothorax lucifer

Mexico: Colibrí Tijereta Altiplanero

RANGE. Southern and eastern Mexico from Jalisco south to Guerrero and east to Puebla. In the United States, the Chisos Mountains of Texas; Brewster County, Texas; Big Bend National Park, Texas; and Fort Bowie, Arizona. U.S. birds withdraw as far as south-central Mexico for the winter.

STATUS. Rare to fairly common.

HABITAT. Arid slopes and tablelands at moderate elevations: 1,250 to 2,500 m (Blake 1953). Nest, built in shrubs 1 to 2 m above the ground, is a cup of soft plant fibers, blossoms, seeds, bits of lichens, etc., bound with spider webs.

SPECIAL HABITAT REQUIREMENTS. Brushy hillsides, cleared land with scattered bushes or scrub.

Further Reading. United States and Canada: Bent 1940b, Robbins et al. 1983, Terres 1980. Mexico: Edwards 1972, Peterson and Chalif 1973.

Ruby-throated Hummingbird

Archilochus colubris

Canada (Quebec): Colibri à Gorge Rubis

Costa Rica: Colibrí Garganta de Rubí

Guatemala: Chupaflor Rubí

Mexico: Colibrí de Paso

Puerto Rico: Zumbadorcito de Garganta Roja

RANGE. Breeds in North America east of the Rocky Mountains, from central Alberta and central Saskatchewan to southern Quebec and New Brunswick south to southern Texas, the Gulf Coast, and Florida. Winters from southern Texas south to Central America, also in southern Florida, but mainly in Mexico and Central America. Rare winter resident in the Bahamas and Cuba. Casual in Hispaniola, Grand Cayman, and Jamaica. Accidental in Puerto Rico. A fairly common migrant and winter visitor in Guatemala. An uncommon to locally common winter resident in north Pacific lowlands of Costa Rica. A rare winter visitor to Panama.

STATUS. Common; long-term and recent increase in eastern North America, although may be declining steadily across its range, perhaps because of declining area or quality of tropical wintering grounds (Dobkin 1992).

HABITAT. Occurs in a variety of wooded habitats, ranging from rather dense to open coniferous and deciduous woodlands, orchards, and shade trees in yards. Also inhabits mixed woodlands, parks, and gardens, often breeding in woodlands near streams or wooded swamps, especially in western portion of range. Places nest 2 to 15 m, typically 3 to 6 m, above the ground, on a fairly level or downward-slanting twig or branch protected from above by larger branches or a leafy canopy. Often locates nest near or directly over water, or near a woodland trail. Uses a variety of trees for nesting, but appears to favor hardwoods over conifers, especially those with rough, lichen-covered bark. May use the same nest site year after year.

SPECIAL HABITAT REQUIREMENTS. Plants that provide tubular nectar-bearing (especially red) flowers such as honeysuckle, lantana, gilia, and trumpet vine.

Further Reading. United States and Canada: Beal and McAtee 1912, De-Graaf and Rudis 1986, Forbush and May 1955, Godfrey 1966, Johnsgard 1979, 1983, Miller and Nero 1983, Terres 1980, Willimont et al. 1988. Mexico and Central America: Edwards 1972, Land 1970, Ridgely and Gwynne 1989, Stiles and Skutch 1989. Caribbean: Bond 1947, Raffaele 1989.

Black-chinned Hummingbird
Archilochus alexandri

Canada (Quebec): Colibri à Gorge Noire

Mexico: Colibrí Gorjinegro

RANGE. Breeds from southwestern British Columbia and northwestern Montana south to Baja California, southern Texas, and northern Mexico, and east to western Wyoming, eastern Colorado, eastern New Mexico, and central Texas. Winters in Mexico, as far south as south-central Mexico; casually to southern Texas, southern Louisiana, northwestern Florida, and southern California.

STATUS. Common; recently declining in the California foothills.

HABITAT. Found most frequently in arid regions, where it inhabits oak and riparian woodlands of canyons and lowlands, small patches of willows along dry washes, chaparral, pine-oak woodlands, orchards, and parks. Prefers sites with a minimal canopy cover. Usually places nest 1 to 2 m (but up to 9 m) above the ground, on a small drooping branch or in a fork of a small tree or shrub, near or overhanging a stream, spring, or dry creek bed. Prefers to nest in oaks, but also nests in alders, cottonwoods, sycamores, laurel, willows, apple and orange trees in orchards, and woody vines and tall herbaceous weeds.

SPECIAL HABITAT REQUIREMENTS. Flowers for nectar.

Further Reading. United States and Canada: Baltosser 1989, Bent 1940b, Godfrey 1966, Harrison 1979, Johnsgard 1983, Phillips et al. 1964, Terres 1980, Verner and Boss 1980. Mexico: Edwards 1972, Peterson and Chalif 1973.

Costa's Hummingbird
Calypte costae

Mexico: Colibrí Coronivioleta Desiértico

RANGE. Breeds from central California, southern Nevada, and southwestern Utah south to southern Arizona and Mexico. Winters from southern California and southern Arizona south to Mexico.

STATUS. Common.

HABITAT. In southwestern deserts, frequents arid washes and hillsides, dry chaparral, and suburban areas where exotic plants have been introduced. In California, inhabits washes, mesas, and hillsides, particularly where sages, ocotillo, yuccas, and cholla cacti are abundant. Relatively independent of water during the breeding season, and thus occupies drier habitats than other hummingbirds. Builds nest in a variety of sites, usually from 1 to 3 m above the ground, on twigs or limbs of oaks, alders, hackberry, willow, palm, citrus trees in open orchards, or other trees; in sage, dead yuccas, branching cacti, and paloverde; or on vines clinging to rock faces.

SPECIAL HABITAT REQUIREMENTS. Flowering plants for nectar.

Further Reading. United States and Canada: Farrand 1983b, Grinnell and Miller 1944, Johnsgard 1983, Terres 1980. Mexico: Edwards 1972, Peterson and Chalif 1973.

Calliope Hummingbird
Stellula calliope

Canada (Quebec): Colibri de Calliope
Mexico: Colibrí Gorjirrayado

RANGE. Breeds in the mountains from central interior British Columbia and southwestern Alberta south to Baja California, and east to northern Wyoming and western Colorado. Winters in western Mexico south to Guerrero.

STATUS. Uncommon.

HABITAT. Frequents meadows and canyons; riparian aspen, willow, and alder thickets; and meadows, burns, and brushy areas within the coniferous forests of western mountains. Occupies a broad vertical range during the breeding season, from 180 m in the northern portions of its range to 3,500 m in the Sierra Nevada. Prefers timber stands with a low to intermediate canopy cover near water. Typically locates nest below a larger branch or canopy of foliage, usually on a branch that has small knots of dead, black or gray mistletoe or pine cones, which nest strongly resembles. Locates nest 1 to 20 m above ground, frequently in a riparian area. May use the same site in subsequent years.

SPECIAL HABITAT REQUIREMENTS. Flowers, preferably red, for nectar.

Further Reading. United States and Canada: Bent 1940b, Calder 1971, Farrand 1983b, Godfrey 1966, Johnsgard 1983, Tamm 1989, Terres 1980, Verner and Boss 1980. Mexico: Edwards 1972, Peterson and Chalif 1973.

Broad-tailed Hummingbird
Selasphorus platycercus

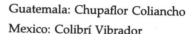

Guatemala: Chupaflor Coliancho
Mexico: Colibrí Vibrador

RANGE. Breeds in the mountains from north-central Idaho, northern Utah, and northern Wyoming south to southeastern California, northeastern Sonora, the state of México, western Texas, and in eastern Chiapas and Guatemala. Winters in Mexico and western Guatemala.

STATUS. Common.

HABITAT. Inhabits meadows, brushy slopes, riparian and montane thickets, and patches of flowers within pine, spruce, fir, and sometimes aspen forests from 1,200 to 3,300 m. Also inhabits gardens in towns and cities, and sometimes ranges eastward onto the Great Plains. Nests are saddled on large horizontal limbs or small twigs in shrubs along moist canyon walls, in Douglas-fir, ponderosa pine, subalpine fir, or other conifer, oak, aspen, alder, willow, or cottonwood. Usually locates nest 1 to 5 m above the ground, and frequently uses the same location for several consecutive summers.

SPECIAL HABITAT REQUIREMENTS. Flowers for nectar.

Further Reading. United States and Canada: Bailey and Niedrach 1965, Bent 1940b, Calder et al. 1983, Johnsgard 1979, 1983, Phillips et al. 1964, Terres 1980, Waser 1976. Mexico: Edwards 1972, Land 1970, Peterson and Chalif 1973.

Rufous Hummingbird

Selasphorus rufus

Canada (Quebec): Colibri Roux

Mexico: Colibrí Colicanelo Rufo

RANGE. Breeds from southern Alaska, southern Yukon, and western and southern British Columbia to western Montana south, primarily in the mountains, to northwestern California, eastern Oregon, and central Idaho. Winters in Mexico south to Guerrero, México, and Veracruz, in small numbers to southern Texas and the Gulf Coast and, rarely, in coastal southern California.

STATUS. Common, but declining.

HABITAT. Occurs in chaparral in meadows, forest edges, and riparian thickets of coniferous woodlands and in northwestern parks and gardens. More abundant in old (>250 yr old) than in young (40–75 yr) Douglas-fir stands

294

in Washington (Manuwal and Huff 1987). During migration, may be found in high mountain meadows as well as in the Pacific lowlands in open areas where flowers are present. In northern latitudes, builds nest close to the ground where it is sheltered from wind and cold; otherwise, builds nest 2 to 15 m above ground. May nest in a variety of sites, sometimes in colonies with up to 20 nests in a small area. Favors the drooping branches of conifers, but also nests in bushes and among vines. Sometimes builds a new nest on top of the previous year's nest.

SPECIAL HABITAT REQUIREMENTS. Flowers (especially red) for nectar.

Further Reading. United States and Canada: Bent 1940b, Calder and Jones 1989, Farrand 1983b, Gass 1979, Godfrey 1966, Johnsgard 1983, Terres 1980. Mexico: Edwards 1972, Peterson and Chalif 1973.

Allen's Hummingbird
Selasphorus sasin

Mexico: Colibrí Colicanelo Sasin

RANGE. Breeds from southwestern Oregon south through coastal California to Santa Barbara County. Resident in southern California in the Channel Islands and on the Palos Verdes Peninsula. Winters mainly in western Mexico.

STATUS. Common, but declining recently.

HABITAT. Found within the Pacific coastal fog belt, inhabiting meadows, moist canyon bottoms, humid woody or brushy ravines, brushy edges of coniferous forest, coastal chaparral, and parks. Usually builds nest on a site with several separate supports, such as a dense tangle of vines. Less frequently, attaches nest to the side of a drooping twig or limb from 1 to 25 m above the ground in trees such as oaks, eucalyptus, and Monterey cypress, or in shrubs in streamside thickets.

SPECIAL HABITAT REQUIREMENTS. Shade, preferably patchy, over the nest site, and flowers for nectar.

Further Reading. United States and Canada: Aldrich 1945, Bent 1940b, Harrison 1979, Terres 1980. Mexico: Edwards 1972, Peterson and Chalif 1973.

Elegant Trogon
Trogon elegans

Costa Rica and Guatemala: Trogón Elegante

Mexico: Trogón Colicobrizo

RANGE. Resident from southern Arizona, primarily in the Chiricahua, Huachuca and Atascosa Mountains, south to Costa Rica. Mostly migratory in northernmost part of range; casual in Arizona, southwestern New Mexico, and southern Texas in winter.

STATUS. Locally fairly common.

HABITAT. Occurs in oak and pine-oak forests in mountain canyons, and in sycamore, walnut, and cottonwood groves along canyon streams. Nests in cavities of large streamside trees such as sycamores or cottonwoods, 4 to 12 m above the ground.

SPECIAL HABITAT REQUIREMENTS. Natural cavities in trees or large, deserted woodpecker holes.

Further Reading. United States and Canada: Bent 1940a, Cottam and Knappen 1939, Farrand 1983b, Oberholser 1974a, Terres 1980. Mexico and Central America: Edwards 1972, Land 1970, Ridgely and Gwynne 1989.

Belted Kingfisher
Ceryle alcyon

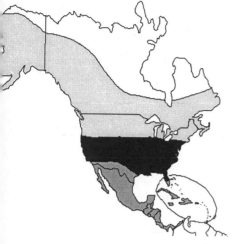

Canada (Quebec): Martin-pêcheur

Costa Rica and Mexico: Martín Pescador Norteño

Guatemala: Martín Pescador, Pescador Norteño

Puerto Rico: Martín Pescador, Pájaro del Rey

Venezuela: Martín Pescador Migratorio

West Indies: Martin-Pecheur, Martín Pescador, Martín Zambullidor, Pájaro del Ray, Pie, Pitirre de Agua, Pitirre de Mangle, Pitirre de Río

RANGE. Breeds from western and central Alaska, central Yukon, and western and south-central Northwest Territories to central Quebec and east-central Labrador south to southern California, southern Texas, the Gulf Coast, and central Florida. Winters from south-coastal and southeastern Alaska, central and southern British Columbia, and western Montana across to Nebraska, the southern Great Lakes, and New England south throughout Central America. Occasional in winter as far south as Guyana, coastal Venezuela, and Colombia. Found throughout the year in the West Indies with some northern birds augmenting local populations.

STATUS. Common; declining in eastern North America.

HABITAT. Occurs in the vicinity of ponds, lakes, rivers, and streams, even rocky seacoasts near areas of exposed vertical ground such as bluffs, road cuts, gravel pits, or sandbanks. Prefers small, clear bodies of water rather than large lakes. Typically excavates a nest burrow 1 to 2 m, but up to 3 m, in a bank with sandy, gravelly, or clay soil with entrance 2 m above level ground or water, and usually within 1 m of the top of the bank. Builds a nest cavity that is an enlarged area at the end of the burrow, often lined with disgorged food pellets. In winter, frequents ice-free waters that allow access to food.

SPECIAL HABITAT REQUIREMENTS. Nests preferably within 1.5 km of water with low turbidity supporting adequate aquatic animal populations, and perches near water to sight prey.

Further Reading. United States and Canada: Brooks and Davis 1987, Cornwall 1963, DeGraaf and Rudis 1986, Godfrey 1966, Johnsgard 1979, Terres 1980, White 1953. Mexico and Central America: Edwards 1972, Land 1970, Ridgely and Gwynne 1989, Stiles and Skutch 1989. South America: Hilty and Brown 1986, Meyer de Schauensee 1966, Meyer de Schauensee and Phelps 1978. Caribbean: Bond 1947, Raffaele 1989.

Yellow-bellied Sapsucker

Sphyrapicus varius
(includes Red-naped Sapsucker,
Sphyrapicus nuchalis)

Canada (Quebec): Pic Maculé

Costa Rica: Carpintero Bebedor

Guatemala: Carpintero Bebedor,
  Picamadero Bebedor

Mexico: Carpintero Aliblanco Común

West Indies: Carpintero de Paso,
  Charpentier

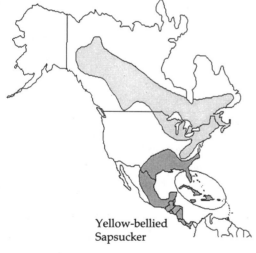

Yellow-bellied
Sapsucker

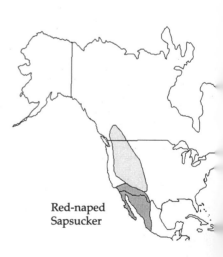

Red-naped
Sapsucker

RANGE. Breeds from eastern Alaska east to central Newfoundland, south to northeastern British Columbia, eastern North Dakota, New Hampshire, and locally in the Appalachians south to eastern Tennessee and western North Carolina; also in the Rocky Mountain region from south-central British Columbia to western Montana south, east of the Cascades, to east-central California and western Texas. Winters from Missouri, the Ohio Valley, and New Jersey south through Texas and the southeastern United States to central Panama; also from southern California, central Arizona, and central New Mexico south to southern Baja California and Jalisco. A transient and winter visitor to the highlands of Guatemala. A transient or winter visitor to Belize, El Salvador, Honduras, and Nicaragua. A very uncommon migrant and winter visitor chiefly in the highlands of Costa Rica, where it is the only migratory woodpecker. A rare and irregular winter visitor to western and central Panama. Not uncommon in most of the West Indies, but rare east of Hispaniola and unknown in the Lesser Antilles south of Dominica.

STATUS. Common.

HABITAT. Breeds in deciduous and mixed deciduous-coniferous forests in the eastern and northern parts of its range, especially in woodlands with aspens. In the Rocky Mountain region, occurs primarily in aspen forests or in coniferous forests where aspen is present. Uses a variety of forests and open woodlands, parks, and orchards in winter. Red-naped Sapsucker occurs in diameter-limit cut stands in Idaho Douglas-fir stands, not in uncut stands (Medin 1985). In western montane riparian habitats, the Red-naped Sapsucker is the most abundant woodpecker and is a key provider of nest sites for secondary cavity nesters (Dobkin and Wilcox 1986). In the central Rocky Mountains, cutting of aspen and subsequent invasion by conifers may reduce sapsucker populations (Finch and Reynolds 1988). In the northern Rockies, retention of deciduous trees in small logging units maintains relative abundance and fledging success (Tobalske 1992).

SPECIAL HABITAT REQUIREMENTS. Dead or live trees with a central decay column. Excavates cavities in snags or in live trees with decay column. Prefers aspens but will nest in ponderosa pine, birch, elm, butternut, cottonwood, alder, willow, beech, maple, and fir. May use the same nest tree for several years, but excavates a new cavity each year.

Further Reading. United States and Canada: Beal 1911, Crockett and Hadow 1975, DeGraaf and Rudis 1986, Godfrey 1966, Howell 1952, Johnsgard 1979, Kilham 1971, Lawrence 1967, Mannan and Meslow 1984, Tate 1973, Thomas et al. 1979. Mexico and Central America: Edwards 1972, Land

1970, Peterson and Chalif 1973, Ridgely and Gwynne 1989, Stiles and Skutch 1989. Caribbean: Bond 1947, Raffaele 1989.

Williamson's Sapsucker
Sphyrapicus thyroideus

Canada (Quebec): Pic de Williamson

Mexico: Carpintero Aliblanco Oscuro

RANGE. Breeds from extreme southern interior British Columbia, Idaho, western Montana, and Wyoming south in the mountains to northern and east-central California, central Arizona, and southern New Mexico. Winters generally from the breeding range south to Baja California, and east to western Texas and western Mexico.

STATUS. Common.

HABITAT. Prefers mixed conifer-hardwood forests in the Rocky Mountain region but also inhabits the subalpine spruce-fir-lodgepole zone, and ponderosa pine, Douglas-fir, and aspen forests. Generally associated with old-growth forests (Mannan and Meslow 1984) and not found in small fragments (Aney 1984).

SPECIAL HABITAT REQUIREMENTS. Dead or live trees infected with *Fomes*, a decay fungus, for cavity nest sites. Nests in different tree species in different regions. In some areas, nests primarily in conifers; in others, prefers aspen, especially those infected with *Fomes*. In Colorado and Arizona, mostly nests in aspen snags or live aspens infected with *Fomes*.

Further Reading. United States and Canada: Bailey and Niedrach 1965, Beal 1911, Bent 1939, Burleigh 1972, Crockett and Hadow 1975, Godfrey 1966, Hubbard 1965, Ligon 1961, Oliver 1970, Packard 1945, Rasmussen 1941, Tatschl 1967. Mexico: Edwards 1972, Peterson and Chalif 1973.

Northern Beardless-Tyrannulet

Camptostoma imberbe
(formerly Beardless Flycatcher)

Costa Rica: Mosquerito Lampiño
Guatemala: Mosquerito Lampiño, Mosquerito Silbador
Mexico: Mosquerito Silbador

RANGE. Breeds from southeastern Arizona, extreme southwestern New Mexico in Guadalupe Canyon, and Kenedy County in Texas south to Costa Rica. Winters in Mexico and Central America, casually northward to southern Arizona. A rare resident found in the lowlands and arid interior of Guatemala. Also resident in Belize, El Salvador, Honduras, Nicaragua, and Costa Rica. An uncommon to fairly common inhabitant of open situations with scrubby growth in either arid or humid regions of Honduras. Northern individuals are migratory and withdraw from the United States to winter from northern Mexico south to Costa Rica.

STATUS. Fairly common to rare.

HABITAT. In Arizona occurs in cottonwoods, dense mesquite thickets, and sycamore–live oak–mesquite associations. Along the lower Rio Grande Valley in Texas inhabits mesquite woodlands, cottonwoods, willows, elms, and great luecaenas. Typically nests far out on a horizontal limb of a bush or tree up to 15 m high, but usually near the ground. Usually nests along the edge of a grove, or among scattered trees in flat, sandy lowlands. Locates nest in a clump of mistletoe or sometimes between the bases of stems of palmetto.

SPECIAL HABITAT REQUIREMENTS. Depending on location: arid scrub thicket, mesquite, open riparian woodland.

Further Reading. United States and Canada: Bent 1942, Harrison 1979, Oberholser 1974a, Phillips et al. 1964, Terres 1980. Mexico and Central America: Edwards 1972, Land 1970, Monroe 1968, Peterson and Chalif 1973, Ridgely and Gwynne 1989, Stiles and Skutch 1989.

Olive-sided Flycatcher
Contopus borealis

Canada (Quebec): Moucherolle à Côtés Olives

Costa Rica: Pibí Boreal

Guatemala: Mosquero Boreal

Mexico: Contopus de Chaleco

Venezuela: Atrapamoscas Boreal

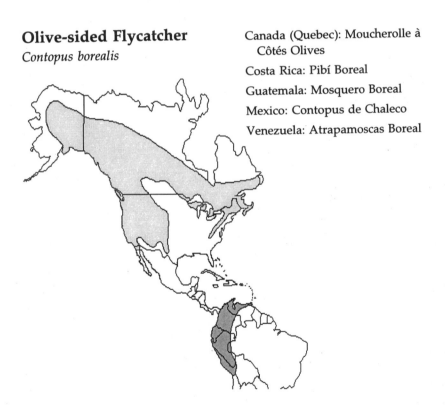

RANGE. Breeds from western and central Alaska and central Yukon to northern Ontario, south-central Quebec, and southern Labrador south to southern California across to western Texas, and east of the Rocky Mountains, to central Saskatchewan, northern Wisconsin, northeastern Ohio, and Massachusetts; also locally in the Appalachians to western North Carolina. Winters from Costa Rica to South America and, casually, in southern California. Breeds in northern Baja California and migrates throughout Mexico and Middle America. Some individuals may overwinter in southern Mexico. A transient and possible winter visitor in Guatemala. An uncommon migrant through Honduras. An uncommon to fairly common fall and spring migrant in Costa Rica, often abundant on the Caribbean coast. A rare winter resident in Costa Rica's foothills and mountains on both slopes. An uncommon migrant throughout Panama except in very open areas. A fairly common migrant and winter resident in Colombia. Winters chiefly in mountains of northern and western South America from Venezuela south to southeastern Peru and northern Bolivia. In Trinidad presumably a migrant from North America and can be regarded as an uncommon visitor to the Arima Valley.

STATUS. Local to fairly common; declining in eastern and western North America.

HABITAT. Inhabits montane and northern coniferous forests up to 3,000 m in elevation, especially in burned-over areas with tall standing dead trees. Prefers forests of tall spruces, firs, balsams, and pines; groves of eucalyptus and Monterey cypress; taiga; subalpine coniferous forests; mixed woodlands near edges and clearings; and wooded streams and borders of northern bogs and muskegs. Prefers stands with a low percentage of canopy cover, and generally responds favorably to opening of coniferous forests. Usually hides nest in a cluster of needles and twigs on a horizontal branch of a conifer, well away from the trunk, usually 5 to 15 m above the ground.

SPECIAL HABITAT REQUIREMENTS. Tall, exposed perches such as snags or high, conspicuous dead branches.

Further Reading. United States and Canada: Beal 1912, Bent 1942, DeGraaf and Rudis 1986, Finch and Reynolds 1988, Forbush and May 1955, Godfrey 1966, Johnsgard 1979, Medin 1985, Medin and Booth 1989, Peterson and Fichtel 1992, Terres 1980. Mexico and Central America: Edwards 1972, Land 1970, Monroe 1968, Peterson and Chalif 1973, Ridgely and Gwynne 1989, Stiles and Skutch 1989. South America: Hilty and Brown 1986, Meyer de Schauensee 1966, Meyer de Schauensee and Phelps 1978. Caribbean: Herklots 1961.

Greater Pewee

Contopus pertinax
(formerly Coues' Flycatcher)

Guatemala: Mosquero Tengofrío, Pibí Tengofrío

Mexico: Contopus José María

RANGE. Breeds from central Arizona and southwestern New Mexico south to Central America. Winters in Mexico and Central America, casually north to southern Arizona. Apparently is not migratory in Honduras—the race resident in Honduras and Nicaragua, *Contopous pertinax minor*, is markedly smaller than the nominate race found from southwestern United States to Guatemala.

STATUS. Fairly common.

HABITAT. Occurs in mountains up to 3,000 m near the U.S.-Mexico border, where it inhabits pine and pine-oak forests with an undergrowth of bushes. Also occurs in sycamore groves along streams in mountain canyons. Locates nest on a horizontal fork 3 to 12 m above the ground in a pine, sycamore, spruce, maple, oak, or other tall tree. Vigorously defends nest against jays, hawks, squirrels, and snakes.

SPECIAL HABITAT REQUIREMENTS. Tall trees for feeding perches and nesting.

Further Reading. United States and Canada: Bent 1942, Godfrey 1966, Harrison 1979, Phillips et al. 1964, Terres 1980. Mexico and Central America: Edwards 1972, Land 1970, Monroe 1968, Ridgely and Gwynne 1989. South America: Hilty and Brown 1986, Meyer de Schauensee 1966.

Western Wood-Pewee
Contopus sordidulus

Canada (Quebec): Pioui de l'Ouest

Costa Rica: Pibí Occidental

Guatemala: Mosquero Occidental, Pibí
Occidental

Mexico: Contopus Occidental

RANGE. Breeds from east-central Alaska, southern Yukon, and southern Northwest Territories to northwestern Minnesota, south to southern Baja California and throuth the highlands of Mexico and Guatemala to Honduras, and east to western South Dakota, western Kansas, and western Texas. Breeding reports from Costa Rica and Panama not verified. Winters from Panama to Peru. A summer resident and transient in Guatemala. All subspecies of the Western Wood-Pewee migrate through Guatemala. An uncommon summer resident and fairly common migrant in the highlands, uncommon in the Pacific lowlands, and rare in the Caribbean lowlands of Honduras. A common migrant but rare winter resident on both slopes of Costa Rica. A transient and occasional winter resident in Panama. Status is uncertain in Colombia because of identification problems, but likely a common transient and winter resident.

STATUS. Common, but long-term decline in western North America.

HABITAT. Occurs in a variety of habitats including open deciduous and coniferous montane forests, pine-oak woodlands, floodplain forests, and
wooded canyons. Found from sea level to the tops of coastal ranges, in cultivated stream valleys, in deciduous trees along borders of lakes and streams,
in cities and towns, and in open, mature pine forests. It is generally adapted
to drier environments than the Eastern Wood-Pewee, and uses areas dominated by conifers. Locates nest on a horizontal limb or fork, dead or live, in
a large variety of trees, generally 5 to 12 m above the ground.

Further Reading. United States and Canada: Beal 1912, Beaver and Baldwin
1975, Godfrey 1966, Harrison 1979, Johnsgard 1979, Terres 1980, Verbeek
1975a, 1975b. Mexico and Central America: Edwards 1972, Land 1970, Monroe 1968, Peterson and Chalif 1973, Ridgely and Gwynne 1989, Stiles and
Skutch 1989. South America: Hilty and Brown 1986.

Eastern Wood-Pewee
Contopus virens

Canada (Quebec): Pioui de l'Est

Costa Rica: Pibí Oriental

Guatemala: Mosquero Norteño, Pibí
Norteño

Mexico: Contopus Verdoso

Venezuela: Atrapamoscas de la Selva

RANGE. Breeds from southeastern Saskatchewan to southern Quebec and New Brunswick, south to Texas, the Gulf Coast, and central Florida, and west to the eastern Dakotas, central Oklahoma, and south-central Texas. Migrates through Mexico and Guatemala. Winters from Nicaragua to South America, mainly from Venezuela to Peru. A common migrant in the Caribbean lowlands and in the islands off the north coast of Honduras. No records for the Pacific Coast. An abundant migrant in Costa Rica during both fall and spring, and a rare winter resident mainly below 1,200 m. A common transient throughout Panama, with a few individuals possibly overwintering. Probably a common transient and winter visitor in Colombia, but sight records are usually uncertain. Migrates through the western islands of the Caribbean.

STATUS. Common. Showing long-term (since 1966) and recent (since early 1980s) declines, especially in the eastern United States and Canada.

HABITAT. Generally associated with deciduous forests; prefers woodlands with a relatively open understory but will use areas with a dense understory if the canopy above is incomplete or sparse. Also inhabits mixed forests, bottomlands, uplands, woodlots, orchards, parks, roadsides, and suburban areas planted to trees. Occurs in floodplain and river-bluff forests at the western edge of its range. Appears to be strongly associated with oaks, and throughout its range probably requires a predominance of hardwoods. Locates nest on a horizontal limb usually well out from the trunk, 3 to 30 m above the ground, often on a dead twig in a living tree. Camouflages nest with spiderwebs and lichens.

SPECIAL HABITAT REQUIREMENTS. Open deciduous or mixed forest or forest edge.

Further Reading. United States and Canada: Beal 1912, DeGraaf and Rudis 1986, Forbush and May 1955, Godfrey 1966, Johnsgard 1979. Mexico and Central America: Edwards 1972, Land 1970, Monroe 1968, Peterson and Chalif 1973, Ridgely and Gwynne 1989, Stiles and Skutch 1989. South America: Hilty and Brown 1986, Meyer de Schauensee 1966, Meyer de Schauensee and Phelps 1978. Caribbean: Bond 1947.

Yellow-bellied Flycatcher
Empidonax flaviventris

Canada (Quebec): Moucherolle à Ventre Jaune

Costa Rica: Mosquerito Vientriamarillo

Guatemala: Mosquerito Olivo, Tontín Olivo, Tontín Vientriamarillo

Mexico: Empidonax Vientriamarillo

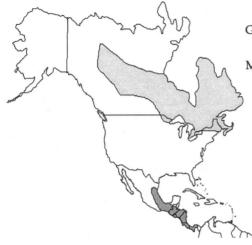

RANGE. Breeds from northern British Columbia and west-central and southern Northwest Territories to southern Labrador and Newfoundland south to central Alberta, northern North Dakota, and northern Minnesota, across to southern Ontario, northeastern Pennsylvania, and Nova Scotia. Winters from Mexico to Panama. Migrant and winter visitor to eastern portions of Mexico, except Yucatán Peninsula.

STATUS. Common.

HABITAT. Prefers predominantly coniferous forests of spruce and fir, frequently low, swampy thickets bordering ponds and streams, spruce, cedar, tamarack and sphagnum bogs, spruce and alder swamps, wet mossy glades, and cool moist mountainsides. Nests on or near the ground, sometimes at the base of a tree or in a cavity formed by upturned roots, but more often beside a hummock or mound and well-hidden in sphagnum moss or other vegetation. May also nest in a damp, mossy crevice of rocks, but always in a quiet, concealed site. In winter, an understory specialist in tall evergreen tropical forest (Tuxtla Mountains, Veracruz), only occasionally found in 10- to 15-m-tall second growth, bordering streams (Rappole et al. 1992, Rappole and Warner 1980). Reported in dense "second growth" in Panama (Wetmore 1972:464) and Costa Rica (Slud 1964:254).

SPECIAL HABITAT REQUIREMENTS. Low, wet areas within coniferous forest.

Further Reading. United States and Canada: Beal 1912, Bent 1942, DeGraaf and Rudis 1986, Forbush and May 1955, Godfrey 1966, Terres 1980. Mexico and Central America: Edwards 1972, Land 1970, Monroe 1968, Peterson and Chalif 1973, Ridgely and Gwynne 1989, Stiles and Skutch 1989.

Acadian Flycatcher
Empidonax virescens

Canada (Quebec): Moucherolle Vert

Costa Rica: Mosquerito Verdoso

Guatemala: Mosquerito Verdoso, Tontín Verdoso

Mexico: Empidonax Verdoso

Venezuela: Atrapamoscas Copete Verde

RANGE. Breeds from southeastern South Dakota, northern Iowa, and extreme southeastern Minnesota to southern New York, Vermont, and Massachusetts, south to central and southern Texas, the Gulf Coast, and central Florida. Winters in Central and South America. Migrates through Middle America to winter in northwestern Venezuela, Colombia east of the Andes, and Ecuador to El Oro. A rare transient in the Bahamas, western Cuba, and Isle of Pines.

STATUS. Common, but long-term decline in the central United States.

HABITAT. Inhabits the lowest tree canopy and understory layers of shady, humid riverbottom forests and wooded swamps. Prefers damp, lowland forests with an understory and uplands with wooded ravines near streams. Favors beech and hemlock forests in the Northeast. Nests on a fork of a horizontal branch well away from the main trunk, usually 3 to 6 m above the ground, often along a stream and sometimes over water. Prefers open space below the nest to approach the nest easily. Favors lower branches of beech, dogwood, and witch-hazel, but also nests in oak, hickory, maple, basswood, and cherry. Occasionally is parasitized by Brown-headed Cowbirds. In winter, occurs only in forest; considered highly vulnerable to deforestation (Powell et al. 1992).

SPECIAL HABITAT REQUIREMENTS. Mature, extensive deciduous forests with tall trees, a closed canopy, and open spaces in understory for feeding.

Further Reading. United States and Canada: Beal 1912, Bent 1942, DeGraaf and Rudis 1986, Forbush and May 1955, Godfrey 1966, Mumford 1964. Mexico and Central America: Edwards 1972, Land 1970, Rappole et al. 1983, Ridgely and Gwynne 1989, Stiles and Skutch 1989. South America: Hilty and Brown 1986, Meyer de Schauensee 1966, Meyer de Schauensee and Phelps 1978.

Alder Flycatcher
Empidonax alnorum

Canada (Quebec): Moucherolle des Aulnes

Costa Rica: Mosquerito de Charral

Mexico: Empidonax Alnoro

RANGE. Breeds from central Alaska and central Yukon to central and eastern Quebec, southern Labrador and southern Newfoundland, south to south-central British Columbia and southern Alberta, across to south-central Minnesota, eastern Pennsylvania, and Connecticut; also in the Appalachians south to western North Carolina. Migrates through Central America. Winters from Colombia and Venezuela south to northern Argentina.

STATUS. Common.

HABITAT. Typically inhabits northern alder swamps, where it occupies a variety of habitats, including thickets of aspen parkland. Usually found near water in dense, low, damp thickets of alders, willows, sumacs, viburnum, elderberries, and red-osier dogwood bordering bogs, swamps, marshes, and along the banks of small streams and shores of ponds. Nests in low trees or shrubs including dogwood, blackberry, hawthorn, viburnum, willow, spiraea, or alder, up to 2 m above the ground, in an upright fork or saddled on a branch.

SPECIAL HABITAT REQUIREMENTS. Forest openings and edges with dense, low shrubs.

Further Reading. United States and Canada: Bent 1942, DeGraaf and Rudis 1986, Farrand 1983b, Mousley 1931, Stein 1958, Terres 1980. Mexico and Central America: Edwards 1972, Ridgely and Gwynne 1989, Stiles and Skutch 1989. South America: Hilty and Brown 1986.

Willow Flycatcher
Empidonax traillii
(formerly Traill's Flycatcher)

Canada (Quebec): Moucherolle des Aulnes

Costa Rica: Mosquerito de Traill

Mexico: Empidonax de Traill

RANGE. Breeds from central British Columbia and southern Alberta east to southern Wisconsin, southern Quebec, central Maine, and Nova Scotia south to southern California, western and central Texas, Arkansas, northern Georgia, and central and eastern Virginia. Winters from Mexico to northwestern Colombia.

STATUS. Common; is generally stable or increasing throughout much of its range, especially in the central United States.

HABITAT. Occurs in a variety of habitats ranging from brushy fields to willows, thickets along streams, prairie woodlots and shelterbelts, shrubby swales, and open woodland edges. Prefers edge habitats that include thickets or groves of small trees and shrubs surrounded by grasslands, savanna-like shrub-willow habitats (in the western United States), as well as the edges of gallery forests along rivers or streams. In areas where its range overlaps that of the Alder Flycatcher, prefers drier, smaller, more open shrubby habitat. In the West, riparian breeding habitat is degraded by livestock grazing and willow control (Taylor and Littlefield 1986). Nests in horizontal forks or upright crotches of shrubs or small trees, 1 to 5 m above the ground, normally about 1 to 2 m. Commonly nests in dogwood, hawthorn, willow, buttonbush, elder, viburnum, and blackberry. Places nest at the outer edge of a shrub or thicket. Commonly parasitized by cowbirds.

SPECIAL HABITAT REQUIREMENTS. Fairly open areas with scattered shrubs.

Further Reading. United States and Canada: Bent 1942, DeGraaf and Rudis 1986, Farrand 1983b, Godfrey 1966, Frakes and Johnson 1982, Holcomb 1972, Johnsgard 1979, King 1955, Sanders and Flett 1989, Sedgewick and Knopf 1989, Stein 1958, Tate and Tate 1982, Walkinshaw 1966. Mexico and Central America: Edwards 1972, Ridgely and Gwynne 1989, Stiles and Skutch 1989. South America: Hilty and Brown 1986, Meyer de Schauensee 1966.

White-throated Flycatcher
Empidonax albigularis

Costa Rica: Mosquerito Gargantiblanco

Guatemala: Mosquerito Gargantiblanco, Tontín Gargantiblanco

Mexico: Empidonax Gorjiblanco

RANGE. Mexico to western Panama. Accidental in the United States.

STATUS. Common.

HABITAT. Mainly highlands, 1,000 to 3,500 m, with brush, semi-open growth, pastures, wet meadows, fields or marshes with scattered shrubs and small trees. The nest is a neat cup of dry grass blades and plant fibers 1 to 2 m high in a shrub.

SPECIAL HABITAT REQUIREMENTS. Brushy or grassy fields, scrub, second growth.

Further Reading. Mexico and Central America: Edwards 1972, Land 1970, Peterson and Chalif 1973, Ridgely and Gwynne 1989, Stiles and Skutch 1989.

Least Flycatcher
Empidonax minimus

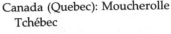

Canada (Quebec): Moucherolle Tchébec

Costa Rica: Mosquerito Chebec

Mexico: Empidonax Mínimo

Guatemala: Mosquerito Chebec, Tontín Chebec

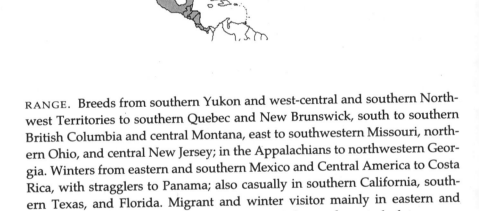

RANGE. Breeds from southern Yukon and west-central and southern Northwest Territories to southern Quebec and New Brunswick, south to southern British Columbia and central Montana, east to southwestern Missouri, northern Ohio, and central New Jersey; in the Appalachians to northwestern Georgia. Winters from eastern and southern Mexico and Central America to Costa Rica, with stragglers to Panama; also casually in southern California, southern Texas, and Florida. Migrant and winter visitor mainly in eastern and southern Mexico, the Yucatán Peninsula, and the south-central plateau.

STATUS. Common; declining in eastern North America, increasing in the western part of range.

HABITAT. Associated with open deciduous forests, where it occurs along forest edges and openings (Breckenridge 1956), burns, and clearings, floodplain forests, open shrublands, wooded margins of lakes and roads, orchards, shelterbelts, overgrown pastures, urban parks, and gardens. In managed hardwood forests, common in shelterwood stands and open stands with understory or midstory layers. Large-scale, intensive logging lowers breeding density (in the Maritime Provinces) (Freedman et al. 1981) and causes this species to retreat into suitable openings in the forest interior

314

(DellaSala and Rabe 1987). Nests in upright crotch or on horizontal fork of deciduous or coniferous trees, usually saplings or small trees, including birch, red pine, cedar, apple, dogwood, oak, sugar maple, willow, and alder. Tends to nest 3 to 5 m above the ground at the edge of a clearing. A habitat generalist on wintering grounds on the Yucatán Peninsula. A scrub-thicket, forest-opening, or forest-edge bird seldom found in closed forest in Tuxtla Mountains, Veracruz (Rappole and Warner 1980). In western Mexico, probably restricted to undisturbed tropical deciduous forest and therefore likely vulnerable there to deforestation in the lowlands (Hutto 1992).

SPECIAL HABITAT REQUIREMENTS. Intermediate openness in the understory of open deciduous woodlands, and some edge habitat for nesting and feeding.

Further Reading. United States and Canada: Beal 1912, Briskie and Sealy 1989, Darveau et al. 1992, DeGraaf and Rudis 1986, DeKirline 1948, Forbush and May 1955, Godfrey 1966, Johnsgard 1979, Sherry and Holmes 1988, Tate and Tate 1982. Mexico and Central America: Edwards 1972, Land 1970, Lynch 1989, Peterson and Chalif 1973, Ridgely and Gwynne 1989, Stiles and Skutch 1989.

Hammond's Flycatcher
Empidonax hammondii

Canada (Quebec): Moucherolle de Hammond

Guatemala: Mosquerito de Hammond, Tontín de Hammond

Mexico: Empidonax de Hammond

RANGE. Breeds from east-central Alaska, southern Yukon, and southwestern Alberta to northwestern Wyoming, south to east-central California, central Utah, northeastern Arizona, and north-central New Mexico. Winters in southeastern Arizona, Mexico, and Central America, casually in southern California.

STATUS. Common.

HABITAT. Inhabits tall, mature, moist, closed-canopy montane conifer forests, sometimes with a deciduous understory. In Colorado, occurs from 2,500 to 3,500 feet in conifer-aspen woodlands; in California, in mature forests of medium to high canopy closure from ponderosa pine up to lodgepole pine forests. In the far north, prefers deciduous forests. In the western United States, is a nesting-habitat specialist, consistently more abundant in intact stands of old-growth or mature (>100 yr) conifer forest (Hejl and Woods 1991, Mannan and Meslow 1984, Sakai and Noon 1991). Nests in a fork of a small tree or on a horizontal branch of a large conifer or deciduous tree, usually 8 to 12 m above the ground. Uses birch, maple, ponderosa pine, western larch, and Douglas-fir for nesting.

SPECIAL HABITAT REQUIREMENTS. Nest sites that are cool and well shaded.

Further Reading. United States and Canada: Beaver and Baldwin 1975, Davis 1954, Farrand 1983b, Godfrey 1966, Manuwal 1970, Terres 1980, Verner and Boss 1980. Mexico and Central America: Edwards 1972, Land 1970, Rappole et al. 1983, Ridgely and Gwynne 1989.

Dusky Flycatcher
Empidonax oberholseri

Canada (Quebec): Moucherolle Sombre

Mexico: Empidonax de Oberholser

RANGE. Breeds from southwestern Yukon, southern Alberta, southwestern Saskatchewan, and western South Dakota, south to southern California, southern Nevada, central Arizona, and central and northeastern New Mexico. Winters in southern Arizona and in Mexico's mountains and plateaus south to Chiapas; casual in southern California.

STATUS. Common.

HABITAT. Prefers shrubby sites or open forests with substantial shrub understory; generally avoids closed-canopy forests. Occurs in a variety of habitats, from montane chaparral to open lodgepole pine forest, including many montane conifer types and aspen; especially favors mixed woodlands or edge of small conifers and brush. In general, prefers drier, more open or patchier forests than Hammond's Flycatcher. More abundant in managed than in old-growth stands in the Pacific Northwest (Hejl and Woods 1991, Mannan and Meslow 1984). Nests on upright or pendant twigs or in crotches of low shrubs or trees in relatively dry sites. Usually nests 1 to 2 m, but up to 12 m, above the ground in willow, alder, aspen, and other trees and shrubs.

SPECIAL HABITAT REQUIREMENTS. Scrub, thickets, and mountain chapparal, especially near water.

Further Reading. United States and Canada: Farrand 1983b, Godfrey 1966, Harrison 1979, Johnsgard 1979, Verner and Boss 1980. Mexico: Edwards 1972, Peterson and Chalif 1973.

Gray Flycatcher
Empidonax wrightii

Mexico: Empidonax de Wright

RANGE. Breeds from south-central Washington and south-central Idaho to central Colorado, south to south-central California, central Arizona, and south-central New Mexico. Winters in central Arizona and Mexico, rarely in southern California; also in Baja California and from Sonora, central Chihuahua, southern Coahuila, and Tamaulipas to south-central Mexico.

STATUS. Fairly common; has been increasing.

HABITAT. Associated with arid woodland and brushy areas, where it inhabits tall sagebrush plains, pinyon-juniper woodlands, and arid, very open pine woods. Nests in a crotch of a thornbush, juniper, or sagebrush, 1 to 2 m above the ground, sometimes in loose colonies. During migration and in winter it occurs in arid scrub and riparian and mesquite woodlands.

SPECIAL HABITAT REQUIREMENTS. Arid woodlands, sagebrush.

Further Reading. United States and Canada: Farrand 1983b, Phillips et al. 1964, Russell and Woodbury 1941, Terres 1980. Mexico: Edwards 1972, Peterson and Chalif 1973.

Pine Flycatcher
Empidonax affinis

Guatemala: Mosquerito Pino, Tontín
de los Pinos

Mexico: Empidonax Afín

RANGE. Central Mexico; winters to Guatemala.

STATUS. Moderately common.

HABITAT. Pine and pine-oak woodlands at high elevations; also humid montane forest.

Further Reading. Edwards 1972, Peterson and Chalif 1973.

Pacific-slope Flycatcher
(Empidonax difficilis)

Cordilleran Flycatcher
(Empidonax occidentalis)
(formerly considered a single
species, "Western" Flycatcher,
Empidonax difficilis)

Canada (Quebec): Moucherolle du
Pacifique

Guatemala: Mosquerito Occidental,
Tontń Occidental

Mexico: Empidonax Difícil (Pacific-
slope Flycatcher), Empidonax
Occidental (Cordilleran Flycatcher)

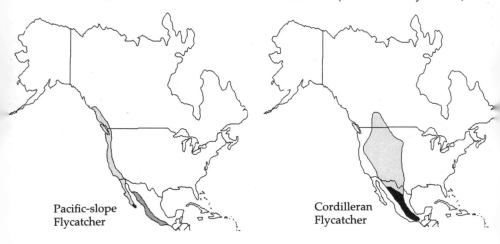

Pacific-slope
Flycatcher

Cordilleran
Flycatcher

RANGE. Pacific-slope Flycatcher: Breeds from extreme southeastern Alaska
and northwestern and central British Columbia (including the Queen Charlotte and Vancouver Islands) south to southwestern California (generally
west of the Cascades and Sierra Nevada) and mountains of northern and
southern Baja California. Winters in southern Baja California and from northwestern Mexico south to the Isthmus of Tehuantepec (Oaxaca). Cordilleran
Flycatcher: Breeds from southeastern Washington, southwestern Alberta,
northern Idaho, western Montana, Wyoming, and western South Dakota
south (generally east of the Cascades and Sierra Nevada, although both species are sympatric in the Siskiyous in northern California [Johnson and Martin 1988]) to northern California, Nevada, and central and southeastern
Arizona, and in the highlands of Mexico to Oaxaca west of the Isthmus of
Tehuantepec, Puebla, and west-central Veracruz, and east to central Colorado, central New Mexico, and western Texas. Winters in southern Baja California and from northern Mexico south through the breeding range; also in
lowlands south of the Isthmus of Tehuantepec.

STATUS. Both species: common.

HABITAT. Pacific-slope Flycatcher: Breeds in warm forests and moist wood-lands, either coastal or lower montane, especially near shaded cliffs, stream banks, and residential dwellings. More abundant in old growth (>250 yr) than in mature (105–165 yr) Douglas-fir stands in the Pacific Northwest (Manuwal and Huff 1987). Winters in mixed woodland and humid lowland forest. Cordilleron Flycatcher: Breeds in cool forest and woodland near cliffs and in shady canyon bottoms, especially along streams in montane coniferous forests. Winters in mixed woodlands and forests. Both species use a variety of sites for nesting, including rock ledges or crevices (often concealed by ferns or clumps of mosses, crotch or tree limb, behind loose bark flap, tree cavities, or old buildings. Nest height ranges from ground level up to 10 m.

SPECIAL HABITAT REQUIREMENTS. Both species: a sheltered nest site, possibly near a water source such as a stream, spring, or seep.

Further Reading. United States and Canada: Beal 1912, Beaver and Baldwin 1975, Davis et al. 1963, Farrand 1983b, Johnsgard 1979, Johnson and Marten 1988, Medin 1985, Verner and Boss 1980. Mexico and Central America: Edwards 1972, Land 1970, Rappole et al. 1983.

Buff-breasted Flycatcher
Empidonax fulvifrons

Guatemala: Mosquerito Leonado, Tontín Pechicanela

Mexico: Empidonax Canelo

RANGE. Breeds very locally from the Huachuca and Chiricahua Mountains of east-central and southeastern Arizona through the highlands of Mexico, Guatemala, and El Salvador to Honduras. Winters in Mexico, Guatemala, El Salvador, and Honduras.

STATUS. Rare and local.

HABITAT. Prefers open stands of pines and riparian trees, but also occurs in mixed pine and oak woods with shrubby undergrowth and on steep canyon slopes, from 1,600 to 2,600 m. Favors trees in open, bare, weedy, or grassy places. Places nest on a branch 3 to 15 m above the ground, often sheltered by an overhanging stub of a branch. Builds nest in pines, oaks, and sycamores.

SPECIAL HABITAT REQUIREMENTS. Forest openings.

Further Reading. United States and Canada: Bent 1942, Cottam and Knappen 1939, Farrand 1983b, Phillips et al. 1964, Terres 1980. Mexico and Central America: Edwards 1972, Land 1970, Monroe 1968, Peterson and Chalif 1973.

Eastern Phoebe
Sayornis phoebe

Canada (Quebec): Moucherolle Phébi
Mexico: Mosquero Fibí

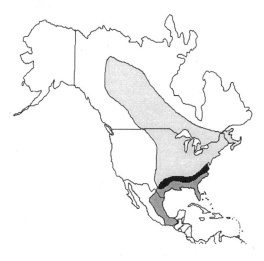

RANGE. Breeds from northeastern British Columbia and west-central and southern Northwest Territories to southwestern Quebec and central New Brunswick, south to southern Alberta, southwestern South Dakota, central New Mexico, and central and northeastern Texas across to northern Georgia and North Carolina. Winters from central Texas, the Gulf States, and Virginia south to Mexico (mainly central and eastern) south to Oaxaca and Veracruz and southern Florida; casually to Oklahoma, southern Ontario, and New England.

STATUS. Common.

HABITAT. Generally occurs near fresh running water in partially wooded habitats; frequents woodland edges, wooded ravines and cliffs, farms, and suburban areas where natural or artificial ledges are available for nesting. Nests on a ledge, usually sheltered above by an overhang, on natural or artificial structures, and generally near lakes or streams. May nest under bridges, culverts, or eaves of buildings, on cliffs, rock bluffs, or in ravines. Frequently uses nests from previous years, but is very adaptable in its nesting habits. Frequently the victim of cowbird parasitism.

SPECIAL HABITAT REQUIREMENTS. Cliffs or ledges at streamside clearings, buildings or other structures in forest openings for nesting. Perches 2 to 5 m high for feeding.

Further Reading. United States and Canada: Beal 1912, DeGraaf and Rudis 1986, Forbush and May 1955, Godfrey 1966, Hespenheide 1971, Johnsgard 1979, Tate and Tate 1982, Weeks 1978. Mexico: Edwards 1972, Peterson and Chalif 1973.

Say's Phoebe
Sayornis saya

Canada (Quebec): Moucherolle à Ventre Roux

Mexico: Mosquero Llanero

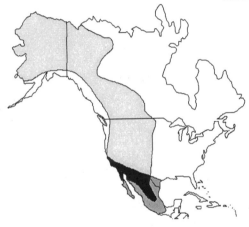

RANGE. Breeds from western and northern Alaska, northern Yukon, northwestern and central Northwest Territories, and central Alberta to southwestern Manitoba, south between coastal ranges and central prairie states to Mexico. Winters from northern California, Arizona, central New Mexico, central Texas, and Mexico south to southern Baja California and Chiapas.

STATUS. Common.

HABITAT. Inhabits open arid and semiarid regions, occurring in shrubsteppe, sagebrush plains, scrubby pine-oak-juniper woodlands, bluffs and cliffs of badlands, dry grasslands, agricultural areas, and open areas near buildings. Unlike the Eastern Phoebe, it is independent of surface water. Prefers to nest in holes, crevices, on ledges, and on other protected horizontal surfaces of cliffs, rimrocks, steep creek banks, and caverns. Frequently nests in abandoned mine shafts, buildings, and under bridges. Also uses old nests of Cliff Swallows, Barn Swallows, and Black Phoebes; often uses the same nest in subsequent years or for successive clutches.

SPECIAL HABITAT REQUIREMENTS. Arid scrub, desert, openings in other arid habitats.

Further Reading. United States and Canada: Beal 1912, Bent 1942, Godfrey 1966, Johnsgard 1979, Ohlendorf 1976, Terres 1980. Mexico: Edwards 1972, Peterson and Chalif 1973.

Vermilion Flycatcher
Pyrocephalus rubinus

Canada (Quebec): Moucherolle Vermillon

Guatemala: Mosquero Colorado, Mosquero Bermellón

Mexico: Mosquero Cardenalito

RANGE. Breeds from southern California, southern Nevada, central Arizona, central New Mexico, and western Oklahoma south to South America. Winters from southern California and southern Nevada to the Gulf Coast, east to south-central Florida, and throughout Mexico south to Nicaragua; also from northern Colombia east across northern Venezuela to Guyana south to northern Chile and central Argentina.

STATUS. Common; long-term decline in eastern part of its range.

HABITAT. In the United States, found in the arid Southwest, occurring almost exclusively near water. Favors wooded groves of cottonwood, willow, oak, mesquite, and sycamore bordering rivers, especially near open, brushy, grassy, or agricultural fields. Elsewhere, arid scrub, desert, savanna, cultivated lands, riparian edge. Also occurs in widely spaced junipers and oaks, and in dry washes on the plains. Builds nest on a small, horizontal forked branch, usually 2 to 8 m above ground, sometimes higher, and usually near a stream or other source of water. Nests in willow, sycamore, mesquite, cottonwood, oak, paloverde, hackberry, and other trees and bushes.

Further Reading. United States and Canada: Bent 1942, Farrand 1983b, Godfrey 1966, Harrison 1979, Johnsgard 1979, Taylor and Hanson 1970, Terres 1980. Mexico and Central America: Edwards 1972, Land 1970, Rappole et al. 1983.

Dusky-capped Flycatcher
Myiarchus tuberculifer
(formerly Olivaceous
Flycatcher)

Costa Rica: Copetón Crestioscuro

Guatemala: Copetón Común, Copetón
Crestioscuro, Mosquero Común

Mexico: Papamoscas Copetón Triste

Venezuela: Atrapamoscas Cresta
Negra

RANGE. Breeds in southeastern Arizona and southwestern New Mexico; also in Central and South America. Winters from Mexico to South America, south to northwestern Argentina.

STATUS. Fairly common.

HABITAT. Generally found below 2,000 m but does occur up to 2,500 m in montane pine-oak woodlands. Prefers dense scrub-oak thickets on hillsides but also occurs along canyon streams where trees grow thick enough to provide deep shade. Builds nest in natural cavities in trees and stumps or in old woodpecker holes, 1 to 15 m above the ground, in oaks, sycamores, or ashes.

SPECIAL HABITAT REQUIREMENTS. Natural cavities or old woodpecker holes in trees for nesting.

Further Reading. United States and Canada: Bent 1942, Cottam and Knappen 1939, Harrison 1979, Phillips et al. 1964, Terres 1980. Mexico and Central America: Edwards 1972, Land 1970, Monroe 1968, Peterson and Chalif 1973, Ridgely and Gwynne 1989, Stiles and Skutch 1989. South America: Hilty and Brown 1986, Meyer de Schauensee 1966, Meyer de Schauensee and Phelps 1978.

Ash-throated Flycatcher
Myiarchus cinerascens

Canada (Quebec): Moucherolle à
  Gorge Cendrée

Costa Rica: Copetón Garganticeniza

Guatemala: Copetón Cenizo,
  Mosquero de Garganta Ceniza

Mexico: Papamoscas Gorjicenizo

RANGE. Breeds from northwestern Oregon and eastern Washington to Colorado and western Kansas, south to Mexico. Winters from southern California and central Arizona south to Guatemala, El Salvador, Honduras, and casually to Costa Rica.

Ash-throated Flycatcher

STATUS. Common; increasing in western part of its breeding range.

HABITAT. Inhabits mesquite and cactus deserts, rocky mesas, shrubby canyons, oak groves on hillsides, mesquite thickets along creek bottoms, open pinyon-juniper woodlands, and open groves of sycamore, oak, willow, or cottonwood along stream courses. Occurs from sea level to 3,000 m in California, but is most frequently found at lower elevations. Uses a variety of cavities for nesting; natural cavities or knotholes in trees and stumps of mesquite, ash, oak, sycamore, juniper, or cottonwood, or old woodpecker holes. May also nest behind loose pieces of bark, in abandoned nests of Cactus Wrens, in cavities in saguaro, in artificial structures, or in stalks of yucca or agave. Usually nests less than 6 m above the ground.

SPECIAL HABITAT REQUIREMENTS. Natural tree cavities or old woodpecker holes for nesting.

Further Reading. United States and Canada: Beal 1912, Bent 1942, Godfrey 1966, Johnsgard 1979, Terres 1980. Mexico and Central America: Edwards 1972, Land 1970, Monroe 1968, Peterson and Chalif 1973, Stiles and Skutch 1989.

Great Crested Flycatcher
Myiarchus crinitus

Canada (Quebec): Moucherolle
Huppé

Costa Rica: Copetón Viajero

Guatemala: Copetón Viajero,
Mosquero Copetón

Mexico: Papamoscas Copetón Viajero

Venezuela: Atrapamoscas Copetón

RANGE. Breeds from east-central Alberta and central and southeastern Saskatchewan to southwestern Quebec and central New Brunswick, south to central and southeastern Texas, the Gulf Coast, and Florida, and west to the eastern Dakotas, western Kansas, and west-central Oklahoma. Winters in central and southern Florida and from Mexico to Colombia and northern Venezuela. Rare visitor in the Bahamas, Cuba, and Puerto Rico.

STATUS. Common.

HABITAT. Prefers fairly extensive hardwood forests but is commonly found in old orchards and woodlots in farming country, and in wooded residential areas. Prefers mature forests with fairly open canopies but will also use second-growth woodlands with scattered large cavity trees. Nests in natural cavities or woodpecker holes in live or dead trees, usually 3 to 6 m, sometimes to 20 m, above the ground. May also use artificial structures such as bird boxes and other hollows; little preference shown for the shape of the opening or the cavity size.

329

SPECIAL HABITAT REQUIREMENTS. Cavities in fairly large trees, preferably in deciduous forests.

Further Reading. United States and Canada: Beal 1912, DeGraaf and Rudis 1986, Forbush and May 1955, Godfrey 1966, Johnsgard 1979, Mousley 1934a. Mexico and Central America: Edwards 1972, Land 1970, Monroe 1968, Peterson and Chalif 1973, Ridgely and Gwynne 1989, Stiles and Skutch 1989. South America: Hilty and Brown 1986, Meyer de Schauensee 1966, Meyer de Schauensee and Phelps 1978. Caribbean: Bond 1947.

Brown-crested Flycatcher

Myiarchus tyrannulus
(formerly Wied's Crested
Flycatcher)

Costa Rica: Copetón Crestipardo

Guatemala: Copetón Colirrufo,
  Mosquero Copetón

Mexico: Papamoscas Copetón
  Tiranillo

RANGE. Breeds from southeastern California, extreme southern Nevada, southwestern Utah, Arizona, and southwestern New Mexico south through southern Mexico and Central America. Resident in South America. Winters in northern Mexico south through the breeding range; rarely in southern Florida. Northern individuals winter as far south as Costa Rica. Individuals breeding in northern Mexico are also migrants. Populations farther south are resident and are augmented by northern migrants.

STATUS. Fairly common.

HABITAT. Inhabits saguaro deserts, riparian deciduous woodlands, and shade trees in urban areas. In Texas, occurs in open woodlands of mesquite, hackberry, and ash; in Arizona, frequents cottonwood, willow, and sycamore woodlands. Builds nest in abandoned woodpecker holes in saguaro or in cavities in cottonwoods, sycamores, mesquite, or old fence posts 1 to 10 m above the ground. Sometimes nests in bird boxes.

SPECIAL HABITAT REQUIREMENTS. Natural tree cavities or abandoned woodpecker holes for nesting.

Further Reading. United States and Canada: Bent 1942, Gambona 1977, Oberholser 1974a, Phillips et al. 1964. Mexico and Central America: Edwards 1972, Land 1970, Monroe 1968, Peterson and Chalif 1973, Ridgely and Gwynne 1989, Stiles and Skutch 1989.

Sulphur-bellied Flycatcher

Myiodynastes luteiventris

Costa Rica: Mosquero Vientriazufrado

Guatemala: Mosquero Cejiblanco, Mosquero Vientriazufrado

Mexico: Papamoscas Rayado Cejiblanco

RANGE. Breeds from southeastern Arizona to Costa Rica. Migrates through Central America, Colombia, and Ecuador. Winters south to Peru and Bolivia.

STATUS. Fairly common.

HABITAT. Inhabits riparian mountain canyons, from 1,600 to 2,400 m in elevation, where sycamore, oak, walnut, Arizona cypress, and pine are common. Usually builds nest in a natural cavity, typically a knothole where a large branch has broken off and a cavity has rotted out, 6 to 16 m above the ground in living sycamores. Occasionally nests in an old flicker hole or a nest box placed high in a tree. Constructs nest on top of a loose platform built inside the cavity.

SPECIAL HABITAT REQUIREMENTS. Natural cavities in trees or abandoned Northern Flicker holes.

Further Reading. United States and Canada: Bent 1942, Cottam and Knappen 1939, Ligon 1971, Phillips et al. 1964. Mexico and Central America: Edwards 1972, Rappole et al. 1983, Stiles and Skutch 1989. South America: Rappole 1989.

Tropical Kingbird
Tyrannus melancholicus

Canada (Quebec): Tyran Mélancolique

Costa Rica: Pecho Amarillo, Tirano Tropical

Guatemala: Chatilla Tropical, Chituri Tropical, Pitirre Chicharrero

Mexico: Tirano Tropical Común

West Indies: Pipiri Jaune

RANGE. Breeds from southeastern Arizona, Sonora, and southern Tamaulipas south on both slopes of Middle America, and in South America from

Colombia, Venezuela, Trinidad and Tobago, and the Guianas south, west of the Andes to central Peru and east of the Andes to central Argentina. Winters from Mexico south to Argentina. Casual in fall and winter along the Pacific Coast to British Columbia, including Vancouver Island, south to southern California; also in southern Texas and Cuba.

STATUS. Uncommon and local.

HABITAT. Inhabits groves of tall trees, especially cottonwoods, next to ponds or flowing streams at low elevations. Frequently occurs with both the Western and Cassin's Kingbirds. Nesting habits similar to those of Western and Cassin's Kingbirds, with which it closely associates.

SPECIAL HABITAT REQUIREMENTS. Savannas, open woodlands, forest edges.

Further Reading. United States and Canada: Farrand 1983b, Godfrey 1966, Terres 1980. Mexico and Central America: Edwards 1972, Land 1970, Rappole et al. 1983, Stiles and Skutch 1989. South America: Rappole 1989. Caribbean: Bond 1947, Herklots 1961, Rappole et al. 1983.

Cassin's Kingbird
Tyrannus vociferans

Guatemala: Chatilla de Cassin, Chituri Gritón, Pitirre Gritón

Mexico: Tirano Gritón

RANGE. Breeds from central California, southern Utah, Colorado, and southeastern Montana south to northern Baja California, and through the Mexican highlands to Michoacán, Oaxaca, and Puebla, and east to western Texas. Winters from southern California and northern Mexico south to central Guatemala, casually to Honduras; irregularly in central California.

Thick-billed Kingbird

STATUS. Fairly common.

HABITAT. Occurs in dry open country such as plains and shrubsteppe, in a variety of habitats from desert riparian areas up to 2,400 m, and in open woodlands in southwestern mountains. Inhabits pinyon-yucca, pinyon-juniper, pine-oak, and ponderosa pine woodlands, canyons of sycamores, and in California, open valley woodlands and grasslands of the foothills among scattered oaks, cottonwoods, and sycamores. Usually nests in fairly tall trees such as pine, oak, cottonwood, walnut, hackberry, or sycamore. Places nest near the end of a horizontal limb 2 to 12 m, but up to 30 m, above the ground. Also places nests in bushes and on posts.

SPECIAL HABITAT REQUIREMENTS. Dry savanna with tall trees for nesting.

Further Reading. United States and Canada: Beal 1912, Bent 1942, Blancher and Robertson 1987, Farrand 1983b, Godfrey 1966, Hespenheide 1964, Johnsgard 1979, Ohlendorf 1974, Terres 1980. Mexico and Central America: Edwards 1972, Land 1970, Rappole et al. 1983, Ridgely and Gwynne 1989.

Thick-billed Kingbird
Tyrannus crassirostris

Guatemala: Chatilla Piquigorda, Chituri Piquigordo, Pitirre Piquigordo

Mexico: Tirano Piquigrueso

RANGE. Breeds from the Patagonia and Guadalupe Mountains in southeastern Arizona, and Guadalupe Canyon in extreme southwestern New Mexico, south to Guatemala. Winters in Mexico and Guatemala.

STATUS. Rare; first discovered in the United States in 1958, this Mexican species has expanded northward since the middle of the twentieth century.

HABITAT. Occurs in riparian habitats dominated by sycamore, cottonwood, willow, savannas, arid scrub, and mesquite. In the United States, nests in streamside sycamores 15 to 20 m above the ground.

Further Reading. United States and Canada: Farrand 1983b, Levy 1959, Oberholser 1974a, Phillips et al. 1964, Terres 1980. Mexico and Guatemala: Edwards 1972, Land 1970, Rappole et al. 1983.

Western Kingbird
Tyrannus verticalis

Canada (Quebec): Tyran de l'Ouest

Guatemala: Chatilla Occidental, Chituri Colinegro, Pitirre Colinegro

Mexico: Tirano Pálido

RANGE. Breeds from southern interior British Columbia to southern Manitoba and western Minnesota south to Baja California, Mexico, and southern and south-central Texas; rarely or sporadically eastward to southern Ontario, Missouri, Arkansas, and Louisiana. Winters in small numbers along the Atlantic and Gulf Coasts from South Carolina to southern Florida and west to southern Louisiana; also from southern Mexico to Costa Rica.

STATUS. Common; long-term increase in eastern part of its breeding range.

HABITAT. Occurs in almost any open habitat with scattered trees at low to moderate elevations, especially in agricultural regions. Commonly occurs near edge habitats such as shelterbelts, woodland borders, orchards, and hedgerows. Builds nest in a variety of sites but prefers trees, when available.

Eastern Kingbird

May nest against the trunk, in a crotch, or on a horizontal branch 2 to 12 m above the ground in cottonwoods, oaks, sycamores, willows, and other trees; if no trees are available, nests in bushes, on utility poles, or on a variety of structures.

SPECIAL HABITAT REQUIREMENTS. Savannas.

Further Reading. United States and Canada: Beal 1912, Bent 1942, Blancher and Robertson 1987, Farrand 1983b, Godfrey 1966, Hespenheide 1964, Johnsgard 1979, Mackenzie and Sealy 1981, Ohlendorf 1974, Terres 1980. Mexico and Central America: Edwards 1972, Land 1970, Rappole et al. 1983, Stiles and Skutch 1989.

Eastern Kingbird

Tyrannus tyrannus

Canada (Quebec): Tyran Tritri

Costa Rica: Tirano Norteño

Guatemala: Chatilla Norteña, Chituri Norteño, Pitirre Norteño

Mexico: Tirano Dorsinegro

RANGE. Breeds from southwestern and north-central British Columbia, southern Northwest Territories, and central Manitoba to southern Quebec and New Brunswick, south to northeastern California, northern Utah, northwestern and central New Mexico, the Gulf Coast, and Florida. Migrates through Central America, Cuba, and the Bahamas. Winters south to Colombia, Venezuela, Guyana, Brazil, Ecuador, Peru, and Bolivia.

STATUS. Common in eastern North America; continuing to decline.

HABITAT. Frequents open areas with scattered trees or tall shrubs; forest edges or hedgerows along pastures, swamps, marshes, fields, or highways; open country around orchards; brushy streamsides; and sometimes open woodlands. Often builds nest over water on a tree limb well away from the trunk, or occasionally in shrubs or on an artificial structure, locating nest 3 to 6 m, sometimes 2 to 20 m, above the ground. Builds nest in the crotch of a tree, on top of a dead stub, or on a fence post if no trees are available. In New England, frequently nests in the upper exposed horizontal limbs of pruned apple trees.

SPECIAL HABITAT REQUIREMENTS. Open habitats with perches for aerial feeding. Heavily dependent on fruit on Costa Rican wintering grounds.

Further Reading. United States and Canada: Beal 1912, Bent 1942, Forbush and May 1955, Johnsgard 1979, Terres 1980. Mexico and Central America: Blake et al. 1990, Edwards 1972, Land 1970, Rappole et al. 1983, Stiles and Skutch 1989. South America: Hilty and Brown 1986, Rappole et al. 1983. Caribbean: Bond 1947, Rappole et al. 1983.

Gray Kingbird
Tyrannus dominicensis

Canada (Quebec): Tyran Gris

Mexico: Tirano Dominicano

Puerto Rico: Chicheri, Chinchiri, Pechari, Pitirre

West Indies: Pestigre, Pipiri, Pipirite, Pitirre, Pitirre Abejero, Titirre

RANGE. Breeds along the Atlantic and Gulf Coasts from southeastern South Carolina south to the Florida Keys, west to southern Alabama and islands off the coast of Mississippi, and south in coastal regions to Venezuela. This species is resident in the West Indies and sporadically from northern Colombia to French Guiana. Northern individuals migrate through the islands west of Hispaniola. Winters south to Panama, Colombia, Venezuela, Guyana, Surinam, and French Guiana; casually in southern Florida.

STATUS. Locally common.

HABITAT. Found within the coastal zone, where it occurs in mangrove swamps, marsh edges, along roadsides, and in woodlands, groves, and yards in urban or rural areas. Nests in a fork or saddle on a horizontal limb of a tree or shrub 1 to 5 m above the ground, often over water. Prefers mangroves for nesting, but also nests in oaks, acacia, sea grape, casuarina, and cabbage palm. Shows a strong attachment to the nesting site, returning yearly to the same tree or shrub.

SPECIAL HABITAT REQUIREMENTS. Habitats within the coastal zone.

Further Reading. United States and Canada: Bent 1942, Farrand 1983b, Godfrey 1966, Harrison 1975, Terres 1980. Central America: Rappole et al. 1983, Ridgely and Gwynne 1989. South America: Hilty and Brown 1986, Rappole et al. 1983. Caribbean: Bond 1947, Raffaele 1989, Rappole et al. 1983.

Scissor-tailed Flycatcher
Tyrannus forficatus

Canada (Quebec): Moucherolle à
Longue Queue

Costa Rica: Tijereta Rosada, Tijerillo

Guatemala: Mosquero Rosado,
Tijereta Rosada

Mexico: Tirano, Tijereta Clara

RANGE. Breeds from southeastern Colorado, southern Nebraska, and north-central Missouri south to northern Nuevo León, western and southern Texas, and western Louisiana; isolated breeding in northeastern Mississippi, central Tennessee, and central Iowa. Winters in southern Florida and in Middle America from Veracruz and Oaxaca south to central Panama, casually in southern Louisiana. Casual north and west of the breeding range from southern British Columbia east across southern Canada and the Great Lakes states; also in the Bahamas, western Cuba, and Puerto Rico.

STATUS. Common.

HABITAT. Occurs on plains, prairies, mesas, and flats, and around pastures, woodland clearings, ranches, and farms. Perches for long periods on tall prairie plants, limbs of dead trees, utility wires, or fences. Typically nests in isolated cottonwoods, elms, or other hardwoods, 2 to 15 m above the ground; occasionally uses fence posts, telephone poles, windmill towers, or buildings for nest sites.

SPECIAL HABITAT REQUIREMENTS. Open habitats with elevated perches.

Further Reading. United States and Canada: Beal 1912, Fitch 1950, Godfrey 1966, Johnsgard 1979, Oberholser 1974a. Mexico and Central America: Edwards 1972, Land 1970, Rappole et al. 1983, Ridgely and Gwynne 1989, Stiles and Skutch 1989.

Rose-throated Becard
Pachyramphus aglaiae

Costa Rica: Cabezón Plomizo

Guatemala: Cabezón Degollado,
Huilo de Garganta Rosada

Mexico: Mosquero Cabezón
Piquigrueso

RANGE. Breeds in southeastern Arizona and in southern Texas (Cameron and Hidalgo Counties) and in Mexico, south to Costa Rica and possibly extreme western Panama. Winters from Mexico south to Costa Rica and casually to western Panama.

STATUS. Rare and local.

HABITAT. Inhabits open groves of mature trees, mostly in arid areas, but also near flowing water, preferably stands of sycamore, cottonwood, and willow. Builds an immense nest of strips of fibrous plant stems, suspending it from twigs at the end of a drooping branch 10 to 20 m above the ground. Often places nest in sycamores but also uses cottonwoods, bald cypress, and willows. Will often build in the same site as the previous year's nest, or very close it.

SPECIAL HABITAT REQUIREMENTS. Open woodlands, plantations.

Further Reading. United States and Canada: Farrand 1983b, Oberholser 1974b, Phillips 1949, Phillips et al. 1964, Terres 1980. Mexico and Central America: Edwards 1972, Land 1970, Rappole et al. 1983, Ridgely and Gwynne 1989, Stiles and Skutch 1989.

Horned Lark
Eremophila alpestris

Mexico: Alondra Cornada

RANGE. Breeds in North America from western and northern Alaska, the arctic coast of northern Canada, Prince Patrick, Devon, and Baffin Islands, and northern Labrador south to Oaxaca, southwestern Louisiana, central Missouri, northern Alabama, and North Carolina; in South America, in the eastern Andes of Colombia. Winters from southern Canada south throughout the breeding range, south to Baja California and Mexico except the Yucatán Peninsula, and locally or irregularly to the Gulf Coast.

STATUS. Locally common; generally declining, especially in central and western North America.

HABITAT. Inhabits a wide variety of open treeless habitats, from coastal dunes and alpine tundra to prairies and deserts. Prefers areas with a minimum of vegetation, such as heavily grazed grasslands, cultivated and plowed fields, golf courses, airports, and other relatively barren areas. In winter, groups in small to enormous flocks on open, barren sites similar to those in its breeding habitat. Nests in a depression on the ground, placed so that the upper edge of the nest is level with the ground surface. Often paves the nest with small pebbles along a portion of the rim. Places nest where there is little or no cover around the nest, or next to a clump of grass or a rock.

341

Further Reading. United States and Canada: Beal and McAtee 1912, Beason and Franks 1974, Cottam and Hanson 1938, DeGraaf and Rudis 1986, Hurley and Franks 1976, Johnsgard 1979, Pickwell 1931, Terres 1980, Verbeek 1967. Mexico and Central America: Edwards 1972, Peterson and Chalif 1973, Ridgely and Gwynne 1989. South America: Meyer de Schauensee and Phelps 1978.

Purple Martin
Progne subis

Canada (Quebec): Hirondelle Pourprée

Costa Rica: Martín Purpúreo, Golondrón

Guatemala: Golondrina Azul, Martín Norteño

Mexico: Golondrina Grande Negruzca

Puerto Rico: Golondrina Purpúrea

Venezuela: Golondrina de Iglesias

West Indies: Golondrina Azul, Golondrina Grande, Golondrina de Iglesias, Hirondelle

RANGE. Breeds from southwestern British Columbia south to Baja California, from northeastern and east-central British Columbia and central Alberta to southern Ontario and New Brunswick and central Nova Scotia south through the Mexican highlands to Michoacán, and throughout the Caribbean, the Gulf Coast, and southern Florida. Local in the Rocky Mountains but avoids most other mountainous areas. Migrates through Central America and the Caribbean. Winters in South America from Colombia, Venezuela, and the Guianas south, east of the Andes, to northern Bolivia and southeastern Brazil.

STATUS. Locally common. Recently declining in many regions of North America.

HABITAT. Inhabits open country near water: grassy river valleys, meadows around ponds, shores of lakes, marsh edges; but also agricultural lands, saguaro deserts, parks, and towns. In the East, breeds almost exclusively in colonial martin houses; in the West, still uses old woodpecker holes to a large extent. Also uses cavities in cliffs or among loose rocks, and crevices in old buildings.

SPECIAL HABITAT REQUIREMENTS. Large, multiroomed martin houses, tree cavities, or abandoned woodpecker holes for nesting, and open spaces for foraging.

Further Reading. United States and Canada: Allen and Nice 1952, Beal 1918, DeGraaf and Rudis 1986, Forbush and May 1955, Godfrey 1966, Johnsgard 1979, Stutchbury 1991, Tate and Tate 1982, Terres 1980. Mexico and Central America: Edwards 1972, Land 1970, Peterson and Chalif 1973, Rappole et al. 1983, Ridgely and Gwynne 1989, Stiles and Skutch 1989. South America: Hilty and Brown 1986, Meyer de Schauensee and Phelps 1978, Rappole et al. 1983. Caribbean: Bond 1947, Raffaele 1989, Rappole et al. 1983.

Gray-breasted Martin
Progne chalybea

Costa Rica: Martín Pechigris

Guatemala: Golondrina Urbana,
Martín Pechigris

Mexico: Golondrina Grande
Pechipálida

Venezuela: Golondrina Urbana

RANGE. Breeds in southeastern Texas (rarely), from Nayarit, Coahuila, and Nuevo León south along both slopes of Middle America, and in South America from Colombia, Venezuela, and the Guianas east of the Andes to Argentina. Winters in the northern parts of the breeding range, but more abundantly and regularly from Costa Rica and Panama south through South America to northern Bolivia and central Brazil. Migratory at both extremes of its range.

STATUS. Fairly common.

HABITAT. Mainly lowlands, foothills, open country (especially near water), clearings, and towns. Usually in colonies around buildings or bridges. The nest is a shallow, loosely built cup of dry grass, weed stems, dead leaves, and twiglets in a cranny in a building, bridge, bird box, or other structure; in rural areas, nests high in an old woodpecker hole or other cavity in a dead tree in a clearing or in a cranny amid rocks.

Further Reading. Mexico and Central America: Edwards 1972, Land 1970, Peterson and Chalif 1973, Rappole et al. 1983, Ridgely and Gwynne 1989, Stiles and Skutch 1989. South America: Hilty and Brown 1986, Meyer de Schauensee and Phelps 1978.

344

Tree Swallow
Tachycineta bicolor

Canada (Quebec): Hirondelle Bicolore

Costa Rica and Guatemala:
Golondrina Bicolor

Mexico: Golondrina Canadiense

Puerto Rico: Golondrina Vientriblanca

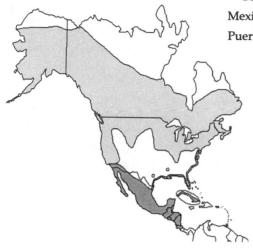

RANGE. Breeds from western and central Alaska and central Yukon to northern Quebec and central Labrador south along the Pacific Coast to southern California and south-central New Mexico; generally sporadic or irregular as a breeder east of the Rocky Mountain states and south of the upper Mississippi and Ohio Valleys, and along the Atlantic Coast south of Massachusetts. Winters from southern California, southwestern Arizona, Texas, the Gulf Coast, and the Atlantic Coast from New York south to Nicaragua and in the Greater Antilles; casual or irregular to Costa Rica, Panama, Colombia, and Guyana.

STATUS. Common. Recently declining in eastern North America.

HABITAT. Prefers open woodlands near ponds, small lakes, or marshes. Occurs around farmlands, river bottomlands, beaver ponds, wooded swamps, and marshes where dead standing trees are in or near water. Prefers to nest in natural cavities and abandoned woodpecker holes, but if nesting holes are scarce, will accept nest boxes placed in open fields or will use crevices in buildings. Uses cavities in the trunk or limb of live or dead trees, especially if the cavity is 1 to 5 m above water. Usually nests singly but is loosely

345

colonial if there are abundant suitable cavities and abundant food. Population can be limited by competition for nest cavities, fall of cavity trees, and the abundance of woodpeckers which excavate nest cavities (Rendell and Robertson 1989). Along forest edges, House Wrens commonly destroy clutches of Tree Swallows (Rendell and Robertson 1990).

SPECIAL HABITAT REQUIREMENTS. Cavities and old woodpecker holes and open feeding areas such as meadows, marshes, or open water.

Further Reading. United States and Canada: Beal 1918, Chapman 1955, DeGraaf and Rudis 1986, Forbush and May 1955, Godfrey 1966, Johnsgard 1979, Scott et al. 1977, Terres 1980, Thomas et al. 1979. Mexico and Central America: Edwards 1972, Land 1970, Rappole et al. 1983, Ridgely and Gwynne 1989, Stiles and Skutch 1989. South America: Hilty and Brown 1986, Rappole et al. 1983. Caribbean: Bond 1947, Raffaele 1989, Rappole et al. 1983.

Violet-green Swallow
Tachycineta thalassina

Canada (Quebec): Hirondelle à Face Blanche

Costa Rica: Golondrina Verde-Violácea

Guatemala: Golondrina Verde-Violeta

Mexico: Golondrina Cariblanca

RANGE. Breeds from central Alaska, central Yukon, and extreme southwestern Northwest Territories south to Mexico, and east to southwestern Saskatchewan, western South Dakota, and western Nebraska. Winters from central coastal and southern California south to Honduras, casually to western Panama.

STATUS. Common.

HABITAT. Inhabits coniferous, deciduous (aspen), and mixed forests, preferring open montane or riparian broken woodlands, or the edges of dense forest. Occurs around towns, in woodland clearings, especially near lakes and streams, and if snags are present, in canyons, and in mountains from the foothills to near timberline. Nests in old woodpecker holes, natural tree cavities, crevices in rocky cliffs, nesting boxes, niches of old buildings, and where cavities are scarce, in old nests of Cliff Swallows and burrows of Bank Swallows. In the central Rocky Mountains, prefers cavities in ponderosa pine but also nests in aspen and other trees.

SPECIAL HABITAT REQUIREMENTS. Cavities or crevices for nesting and open terrain or forest openings for feeding.

Further Reading. United States and Canada: Bailey and Niedrach 1965, Beal 1918, Bent 1942, Combellack 1954, Godfrey 1966, Johnsgard 1979, Scott et al. 1977, Terres 1980. Mexico and Central America: Edwards 1972, Land 1970, Rappole et al. 1983, Ridgely and Gwynne 1989, Stiles and Skutch 1989.

Northern Rough-winged Swallow

Stelgidopteryx serripennis
(includes Southern Rough-winged Swallow, *Stelgidopteryx ruficollis*)

Canada (Quebec): Hirondelle à Ailes Hérissées

Costa Rica: Golondrina Alirrasposa Norteña

Guatemala: Golondrina Alirrasposa, Golondrina de Sierra

Mexico: Golondrina Gorjicafé

Puerto Rico: Golondrina de Garganta Pálido

RANGE. Breeds from southeastern Alaska, central British Columbia, and southern Alberta to southwestern Quebec and Maine south through Middle and South America to southern Chile and central Argentina. Winters from Louisiana, Texas, and southern Florida south through the Greater Antilles, Middle America, and northern South America. Populations in Argentina, Uruguay, and southern Brazil are austral migrants.

STATUS. Fairly common.

HABITAT. Inhabits open country near water, wherever a suitable nest site can be found. In the East, frequents rocky gorges, shale banks, stony road cuts, railroad embankments, river valleys, and stream banks. In the Midwest and West, often found around sand banks, gravel pits, and stream banks. Excavates nests in banks of clay, sand, or gravel or uses abandoned Bank Swallow or kingfisher burrows. Sometimes uses natural rock crevices, drain-pipes, culverts, cracks in bridges, and crevices in buildings. Commonly nests singly, or in Bank Swallow colonies; may be somewhat colonial in the west-ern part of its breeding range.

SPECIAL HABITAT REQUIREMENTS. Suitable nest sites preferably near, but up to 1 km from, water.

Further Reading. United States and Canada: Beal 1918, DeGraaf and Rudis 1986, Godfrey 1966, Johnsgard 1979, Lunk 1962, Ricklefs 1972. Mexico and Central America: Edwards 1972, Land 1970, Rappole et al. 1983, Ridgely and Gwynne 1989, Stiles and Skutch 1989.

Bank Swallow
Riparia riparia

Canada (Quebec): Hirondelle des
Sables

Costa Rica: Golondrina Ribereña

Guatemala: Avión Zapador,
Golondrina Ribereña

Mexico: Golondrina Pechifajada

Puerto Rico and Venezuela:
Golondrina Parda

RANGE: Breeds in North America from western and central Alaska and central Yukon to central Quebec and southern Labrador, south to southern California, western Nevada, southern New Mexico, southern Texas, northern Alabama, eastern Virginia, and casually, northwestern North Carolina and south-central South Carolina. Migrates through Central America and the Caribbean. Winters in South America from Colombia, Venezuela, and the Guianas south, essentially east of the Andes, to Peru, northern Argentina, and Paraguay.

STATUS. Locally common.

HABITAT. Prefers grasslands and cultivated fields but uses a variety of open habitats, usually near water and suitable nest sites. Nests in riverbanks, burrow pits, gravel pits, road cuts, sand banks, and other exposed banks of sand, gravel, or clay. Excavates a burrow (or repairs an existing burrow) ranging from 0.25 to 2 m, but generally about 0.5 m, in length near the top

350

of a steep, stable (e.g., grass-topped) bank. Forms dense colonies, with up to several hundred nests in a bank.

SPECIAL HABITAT REQUIREMENTS. Stabilized steep banks of sand, gravel, or clay in an open habitat, preferably near flowing water.

Further Reading. United States and Canada: Allen 1933, Beal 1918, Beyer 1938, DeGraaf and Rudis 1986, Forbush and May 1955, Freer 1979, Godfrey 1966, Johnsgard 1979, Peterson 1955, Tate and Tate 1982. Mexico and Central America: Edwards 1972, Land 1970, Rappole et al. 1983, Ridgely and Gwynne 1989, Stiles and Skutch 1989. South America: Hilty and Brown 1986, Rappole et al. 1983. Caribbean: Bond 1947, Raffaele 1989, Rappole et al. 1983.

Cliff Swallow
Hirundo pyrrhonota

Canada (Quebec): Hirondelle à Front Blanc

Costa Rica, Mexico, and Venezuela: Golondrina Risquera

Guatemala: Golondrina Risquera, Golondrina de las Rocas

Puerto Rico: Golondrina de Peñasco

RANGE. Breeds from western and central Alaska and central Yukon to northern Ontario, southern Quebec, and New Brunswick south to Mexico, southwestern Louisiana, northern portion of the Gulf States, and southern North Carolina; also in the Lake Okeechobee region of southern Florida. Migrates through Central America, the Caribbean, and northern South America. Winters in Brazil, Paraguay, Uruguay, and Argentina.

STATUS. Common in the western United States, locally fairly common in the East; overall populations are stable or increasing, except in some northeastern states where the species is declining.

HABITAT. Originally restricted to the vicinity of cliffs and banks; now occurs over open country around farmlands, towns, bridges, dams, freeway overpasses, and other areas near mud supplies and potential nest sites. Originally nested on bluffs, cliffs, and deep gorges in mountains, usually near water and sometimes on the side of large pine trees and in caves; has adapted to building its gourdlike mud nests under the eaves of, or in, buildings, under bridges, in culverts, on the faces of dams, and under freeway overpasses. Forms colonies of up to several hundred nests in favorable locations.

SPECIAL HABITAT REQUIREMENTS. A vertical substrate with an overhang for nest attachment, a supply of mud suitable for nest construction, fresh water with a smooth surface for drinking, and an open foraging area near the nest site.

Further Reading. United States and Canada: Beal 1918, Bent 1942, DeGraaf and Rudis 1986, Forbush and May 1955, Godfrey 1966, Johnsgard 1979, Mayhew 1958, Samuel 1971, Tate and Tate 1982, Withers 1977. Mexico and Central America: Edwards 1972, Land 1970, Rappole et al. 1983, Stiles and Skutch 1989. South America: Hilty and Brown 1986, Meyer de Schauensee and Phelps 1978, Rappole et al. 1983. Caribbean: Bond 1947, Herklots 1961, Raffaele 1989, Rappole et al. 1983.

Cave Swallow
Hirundo fulva

Mexico: Golondrina Fulva

RANGE. Breeds from Carlsbad Caverns in southeastern New Mexico and in western and south-central Texas south through Mexico to central Chiapas and the state of México. Winter range of northern populations is unknown. Isolated resident population in South America in southwestern Ecuador and northwestern Peru.

STATUS. Locally fairly common; range is expanding as the species adapts to human-altered environments.

HABITAT. Originally restricted to open country in the vicinity of limestone caves and sinkholes; has adapted its nesting habits to artificial structures. In the northern part of its range, now also nests in culverts and on bridges where water and mud are available. Forms colonies. In caves, tends to build its mud nests in isolated crevices and pockets, or under overhanging ledges. Also nests in sinkholes, in highway culverts, and under bridges. May reuse nest year after year and will sometimes share the same nest site with Barn Swallows.

SPECIAL HABITAT REQUIREMENTS. Caves, sinkholes. Roughened or pitted surfaces for nesting (culverts), water for drinking, and mud suitable for nest construction.

Further Reading. United States and Canada: Martin 1974, Selander and Baker 1957, Wauer and Davis 1972. Mexico: Edwards 1972, Rappole et al. 1983.

Barn Swallow
Hirundo rustica

Canada (Quebec): Hirondelle des
Granges

Costa Rica and Mexico: Golondrina
Tijereta

Guatemala: Golondrina Tijereta,
Golondrina Tijerilla

Puerto Rico and Venezuela:
Golondrina de Horquilla

RANGE. Breeds from south-coastal and southeastern Alaska and southern Yukon across to central Manitoba, northern Ontario, and southern Quebec south to Mexico, the Gulf Coast, north-central Florida, and southern North Carolina. Migrates through the Caribbean. Winters throughout Central and South America, casually north to the southwestern United States and throughout southern Florida.

STATUS. Common; but declining over much of its breeding range.

HABITAT. Occurs virtually throughout the United States wherever suitable nest sites are found, but favors farmlands, rural and suburban areas, and sometimes open forests. Originally nested on cliffs and in caves and rock crevices in mountains, along rocky coasts, and on high shores of lakes and rivers. Still uses such sites in the far north and on the Pacific Coast, but in other areas nests on horizontal beams or ledges inside barns or other buildings, or under bridges, culverts, or wharves. Usually nests colonially.

SPECIAL HABITAT REQUIREMENTS. Overhead protection, especially buildings, for nesting.

Further Reading. United States and Canada: DeGraaf and Rudis 1986, Forbush and May 1955, Godfrey 1966, Johnsgard 1979, Samuel 1971, Shields 1984, Snapp 1976. Mexico and Central America: Edwards 1972, Land 1970, Rappole et al. 1983, Stiles and Skutch 1989. South America: Meyer de Schauensee and Phelps 1978, Rappole et al. 1983. Caribbean: Raffaele 1989, Rappole et al. 1983.

Rock Wren
Salpinctes obsoletus

Costa Rica: Soterrey Roquero
Mexico: Troglodita Saltarroca

RANGE. Breeds from south-central British Columbia and southern Alberta to the western Dakotas, south (east of the coast ranges in Washington,

Oregon, and northern California) to Baja California and the highlands of Middle America south to northwestern Costa Rica, and east to western Nebraska and central and southern Texas. Winters from northern California, southern Nevada, and southern Utah to north-central Texas south through the southern portions of the breeding range, wandering to lower elevations.

STATUS. Fairly common.

HABITAT. In widely scattered pairs, inhabits arid and semiarid open, rocky areas such as rock outcrops, canyons, fractured cliff faces, talus slopes, arroyos, and badlands. May be found up to 3,000 m in the Rocky Mountains. Shows no preference for areas with water throughout its range. Typically locates nest on slopes of loose rocks and boulders, in crevices of canyon walls, or sometimes in rodent cavities in banks or in tree holes. Builds a well-hidden nest, often with a small runway of stones, leading to the nest. Locally common resident along Pacific slope of Cordillera de Guanacaste, south to Volcán Miravalles, 500 to 1,600 m. Restricted to high barren slopes, often on volcanoes.

SPECIAL HABITAT REQUIREMENTS. Rough, rocky canyons; concrete or stone structures with crevices for foraging and cover.

Further Reading. United States and Canada: Bent 1948, Farrand 1983b, Johnsgard 1979, Verner and Boss 1980, Wolf et al. 1985. Mexico and Central America: Edwards 1972, Stiles and Skutch 1989, Ridgely and Gwynne 1989.

Bewick's Wren
Thryomanes bewickii

Canada (Quebec): Troglodyte de Bewick

Mexico: Troglodita Colinegro

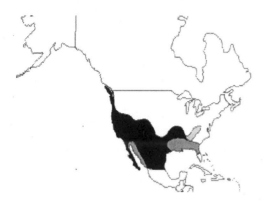

RANGE. Breeds from southwestern British Columbia, Wyoming, Nebraska, Minnesota, southern Ontario, and New York south to Mexico, Texas, the northern portions of the Gulf States, Georgia, and South Carolina. Resident in much of the western United States and northern Mexico. Winters from Kansas, the lower Ohio Valley, and North Carolina to Mexico, the Gulf Coast, and Florida.

STATUS. Scarce and local throughout the eastern portion of the breeding range; since the early 1980s, declining in the western United States.

HABITAT. Generally associated with dense, brushy habitats such as thickets of mesquite, oaks, and cacti; chaparral; mixtures of pine, junipers, and oaks; dense growths of alder, cottonwood, and willow. In the Southwest, occurs in mountain canyons to 2,000 m elevation. In the Southeast, seems to favor small brushy openings in extensive forest. Nests near the ground in secluded natural tree cavities, old woodpecker holes, rock crevices, deserted buildings, nest boxes placed low, or in almost any cavity where a nest could be built. Nest and nesting sites are like those of the House Wren, and the two species usually compete when in the same area.

SPECIAL HABITAT REQUIREMENTS. A brushy understory and cavities for nesting.

358

Further Reading. United States and Canada: Beal 1907, Bent 1942, Godfrey 1966, Hunter 1990, Johnsgard 1979, Miller 1941, Tate and Tate 1982, Verner and Boss 1980. Mexico: Edwards 1972.

House Wren
Troglodytes aedon

Canada (Quebec): Troglodyte Familier
Mexico: Troglodita Continental

RANGE. Breeds from southern and east-central British Columbia and northern Alberta east to southwestern Quebec and New Brunswick, and south to northern Baja California, southern Arizona and New Mexico, western and northern Texas, central Arkansas, southern Tennessee, and North Carolina. Winters from southern California to northern Texas, the northern portion of the Gulf States, and coastal Maryland south to southern Baja California, the

Gulf Coast, Florida, and throughout Mexico to Oaxaca and Veracruz. Resident from Oaxaca through central America to southern South America.

STATUS. Common; increasing throughout its breeding range.

HABITAT. Originally associated with deciduous forests and open woods, but has adapted to woody vegetation in cities, towns, and around farms. Frequents edges of woodlands, open forests, clearings, swampy woodlands, orchards, farmlands, and suburban gardens. Ranges from the plains to near timberline in western riparian forests but avoids high elevations in the East. Uses almost any type of cavity as a nest site, including natural cavities in trees, fence posts, or stumps, woodpecker holes, and bird boxes or other artificial cavities with openings preferably about 2.5 cm in diameter. Typically chooses a nest site less than 3 m above the ground. Commonly destroys clutches of other passerines near the nest.

SPECIAL HABITAT REQUIREMENTS. Thickets and cavities for nesting.

Further Reading. United States and Canada: Bent 1948, DeGraaf and Rudis 1986, Finch 1990, Forbush and May 1955, Godfrey 1966, Guinan and Sealy 1989, Johnsgard 1979, Kendeigh 1941. Mexico and Central America: Peterson and Chalif 1973.

Sedge Wren

Cistothorus platensis
(formerly Short-billed Marsh
Wren)

Canada (Quebec): Troglodyte à Bec
   Court

Mexico: Troglodita Pantanero
   Piquicorto

RANGE. Breeds in North America from extreme east-central Alberta and central Saskatchewan east to northern Michigan and southern New Brunswick, south to east-central Arkansas, central Kentucky, and southeastern Virginia, and west to central North Dakota and eastern Kansas. Resident locally in Middle America and in South America locally in the Andes from Colombia to Chile and from Brazil to Tierra del Fuego. Winters from western Tennessee and Maryland to northeastern Mexico, Texas, the Gulf Coast, and Florida.

STATUS. Scarce and local; populations have been increasing in the eastern part of the breeding range in the United States and Canada.

HABITAT. Inhabits wet meadows and the damp upper margins of marshes and sphagnum bogs. In the Northeast, commonly inhabits sedge meadows, shallow sedge marshes with scattered shrubs and little or no standing water, and coastal brackish marshes of marshhay cordgrass with scattered low shrubs and herbs. In the Midwest, prefers wet meadows dominated by sedges, cottongrass, mannagrass, and reed grass, but also frequents emergent vegetation associated with marshes, and retired croplands and fields. May nest in loose colonies in good habitat, otherwise nests singly. Builds nest over land or water in dense vegetation such as canarygrasses, sedges, or bulrushes, but shuns cattails; usually places nest, interwoven with live grasses, less than 1 m above the substrate. Males build many unlined dummy nests, but few are used by females. Nomadic breeders, Sedge Wrens shift nesting areas from year to year.

SPECIAL HABITAT REQUIREMENTS. Wet meadows or drier edges of marshes; in South America, dry grasslands.

Further Reading. United States and Canada: Crawford 1977, DeGraaf and Rudis 1986, Farrand 1983b, Forbush and May 1955, Godfrey 1966, Johnsgard 1979, Mousley 1934b, Picman and Picman 1980, Tate and Tate 1982, Walkinshaw 1935. Mexico: Edwards 1972, Rappole et al. 1983.

Marsh Wren
Cistothorus palustris
(formerly Long-billed Marsh
Wren)

Canada (Quebec): Troglodyte des
Marais

Mexico: Troglodita Pantanero
Piquilargo

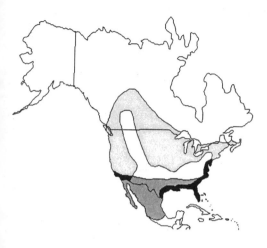

RANGE. Breeds from southwestern and east-central British Columbia and northern Alberta east to northern Michigan and eastern New Brunswick, south to Baja California, southwestern Arizona, extreme western and southern Texas, the Gulf Coast, and east-central Florida. Generally very local in interior North America. Winters in coastal areas throughout the breeding range, and in the interior from the southern United States to Mexico.

STATUS. Locally common; has been increasing in the western United States, but has recently declined in the East.

HABITAT. Prefers large fresh or brackish marshes with an abundance of tall emergent vegetation such as cattails, loosestrife, sedges, or rushes. Also frequents prairie sloughs, pond and sluggish river shores, marsh-fringed lakes, and the banks of tidal rivers bordered with tall emergent vegetation. Prefers large marshes grown with narrow-leaved cattails to those with broad-leaved cattails. Builds domed elliptical nest, preferably in cattail stands of moderate density, 1 to 2 m above the marsh substrate, which is generally shallow water. Usually attaches nest to cattails or other tall emergent vegetation, but often places it in small bushes or trees. Constructs many dummy nests and uses some for roosting.

SPECIAL HABITAT REQUIREMENTS. Marshy habitats with tall emergent vegetation.

Further Reading. United States and Canada: Bent 1948, DeGraaf and Rudis 1986, Farrand 1983b, Forbush and May 1955, Godfrey 1966, Johnsgard 1979, Kroodsma 1989, Leonard and Picman 1986, 1987, Low and Mansell 1983, Metz 1991, Verner 1965, Verner and Engelsen 1970, Welter 1935. Mexico: Edwards 1972, Rappole et al. 1983.

Golden-crowned Kinglet
Regulus satrapa

Canada (Quebec): Roitelet à Couronne Dorée

Guatemala: Reyezuelo de Coronilla Dorada, Reyezuelo Monidorado

Mexico: Reyezuelo Cabecirrayado

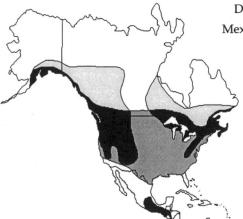

RANGE. Breeds from southern Alaska to northern Alberta, southern Quebec, and Newfoundland south in the coastal and interior mountains to southern and eastern California, southern Utah, south-central New Mexico, Mexico, Guatemala, and east of the Rockies to southern Manitoba, north-central Michigan, New York, eastern Tennessee, western North Carolina, northern New Jersey, and southern Maine. Winters from south-coastal Alaska and southern Canada south to northern Baja California, through the breeding range to Guatemala, the Gulf Coast, and central Florida.

STATUS. Common in parts of its range; has declined in western regions.

HABITAT. Breeds primarily in dense coniferous forests, especially where spruce is present. More abundant in old growth (>200 yr old) than in managed (85 yr) conifer stands in the northwestern United States, and is sensitive to forest cutting. Winters in coniferous forests and occasionally in deciduous woodland scrub and brush. Builds a globular nest with entrance at the top, woven into the twigs of a horizontal limb of a conifer.

SPECIAL HABITAT REQUIREMENTS. Dense conifer stands.

Further Reading. United States and Canada: Bent 1949, DeGraaf and Rudis 1986, Forbush and May 1955, Godfrey 1966, Mannon and Meslow 1984, Medin 1985, Tate and Tate 1982, Terres 1980. Mexico and Guatemala: Edwards 1972, Land 1970.

Ruby-crowned Kinglet
Regulus calendula

Canada (Quebec): Roitelet à Couronne Rubis

Guatemala: Reyezuelo de Coronilla Colorada, Reyezuelo Monicolorado

Mexico: Reyezuelo Sencillo

RANGE. Breeds from northwestern and north-central Alaska, northern Saskatchewan, northern Ontario, and Newfoundland south to southern Alaska, in the mountains to southern California, southern Arizona, south-central New Mexico, and east-central Colorado, and east of the Rockies to central Alberta, southern Manitoba, northeastern Minnesota, northern Michigan, northern New York, northern Maine, and Nova Scotia. Winters from southern British Columbia, Idaho, northern Arizona, Nebraska, southern Ontario, and New Jersey south to Baja California, southern Texas, southern Florida, and through Mexico to Guatemala.

STATUS. Locally common. Had been declining in eastern North America; recently has increased.

HABITAT. Generally inhabits coniferous forests or coniferous-deciduous woodlands during the breeding season. Avoids pure or nearly pure deciduous forest in the central Rockies (Scott and Crouch 1988). More abundant in managed than in old-growth conifer stands in the northwestern United States (Mannan and Meslow 1984). In migration and during winter also found in deciduous forests, open woodlands, brush, and scrub. Usually attaches nest to pendant twigs beneath a horizontal conifer branch (commonly spruce, occasionally fir or pine), generally from 5 to 20 m above the ground.

SPECIAL HABITAT REQUIREMENTS. Breeding: coniferous or mixed forest, muskeg. Winter: also deciduous woodland, brush, scrub.

Further Reading. United States and Canada: Beal and McAtee 1912, DeGraaf and Rudis 1986, Forbush and May 1955, Godfrey 1966, Terres 1980. Mexico and Guatemala: Edwards 1972, Land 1970, Rappole et al. 1983.

Blue-gray Gnatcatcher
Polioptila caerulea

Canada (Quebec): Gobe-mouches Gris-Bleu
Guatemala: Perlita Grisilla
Mexico: Perlita Gris
West Indies: Rabuita

RANGE. Breeds from southern Oregon, northern California, southern Idaho, central Utah, Colorado, Nebraska, western Iowa, southeastern Minnesota, Michigan, southwestern Quebec, central New York, central Vermont, and southern Maine south to Baja California, southeastern Texas, the Gulf Coast, and southern Florida, throughout Mexico to Central America. Winters from southern California, southern Nevada, western and central Arizona, central Texas, the southern portions of the Gulf States, and on the Atlantic Coast from Virginia south through Mexico to Honduras, and in the Bahamas and Cuba.

STATUS. Common in parts of its range; increasing in eastern North America.

HABITAT. In the Southeast, inhabits forested river bottoms and upland pine woods with an understory of oaks. In other areas, may inhabit open scrub and woodlands, or tall trees with closed canopies along river floodplains. Throughout the West, breeds in oaks, pinyon-juniper, and less frequently in chaparral. Nest is saddled on a horizontal limb 1 to 20 m high (average 8 m), in a deciduous tree or conifer.

SPECIAL HABITAT REQUIREMENTS. An abundant supply of arthropods.

Further Reading. United States and Canada: Forbush and May 1955, Godfrey 1966, Root 1967, 1969, Terres 1980. Mexico and Central America: Edwards 1972, Land 1970, Rappole et al. 1983. Caribbean: Bond 1947, Rappole et al. 1983.

Eastern Bluebird
Sialia sialis

Canada (Quebec): Merle bleu à Poitrine Rouge

Guatemala: Azulejo, Azulillo

Mexico: Azulejo Gorjicanelo

RANGE. Breeds from southern Saskatchewan, southern Quebec, and western Nova Scotia south to southern Texas and southern Florida, and west to the Dakotas, western Kansas, Texas, and southeastern New Mexico; also in southeastern Arizona and through the highlands of Mexico to Nicaragua. Winters from the middle portions of the eastern United States south throughout the breeding range, and on Cuba. Individuals from Guatemala south are resident.

STATUS. Populations low but increasing throughout the breeding range.

HABITAT. Inhabits fields, forest edges, open woodlands, and open country with scattered trees, orchards, shelterbelts, and riparian woodlands. Nests in old woodpecker holes, hollows of decayed trees, crevices of rocks, and hollows in wooden fence posts when available. Many now nest in nest boxes placed in open areas or at the edge of a forest.

SPECIAL HABITAT REQUIREMENTS. Low cavities for nesting and perches for foraging.

Further Reading. United States and Canada: Beal 1915a, Forbush and May 1955, Godfrey 1966, Hartshorne 1962, Pearson 1936, Rustad 1972, Sauer and

368

Droege 1990, Tate and Tate 1982, Thomas 1946. Mexico and Central America: Edwards 1972, Land 1970, Rappole et al. 1983. Caribbean: Bond 1947, Rappole et al. 1983.

Mountain Bluebird
Sialia currucoides

Canada (Quebec): Merle Bleu des Montagnes

Mexico: Azulejo Pálido

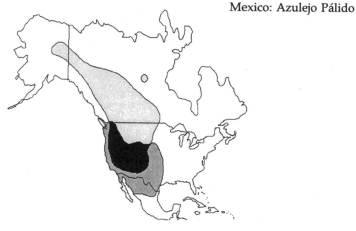

RANGE. Breeds from east-central Alaska, southern Yukon, and western Manitoba south in the mountains to southern California, central and southeastern Nevada, northern and east-central Arizona, and southern New Mexico, and east to northeastern North Dakota, western South Dakota, and central Oklahoma. Winters from southern British Columbia and western Montana south to Baja California, northern Mexico, and southern Texas, and east to eastern Kansas, western Oklahoma, and central Texas.

STATUS. Populations are low; has been declining in eastern part of breeding range.

HABITAT. Nests in open woodlands of nearly all forest types of the Rocky Mountain region, usually from 2,200 to 3,500 m in open and logged forests or near forest edges. Usually nests in old woodpecker holes or natural cavities in dead trees in open areas or near forest edges. Will also use nest boxes. During migration and in winter, also frequents grasslands, open brushy country, and agricultural lands.

SPECIAL HABITAT REQUIREMENTS. Cavity nests and feeding perches.

Further Reading. United States and Canada: Beal 1915a, Burleigh 1972, Godfrey 1966, Herlugson 1981b, Power 1966, 1980, Scott et al. 1977, Tate and Tate 1982. Mexico and Central America: Edwards 1972, Rappole et al. 1983.

Western Bluebird
Sialia mexicana

Canada (Quebec): Merle Bleu à Dos Marron

Mexico: Azulejo Gorjiazul

RANGE. Resident from southern British Columbia, western and south-central Montana, and north-central Colorado south through the mountains to Baja California, western and southern Nevada, southern Utah, western and southeastern Arizona, central New Mexico, western Texas, and the highlands of Mexico. Wanders in winter to lowland areas throughout the breeding range, and to islands off California and Baja California.

STATUS. Overall, populations are low but stable; moderately common.

HABITAT. Mostly inhabits open ponderosa pine forests of the transition zone but is also found in other open coniferous, deciduous (aspen), and mixed forests, partly open country with scattered trees, savanna, and riparian woodlands. In Mexico, open country with only a few scattered trees or other nesting site, or woodland edges where large open areas are nearby. Usually nests in old woodpecker holes, but also uses natural cavities and nest boxes. Commonly locates nests in rather open forests or at forest edges.

SPECIAL HABITAT REQUIREMENTS. Cavities for nesting and perches for feeding.

Further Reading. United States and Canada: Beal 1915a, Herlugson 1981a, Mock 1991, Tate and Tate 1982. Mexico: Edwards 1972.

Townsend's Solitaire
Myadestes townsendi

Canada: Solitaire de Townsend
Mexico: Clarín Norteño

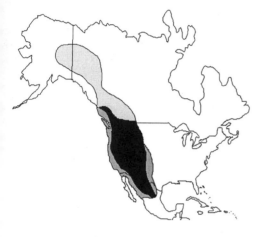

RANGE. Breeds from east-central and southeastern Alaska and western Northwest Territories south to the mountains of southern California, northern and east-central Arizona, central New Mexico, and northern Mexico, and east to southwestern Alberta, western and southern Montana, southwestern South Dakota, and northwestern Nebraska. Winters from southern British Columbia, southern Alberta, Montana, and South Dakota south to Baja California, the southern limit of the breeding range in Mexico, and east to western Missouri, western Oklahoma, and central Texas.

STATUS. Common within parts of its range.

HABITAT. Breeds in open montane and subalpine coniferous forests and in thickets and brushy areas on steep slopes or adjacent to rocky cliffs up to 3,900 m in elevation. More abundant in Douglas-fir stands that have been partially opened by logging than in uncut stands. Builds nest on the ground, partly concealed at the base of a pine or fir, under overhanging banks, or among the roots of a fallen tree. Winters in open woodland, pinyon-juniper associations, riparian woodlands, chaparral, and desert.

SPECIAL HABITAT REQUIREMENTS. Juniper berries for winter food.

Further Reading. United States and Canada: Beal 1915b, Godfrey 1966, Lederer 1977, Medin 1985, Poddar and Lederer 1982, Salomonson and Balda 1977, Terres 1980, Verner and Boss 1980. Mexico and Central America: Edwards 1972, Rappole et al. 1983.

Veery
Catharus fuscescens

Canada (Quebec): Grive Fauve
Costa Rica: Zorzal Dorsirrojizo
Guatemala: Tordo de Espalda Café, Zorzal Dorsicafé
Mexico: Zorzalito de Wilson
Venezuela: Paraulata Cachetona

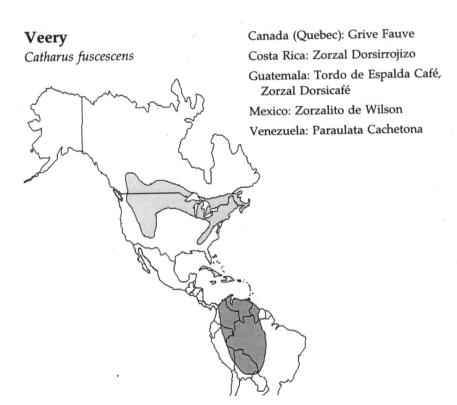

RANGE. Breeds from south-central and southeastern British Columbia to New Brunswick and southwestern Newfoundland, south to central Oregon, southern Idaho, northeastern South Dakota, northern Illinois, and northern Ohio, in the mountains through West Virginia, western and central Maryland, eastern Kentucky, western and central Virginia, eastern Tennessee, and western North Carolina to northwestern Georgia, and in the Atlantic region to eastern Pennsylvania, central New Jersey, and the District of Columbia. Also in east-central Arizona. Migrates through Central America, the Bahamas, and Cuba. Winters in Colombia, Venezuela, Guyana, Ecuador, Peru, and Bolivia.

STATUS. Common, but generally declining.

HABITAT. Inhabits low or moist deciduous woods, bottomland forests, wooded swamps, and damp ravines; prefers sapling stands of deciduous second-growth or open woods with fairly dense undergrowth of ferns, shrubs, and trees. Builds a bulky nest on or near the ground at the base of a shrub, on a mossy stump, in a clump of weeds, or occasionally in a low shrub or tree.

SPECIAL HABITAT REQUIREMENTS. Moist woodlands with understory of low trees or shrubs.

Further Reading. United States and Canada: Beal 1915b, Bertin 1977, DeGraaf and Rudis 1986, Forbush and May 1955, Godfrey 1966, Holmes and Robinson 1988, Noon 1981, Terres 1980. Mexico and Central America: Edwards 1972, Land 1970, Rappole et al. 1983, Stiles and Skutch 1989. South America: Meyer de Schauensee and Phelps 1978, Rappole et al. 1983. Caribbean: Rappole et al. 1983.

Gray-cheeked Thrush

Catharus minimus
(includes Bicknell's Thrush,
Catharus bicknelli)

Canada (Quebec): Grive à Joues Grises

Costa Rica: Zorzal Carigrís

Guatemala: Tordo Carigrís, Zorzal Carigrís

Mexico: Zorzalito Carigrís

Puerto Rico: Zorzal de Mejilla Gris

Venezuela: Paraulata de Cara Gris

RANGE. Gray-cheeked Thrush breeds from northern Alaska, southeastern Northwest Territories, and Newfoundland south to southern Alaska, northwestern British Columbia, northeastern Saskatchewan, Nova Scotia, eastern New York, Massachusetts (Mt. Greylock), central Vermont, northern New Hampshire, and central Maine. Migrates through Central America and the Bahamas, Cuba, Hispaniola, and Trinidad and Tobago. Winters in the Guianas, Venezuela, Colombia, Ecuador, Peru, and Bolivia; also on Hispaniola.

STATUS. Common in portions of its range.

HABITAT. During summer, inhabits coniferous forests (primarily spruce) and tall shrubby areas in taiga. During migration and in winter, also found in deciduous forests and open woodlands. Generally inhabits krummholz on mountain tops in the Northeast. Usually builds its nest in willows, alders, or spruces, from ground level to 5 to 6 m, but usually about 2 m, above ground on divergent branches close to the trunk.

374

SPECIAL HABITAT REQUIREMENTS. Coniferous forests.

Further Reading. United States and Canada: Beal 1915b, DeGraaf and Rudis 1986, Ouellet 1993, Terres 1980. Mexico and Central America: Edwards 1972, Land 1970, Noon 1981, Rappole et al. 1983, Stiles and Skutch 1989. South America: Meyer de Schauensee and Phelps 1978, Rappole et al. 1983. Caribbean: Raffaele 1989, Rappole et al. 1983.

Swainson's Thrush
Catharus ustulatus

Canada (Quebec): Grive à Dos Olive

Costa Rica: Zorzal de Swainson

Guatemala: Tordo Aceitunado, Zorzal de Swainson

Mexico: Zorzalito de Swainson

Venezuela: Paraulata Lomiaceituna

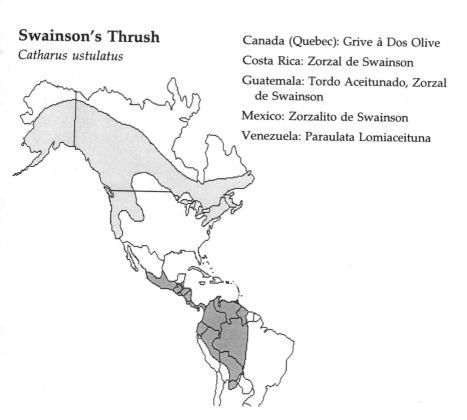

RANGE. Breeds from western and central Alaska, northern Saskatchewan, central Quebec, and Newfoundland south to southern Alaska, southern and east-central California, central Utah, north-central New Mexico, extreme northern Nebraska, eastern Montana, southern Manitoba, northern Minnesota, southern Ontario, northern Pennsylvania, and southern Maine. Also in eastern West Virginia, western Virginia, and western Maryland. Winters from Nayarit and southern Tamaulipas south through Middle America to

Guyana, western Brazil, Peru, Bolivia, and northern Argentina. Not found in the Caribbean.

STATUS. Rare to locally common; long-term and recent declines in eastern North America.

HABITAT. In summer, inhabits dense coniferous forests (especially spruce), dense tall understory vegetation in deciduous or mixed forests, or (especially in New England) recent clearcuts with remaining thickets in low damp areas. In western parts of range, prefers mixed-conifer old growth (Finch and Reynolds 1988, Mannan and Meslow 1984), especially on moist slopes, aspen forests with dense understories, and willow or alder thickets. In winter, frequents deciduous forests. Builds a bulky cup nest, usually near the trunk, on a horizontal branch of a conifer 0.5 to 5 m above the ground. In winter, one of the most abundant birds in young second-growth forests in Costa Rica. Some population declines have been related to losses of winter habitat in Central America (Marshall 1988, Morton 1992).

SPECIAL HABITAT REQUIREMENTS. Damp forests.

Further Reading. United States and Canada: Beal 1915b, DeGraaf and Rudis 1986, Forbush and May 1955, Godfrey 1966, Holmes and Robinson 1988, Noon 1981, Sealy 1974, Terres 1980, Verner and Boss 1980, Winker et al. 1992. Mexico and Central America: Blake et al. 1990, Edwards 1972, Land 1970, Rappole et al. 1983, Stiles and Skutch 1989. South America: Meyer de Schauensee and Phelps 1978, Rappole et al. 1983.

Hermit Thrush
Catharus guttatus

Canada (Quebec): Grive Solitaire

Guatemala: Tordo de Cola Rojiza, Zorzal Colirrojizo

Mexico: Zorzalito Colirrufo

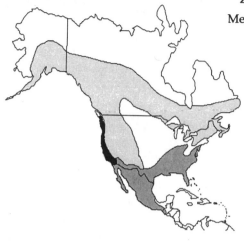

RANGE. Breeds from western and central Alaska, northern Saskatchewan, and Newfoundland south to southern Alaska, in the mountains to southern California, southern Nevada, southern New Mexico, and western Texas, and east of the Rockies to central Alberta, central Wisconsin, southern Ontario, central Pennsylvania, western Virginia, western Maryland, southern New York, and in the Black Hills in southwestern South Dakota. Winters from southern British Columbia and the northeastern United States south through Mexico (except Yucatán Peninsula) to Guatemala, southern Texas, and Florida.

STATUS. Common, generally increasing.

HABITAT. In summer, generally inhabits moist coniferous or mixed forests that have nearly closed or closed canopies. In the western part of range, also occurs in drier conifer types such as ponderosa pine. Is sensitive to intensive tree harvests, especially cuts that fragment mature stands (Keller and Anderson 1992). Usually nests in a depression on the ground, under rock ledges, or under low overhanging limbs. Sometimes locates nest in a shrub or small tree near the ground, especially in the western part of the breeding range. Almost always nests in small white firs in central Arizona (Martin and Roper

1988). During migration and in winter, also inhabits chaparral, riparian woodlands, arid pine-oak associations, and desert scrub.

SPECIAL HABITAT REQUIREMENTS. Relatively undisturbed, rather dense coniferous or mixed forests.

Further Reading. United States and Canada: Beal 1915b, DeGraaf and Rudis 1986, Forbush and May 1955, Godfrey 1966, Holmes and Robinson 1988, Martin and Roper 1988, Noon 1981, Sealy 1974, Szaro and Balda 1982, Terres 1980, Verner and Boss 1980. Mexico and Central America: Edwards 1972, Land 1970, Rappole et al. 1983.

Wood Thrush
Hylocichla mustelina

Canada (Quebec): Grive des Bois

Costa Rica: Zorzal del Bosque

Guatemala: Tordo Maculado, Zorzal Maculado

Mexico: Zorzalito Maculado

RANGE. Breeds from southeastern North Dakota, northern Michigan, northern Vermont, southwestern Maine, and Nova Scotia south to east-central Texas, the Gulf Coast, and northern Florida, and west to eastern South Dakota, central Kansas, and eastern Oklahoma. Migrates through Cuba and the Bahamas. Winters from southern Texas south to Panama.

STATUS. Declining across the breeding range. This species' natural history is the best known among all Neotropical migrants.

HABITAT. Inhabits interiors and edges of cool, mature, lowland deciduous or mixed forests, particularly damp woodlands near swamps or water. In New England, also found on wooded slopes. May be somewhat tolerant of forest fragmentation on breeding grounds, commonly found in woodlots 1 to 5 ha in size (Whitcomb et al. 1981). Builds a compact cup nest on a horizontal limb, in a fork of a sapling or tree, or well hidden in dense shrubbery, generally 2 to 15 m (average 3 m) above the ground. A moist tropical forest habitat specialist during winter on the Yucatán Peninsula. Common in tall evergreen forests (the *selva alta perenifolia* [Pennigton and Sarukhan 1968:5–11]) of the Caribbean lowlands of Middle America. Occurs primarily within the forest interior in Costa Rica but also found in early successional habitats during migration. Common in wet, lowland tropical forests from Veracruz, Mexico, to Panama. All tropical studies of the Wood Thrush show that it is found in moist tropical forest understory (Morton 1992). Second growth is apparently an important winter habitat in Belize (Petit et al. 1992), Costa Rica (Blake and Loiselle 1992), and Panama (Martin and Karr 1986). In addition to territorial birds, there are floaters which use patches of abundant fruit in the canopy as an alternative foraging habitat when forests are saturated with territorial birds (Rappole et al. 1992, Winker et al. 1990). Floaters may be mostly younger birds (Morton 1992) and suffer higher mortality from birds and mammals than do sedentary birds (Rappole et al. 1989). Carrying capacity of tropical wintering grounds is potentially reduced by conversion of large amounts of primary tropical forest habitat. In the Tuxtla Mountains of Veracruz, the Wood Thrush declined 70% from 1960 to 1985 because of deforestation. It may be breeding farther south in Central America; uncommon in the 1970s, the Wood Thrush is now common in Parque Nacional Soberania in Panama (Morton 1992).

SPECIAL HABITAT REQUIREMENTS. Mature moist deciduous or mixed forests with closed canopies.

Further Reading. United States and Canada: Beal 1915b, Brackbill 1943, DeGraaf and Rudis 1986, Forbush and May 1955, Terres 1980. Mexico and Central America: Edwards 1972, Land 1970, Lynch 1989, 1992, Rappole et al. 1983, 1989, Stiles and Skutch 1989.

American Robin

Turdus migratorius

Canada (Quebec): Merle Américain

Guatemala: Mirlo Migratorio,
Sensontle Migratorio

Mexico: Zorzal Pechirrojo

Puerto Rico: Mirlo Norteamericano

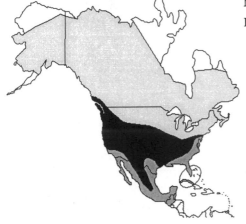

RANGE. Breeds from western and northern Alaska, southeastern Northwest Territories, northern Quebec, Labrador, and Newfoundland south to southern California, central and southeastern Arizona, Mexico, southern Texas, and central Florida. Winters from southern Alaska, the northern United States, and Newfoundland south to Baja California, southern Texas, and southern Florida, throughout Mexico to Guatemala, the Bahamas, and Cuba.

STATUS. Abundant.

HABITAT. Very widely distributed. Breeds in forests, open woodlands, scrub, farmlands, grasslands with scattered trees, from tree limit in sparsely wooded barrens up to 3,800 m in the mountains of western North America, along forest borders, hedges, orchards, gardens, city parks, and in suburban yards; production low in urban/suburban habitats. Extended its range westward throughout the Great Plains with settlement. Places nest, constructed of mud and vegetation, on almost any substantial support, usually in a fork or on a horizontal branch of a shrub or tree; rarely on the ground.

SPECIAL HABITAT REQUIREMENTS. Mud for nest building.

Further Reading. United States and Canada: Beal 1915b, Forbush and May 1955, Knupp et al. 1977, Terres 1980, Wheelwright 1986, Young 1955. Mexico and Guatemala: Edwards 1972, Land 1970, Rappole et al. 1983. Caribbean: Rappole et al. 1983.

Gray Catbird
Dumetella carolinensis

Canada (Quebec): Moqueur-chat

Costa Rica: Pájaro-Gato Gris

Guatemala: Maullador, Pájaro-Gato Norteño

Mexico: Mímido Gris

West Indies: Zorzal Gato

RANGE. Breeds from southern British Columbia, southern Ontario, and Nova Scotia, south to central New Mexico and northern Florida, and west to northern and south-central Washington, south-central and eastern Oregon, north-central Utah, and central and northeastern Arizona. Winters from north-central and eastern Texas, the central portions of the Gulf States, and the Atlantic Coast lowlands from Long Island, New York, south along the Gulf-Caribbean slope of Central America to Panama. Also winters in the Bahamas, Cuba, Jamaica, and Hispaniola.

STATUS. Common, but generally declining.

HABITAT. Prefers dense thickets of shrubby edge habitat, but also inhabits shrubs, briars, vines along woodland borders, regenerating hardwoods, dry marsh edges, roadside shrubs, old house sites, abandoned fields, shelterbelts, and fence rows; in drier parts of the West, low moist thickets on forest edges or open places, riparian thickets, aspen forests with dense understories. Not found in coniferous forests. Nests about 1 to 3 m above the ground in almost any dense woody vegetation such as multiflora rose, barberry, osage orange, hedges, or occasionally a coniferous tree. A habitat generalist on wintering grounds on the Yucatán Peninsula. Found in dense shrubs in mature forest understory, second growth, or scrub thickets in Tuxtla Mountains, Veracruz (Rappole and Warner 1980, Rappole et al. 1992).

SPECIAL HABITAT REQUIREMENTS. Low, dense, shrubby vegetation.

Further Reading. United States and Canada: Darley et al. 1977, DeGraaf and Rudis 1986, Forbush and May 1955, Godfrey 1966, Johnson and Best 1982, Nickell 1965, Terres 1980, Yahner 1991. Mexico and Central America: Edwards 1972, Land 1970, Lynch 1989, Rappole et al. 1983, Ridgely and Gwynne 1989, Stiles and Skutch 1989. Caribbean: Bond 1947, Rappole et al. 1983.

Sage Thrasher
Oreoscoptes montanus

Canada (Quebec): Moqueur des Armoises

Mexico: Mímido Pinto

RANGE. Breeds largely between the Sierra Nevada-Cascade Mountain axis and the Rocky Mountains from southern British Columbia (very locally), Montana, southwestern Saskatchewan (very locally), and Wyoming south to east-central California, southern Nevada, northern Arizona, New Mexico, and northwestern Texas. Winters from northern Arizona and central Texas south to Chihuahua and Tamaulipas, Mexico.

STATUS. Common throughout most its range.

HABITAT. Mainly limited to semiarid sagebrush plains, but may extend into lower-elevation junipers and mountain-mahogany habitats near sagebrush. Sometimes nests on the ground under sagebrush, but usually in branches near the main stem of the tallest, densest sagebrush plants (Peterson and

Best 1991), usually less than 1 m above ground. May also nest in other low-growing shrubs such as greasewood, horsebrush, rabbitbrush, and saltbush.

SPECIAL HABITAT REQUIREMENTS. Sagebrush.

Further Reading. United States and Canada: Farrand 1983c, Godfrey 1966, Reynolds 1981, Reynolds and Rich 1978, Rich 1978, Rotenberry and Wiens 1989, Terres 1980. Mexico: Edwards 1972, Rappole et al. 1983.

Bendire's Thrasher
Toxostoma bendirei

Mexico: Cuitlacoche Piquicorto

RANGE. Breeds from southeastern California, southern Nevada, southern Utah, and western New Mexico south to southern Sonora, Mexico. Winters from central-southern Arizona to Sinaloa, Mexico. Rare in fall north to the southern coast of California.

STATUS. Locally common.

HABITAT. Inhabits open desert habitats, especially areas with tall vegetation, cholla, creosote bush, and yucca. May also inhabit pinyon-juniper-sage communities, but tends to avoid large areas of continuous, dense brushy cover and grasslands. Builds a compact nest in almost any of the shrubby vegetation found within its habitat, usually 1 to 3 m above the ground.

SPECIAL HABITAT REQUIREMENTS. Desert communities.

Further Reading. United States and Canada: Bent 1948, Farrand 1983c, Phillips et al. 1964, Terres 1980. Mexico: Edwards 1972.

American Pipit
Anthus rubescens

Canada (Quebec): Pipit Commun

Guatemala: Bisbita de Agua, Cachirla de Agua

Mexico: Bisbita Americana

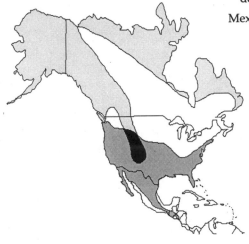

RANGE. Breeds on the arctic tundra and in the mountains of western North America and in Maine. Winters on the Pacific Coast from British Columbia south, and from the central United States south through Mexico to El Salvador.

STATUS. Common.

HABITAT. Breeds in arctic tundra and alpine tundra and meadows with rough features such as tussocks, tilted rocks, or eroded spots for nest sites. Requires nesting habitat that is free of snow early in the breeding season; prefers moss-grown slopes with southern exposures. In migration, occurs in grasslands. In winter, found in bare fields and grasslands, lake shores, and pastures or other completely open terrain. Builds nest on the ground in the shelter of a rock or bank, beside a mossy hummock, or at the base of a tussock.

SPECIAL HABITAT REQUIREMENTS. Alpine or arctic tundra with some rough features.

Further Reading. United States and Canada: Forbush and May 1955, Gibb 1956, Godfrey 1966, Miller and Green 1987, Sutton and Parmalee 1954, Terres 1980, Verbeek 1970. Mexico and Central America: Edwards 1972, Land 1970, Rappole et al. 1983.

384

Sprague's Pipit
Anthus spragueii

Canada (Quebec): Pipit des Prairies
Mexico: Bisbita Llanera

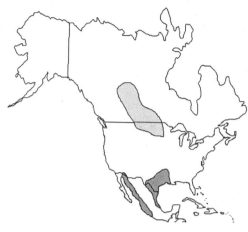

RANGE. Breeds from north-central Alberta, central Saskatchewan, and west-central and southern Manitoba south to Montana, South Dakota, and Minnesota. Winters from Arizona south along the Pacific states of Mexico to Guerrero and from Mississippi, Louisiana, Arkansas, and Texas south in Mexico to Veracruz.

STATUS. Common; long-term and recent decline in the western part of the breeding range.

HABITAT. Primarily inhabits extensive areas of native shortgrass prairie or other grasslands dominated by grasses of short to medium height. Also inhabits large alkaline meadows and meadow zones of large alkali lakes. In winter usually found walking about in fields of tall grass or other treeless areas with considerable grass cover. Constructs nest of grasses in hollows on the ground, and in clumps of grasses or sedges.

SPECIAL HABITAT REQUIREMENTS. Extensive prairie; in migration and winter, also pastures and weedy fields.

Further Reading. United States and Canada: Bent 1950, Forbush and May 1955, Godfrey 1966, Johnsgard 1979, Terres 1980. Mexico: Edwards 1972, Rappole et al. 1983.

Cedar Waxwing
Bombycilla cedrorum

Canada (Quebec): Jaseur des Cèdres

Costa Rica: Ampelis Americano

Guatemala: Ampelis Americano, Capuchino

Mexico: Ampelis Americano, Ampelis Chinito

Puerto Rico: Picotera

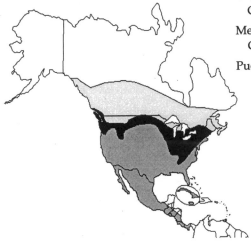

RANGE. Breeds from southeastern Alaska, central British Columbia, Alberta, Saskatchewan, northern Manitoba, Ontario, central Quebec, and Newfoundland south to northern California, Nevada, Utah, Colorado, South Dakota, central Missouri, Illinois, Indiana, northern Georgia, western North Carolina, and Virginia. Winters from southern British Columbia, Montana, Saskatchewan, Manitoba, Ontario, New York, and New England south to Panama, and throughout the Greater and Lesser Antilles.

STATUS. Locally common; generally increasing.

HABITAT. Inhabits a wide variety of habitats that provide berries: open coniferous and deciduous forests, forest and wooded marsh edges, farmsteads, shelterbelts, parks, residential areas, and open areas with scattered trees, but absent from dense forests. Nests semicolonially in dense coniferous thickets (often cedar) but will use a variety of deciduous trees and shrubs. Places nest on a horizontal limb, or in a crotch next to the main trunk, 2 to 16 m above the ground. Parasitized by cowbirds. During winter, found in flocks where trees and shrubs with persistent fruits are present.

SPECIAL HABITAT REQUIREMENTS. Fruit- and berry-producing trees and shrubs.

Further Reading. United States and Canada: DeGraaf and Rudis 1986, Farrand 1983c, Forbush and May 1955, Godfrey 1966, Lea 1942, Leck and Cantor 1979, McPherson 1987, Putnam 1949, Terres 1980. Mexico and Central America: Edwards 1972, Land 1970, Rappole et al. 1983, Stiles and Skutch 1989. Caribbean: Raffaele 1989, Rappole et al. 1983.

Phainopepla

Phainopepla nitens

Mexico: Capulinero Negro

RANGE. Breeds from central California, southern Nevada, southern Utah, southern New Mexico, and western Texas, south to Baja California and into Mexico. Winters from southern California, southern Nevada, central Arizona, southern New Mexico, and western and southern Texas south in Mexico to Puebla and Veracruz.

STATUS. Locally common to uncommon or rare.

HABITAT. In deserts, primarily inhabits washes, riparian areas, and other habitats that support a brushy growth of mesquite and paloverde. In more northern and coastal areas, inhabits oak chaparral and riparian oak woodlands. Nest (built almost exclusively by the male) is in a forked limb of a mesquite, cottonwood, hackberry, willow, sycamore, oak, or citrus tree, often in a clump of mistletoe, 2 to 16 m above the ground.

SPECIAL HABITAT REQUIREMENTS. Trees or shrubs and berries (especially mistletoe).

Further Reading. United States and Canada: Farrand 1983c, Phillips et al. 1964, Terres 1980, Verner and Boss 1980. Mexico: Edwards 1972.

Loggerhead Shrike
Lanius ludovicianus

Canada (Quebec): Pie-grièche Migratrice

Mexico: Verdugo Americano

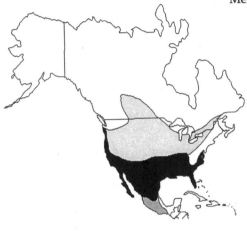

RANGE. Breeds from central Alberta, central Saskatchewan, southern Manitoba, Minnesota, central Wisconsin, central Michigan, and southeastern Ontario, south to Mexico and the Gulf Coast. Very rare or absent from most of the Appalachians, Pennsylvania, New York, and New England. Winters in the southern half of the United States and south to Oaxaca, Mexico.

STATUS. A species of concern in the East; now also declining in parts of the western and central United States. The San Clemente subspecies, *mearnsi*, is endangered. Habitat loss and pesticides may be responsible, but declines are poorly understood (Dobkin 1992).

HABITAT. Inhabits open country with scattered shrubs or small trees such as shelterbelts, cemeteries, farmsteads, or hedgerows in the Great Plains and Midwest. In the West, breeds in savanna, pine-oak woodlands, pinyon-juniper woodland, shrubsteppe, and scrub and chapparal; prefers very open stands. Builds a bulky, cup-shaped nest in a variety of shrubs and low, dense trees, rarely less than 1 m or more than 8 m above the ground. Hides the nest well below the crown of the bush or tree. Sometimes remodels old nests or nests of other passerines. Maintenance of short-grass habitats and shelterbelts may be important to Loggerhead Shrikes in the southeastern United States. Survival on the wintering ground may be responsible for declines of breeding populations in the Midwest (Brooks and Temple 1990) and the Southeast (Gawlik and Bildstein 1990). Low rates of return, attributable to low breeding site fidelity, make assessment of winter mortality difficult (Haas and Sloane 1989).

Further Reading. United States and Canada: Bartgis 1992, Beal and McAtee 1912, Farrand 1983c, Godfrey 1966, Hunter 1990, Johnsgard 1979, Kridelbaugh 1983, Miller 1931, Morrison 1980, 1981, Porter et al. 1975, Tate and Tate 1982. Mexico: Edwards 1972, Rappole et al. 1983.

White-eyed Vireo
Vireo griseus

Canada (Quebec): Viréo aux Yeux Blancs

Costa Rica, Guatemala, and Mexico: Vireo Ojiblanco

Puerto Rico: Julián Chibi Ojiblanco

RANGE. Breeds from southeastern Nebraska and central Iowa to southern Michigan, southern Ontario, and southern Massachusetts, south through eastern Texas and Florida to San Luis Potosí, northern Hidalgo, northern Veracruz, and Tamaulipas. Winters from southern Texas south to Honduras and east across the Gulf Coast to Florida, north to central coastal North Carolina. Also winters on Cuba and in the Bahamas.

STATUS. Common; declining in the central United States.

HABITAT. Inhabits deciduous thickets, woodland edges, brambles, woodland undergrowth, hedgerows, and the dense understory of bottomland forests, generally favoring open woodlands or thickets near water. Builds a cone-shaped cup nest that is suspended from forked twigs of a low shrub or tree, usually well concealed, 0.5 to 2 m above the ground. Found in thickets, scrub, and brushy woodland during winter on the Yucatán Peninsula.

Common in mature evergreen forest in Tuxtla Mountains, Veracruz (Rappole and Warner 1980).

SPECIAL HABITAT REQUIREMENTS. Low shrubby vegetation, scrub, brushy woodland.

Further Reading. United States and Canada: Chapin 1925, Farrand 1983c, Forbush and May 1955, Johnsgard 1979, McComb et al. 1989, Nolan and Wooldridge 1962. Mexico and Central America: Edwards 1972, Land 1970, Lynch 1989, 1992, Rappole et al. 1983. Caribbean: Raffaele 1989, Rappole et al. 1983.

Bell's Vireo
Vireo bellii

Guatemala and Mexico: Vireo de Bell

RANGE. Breeds from southern California (local and rare), southern Nevada, Arizona, and southern New Mexico north into the Midwest (east of the Rocky Mountains) to North Dakota and east to Illinois and south to southwestern Tennessee, Arkansas, northwestern Louisiana, Texas, northern Baja California, and southern Sonora to southern Tamaulipas. Winters from southern Baja California, southern Sonora, and Veracruz along both slopes of Middle America south to Nicaragua.

STATUS. Rare and local in interior California (Tehama County), widespread decline throughout breeding range.

HABITAT. Throughout most its range, inhabits dense riparian thickets, especially willows. In the arid Southwest, found along water courses and marshes where mesquite is mixed with cottonwood, saltcedar, elderberry, and desert hackberry. In the Great Plains, generally associated with thickets near streams and rivers, or with second-growth scrub, forest edges, and brush patches. Generally found in the same habitats through the year. Builds a small, basketlike cup nest attached to a forked branch of mesquite, hackberry, catclaw, oak, willow, ash, cottonwood, or low shrub, usually near water and seldom more than 2 m above the ground. Widespread declines have been linked closely to destruction of riparian habitats by channelization and livestock grazing, and to cowbird parasitism (Dobkin 1992).

SPECIAL HABITAT REQUIREMENTS. Dense brush, riparian thickets.

Further Reading. United States and Canada: Baird and Rieger 1989 (California), Barlow 1962, Chapin 1925, Farrand 1983c, Forbush and May 1955, Greaves 1989 (California), Hendricks and Rieger 1989 (California), Johnsgard 1979, Tate and Tate 1982, Verner and Boss 1980. Mexico and Central America: Edwards 1972, Land 1970, Rappole et al. 1983.

Black-capped Vireo
Vireo atricapillus

Mexico: Vireo Gorrinegro

RANGE. Breeds in central Oklahoma locally through central and western Texas to north-central Mexico. Winters mainly in western Mexico.

STATUS. Fairly common to uncommon.

HABITAT. Generally inhabits the dense, low, ragged-topped thickets growing in hot, rocky hillsides, including stands of oaks, mescalbean, sumac, cedar, or other chaparral brush; prefers scrub oaks. May also be found in prairie ravines and early successional stages of brushlands. Prefers winter habitat arranged in rectangular or circular rather than linear shape (small amount of edge relative to area). Builds a deep, cuplike nest suspended from a fork of slender twigs in trees or shrubs, usually 1 to 2 m above the ground.

Prefers oaks in parts of its range, but uses other tree and shrub species. In winter in western Mexico (Sinaloa south to Chiapas) probably restricted to undisturbed tropical deciduous forest and therefore likely sensitive to deforestation in the lowlands (Hutto 1992).

SPECIAL HABITAT REQUIREMENTS. Low, dense shrubs or trees.

Further Reading. United States and Canada: Farrand 1983c, Graber 1961, Johnsgard 1979, Oberholser 1974b. Mexico: Edwards 1972, Rappole et al. 1983.

Gray Vireo
Vireo vicinior

Mexico: Vireo Gris

RANGE. Breeds locally from southern California, southern Nevada, southern Utah, and northwestern and central New Mexico south to Baja California, central and southeastern Arizona, southern New Mexico, western Oklahoma, western Texas, and central Mexico. Winters in southern Arizona, western Texas, and western Mexico.

STATUS. Rare to locally common.

HABITAT. Inhabits thorn scrub, oak-juniper woodland, pinyon-juniper, dry chaparral, mesquite, and riparian willow habitats. Favors dry chaparral that forms a continuous thicket of twigs up to 2 m above ground. In winter found in arid open country with scattered small trees and thickets, or even in rather dense but scrubby vegetation. Builds a basketlike cup nest suspended 1 to 3 m above the ground, from the forks of twigs in a variety of low, thorny shrubs and small trees.

SPECIAL HABITAT REQUIREMENTS. Low, dense scrubs.

Further Reading. United States and Canada: Chapin 1925, Farrand 1983c, Johnsgard 1979, Terres 1980. Mexico: Edwards 1972, Rappole et al. 1983.

Solitary Vireo
Vireo solitarius

Canada (Quebec): Viréo à Tête Bleue
Costa Rica: Vireo Solitario
Guatemala: Vireo Anteojillo
Mexico: Vireo Anteojillo

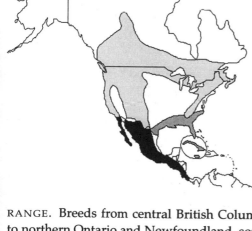

RANGE. Breeds from central British Columbia east through central Canada to northern Ontario and Newfoundland, southwest of and through the Rockies to southern California and west Texas, south through Mexico to Honduras, and east of the Rockies to North Dakota, Illinois, and Massachusetts; in the Appalachian and Piedmont regions to eastern Tennessee, Alabama, Georgia, South Carolina, North Carolina, Virginia, and Maryland. Winters from southern California, central Texas, the northern portions of the Gulf States, and North Carolina south to Costa Rica, and also on Cuba and Jamaica. Occurs as a migrant in the Bahamas.

STATUS. Common; increasing in eastern and western parts of the breeding range.

HABITAT. Usually inhabits coniferous or mixed forests, especially spruce and tamarack swamps in eastern parts of its range, montane woodlands in the West. Seems to prefer open mixed forests with considerable undergrowth. More abundant in rotation-age than in old-growth forest in the West (Hejl and Woods 1991). In winter found generally in open or dense woodland or woodland edges. Usually 5 to 10 conifers per hectare are sufficient for the species to occur in hardwood stands in the Northeast. Builds a deep cup nest that is suspended from the fork of a horizontal branch, generally 1 to 6 m above the ground, often about midway in a small conifer, but occasionally in a small deciduous tree or shrub. Commonly parasitized by cowbirds, especially in parts of the West (Marvil and Cruz 1989).

Further Reading. United States and Canada: DeGraaf and Rudis 1986, Farrand 1983c, Forbush and May 1955, Godfrey 1966, Harrison 1975, Johnsgard 1979, Terres 1980. Mexico and Central America: Edwards 1972, Land 1970, Rappole et al. 1983, Stiles and Skutch 1989. Caribbean: Rappole et al. 1983.

Yellow-throated Vireo
Vireo flavifrons

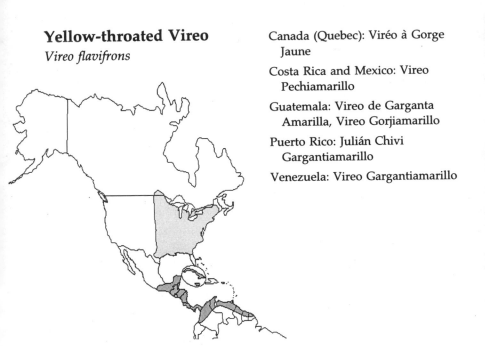

Canada (Quebec): Viréo à Gorge Jaune

Costa Rica and Mexico: Vireo Pechiamarillo

Guatemala: Vireo de Garganta Amarilla, Vireo Gorjiamarillo

Puerto Rico: Julián Chivi Gargantiamarillo

Venezuela: Vireo Gargantiamarillo

RANGE. Breeds from southern Manitoba, Minnesota, southern Ontario, New Hampshire, and southwestern Maine south to eastern Texas, the Gulf Coast, and central Florida and west to the Dakotas, Nebraska, Kansas, Oklahoma, and west-central Texas. Winters mainly in Mexico and Central and South America from Colombia to French Guiana, including Trinidad and Tobago; a few winter in Florida. Some also winter in the Bahamas, Cuba, and Jamaica.

STATUS. Rather uncommon to common; increasing in the eastern and central United States.

HABITAT. In summer, inhabits mature, moist deciduous forests, especially river-bottom forests, riparian woodlands, or (in southern parts of range) north-facing slopes; prefers open woodlands with partially open canopies. Found less frequently in wooded residential areas, seldom in dense deciduous forests, and rarely in coniferous forests. Suspends nest between the forks of a slender branch, usually near the trunk of a deciduous tree, preferably a large oak, normally more than 10 m above the ground (range 1 to 20 m). Extremely vulnerable to deforestation of tropical wintering grounds (Morton 1992). In winter, occurs in seasonal deciduous, broadleaf evergreen, and gallery forests (Rappole et al. 1983).

SPECIAL HABITAT REQUIREMENTS. Open, moist, mature deciduous woodlands.

Further Reading. United States and Canada: Chapin 1925, Forbush and May 1955, Godfrey 1966, James 1976, Johnsgard 1979. Mexico and Central America: Edwards 1972, Land 1970, Rappole et al. 1983, Stiles and Skutch 1989. South America: Meyer de Schauensee and Phelps 1978, Rappole et al. 1983. Caribbean: Raffaele 1989, Rappole et al. 1983.

Warbling Vireo
Vireo gilvus

Canada (Quebec): Viréo Mélodieux
Costa Rica: Vireo Canoro
Guatemala: Vireo Chipe
Mexico: Vireo Gorjeador Norteño

RANGE. Breeds locally from southeastern Alaska, northern British Columbia, and southern Northwest Territories southeast to southern Ontario and New Brunswick south through Middle America and parts of northern and western South America. Winters from central Mexico south through the breeding range. Resident populations in Middle and South America are considered to be a separate species, *Vireo leucophrys*, by some authors.

STATUS. Common and widespread, generally increasing.

HABITAT. Inhabits mature riparian woodlands and floodplain forests; also open deciduous and mixed deciduous-coniferous forests, wooded farmsteads, shelterbelts, deciduous groves, patches within coniferous forest, scrubby hillside trees, and street trees in residential areas. In mixed forests, generally associated with deciduous trees, and prefers forests with a sub-

396

stantial herb or shrub layer and low to intermediate canopy cover. More abundant in young (40–75 yr old) than in older (>105 yr old) Douglas-fir stands in the Pacific Northwest (Manuwal and Huff 1987). Present only in partially opened stands in Idaho (Medin 1985). Builds a cup nest that is usually suspended from a horizontal branch of a deciduous tree, generally in branches well away from the trunk and higher than those of other vireos (6 to 30 m above the ground). Commonly parasitized by cowbirds.

SPECIAL HABITAT REQUIREMENTS. Scattered deciduous trees or wooded streamsides.

Further Reading. United States and Canada: Chapin 1925, DeGraaf and Rudis 1986, Forbush and May 1955, Godfrey 1966, Harrison 1975, James 1976, Johnsgard 1979. Mexico and Central America: Edwards 1972, Land 1970, Rappole et al. 1983, Stiles and Skutch 1989.

Philadelphia Vireo
Vireo philadelphicus

Canada (Quebec): Viréo de
   Philadelphie
Costa Rica: Vireo Amarillento
Guatemala: Vireo Gris-oliva
Mexico: Vireo Filadélfico

RANGE. Breeds from east-central British Columbia to central Manitoba and southwestern Newfoundland, south to south-central Alberta, north-central North Dakota, northeastern Minnesota, southern Ontario, northern New

Hampshire, northern Vermont, and Maine. Winters to western Panama and casually to Colombia west of the Andes.

STATUS. Uncommon to rare; may be increasing in the eastern parts of breeding range.

HABITAT: Inhabits open, early-successional deciduous, coniferous, or mixed forests, woodland edges, burned or cutover areas with young deciduous regeneration, and willow and alder thickets along streams. In aspen groves in western parts of its range. In winter found in woodland or woodland edges, or in open country with scattered trees. Builds a deep cup nest that is suspended from a horizontal, forked branch of a deciduous tree or shrub, usually 3 to 12 m above the ground.

SPECIAL HABITAT REQUIREMENTS. Early successional forest.

Further Reading. United States and Canada: Chapin 1925, DeGraaf and Rudis 1986, Johnsgard 1979. Mexico and Central America: Edwards 1972, Land 1970, Rappole et al. 1983, Stiles and Skutch 1989. South America: Hilty and Brown 1986, Rappole et al. 1983.

Red-eyed Vireo
Vireo olivaceus

Canada (Quebec): Viréo aux Yeux Rouges

Costa Rica and Guatemala: Vireo Ojirrojo

Mexico: Vireo Ojirrojo Norteño

Puerto Rico and Venezuela: Julián Chivi Ojirrojo

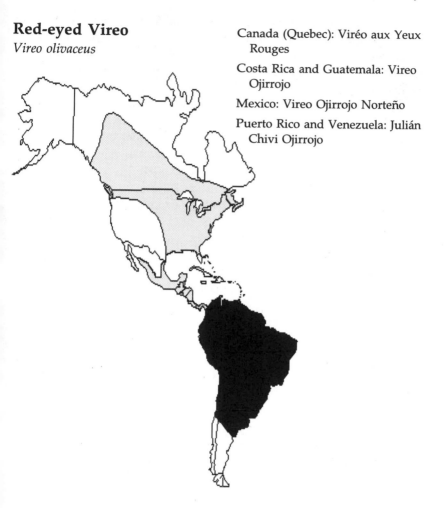

RANGE. Breeds from southwestern British Columbia and southern Northwest Territories southeast to central Ontario and the Maritime Provinces, south to Oregon, Colorado, Oklahoma, Texas, the Gulf Coast, and Florida, south through Middle America (except northern Mexico) and the northern two-thirds of South America. Winters in South America. Middle and South American breeding populations are recognized as separate species by some authors, *Vireo flavoviridis* and *Vireo chivi*, respectively.

STATUS. Common; increasing in eastern part of breeding range.

HABITAT. Inhabits open deciduous and mixed forests with moderate to dense understories, wooded clearings, borders of burns. The best habitats in the mixed mesophytic forest may be areas with high basal area and low understory density. Found in both upland and river-bottom forests, and sometimes in residential areas where abundant shade trees provide a fairly continuous canopy. Seldom found where conifers make up 75% or more of the basal area. Builds nest in deciduous or coniferous trees or shrubs. Suspends deep cup nest from a horizontal fork of a slender branch, usually in dense foliage 2 to 3 m above the ground, but sometimes as high as 20 m. In winter, commonly found in sparsely wooded habitats.

SPECIAL HABITAT REQUIREMENTS. Deciduous trees with dense understory.

Further Reading. United States and Canada: Chapin 1925, Darveau et al. 1992, Forbush and May 1955, Graham 1988, James 1976, Johnsgard 1979, Lawrence 1953a, McComb 1985, Robinson 1981, Williamson 1971. Mexico and Central America: Edwards 1972, Land 1970, Rappole et al. 1983, Stiles and Skutch 1989. South America: Meyer de Schauensee and Phelps 1978, Rappole et al. 1983. Caribbean: Raffaele 1989, Rappole et al. 1983.

Yellow-green Vireo
Vireo flavoviridis

Costa Rica: Cazadora, Fraile, Vireo Cabecigris
Guatemala: Vireo Cabecigris
Mexico: Vireo Ojirrojo Tropical

RANGE. Breeds from southern Texas and western Mexico to central Panama. Winters in the western Amazon basin.

STATUS. Common to abundant during the breeding season in Guatemala, Costa Rica, and Panama.

HABITAT. Open woodlands, shrubby clearings with scattered trees, coastal scrub and mangroves, savannas, and in gardens at lower elevations. Found in canopy and middle levels of both deciduous and evergreen forests. Builds a cup nest attached by its rim to the arms of a stout forked twig, of varied materials, including grass blades, fragments of broad leaves, strips of papery epidermis and bark, and lined with slender, curved rachises, 1 to 10 m above the ground in a shrub or a tree, usually below 4 m.

Further Reading. Mexico and Central America: Ridgely and Gwynne 1989, Stiles and Skutch 1989.

Black-whiskered Vireo
Vireo altiloquus

Cuba: Bien-te-veo

Dominican Republic and Puerto Rico: Bien-te-veo, Chavos-por-él, Juan Chivi, Julián Chivi, Julián Chivi Bigotinegro

Haiti: Oiseau Canne, Petit Panache

Lesser Antilles: Choueque, Père Gris, Piade

Mexico: Vireo Bigotinegro

Venezuela: Julián Chivi Bigotinegro

RANGE. Breeds in central and southern Florida and on islands in the Caribbean; winters in Colombia, Venezuela, Guyana, Surinam, and Brazil.

STATUS. Common within much of its limited U.S. range.

HABITAT. In the United States found mostly in mangroves along the coasts of Florida and the Florida Keys. In winter will utilize palms as well as mangroves. Suspends deep cup nest between twigs (usually in red mangroves but occasionally in other trees or shrubs) as high as 5 m above the ground or water.

SPECIAL HABITAT REQUIREMENTS. Dense mangroves.

Further Reading. United States and Canada: Bent 1950, Chapin 1925, Farrand 1983c, Terres 1980. South America: Meyer de Schauensee and Phelps 1978, Rappole et al. 1983. Caribbean: Bond 1947, Raffaele 1989, Rappole et al. 1983.

Bachman's Warbler
Vermivora bachmanii

Mexico: Chipe de Bachman

RANGE. Formerly bred in northeastern Arkansas, southeastern Missouri, south-central Kentucky, central Alabama, and southeastern South Carolina. May still breed in South Carolina. Formerly migrated through Florida and the Keys to winter in western Cuba and possibly the Isle of Pines.

STATUS. Rarest songbird in North America; endangered, possibly extinct; no breeding population known. Only two observations in Florida between 1949 and 1971 (Stevenson 1972). Reasons for its decline are not clear. May have been associated with abandoned rice-growing areas in the Southeast (Hooper and Hamel 1977).

HABITAT. Inhabited low, moist, secondary successional openings in deciduous woodlands and swamps of the southern coastal plain, where it probably occurred primarily in sweet bay–swamp tupelo–red maple associations of headwater swamps, sweet gum–willow oak associations of wet flats, and in openings in bottomland hardwoods. Appeared to use forested bottomlands and headwater swamps that were inundated with water for relatively short periods of time. Located nest in canebrakes and thickets about 1 m above the ground in and along the margins of low, wet forested habitats. Wintering habitat is now limited; the area is agricultural, mostly planted to sugarcane since the 1500s, though Bachman's Warbler was fairly abundant on migration in Florida in the 1880s to 1890 (Brewster 1891). Decline may have been due to loss of winter habitat (Terborgh 1974) and a population bottleneck from which the species could not recover because of resultant limited genetic variability (Rappole et al. 1983). Decline of cane breeding habitat also paralleled the bird's decline (Remsen 1986). See Morse 1989 for concise review of the bird's decline and possible causes.

SPECIAL HABITAT REQUIREMENTS. Disturbed areas in southern deciduous swamp forests (Morse 1989:257); perhaps (former) extensive stands of cane (Meanley 1971).

Further Reading. United States and Canada: Farrand 1983c, Griscom and Sprunt 1979, Hamel 1986, Harrison 1984, Hooper and Hamel 1977, Mackenzie 1977, Meanley and Mitchell 1958, Morse 1989, Stevenson 1972. Caribbean: Bond 1947, Rappole et al. 1983.

Blue-winged Warbler
Vermivora pinus

Canada (Quebec): Fauvette à Ailes Bleues

Costa Rica: Reinita Aliazul

Guatemala: Chipe Aliazul, Gusanero Aliazul, Reinita Aliazul

Mexico: Chipe Aliazul

RANGE. Breeds from eastern Nebraska and southeastern Minnesota east to southern Vermont and southern Maine, and south to northwestern Arkansas, northern Alabama, northern Georgia, western South Carolina, and Delaware. Range is expanding in the Northeast. Migrates through the Bahamas and Cuba. Rare elsewhere in the West Indies. Winters in Mexico and northern Central America, casually to Panama.

STATUS. Uncommon.

HABITAT. Inhabits rank growth near the borders of swamps or streams, woodland edges, brushy overgrown fields and pastures, thickets, and second-growth woods. Prefers old fields with saplings greater than 3 m tall. Prefers moister habitats than the Golden-winged Warbler, a closely related species with which Blue-winged hybridizes. (The Brewster's Warbler is thought to result from the original cross of a Blue-winged and Golden-winged, and is more common than the rarer Lawrence's, which represents a combination of recessive characteristics from crosses among hybrids.) Builds nest on the ground, attached to upright stems of weeds or grass clumps among bushes, ferns, tangles of vines, or grasses. Usually nests in loose aggregations or colonies within a larger area of similar habitat (Ficken and Ficken 1968, Gill 1980, Morse 1989:100). In winter prefers second growth, semi-open areas, and hedgerows (Robbins et al. 1983). A tropical forest specialist on the Yucatán Peninsula in winter (Lynch 1989). On spring migration, common in crowns of mature forest trees feeding on caterpillars with other warblers, especially Chestnut-sided.

SPECIAL HABITAT REQUIREMENTS. Brushy open habitats, commonly near water.

Further Reading. United States and Canada: Confer and Knapp 1981, DeGraaf and Rudis 1986, Godfrey 1966, Griscom and Sprunt 1979, Harrison 1975, Harrison 1984, Johnsgard 1979. Mexico and Central America: Edwards 1972, Land 1970, Rappole et al. 1983, Stiles and Skutch 1989. Caribbean: Rappole et al. 1983.

Golden-winged Warbler
Vermivora chrysoptera

Canada (Quebec): Fauvette à Ailes Dorées

Costa Rica, Puerto Rico, and Venezuela: Reinita Alidorado

Guatemala: Chipe Alidorado, Gusanero Alidorado, Reinita Alidorado

Mexico: Chipe Alidorado

RANGE. Breeds from southern Manitoba, central Minnesota, and northern Wisconsin east to southern Vermont and eastern Massachusetts, and south to southeastern Iowa, southern Ohio, and southern Connecticut, and in the Appalachian Mountains south to northern Georgia. Breeding range in the Northeast and Appalachians has been decreasing since the late 1970s, partly as a result of displacement by Blue-winged Warblers and loss of early-successional habitat. Migrants pass through the Bahamas and Cuba. Winters in Central America, Colombia, and Venezuela; rarely in the West Indies.

STATUS. Locally common; has been declining.

HABITAT. Inhabits short-lived openings in deciduous forests or old fields where there is a dense understory of forbs, grasses, or ferns. Also inhabits damp fields heavily vegetated with thick grass, overgrown pastures, dense scrubby thickets, second-growth woods, and brush-bordered lowland areas. Generally occupies higher and earlier-successional drier habitats than the Blue-winged Warbler, although there is a broad overlap in habitats. Nests in loose aggregations or colonies on or close to the ground. Usually locates nest within the shade of a forest edge, supported by weed stalks such as goldenrod, or by tufts of grass, or on a substrate of dead leaves. In winter prefers wooded areas, especially forest canopy and edges or gaps, in adjacent tall second growth or semi-open areas. In migration it also may occur in low scrub. Unlike some other warblers, may be an obligate insectivore on the wintering grounds (sensu Morse 1989:222).

SPECIAL HABITAT REQUIREMENTS. Brushy edge habitats or openings in cover with saplings, forbs, and grasses.

Further Reading. United States and Canada: Confer 1992, Confer and Knapp 1981, DeGraaf and Rudis 1986, Eyer 1963, Farrand 1983c, Godfrey 1966, Griscom and Sprunt 1979, Harrison 1984, Johnsgard 1979, Morse 1989, Tate and Tate 1982. Mexico and Central America: Edwards 1972, Land 1970, Rappole et al. 1983, Stiles and Skutch 1989. South America: Meyer de Schauensee and Phelps 1978, Rappole et al. 1983. Caribbean: Raffaele 1989, Rappole et al. 1983.

Tennessee Warbler
Vermivora peregrina

Canada (Quebec): Fauvette Obscure

Costa Rica: Reinita Verdilla

Guatemala: Chipe Verdillo, Gusanero Verdillo, Reinita Verdilla

Mexico: Chipe Peregrino

Venezuela: Reinita Gorro Gris

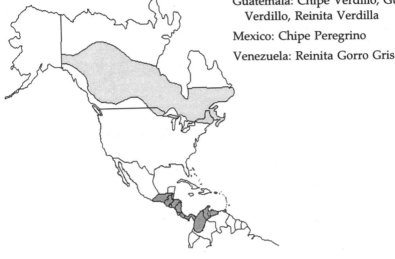

RANGE. Breeds from southeastern Alaska and southern Yukon across Canada to north-central Quebec and southern Labrador, and south to south-central British Columbia, northwestern Montana, northern Minnesota, northeastern New York, and southern Maine. Some individuals migrate through the Bahamas and Cuba. Winters from Mexico to Colombia and Venezuela.

STATUS. Common; declining since the early 1980s in western parts of breeding range. Responds rapidly to infestations and crashes of spruce budworm, producing dramatic regional population fluctuations.

HABITAT. In northern coniferous, mixed, and deciduous woodlands, inhabits forest openings with grasses, dense shrubs, and scattered clumps of young trees, open spruce and tamarack or white-cedar bogs where sphagnum moss is abundant, brushy hillsides, and occasionally dry pine sites. Drawn to areas of spruce-budworm outbreaks (Erskine 1977, Kendeigh 1947). Also found in recently logged areas, especially where a few small trees are left. Nests singly or sometimes in loose colonies. Conceals nest on moist ground, typically in sphagnum-covered hummocks or among grasses, protected by plants overhead, or less frequently, on dry hillsides under the cover of shrubs or saplings. During winter this species prefers semi-open condi-

406

tions, second growth, coffee plantations, and gardens; regularly found in the canopy and forest edge. Shifts wintering habitat (Panama) and diet from insects to fruit and nectar as the latter become available (Morton 1980b). In migration (late spring) it may be found almost anywhere.

SPECIAL HABITAT REQUIREMENTS. Brushy, semi-open habitat in coniferous or mixed forests. The Tennessee Warbler is a spruce-budworm specialist—populations become very high during major outbreaks—and it may not breed in the area when the outbreak is over (Brewster 1938, Erskine 1984, Morse 1989:99).

Further Reading. United States and Canada: Bent 1953a, DeGraaf and Rudis 1986, Godfrey 1966, Griscom and Sprunt 1979, Harrison 1984, Johnsgard 1979, Morse 1989. Mexico and Central America: Edwards 1972, Land 1970, Rappole et al. 1983, Stiles and Skutch 1989. South America: Meyer de Schauensee and Phelps 1978, Rappole et al. 1983. Caribbean: Rappole et al. 1983.

Orange-crowned Warbler
Vermivora celata

Canada (Quebec): Fauvette Verdâtre

Costa Rica: Reinita Olivada

Guatemala: Chipe de Coronilla Naranjada, Gusanero Deslustrado, Reinita Deslustrada

Mexico: Chipe Celato

RANGE. Breeds from western and central Alaska and central Yukon across Canada to northern Ontario, central Quebec, and southern Labrador south to southwestern and central California, central Utah, southern New Mexico,

407

and extreme western Texas, and east of the Rockies, to southern Saskatchewan, central Ontario, and south-central Quebec. Winters from coastal and southern California, Baja California, central Arizona, Texas, the southern portion of the Gulf States, and South Carolina south to Guatemala and Belize.

STATUS. Common in the western United States and Canada.

HABITAT. Occurs in a variety of woodland and brushy habitats, especially sites with considerable shrub cover. Prefers chaparral, brushy open woods, woodland edges of low deciduous growth, shrubby burns, overgrown pastures, riparian thickets, and the edges of clearings. In the Pacific Northwest, found in the mountains up to about 2,000 m, inhabiting dense mixed groves of aspen, alder, willow, and pine in meadows of subalpine parks. Conceals nest on the ground in a bramble tangle, hummock, at the base of a bush or stump, or occasionally up to 1 m high in low, dense bushes. Winter habitats of Orange-crowned Warblers are similar to those of Common Yellow-throats, Yellow-breasted Chats, and Tennessee Warblers—semi-open second growth and scrub thickets—in which these species are territorial (Rappole and Morton 1985). Seldom in forest (Rappole and Warner 1980).

SPECIAL HABITAT REQUIREMENTS. Dense shrubs for nesting.

Further Reading. United States and Canada: Bent 1953a, Godfrey 1966, Griscom and Sprunt 1979, Harrison 1979, 1984, Morrison 1981, Morse 1989, Verner and Boss 1980. Mexico and Central America: Edwards 1972, Land 1970, Rappole et al. 1983, Stiles and Skutch 1989.

Nashville Warbler
Vermivora ruficapilla

Canada (Quebec): Fauvette à Joues Grises

Costa Rica: Reinita Cachetigris

Guatemala: Chipe de Cabeza Gris, Gusanero Capigris, Reinita Capigris

Mexico: Chipe de Nashville

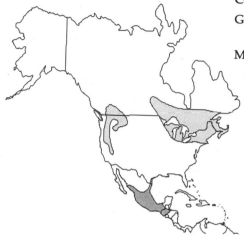

RANGE. Breeds from southern interior British Columbia and northwestern Montana south to northwestern and south-central California and extreme west-central Nevada; and from central Saskatchewan to southern Quebec, Nova Scotia, and New Brunswick, south to southern Manitoba, southern Wisconsin, southern Michigan, northern New Jersey, and Rhode Island. Winters from southern Texas south to Mexico, Guatemala, and Belize, rarely in California and southern Florida.

STATUS. Common.

HABITAT. Prefers brushy sphagnum bogs and open second-growth deciduous woodlands. Also occurs in regenerating areas that have been burned or cut; overgrown pastures and fields; woodland edges, swales, slashings, and undergrowth of open mixed forests; and shallow-to-bedrock dry openings in the forest, especially those containing aspen or birch. Sometimes found on poor sites within mesic woodlands. Conceals nest on the ground in a small depression, sometimes in a sphagnum hummock, often with an overhead cover of ferns or other overhanging vegetation. In Costa Rica evidently prefers dense shrubbery along forest edges and roadsides.

SPECIAL HABITAT REQUIREMENTS. Second-growth deciduous forest; scattered trees interspersed with brush.

Further Reading. United States and Canada: DeGraaf and Rudis 1986, Farrand 1983c, Godfrey 1966, Griscom and Sprunt 1979, Harrison 1984, Johnsgard 1979, Johnson 1976, Lawrence 1948, Morse 1989. Mexico and Central America: Edwards 1972, Land 1970, Rappole et al. 1983, Stiles and Skutch 1989.

Virginia's Warbler
Vermivora virginiae

Canada (Quebec): Fauvette de Virginia

Mexico: Chipe de Virginia

RANGE. Breeds from east-central California, central Nevada, southeastern Idaho, and southern Wyoming south to south-central California, central and southeastern Arizona, southern New Mexico, and extreme western Texas. Winters in Mexico.

STATUS. Common.

HABITAT. Inhabits arid montane woodlands from 2,000 to 3,000 m, preferring scrubby brush interspersed with pinyon-juniper and yellow pine. Frequents dense growths of low scrub oaks in canyons, mountain-mahogany, and chokecherry, rocky steep slopes and ravines, chaparral in foothills, pinyon-juniper brushlands, riparian willow and alder thickets, and open spruce and fir forests near scrubby thickets. Builds nest on the ground, among dead leaves or embedded in loose soil, sometimes at the base of a bush, or hidden under a tussock of grass, but usually concealed by overhanging vegetation.

SPECIAL HABITAT REQUIREMENTS. Scrubby vegetation for nesting.

Further Reading. United States and Canada: Bent 1953a, Farrand 1983c, Godfrey 1966, Griscom and Sprunt 1979, Harrison 1984, Van Tyne 1936. Mexico: Edwards 1972, Rappole et al. 1983.

Colima Warbler

Vermivora crissalis

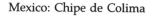

Mexico: Chipe de Colima

RANGE. Breeds in the Chisos Mountains in extreme western Texas and northern Mexico. Winters in Mexico.

STATUS. Rare.

HABITAT. Inhabits forested canyons and slopes above 2,000 m, where it frequents thickets of young maples and oaks along dry stream beds, clumps of small oaks along mountain slopes, and mixed woods of maple, oak, Arizona cypress, and yellow pine. Nests on the ground among fallen leaves and vines, which may partly or completely conceal the nest. Locates nest on rocky slopes, or adjacent to dry stream beds among small rocks and leaves where there are clumps of small oaks.

SPECIAL HABITAT REQUIREMENTS. Oak thickets above 2,000 m.

Further Reading. United States and Canada: Blake 1949, Griscom and Sprunt 1979, Harrison 1984, Oberholser 1974b, Van Tyne 1936. Mexico: Edwards 1972, Rappole et al. 1983.

Lucy's Warbler   Mexico: Chipe de Lucy
Vermivora luciae

RANGE. Breeds from southeastern California, southern Nevada, and Utah south to southern Arizona, northern Mexico, and extreme western Texas. Winters in western Mexico south to Jalisco and Guerrero.

STATUS. Common.

HABITAT. Found in southwestern deserts, generally wherever there are large mesquites and cottonwoods, especially along main watercourses. Also occurs in sycamores and live oaks in mountain foothills. The only cavity-nesting warbler besides the Prothonotary; generally places nest in four types of cavities: natural cavities in trees (usually mesquite), where the entrance is in a sheltered spot; under loose bark; in abandoned woodpecker holes; and in deserted Verdin nests. Generally locates nest less than 2 m above the ground, but ranges up to 5 m. Occasionally nests in holes in banks, in yuccas, willows, sycamores, or elderberries.

SPECIAL HABITAT REQUIREMENTS. Cavities for nesting.

Further Reading. United States and Canada: Bent 1953a, Griscom and Sprunt 1979, Harrison 1984. Mexico: Edwards 1972, Rappole et al. 1983.

Northern Parula

Parula americana
(includes Tropical Parula,
Parula pitiayumi)

Canada (Quebec): Fauvette Parula

Costa Rica: Parula Norteña

Guatemala: Chipe Azulado Norteño,
Parula Norteña, Reinita
Pechidorada

Mexico: Chipe Azul-Olivo Norteño

Puerto Rico: Reinita Pechidorada

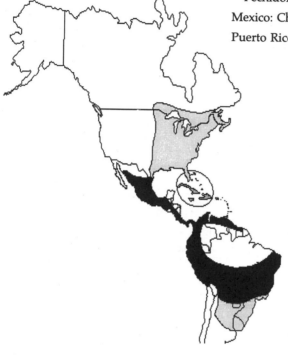

RANGE. Breeds from southeastern Manitoba and central Ontario east to New Brunswick south to south-central and southern Texas, the Gulf Coast, and Florida, and west to the eastern edge of the Great Plains states. Winters in Florida and the West Indies and from Veracruz and Oaxaca south through Mexico to Costa Rica.

STATUS. Common.

HABITAT. Primarily associated with swampy woods, especially in the Southeast, where it inhabits mature coniferous and deciduous woodlands where mosslike lichens or Spanish moss are found. Now common throughout much or all of Bachman's Warbler's former breeding range (Morse 1989: 265). In the north, found at edges of spruce forests, spruce-covered islands

413

off the Maine coast, and bogs with abundant bearded lichens. Occasionally occurs in woodlands without moss or lichens. Typically suspends cup nest near the tip of a tree limb that is covered with bearded lichen or Spanish moss at heights averaging 3 m but ranging from 2 to 30 m above the ground. Conceals nest with moss or lichen, and constructs it of these materials as well. In areas where these are not available, uses other materials. (Tropical Parulas also hide nests; see Morse 1989:316–317 for review.) In winter Northern Parula can be found in semi-open habitats, including arboreal agriculture, sun coffee, and tall second growth, usually from middle heights to high in the canopy. It is a habitat specialist on the wintering grounds in the Yucatán Peninsula (Lynch et al. 1985); has exhibited sexual segregation in Quintana Roo with more males in taller forest and more females in scrub habitats (Morton et al. 1987).

SPECIAL HABITAT REQUIREMENTS. Bearded lichen or Spanish moss for nesting material or for nest sites.

Further Reading. United States and Canada: Bent 1953a, DeGraaf and Rudis 1986, Godfrey 1966, Graber and Graber 1951, Griscom and Sprunt 1979, Johnsgard 1979, Morse 1989. Mexico and Central America: Edwards 1972, Land 1970, Rappole et al. 1983, Stiles and Skutch 1989. Caribbean: Raffaele 1989, Rappole et al. 1983.

Yellow Warbler

Dendroica petechia

Canada (Quebec): Fauvette Jaune

Costa Rica: Reinita Amarilla

Guatemala: Chipe Amarillo, Chipe Canario, Reinita Amarilla

Mexico: Chipe Amarillo Norteño

Puerto Rico and Venezuela: Canario de Mangle

West Indies: Canario, Canario de Mangle, Oiseau Jaune, Petit-jaune Didine, Sucrier Barbade, Sucrier Mangle

RANGE. Breeds from northwestern and north-central Alaska and northern Yukon to northern Ontario, central Quebec, and southern Labrador south through Mexico and Central America to northern South America, central and northeastern Texas, northern Arkansas, central Georgia, and central South Carolina. Migrates and winters throughout the Caribbean. Winters from southern California, southwestern Arizona, Mexico, and southern Florida south to Venezuela, Brazil, Peru and the Galápagos. Also resident in southern Florida, the Greater Antilles, the Virgin Islands, and other islands in the West Indies; also in Central America, coastal Colombia, Ecuador, and Venezuela.

STATUS. Common, some declines in the western United States due to losses of riparian thickets because of channelization and livestock grazing (Dobkin 1992); populations enhanced by willow restoration and protection of riparian habitats (Taylor and Littlefield 1986).

HABITAT. Prefers moist habitats with dense understories such as willow- and alder-lined streams and ponds, brushy bogs, and the edges of marshes, swamps, or creeks. Also occurs in dry sites such as hedgerows, roadside

415

thickets, orchards, farmlands, forest edges, and suburban yards and gardens. Common in the aspen parkland in the prairie provinces. Uses only deciduous habitats except on islands off the Maine coast, where coniferous habitats are used (Morse 1989:110). Usually builds nest in an upright fork or crotch of a tree or bush, typically 1 to 2 m, occasionally up to 12 m, above the ground. May nest colonially in ideal habitats. One of the species most frequently parasitized by cowbirds; builds up the nest lining to cover the cowbird eggs. Generally occurs wherever patches of trees or shrubs grow, but avoids heavy forests. In the Florida Keys, only inhabits coastal mangroves. Winter habitats include mangroves, freshwater swamps, dry scrub, and shade trees near water. Coastal scrub is the only winter habitat used on the northern Yucatán Peninsula (Lynch 1992).

SPECIAL HABITAT REQUIREMENTS. Small scattered trees or dense shrubbery, commonly near water.

Further Reading. United States and Canada: Bent 1953a, Busby and Sealy 1979, DeGraaf and Rudis 1986, Farrand 1983c, Godfrey 1966, Graham 1988, Griscom and Sprunt 1979, Harrison 1984, Johnsgard 1979, Morse 1989, Schrantz 1943, Tate and Tate 1982. Mexico and Central America: Edwards 1972, Land 1970, Rappole et al. 1983, Stiles and Skutch 1989, Wiedenfeld 1992. South America: Meyer de Schauensee and Phelps 1978, Rappole et al. 1983. Caribbean: Bond 1947, Raffaele 1989, Rappole et al. 1983.

Chestnut-sided Warbler
Dendroica pensylvanica

Canada (Quebec): Fauvette à Flancs Marron

Costa Rica: Reinita de Costillas Castañas

Guatemala: Chipe Pardoblanco, Reinita Pardoblanca

Mexico: Chipe Gorriamarillo

Puerto Rico: Reinita Costadicastaña

Venezuela: Reinita Lados Castaños

RANGE. Breeds from east-central Alberta and central Saskatchewan to southern Quebec and New Brunswick south to eastern Colorado, Iowa, central Ohio, and Connecticut; in the Appalachians south to north-central Georgia and northwestern South Carolina. Winters from Oaxaca, Chiapas, and Guatemala to Panama, casually to Colombia, Ecuador, and Venezuela. Also in the Greater Antilles east to the Virgin Islands.

STATUS. One of the most common breeding birds in eastern North America. Apparently extremely rare 150 years ago; Audubon and Wilson saw it once and twice, respectively (Bent 1953), probably after fire or in regenerating beaver meadows. Dramatically increased as New England's abandoned farmlands reached brushy stage (see Morse 1989:262 for review).

HABITAT. Inhabits rather open and dry areas having some deciduous woody vegetation in the form of shrubs and small trees, preferring the brushy hardwood regeneration of clear-cut forests or regenerating shelterwood cuts. Also abandoned pastures and fields, second-growth woodland edges and clearings, low shrubbery, briar thickets, brushy hillsides and brooksides, and roadside thickets. Conceals nest from 0.5 to 1 m above the ground in briar thickets, bushes, saplings, or vines, such as spirea, raspberry, red-osier dogwood, azalea, laurel, gooseberry, meadow-rue, and hazel. In winter, territorial (Greenberg 1984) in middle and upper levels of the forest, freely descending to shrub level in second growth, edges, forest gaps, and shady gardens.

SPECIAL HABITAT REQUIREMENTS. Early second-growth deciduous wood-lands with dense vegetation about 1 to 3 m tall to provide nest sites and foraging habitat.

Further Reading. United States and Canada: Cripps 1966, Farrand 1983c, Godfrey 1966, Griscom and Sprunt 1979, Harrison 1984, Lawrence 1948, Morse 1989, Tate 1973, Terres 1980. Mexico and Central America: Edwards 1972, Land 1970, Rappole et al. 1983, Ridgely and Gwynne 1989, Stiles and Skutch 1989. South America: Meyer de Schauensee 1966, Meyer de Schauensee and Phelps 1978, Rappole et al. 1983. Caribbean: Bond 1947, Raffaele 1989, Rappole et al. 1983.

Magnolia Warbler
Dendroica magnolia

Canada (Quebec): Fauvette à Tête Cendrée

Costa Rica: Reinita Colifajeada

Guatemala: Chipe Pechirrayado, Reinita Colifajeada

Mexico: Chipe Colifajeado

Puerto Rico: Reinita Manchada

RANGE. Breeds from northeastern British Columbia, west-central and southern Northwest Territories east to north-central Manitoba and south-central and eastern Quebec, and south to south-central British Columbia, central Saskatchewan, northeastern Minnesota, central Michigan, western Virginia, northwestern New Jersey, and Connecticut. Winters from Oaxaca and central Veracruz to Panama, occasionally along the Gulf Coast and West Indies east to the Virgin Islands.

STATUS. Common.

HABITAT. Inhabits a wide range of coniferous or mixed coniferous-deciduous woodlands, from open clearcuts to mature stands (Titterington et al. 1979), preferring spruce and fir forests with low trees and coniferous bogs. Also inhabits dense thickets of spruce and fir, old clearings with small coniferous saplings, second growth following logging, woodland edges, and coniferous thickets along roadsides and balsam fir understory of wet-site aspen stands. Conceals nest in a small conifer, in foliage near the top of a horizontal branch, generally up to 2 m, but typically about 1 m, above the ground. Prefers spruce, fir, and hemlock for nesting, but may use hardwoods. In winter frequents open groves, arboreal agriculture (cacao, citrus, shade coffee [Robbins et al. 1992]), thickets, and gallery woodlands. No permanent resident on the wintering grounds forages in the same portion of the canopy—the upper third of the canopy where the leaves are small—as the Magnolia Warbler; its closest competitor there is Wilson's Warbler, another migrant (Rappole and Warner 1980). A habitat generalist on the wintering grounds in the Yucatán Peninsula (Lynch 1989). Sexual segregation by habitat in Quintana Roo, Mexico, with males predominant in forest and females predominant in scrub forest (Ornat and Greenberg 1990). Common in mature evergreen forest in Tuxtla Mountains of Veracruz (Rappole and Warner 1980).

SPECIAL HABITAT REQUIREMENTS. Young conifer stands for nesting.

Further Reading. United States and Canada: Bent 1953a, DeGraaf and Rudis 1986, Farrand 1983c, Godfrey 1966, Griscom and Sprunt 1979, Harrison 1984, Johnsgard 1979, Morse 1989. Mexico and Central America: Edwards 1972, Land 1970, Rappole et al. 1983, Stiles and Skutch 1989. Caribbean: Raffaele 1989, Rappole et al. 1983.

Cape May Warbler
Dendroica tigrina

Canada (Quebec): Fauvette Tigrée
Costa Rica: Reinita Tigrina
Mexico: Chipe Tigrino
Puerto Rico: Reinita Tigre

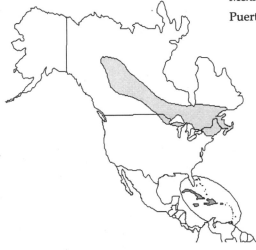

RANGE. Breeds from southwestern and south-central Northwest Territories and northeastern British Columbia east to central Ontario, southern Quebec, and New Brunswick south to central Alberta, southeastern Manitoba, northern Wisconsin, southern Ontario, northeastern New York, and east-central Maine. Winters in central and southern Florida and the West Indies, and casually from the Yucatán Peninsula to Panama.

STATUS. Uncommon.

HABITAT. Inhabits fairly open coniferous forests, especially those with a high percentage of mature spruce. Also frequents dense spruce forests with a scattering of tall trees extending above the general canopy level; the edges of coniferous forests, especially if birches or hemlocks are present; and more open land with small trees. A spruce-budworm specialist; proliferates in areas that are heavily infested, and may not occur after the outbreak has subsided (Brewster 1938, Morse 1989:99). Builds nest invariably in the uppermost clump of thick foliage near the top of tall conifers, generally invisible from below. Little is known about the nesting biology of this species, mainly because it tends to nest so high in conifers and because females tend to land near the tree base and move up through the tree rather than fly to

420

the nest. On the wintering grounds, known to defend discrete food source when the widely spaced century plants bloom in the Bahamas, providing insects and nectar (Emlen 1973, Morse 1989:227).

SPECIAL HABITAT REQUIREMENTS. Tall conifer trees, especially spruce and especially in areas of spruce-budworm outbreaks, for nesting.

Further Reading. United States and Canada: Bent 1953a, DeGraaf and Rudis 1986, Farrand 1983c, Griscom and Sprunt 1979, Harrison 1984, Johnsgard 1979, Morse 1989. Mexico and Central America: Edwards 1972, Rappole et al. 1983, Stiles and Skutch 1989. Caribbean: Raffaele 1989, Rappole et al. 1983.

Black-throated Blue Warbler

Dendroica caerulescens

Canada (Quebec): Fauvette Bleue à Gorge Noire

Costa Rica: Reinita Azul y Negra

Guatemala: Chipe Azulado Gargantinego, Reinita Azul y Negra

Mexico: Chipe Azul Pizarra

Puerto Rico: Reinita Azul

RANGE. Breeds from western and central Ontario to New Brunswick, south to northeastern Minnesota, central Michigan, northeastern Pennsylvania, and southern New England, and in the Appalachians to northeastern Georgia and northwestern South Carolina. Winters primarily from southern Florida to the West Indies, casually in Central America and Colombia.

STATUS. Common.

HABITAT. Prefers northern hardwood forests with a dense understory of saplings—coniferous or deciduous shrubs. Also inhabits mature coniferous-deciduous forests, especially those with an undergrowth of creeping yew, laurel, hazel, maple, or brushy saplings. In the southern Appalachians, often inhabits dense mountain-laurel thickets. Constructs well-concealed nest near the ground, generally not higher than 1 m above ground, in fallen tree tops, seedlings, small trees, or shrubs. Partial to nesting in rhododendron, laurel, hemlock, small spruce, fir, and maple. In winter found mainly in forest canopy, edges, and semi-open areas in hilly country. Both sexes occupy territories on wintering grounds in Jamaica, and show more site fidelity there in winter than in summer in New Hampshire, returning precisely to specific wintering ground sites (Holmes and Sherry 1992). In Puerto Rico, females are more common in young, shrubby forest at high altitude of montane forest; males more common in older, taller forests of lower altitude (Wunderle 1992). Common in shade coffee in Puerto Rico (Robbins et al. 1992).

SPECIAL HABITAT REQUIREMENTS. Woodlands with shrubby undergrowth.

Further Reading. United States and Canada: Bent 1953a, DeGraaf and Rudis 1986, Farrand 1983c, Godfrey 1966, Griscom and Sprunt 1979, Harding 1931, Harrison 1984, Johnsgard 1979, Morse 1989. Mexico and Central America: Edwards 1972, Land 1970, Rappole et al. 1983, Stiles and Skutch 1989. Caribbean: Holmes et al. 1989, Raffaele 1989, Rappole et al. 1983.

Yellow-rumped Warbler
Dendroica coronata

Canada (Quebec): Fauvette à Croupion Jaune

Costa Rica: Reinita Lomiamarilla

Guatemala: Chipe Coronado, Reinita Coronada

Mexico: Chipe Grupidorado

Puerto Rico: Reinita Coronada

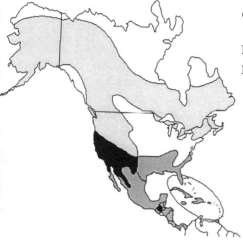

RANGE. The northern and eastern race (*Coronata* group) breeds from western Alaska and central Northwest Territories to north-central Labrador south to northern British Columbia, southeastern Saskatchewan, central Michigan, and Massachusetts, and in the Appalachians to eastern West Virginia. The western race (*Auduboni* group) breeds from central British Columbia and southwestern Saskatchewan south to southern California and northern Baja California, east to western Texas and south to western Chihuahua. The *Coronata* group winters from southwestern British Columbia through the Pacific States, southern Arizona, and Colorado, and from Kansas east across the central United States to New England, south to Panama and throughout the West Indies, where it is one of the last warblers to arrive in the Caribbean. The *Auduboni* group winters from southwestern British Columbia and Idaho to southern Baja California, through Mexico to Guatemala and western Honduras.

STATUS. Common, perhaps the most abundant warbler in North America.

HABITAT. Generally inhabits coniferous forests throughout its range, but also found in aspen forests in the Rocky Mountains. In the West, prefers open montane coniferous forest, and forest edges such as those around meadows or lakes. More abundant in selection-cut Douglas-fir stands in

423

Idaho (Medin and Booth 1989), and in old-growth rather than in rotation-age stands (Hejl and Woods 1991), probably because of the more open conditions in old growth. In the East, prefers spruce-fir woodlands, but also frequents young coniferous growth near the edges of woods, mixed woods, and evergreen plantations. Builds nest well out on a horizontal branch of a coniferous tree, screened from above by clumps of needles. Typically locates nest 5 to 6 m above the ground, but sometimes 1 to 16 m. In winter frequents a variety of habitats, many at least partly deciduous: pastures, savannas, roadsides, low scrub, and other open places; in the Caribbean, forests, open thickets, mangroves.

Winters farther north than any other warbler, being confined to areas where bayberry grows (Morse 1989:5, Place and Stiles 1992). In isolated cloud forests in Chiapas, Mexico, *Auduboni* group, together with Townsend's and Wilson's Warblers, dominates the autumn avifauna (Vidal-Rodriguez 1992).

SPECIAL HABITAT REQUIREMENTS. Coniferous trees for nesting.

Further Reading. United States and Canada: Beal and McAtee 1912, Bent 1953a, DeGraaf and Rudis 1986, Farrand 1983c, Godfrey 1966, Griscom and Sprunt 1979, Harrison 1984, Johnsgard 1979, Morse 1989, Verner and Boss 1980. Mexico and Central America: Edwards 1972, Land 1970, Rappole et al. 1983, Stiles and Skutch 1989. Caribbean: Raffaele 1989, Rappole et al. 1983.

Black-throated Gray Warbler

Dendroica nigrescens

Canada (Quebec): Fauvette Grise à
  Gorge Noire

Guatemala: Chipe Negruzco, Reinita
  Negruzca

Mexico: Chipe Negrigrís

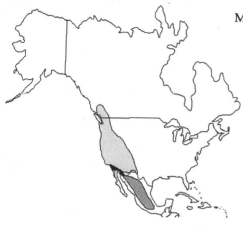

RANGE. Breeds from southwestern British Columbia and western Washington to northern Utah and southern Wyoming south, primarily in mountains, to northern Baja California, central and southeastern Arizona, extreme western Texas, and Mexico. Winters from coastal southern California and southern Arizona south to Mexico and casually to Guatemala.

STATUS. Common.

HABITAT. In the northern portion of its range, inhabits conifer forests that are open and interspersed with shrubs or forest edges. Farther south, seems to prefer shrubby stands of oaks, pinyon, juniper, and manzanita. Throughout its range, prefers and is perhaps limited to dry slopes. In Washington, seems to nest exclusively in fir trees, placing nest on a horizontal branch 2 to 16 m above the ground. In habitats farther south, builds nest in shrubs such as manzanita, oak, or ceanothus, or in large white oaks and sycamores.

Further Reading. United States and Canada: Bent 1953a, Godfrey 1966, Griscom and Sprunt 1979, Harrison 1984, Morse 1989, Phillips et al. 1964. Mexico and Central America: Edwards 1972, Land 1970, Rappole et al. 1983.

Townsend's Warbler
Dendroica townsendi

Canada (Quebec): Fauvette de Townsend

Costa Rica: Reinita de Townsend

Guatemala and Mexico: Chipe de Townsend, Reinita Negriamarilla

RANGE. Breeds from east-central Alaska, southern Yukon, and coastal British Columbia, as well as parts of Alberta and Saskatchewan, south along the Pacific Coast into northwestern Washington, and inland to central and southeastern Washington, central and northeastern Oregon, northern Idaho, northwestern and south-central Montana, and northwestern Wyoming. In general, breeding distribution follows that of conifers and mixed coniferous-deciduous forests. Winters in the coastal region of Oregon and California south through the highlands of Mexico and Central America to Costa Rica, and casually to western Panama.

STATUS. Common in coniferous forests of the Pacific Northwest. Characteristic of multiple canopy, mature, and old-growth true fir and mixed conifer stands.

HABITAT. During breeding season, primarily inhabits mature coniferous and mixed forests of the Pacific Northwest, most commonly in the more mountainous regions. Usually forages high in the crowns of tall trees. More abundant in old growth than in younger forests (Hejl and Woods 1984, Mannan and Meslow 1984). Nests in conifers, particularly firs and Douglas-fir, generally on the limb of the tree and not in the fork or crotch. Constructs nest of bark and slender twigs woven together and usually places it near the crowns of trees. In shorter trees, places nest within 3 or 4 m of the ground. During migration and winter, moves into humid forests as well as pine-oak associations, open woodlands, and second-growth scrub forests. Found in disturbed forests, edges, second growth, and clearings; in Costa Rican highlands in winter. In an isolated cloud forest in Chiapas, Mexico, this species is a canopy specialist which, together with Wilson's and *Auduboni* group of Yellow-rumped Warblers, dominates the autumn avifauna (Vidal-Rodriguez 1992).

SPECIAL HABITAT REQUIREMENTS. Tall, coniferous forests of the north Pacific Coast.

Further Reading. United States and Canada: Bent 1953a, Godfrey 1966, Griscom and Sprunt 1979, Harrison 1984, Morse 1989. Mexico and Central America: Edwards 1972, Land 1970, Rappole et al. 1983, Stiles and Skutch 1989.

Hermit Warbler
Dendroica occidentalis

Costa Rica: Reinita Cabecigualda

Guatemala: Chipe Cabeciamarillo, Reinita Cabeciamarilla

Mexico: Chipe Occidental

RANGE. Breeds in coniferous forests from southern British Columbia south through the coastal ranges and Sierra Nevada to southern California. Winters from Sinaloa and Durango south through the highlands of Central America (except Belize) to Nicaragua. In general, distribution coincides with that of tall coniferous forests.

STATUS. Uncommon and local.

HABITAT. Uses very tall conifers, especially Douglas-fir and cedar. Prefers scattered groups of tall trees (sometimes 70 or more m tall) that tower above the rest of the forest. Generally found in the upper canopies of such trees, and though very active, is often difficult to identify. Builds a well-concealed nest, generally supported by needles on the scraggly limbs of conifers, often up to 12 or 16 m above the ground. Generally forages on the upper portions of trees, whereas nest is built in the midcanopy of tall trees. Little known in Costa Rica; seen mostly in hedgerows or at forest edges.

SPECIAL HABITAT REQUIREMENTS. Tall coniferous trees.

Further Reading. United States and Canada: Bent 1953a, Godfrey 1966, Griscom and Sprunt 1979, Harrison 1984, Morse 1989. Mexico and Central America: Edwards 1972, Land 1970, Rappole et al. 1983, Stiles and Skutch 1989.

Black-throated Green Warbler

Dendroica virens

Canada (Quebec): Fauvette Verte à Gorge Noire

Costa Rica: Reinita Cariamarilla

Guatemala: Chipe Verde Gargantinegro, Reinita Gorjinegra

Mexico: Chipe Negriamarillo Dorsiverde

Puerto Rico: Reinita Verdosa

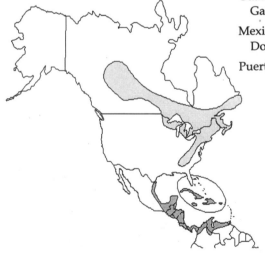

RANGE. Breeds from east-central British Columbia and northern Alberta to central Ontario and Newfoundland south to central Alberta, southern Manitoba, Minnesota, Pennsylvania, and northern New Jersey, and south in the Appalachians to northern Alabama and Georgia. Winters from south Texas and the Greater Antilles south through eastern Mexico and Central America to northern South America (Colombia and Venezuela).

STATUS. Common.

HABITAT. Inhabits large stands of mature open mixed woodlands (especially northern hardwood-hemlock stands), northern coniferous forests with large trees, and larch bogs. Less often, inhabits second-growth hardwoods and pastures with cedars. It occurs rather commonly in pine barrens in Maine and jack pines in Michigan. Sensitive to heavy (>50% volume) but not light (<25% volume) logging in northern hardwoods (Webb et al. 1977). May decline in areas of spruce-budworm outbreak (Kendeigh 1947, Morris et al. 1958). Builds a compact, deep cup nest, usually placed on a branch or in a fork of a conifer tree, 1 to 25 m above the ground; occasionally uses a deciduous tree. In winter prefers forest canopy and edges, pasture trees, and semi-open areas, but occasionally found foraging low in scrubby second growth.

SPECIAL HABITAT REQUIREMENTS: Mixed mature stands in breeding season.

Further Reading. United States and Canada: DeGraaf and Rudis 1986, Godfrey 1966, Griscom and Sprunt 1979, Harrison 1975, 1984, Morse 1989, Pitelka 1940, Terres 1980. Mexico and Central America: Edwards 1972, Land 1970, Rappole et al. 1983, Stiles and Skutch 1989.

Golden-cheeked Warbler
Dendroica chrysoparia

Guatemala: Chipe Cariamarillo

Mexico: Chipe Negriamarillo
Dorsinegro

RANGE. Breeds from central Texas south to the Edward's Plateau region into Medina and Bexar Counties and west to Real and Kerr Counties. Winters in the highlands of Central America from Mexico to Nicaragua. Occasionally visits the Farallon Islands of California and isolated areas of Florida.

STATUS. Uncommon and very local. Endangered.

HABITAT. An extreme habitat specialist; prefers to nest in stands of mature Ashe juniper (locally called *cedar*) on rough woody hillsides in canyons or on ridges that separate headwaters of streams on the Edwards Plateau of central Texas (Pulich 1976). Also may need presence of Bigelow oak among Ashe juniper (Kroll 1980). Range and populations have been reduced by eradication of "cedar brakes" in the 1940s (see Morse 1989:254). Mature stands of Ashe juniper ranging from 100 to several hundred hectares or more are necessary to ensure habitat for this species. Builds nests of Ashe juniper bark interspersed with webs, often from spiders. Nest is fastened to limbs in the midcanopy level of Ashe juniper and is difficult to locate because it resembles the bark of the tree. Often parasitized by cowbirds; parasitized nests deserted, second nest often more successful than first attempt (Morse 1989:255).

SPECIAL HABITAT REQUIREMENTS. Breeds in mature Ashe juniper (and perhaps Bigelow oak) in canyons or draws within 1 to 2 km of water. In migration, open woodland, scrub, thickets. In winter, montane oak-pine woodland.

Further Reading. United States and Canada: Griscom and Sprunt 1979, Harrison 1984, Morse 1989, Oberholser 1974b, Pulich 1976. Mexico and Central America: Edwards 1972, Land 1970, Rappole et al. 1983.

Blackburnian Warbler
Dendroica fusca

Canada (Quebec): Fauvette à Gorge Orangée

Costa Rica: Reinita Gorjinaranja

Guatemala: Chipe de Garganta Naranjada, Reinita Gorjinaranjada

Mexico: Chipe Gorjinaranja

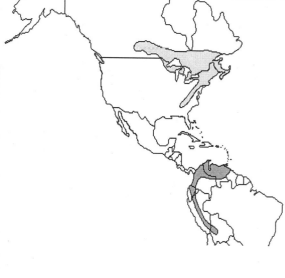

RANGE. Breeds from central Alberta east to southern Quebec and Nova Scotia south to southern Manitoba, northeastern Ohio, Pennsylvania, and southeastern New York, and in the Appalachians to South Carolina and northern Georgia. Migrates through Cuba and the Bahamas. Winters from mid-Central America to Venezuela, Colombia, Ecuador, Peru, and Bolivia.

STATUS. Common.

HABITAT. Throughout most of its breeding range favors mature conifer forests with few deciduous trees. May be reduced in areas of spruce-budworm outbreak (Kendeigh 1947, Morris et al. 1958). Also inhabits climax stands of conifers with sparse understory and with deciduous trees and shrubs around the edges. In the Appalachians, inhabits oak forests along ridges. Builds a deeply cupped nest saddled to an upper horizontal branch of a large tree. Nests in conifers throughout its range but also uses deciduous trees in the South. In winter prefers forest canopy and edges, semi-open areas, and tall

431

second growth. In migration also found in low scrub, hedgerows, and gardens.

SPECIAL HABITAT REQUIREMENTS. Tall coniferous forests; oaks in the Appalachians.

Further Reading. United States and Canada: Bull and Farrand 1977, De-Graaf and Rudis 1986, Godfrey 1966, Griscom and Sprunt 1979, Harrison 1984, Lawrence 1953b, Morse 1989, Terres 1980. Mexico and Central America: Edwards 1972, Land 1970, Rappole et al. 1983, Stiles and Skutch 1989. South America: Meyer de Schauensee and Phelps 1978, Rappole et al. 1983.

Yellow-throated Warbler
Dendroica dominica

Canada (Quebec): Fauvette à Gorge Jaune

Costa Rica: Reinita Gorjiamarilla

Guatemala: Chipe de Garganta Amarilla, Reinita Gorjiamarilla

Mexico: Chipe Domínico

RANGE. Breeds from southeastern Kansas, central Ohio, and central New Jersey south to south-central and eastern Texas, the Gulf Coast, central Florida, and the northern Bahama Islands. Wanders north to the Great Lakes and the Maritime Provinces. Winters from southeastern Texas, the Gulf Coast, and South Carolina south to Costa Rica.

STATUS. Common in southeastern United States.

HABITAT. Generally inhabits large trees along river banks, swamps, and bottomlands, as well as open stands of pines, live oaks, and mixed forests. In the South, prefers forests with abundant Spanish moss. Tends to utilize the upper canopy level of the forests. In coastal areas, nearly always builds nest in clumps of Spanish moss. In areas with no Spanish moss, saddles nest on a horizontal branch. Locates nests 3 to 30 m above the ground (average 10 m), generally far out from the tree trunk. In winter prefers semi-open areas, old second growth, thinned woodlands, and sometimes suburban gardens. Coastal scrub is the only winter habitat utilized in the northern Yucatán Pensinsula (Lynch 1992).

Further Reading. United States and Canada: Bull and Farrand 1977, Farrand 1983c, Griscom and Sprunt 1979, Harrison 1975, 1984, Johnsgard 1979, Morse 1989, Terres 1980. Mexico and Central America: Edwards 1972, Land 1970, Rappole et al. 1983, Stiles and Skutch 1989.

Grace's Warbler
Dendroica graciae

Guatemala: Chipe de Grace, Reinita Pinera

Mexico: Chipe Pinero Gorjiamarillo

RANGE. Breeds from southern Nevada, southern Utah, southwestern Colorado, northern New Mexico, and western Texas (Guadalupe and Davis Mountains) south through the mountains of western Mexico to north-central Nicaragua. Also in lowland pine savanna in Belize and eastern Honduras. Winters from Sonora and Chihuahua south through the breeding range. Resident from central Mexico to southern limits of breeding range.

STATUS. Locally common in pine-oak forests above 2,200 m.

HABITAT. Typically inhabits southwestern and tropical hard-pine forests, usually the upper portions of yellow pines (Howell 1971). Sometimes inhabits hemlock and fir forests, and occasionally oak thickets, generally from 2,000 to 3,000 m. Locates nest on limbs of pines up to 20 m above the ground, usually in clumps of needles at the ends of a brush. Nest is very compact and composed of a variety of grass fibers, hair, vegetable material, and insect webbing, often well hidden from view.

SPECIAL HABITAT REQUIREMENTS. Pine forests approximately 2,300 m in elevation.

Further Reading. United States and Canada: Griscom and Sprunt 1979, Harrison 1984, Morse 1989. Mexico and Central America: Edwards 1972, Land 1970, Rappole et al. 1983.

Pine Warbler

Dendroica pinus

Canada (Quebec): Fauvette des Pins

Mexico: Chipe Nororiental

West Indies: Petit-Chitte de Bois Pin, Siguita del Pinar

RANGE. Breeds from southern Manitoba, western Ontario, southwestern Quebec, and central Maine south to eastern Texas, the Gulf Coast, southern Florida, and the Bahamas. Rare or absent in the upper Mississippi and Ohio River Valleys. Winters in the southeastern United States south to southern Texas, the Gulf Coast, southern Florida, and the Bahamas.

STATUS. Locally common; increasing.

HABITAT. Inhabits open pine forests and pine barrens, especially jack pine in Minnesota and upland southern pines. Abundant in loblolly pine plantations (Morse 1989:103). Generally avoids tall, moist, and dense coniferous forests. Builds nest saddled on horizontal limbs of conifers 2 to 25 (usually 10 to 15 m) above the ground, usually far out from the tree trunk and well concealed in foliage. Apparently rarely parasitized by cowbirds. In winter frequents semi-open areas and old second growth. Highly aggressive in winter in southern United States, attacking conspecifics (Morse 1967, 1970), Brown-headed Nuthatches (Morse 1967), and Yellow-throated Warblers (Gaddis 1983). See Morse 1989:227–228.

SPECIAL HABITAT REQUIREMENTS. Open pine forests.

Further Reading. United States and Canada: Bull and Farrand 1977, De-Graaf and Rudis 1986, Godfrey 1966, Griscom and Sprunt 1979, Harrison 1975, 1984, Morse 1989. Caribbean: Bond 1947, Rappole et al. 1983.

Kirtland's Warbler
Dendroica kirtlandii

Canada (Quebec): Fauvette de Kirtland

Mexico: Chipe de Kirtland

RANGE. Breeds only in extensive tracts of small jack pines in a few counties in the northern half of Michigan's Lower Peninsula, from Otsego, extreme southwestern Presque Isle, and Alpena Counties south to Kalkaska, northwestern Clare, Roscommon, Ogemaw, and Iosco Counties. Occasionally individuals stray to similar habitats in Wisconsin, southern Ontario, and southern Quebec. Winters throughout the Bahamas.

STATUS. Endangered. Approximately 1,000 birds in existence.

HABITAT. Breeds in a very specific habitat: large stands (25+ ha—ideally 65 ha) of young jack pine that are 2 to 6 m tall and have living pine branches near the ground; the ground beneath the stand must be porous and sandy with low ground cover. Usually moves into burned-over jack pine forests 6 to 13 years after fire and inhabits these young stands for 10 to 12 years. Tends to nest in loose colonies. Conceals nest under low vegetation (particularly bluestem grass and blueberry) near the base of a small jack pine on flat, dry, porous soil, usually depressed below ground level; about half of the nests are parasitized by Brown-headed Cowbirds. Cowbirds likely did not encounter Kirtland's Warblers until cowbirds expanded in the Kirtland's range in the late 1880s. Kirtland's Warblers have no known defense against cowbird parasitism, such as nest abandonment and renesting.

SPECIAL HABITAT REQUIREMENTS. Dense stands of young jack pine.

Further Reading. United States and Canada: Godfrey 1966, Griscom and Sprunt 1979, Harrison 1984, Mackenzie 1977, Mayfield 1960, Morse 1989, Walkinshaw 1983. Caribbean: Rappole et al. 1983.

Prairie Warbler
Dendroica discolor

Canada (Quebec): Fauvette des Prés

Costa Rica: Reinita Galana

Guatemala: Chipe Galano, Reinita Galana

Mexico: Chipe Galán

RANGE. Breeds from eastern Nebraska, central Missouri, northern Illinois, central Wisconsin, northern Michigan, southern Ontario, southeastern New York, and New Hampshire south to eastern Texas, the Gulf Coast, and southern Florida. Winters from central Florida south to the West Indies and Central America and occasionally to northern South America.

STATUS. Common, but declining over much of the breeding range as forests reclaim old fields.

HABITAT. Generally inhabits old fields, open brushy lands, often mixed pine and scrub-oak woodlands or dry sites such as gray birch, understories of barrens, pitch pine, scrub oaks on glacial outwash plains, and partially cut woodlands (in the interior southeastern United States). Also inhabits southern pine forests, sand dunes, mangroves, and jack pine plains but avoids dense forests. Suitable habitats have increased on abandoned farms, unmowed orchards, strip-mine lands, Christmas-tree plantations, and burned and grazed woodlands. In winter evidently prefers scrubby woodland edges and second growth; also found in open groves. Sometimes nests in loose colonies. Attaches well-concealed cup nest to stems and branches of a variety of shrubs and trees, usually about 1 m above the ground. Regularly parasitized by Brown-headed Cowbirds. In winter, formerly common in tropical deciduous forest in southwestern Puerto Rico, but declined 1972–1990 and continues to be rare (Faaborg and Arendt 1992).

SPECIAL HABITAT REQUIREMENTS. Low trees and shrubs; tends to favor areas with some conifers.

Further Reading. United States and Canada: DeGraaf and Rudis 1986, Godfrey 1966, Griscom and Sprunt 1979, Harrison 1984, Morse 1989, Nolan 1978. Mexico and Central America: Edwards 1972, Land 1970, Rappole et al. 1983, Ridgely and Gwynne 1989, Stiles and Skutch 1989. Caribbean: Rappole et al. 1983.

Palm Warbler
Dendroica palmarum

Canada (Quebec): Fauvette à
Couronne Rousse

Costa Rica: Reinita Coronicastaña

Mexico: Chipe Playero

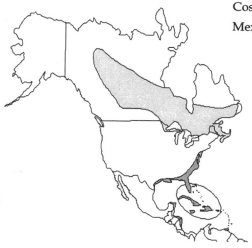

RANGE. Breeds from southern Northwest Territories and northern Alberta to central Quebec and southern Newfoundland south to northeastern British

Columbia, central Alberta, northern Minnesota, southern Quebec, Maine, and Nova Scotia. Winters mostly from north-central Texas to North Carolina south to southern Texas, the Gulf Coast, southern Florida, and the West Indies.

STATUS. Fairly common.

HABITAT. Inhabits boggy areas dominated by tamarack, black spruce, and white cedar, and dry, open forests of spruce or jack pine. Sometimes nests in loose colonies. Usually builds nest on the ground, nearly buried in sphagnum mosses, but may nest in the low branches of conifer saplings. In winter prefers open areas near water, including wet pastures, flooded fields with short grass, roadside puddles, lawns, and saline dikes. Exhibits strong site fidelity in wintering grounds (Stewart and Conner 1980). Coastal scrub is the only winter habitat utilized on the Yucatán Peninsula (Lynch 1992).

SPECIAL HABITAT REQUIREMENTS. Breeding: bogs, open boreal coniferous forest; winter and in migration: second-growth woodlands, thickets, and mangroves.

Further Reading. United States and Canada: Bent 1953b, Godfrey 1966, Griscom and Sprunt 1979, Harrison 1984, Morse 1989. Mexico and Central America: Edwards 1972, Rappole et al. 1983, Stiles and Skutch 1989. Caribbean: Rappole et al. 1983.

Bay-breasted Warbler
Dendroica castanea

Canada (Quebec): Fauvette à Poitrine Baie

Costa Rica and Puerto Rico: Reinita Castaña

Guatemala: Chipe Castaño, Reinita Castaña

Mexico: Chipe Pechicastaño

Venezuela: Reinita Pecho Bayo

RANGE. Breeds from southwestern Northwest Territories to north-central Saskatchewan and Newfoundland south to northeastern British Columbia, southern Manitoba, and northeastern Minnesota to southern Maine. Migrates through Central America and the Caribbean. Winters in northern South America—northern Colombia and western Venezuela—and occasionally south to Peru and Brazil.

STATUS. Fairly common. Has declined since the early 1980s, especially in the eastern part of the breeding range.

HABITAT. Inhabits northern coniferous forests or mixed forests, favoring balsam fir. Like Cape May and Tennessee Warblers, populations increase dramatically in response to spruce-budworm outbreaks in northern New England and northeastern Canada. May be absent from the area after the outbreak has run its course (Brewster 1938, Morse 1989:99). Usually saddles nest on a horizontal branch of a conifer, 1 to 12 m above the ground, well out from the tree trunk. Also found in woodlands bordering streams, along fence rows and highways, and in mixed woods around ponds and lakes. During migration, uses all types of woodlands but prefers conifers. In winter prefers forest edge and semi-open areas, sometimes forest canopy or second growth. An omnivorous foliage gleaner, this species is found virtually any-where in fall, from interior of forest to pasture, and moves regionally on the wintering grounds (Morse 1989:220).

SPECIAL HABITAT REQUIREMENTS. Early stages of second-growth conifer forests.

Further Reading. United States and Canada: DeGraaf and Rudis 1986, God-frey 1966, Griscom and Sprunt 1979, Harrison 1975, 1984, Mendall 1937, Morse 1978, 1989. Mexico and Central America: Edwards 1972, Land 1970, Rappole et al. 1983, Stiles and Skutch 1989. South America: Meyer de Schauensee and Phelps 1978, Rappole et al. 1983. Caribbean: Raffaele 1989, Rappole et al. 1983.

Blackpoll Warbler

Dendroica striata

Canada (Quebec): Fauvette Rayée

Mexico: Chipe Gorrinegro

Puerto Rico and Venezuela: Reinita Rayado

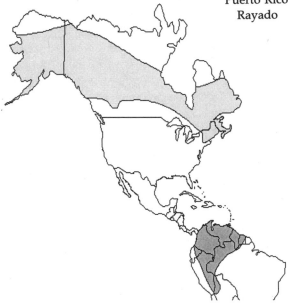

RANGE. Breeds from western and north-central Alaska throughout most of central Canada around the lower part of Hudson Bay to the Atlantic Coast, coincidental with boreal forests. Migrates south across the eastern United States and through the Caribbean, and winters in South America south to Peru and Brazil. May stray to Costa Rica when migrating through the Caribbean. This species has the longest migration of any New World land bird— it has the most northerly breeding range and the most southerly wintering ground of any warbler nesting in North America (Morse 1989:27).

STATUS. Common throughout the north-central boreal forests; recent decline in eastern North America.

HABITAT. Inhabits northern coniferous forests, favoring young or medium-sized conifers, and especially the upper canopy. Occurs on mountains in the southern parts of the breeding range—common in high-elevation stunted forests of northern New England. Builds a generally bulky nest of a variety of twigs, bark, and sometimes moss and grass, lined usually with plant fiber, grass, hair, and rootlets. Often locates nest 1 to 2 m above the ground, usually near the trunk of a tree, supported by horizontal branches, and quite well concealed. Migrates primarily at night and is often attracted to bright lights. A transient migrant through the Caribbean, moves en masse on the way to South American wintering grounds (Nisbet 1970). Probably one of the most common warbler species seen in migration, when large numbers are found in gardens and parks. A few reported from the British Isles in fall (Simms 1985). In Costa Rica prefers edges and second growth in winter, but much of the wintering grounds are in areas that few other northern warblers penetrate—Amazonia and subtropical Brazil (Sick 1971).

SPECIAL HABITAT REQUIREMENTS. Northern coniferous forests. Requires conifers for nesting and prefers spruce.

Further Reading. United States and Canada: DeGraaf and Rudis 1986, Godfrey 1966, Griscom and Sprunt 1979, Harrison 1984, Morse 1989. Mexico and Costa Rica: Edwards 1972, Rappole et al. 1983, Stiles and Skutch 1989. South America: Meyer de Schauensee and Phelps 1978, Rappole et al. 1983. Caribbean: Raffaele 1989, Rappole et al. 1983.

Cerulean Warbler

Dendroica cerulea

Canada (Quebec): Fauvette Azurée

Costa Rica and Venezuela: Reinita Cerúlea

Guatemala: Chipe Cerúleo, Reinita Cerúlea

Mexico: Chipe Cerúleo

RANGE. Breeds from southeastern Minnesota, southern Wisconsin, southern Michigan, Ontario, New York, and western New England south to northeastern Texas, southeastern Louisiana, Mississippi, Alabama, and central North Carolina. Migrates principally along Mississippi River drainage through Central America, the Bahamas, Cuba, and Jamaica. Winters in Colombia, Venezuela, Ecuador, Peru, and Bolivia.

STATUS. Declining over much of the breeding range. From 1966 to 1982 showed the most precipitous decline of any North American warbler, and decline may be continuing.

HABITAT. Prefers extensive mature floodplain and bottomland forests and shady, mature upland woods with sparse understories. Sensitive to forest fragmentation; rarely found in tracts less than 250 ha in size. A canopy gleaner and breeding habitat specialist—prefers large tracts of semi-open deciduous forests with tall trees and little undergrowth (Robbins et al. 1992). Usually builds nest 5 to 30 m above ground in tall trees, well away from the trunk, on horizontal branches that are free from vegetation below. Frequently parasitized by cowbirds. In the midwestern United States, favors elms for nesting, but will nest in oaks, maples, basswood, and yellow-poplar. One of the earliest fall migrant passerines to reach the northern Gulf Coast (only the Louisiana Waterthrush arrives there sooner). A habitat specialist on the wintering grounds, it occurs strictly in primary humid evergreen forest along

442

an extremely narrow elevational zone between 620 and 1300 m in the Andean foothills. This zone is among the most intensively logged and cultivated regions in the Neotropics (Robbins et al. 1992).

SPECIAL HABITAT REQUIREMENTS. In breeding season, tall deciduous trees; extensive forest. In winter, mature humid evergreen forest at 620 to 1300 m in Andean foothills.

Further Reading. United States and Canada: Bent 1953a, DeGraaf and Rudis 1986, Farrand 1983c, Godfrey 1966, Griscom and Sprunt 1979, Hamel 1992, Harrison 1984, Johnsgard 1979. Mexico and Central America: Edwards 1972, Land 1970, Rappole et al. 1983, Stiles and Skutch 1989. South America: Meyer de Schauensee and Phelps 1978, Rappole et al. 1983. Caribbean: Rappole et al. 1983.

Black-and-white Warbler
Mniotilta varia

Canada (Quebec): Fauvette Noire et Blanche

Costa Rica, Puerto Rico, and Venezuela: Reinita Trepadora

Guatemala: Chipe Rayado, Pepino, Reinita Trepadora

Mexico: Chipe Trepador

West Indies: Bijirita Trepadora, Madras, Mi-Dueil, Reinita Trepadora

RANGE. Breeds from west-central Northwest Territories, northern Alberta, and central Saskatchewan to southern Quebec and Newfoundland, south to eastern Montana, central Texas, Louisiana, Alabama, Georgia, and North Carolina. Winters from southern Texas and Florida through Mexico and Cen-

443

tral America to northern South America. Also migrates and winters throughout the Caribbean.

STATUS. Common.

HABITAT. Generally associated with somewhat open deciduous or mixed forest; usually not abundant in coniferous forest. Found in mature and second-growth forests, especially those composed of immature or scrubby trees. Nests in a slight depression in the ground, usually at the base of a tree or stump, beside a log, or sometimes in the roots of a fallen tree. Commonly parasitized by Brown-headed Cowbirds. Found in almost any tall scrub during migration, but in winter prefers mature forest, semi-open areas, occasionally old second growth, parks, and gardens—forages on large limbs and trunks as in breeding season (Tramer and Kemp 1980). Much more abundant in moist forest than in dry evergreen woodland on U.S. Virgin Islands (Askins et al. 1992). Highly vulnerable to tropical deforestation (Morton 1992).

SPECIAL HABITAT REQUIREMENTS. Breeding: deciduous and mixed forests and woodlands; migration and winter: forest, second growth, and scrub.

Further Reading. United States and Canada: Bent 1953a, DeGraaf and Rudis 1986, Forbush and May 1955, Godfrey 1966, Griscom and Sprunt 1979, Harrison 1984, Morse 1989, Smith 1934. Mexico and Central America: Edwards 1972, Land 1970, Rappole et al. 1983, Stiles and Skutch 1989. South America: Meyer de Schauensee and Phelps 1978, Rappole et al. 1983. Caribbean: Bond 1947, Raffaele 1989, Rappole et al. 1983.

American Redstart
Septophaga ruticilla

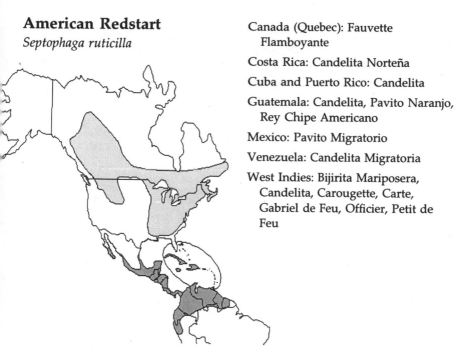

Canada (Quebec): Fauvette Flamboyante

Costa Rica: Candelita Norteña

Cuba and Puerto Rico: Candelita

Guatemala: Candelita, Pavito Naranjo, Rey Chipe Americano

Mexico: Pavito Migratorio

Venezuela: Candelita Migratoria

West Indies: Bijirita Mariposera, Candelita, Carougette, Carte, Gabriel de Feu, Officier, Petit de Feu

RANGE. Breeds from southeastern Alaska east to Labrador and Newfoundland, south to Utah, southeastern Oklahoma, and East Texas east to South Carolina. Absent as a breeding bird through most of the Great Plains region. Migrates through Central America and the Caribbean. Winters from Baja California, southern Texas, and central Florida south to Brazil. Also winters throughout the Caribbean.

STATUS. Common.

HABITAT. Prefers open deciduous forest with a well-developed understory of shrubs and young trees, but is very adaptable. Frequently breeds in mixed coniferous-deciduous forests, sapling stands of regenerating hardwoods, aspen groves, shade trees and shrubbery around farms, orchards, and willow and alder thickets bordering ponds and streams. Normally builds its nest 3 to 6 m above the ground in a crotch or on a horizontal limb of a second-growth deciduous tree. Frequently parasitized by Brown-headed Cowbirds when nesting outside of woodlands. The size of some breeding populations may be most limited by nest predators and brood parasites (Sherry and

445

Holmes 1992); other populations may be most limited by conditions on the wintering grounds (Bennett 1980). In winter frequents middle and upper levels of forests, open woods, semi-open areas, tall second growth, and a variety of arboreal agricultural habitats including sun and shade coffee (Robbins et al. 1992). Occurs from the mountains to the mangroves in Puerto Rico and the Virgin Islands. Much more abundant in moist forest than in dry evergreen woodland on U.S. Virgin Islands (Askins et al. 1992). Both sexes occupy territories on wintering grounds in Jamaica (Holmes and Sherry 1992), but habitats sexually segregated in Quintana Roo, Mexico, with more males in forest and more females in scrub (Ornat and Greenberg 1990).

SPECIAL HABITAT REQUIREMENTS. Breeding: open deciduous or mixed woodland, second-growth deciduous forest; migration: forests, thickets, scrub.

Further Reading. United States and Canada: Baker 1944, DeGraaf and Rudis 1986, Farrand 1983c, Godfrey 1966, Griscom and Sprunt 1979, Harrison 1984, Lemon et al. 1992, Morse 1989. Mexico and Central America: Edwards 1972, Land 1970, Rappole et al. 1983, Stiles and Skutch 1989. South America: Meyer de Schauensee and Phelps 1978, Rappole et al. 1983. Caribbean: Bond 1947, Holmes et al. 1989, Raffaele 1989, Rappole et al. 1983.

Prothonotary Warbler
Protonotaria citrea

Canada (Quebec): Fauvette Orangée

Costa Rica: Reinita Cabecidorada

Guatemala: Chipe Anaranjado,
Reinita Anaranjada

Mexico: Chipe Cabecidorado

Puerto Rico: Reinita Anaranjada

Venezuela: Reinita Protonotaria

RANGE. Breeds from east-central and southeastern Minnesota, south-central Wisconsin, southern Michigan, southern Ontario, central New York, and northern New Jersey south to south-central and eastern Texas, the Gulf

Coast, and central Florida, and west to eastern Oklahoma, eastern Kansas, and central Oklahoma. Winters, from the Yucatán Penninsula south through Central America and in South America from Colombia and northern Ecuador east to northern Venezuela (also Netherlands Antilles, Trinidad and Tobago), casually to Surinam.

STATUS. Uncommon; generally declining in the eastern United States.

HABITAT. Generally associated with moist bottomland or swampy deciduous woods, including woods that are frequently flooded, and willow-lined streamsides. Nests in natural cavities and old woodpecker (especially Downy Woodpecker) and chickadee holes in stumps or stubs that are standing in or near water. Will occasionally nest in nest boxes. Usually places nest low, about 1.5 m above the ground. The only hole-nesting eastern wood warbler. In winter prefers thickets adjoining rivers, streams, ponds, and lagoons, and also mangroves.

SPECIAL HABITAT REQUIREMENTS. Moist woodlands with cavities, usually in a rotted stub, for nesting. In winter, wooded swamps, scrub thickets, mangroves.

Further Reading. United States and Canada: Bent 1953a, Forbush and May 1955, Godfrey 1966, Griscom and Sprunt 1979, Harrison 1984, Morse 1989, Pearson 1936, Walkinshaw 1953. Mexico and Central America: Edwards 1972, Land 1970, Rappole et al. 1983, Stiles and Skutch 1989. South America: Meyer de Schauensee and Phelps 1978, Rappole et al. 1983. Caribbean: Raffaele 1989, Rappole et al. 1983.

Worm-eating Warbler
Helmitheros vermivorus

Canada (Quebec): Fauvette Vermivore

Costa Rica and Puerto Rico: Reinita Gusanera

Guatemala: Chipe Come-gusano, Pulgonero, Reinita Gusanera

Mexico: Chipe Vermívoro

RANGE. Breeds from southeastern Nebraska, southeastern Iowa to southern and east-central Ohio, and southeastern Massachusetts, south to southeastern Oklahoma and northeastern Texas to northwestern Florida. Winters in the West Indies, Veracruz, Chiapas, and the Yucatán Penninsula south through Central America (primarily on the Caribbean slope) to Panama. A few winter in southern Florida.

STATUS. Locally common to rare.

HABITAT. Inhabits extensively wooded hillsides and ravines with stands of deciduous trees and dense undergrowth, often near streams or swampy bogs rimmed by shrubs and vines. Nests on the ground at the base of a tree or sapling, usually well concealed under dead leaves. Generally locates nest on a hillside or bank of a ravine, but sometimes under an overhanging bank or under shrubbery. Sensitive to forest fragmentation and mammalian predation (Morse 1989:283). In winter frequents forest undergrowth and thickets, preferring evergreen forests such as those that can be found in the dry northwest of Costa Rica. Occasionally in second growth.

SPECIAL HABITAT REQUIREMENTS. Ravines with dense undergrowth.

Further Reading. United States and Canada: DeGraaf and Rudis 1986, Godfrey 1966, Griscom and Sprunt 1979, Harrison 1984, Morse 1989. Mexico and Central America: Edwards 1972, Land 1970, Rappole et al. 1983, Stiles and Skutch 1989. Caribbean: Raffaele 1989, Rappole et al. 1983.

Swainson's Warbler
Limnothlypis swainsonii

Mexico: Chipe Coronicafé
Puerto Rico: Reinita de Swainson

RANGE. Breeds locally from northeastern Oklahoma, southern Missouri, southern Illinois, southwestern Indiana, southwestern and eastern Kentucky, southern Ohio, western West Virginia, western and southern Virginia, and southern Delaware south to east-central Texas, the Gulf Coast, and northern Florida. Breeding populations disjunct—in coastal-plain swamps, floodplains, and southern Appalachians to 1,000 m. Winters in the northern Bahamas, Cuba, Cayman Islands, Jamaica, Yucatán Peninsula, and Belize.

STATUS. Uncommon.

HABITAT. Generally inhabits rich, damp woodlands with deep shade and dense undergrowth, including wooded swamps and canebrakes of lowlands (often occupying only contiguous territories within extensive, unoccupied areas of similar habitat [Meanley 1971]) and, locally, rhododendron thickets of the southern Appalachian mountains. Builds a large bulky nest, usually 1 to 2 m above the ground. In coastal lowlands, commonly nests in cane or palmetto; in highlands, nests in shrubs, small trees, vines, briars, rhododendron, or laurel. Sometimes parasitized by cowbirds. Unlike Bachman's Warbler, Swainson's winter on the mainland (Yucatán Peninsula), a factor that may be important to its survival since much of the lowland habitat on Cuba has been destroyed by agriculture (Rappole et al. 1983).

SPECIAL HABITAT REQUIREMENTS. Breeding: dense underbrush in moist forest, canebrakes; winter: lowland scrub, thickets, mangroves.

Further Reading. United States and Canada: Harrison 1975, 1984, Meanley 1966, 1971, Morse 1989. Mexico: Edwards 1972, Rappole et al. 1983. Caribbean: Raffaele 1989, Rappole et al. 1983.

Ovenbird
Seiurus aurocapillus

Canada (Quebec): Fauvette Couronnée

Costa Rica and Venezuela: Reinita Hornera

Guatemala: Chipe de Tierra, Chipe-Tordo Raicero, Pizpita Hornero, Pizpita Raicera

Mexico: Chipe Suelero Coronado

Puerto Rico: Pizpita Dorada

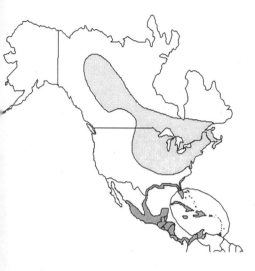

RANGE. Breeds from northeastern British Columbia, southern Northwest Territories, northern Alberta, across southern Canada to Newfoundland, south to eastern Colorado, eastern Oklahoma, northern Arkansas, and the Middle Atlantic States to northern Georgia. Winters in coastal South Carolina, Florida, the Gulf States, coastal Texas, the West Indies (fairly common visitor to Puerto Rico and St. John, which have extensive forest), Mexico, Central America, Venezuela, Colombia, Ecuador, and Peru.

STATUS. Common, but long-term decline in central North America, increase in eastern North America.

HABITAT. Usually inhabits extensive open, mature, dry deciduous or mixed forests without thick brush and tangles, preferring areas with an abundance of fallen leaves, logs, and rocks. Occasionally inhabits floodplain or even swampy forests; in the north, inhabits jack pine and spruce forests. Males occupying territories in large forest patches were more likely to obtain mates than males in small forest patches in Missouri (Gibbs and Faaborg 1990). Locates nest on or in a slight depression in the ground. Uses almost any available vegetation to construct an arched nest resembling a clay oven, with the entrance hole at or near ground level. Frequently parasitized by cowbirds in fragmented forests; also vulnerable to mammalian predators, especially chipmunks and red squirrels (Reitsma et al. 1990). In winter prefers shady understory of forest with well-developed shrub layer and deep leaf litter; occasionally in second growth and tall scrub (Tuxtla Mountains, Veracruz), even mangroves; common in young second-growth forest in Costa Rica and on migration in shaded gardens. In western Mexico, probably restricted to undisturbed tropical deciduous (lowland) forest (Hutto 1992); highly vulnerable to tropical deforestation (Morton 1992). Uncommon in agricultural habitats (Robbins et al. 1992).

Further Reading. United States and Canada: DeGraaf and Rudis 1986, Farrand 1983c, Godfrey 1966, Griscom and Sprunt 1979, Hann 1937, Harrison 1984, Holmes and Robinson 1988, Morse 1989, Smith and Shugart 1987, Stenger 1958, Sweeney and Dijak 1985. Mexico and Central America: Edwards 1972, Land 1970, Rappole et al. 1983, Stiles and Skutch 1989, Zach and Falls 1975. South America: Meyer de Schauensee and Phelps 1978, Rappole et al. 1983. Caribbean: Raffaele 1989, Rappole et al. 1983.

Northern Waterthrush

Seiurus noveboracensis

Canada (Quebec): Fauvette des Ruisseaux

Costa Rica: Reinita Acuática Norteña

Guatemala: Chipe de Agua Norteña, Chipe-Tordo Norteño, Pizpita de Agua

Mexico: Chipe Suelero Gorjijaspeado

Puerto Rico: Pizpita de Mangle

Venezuela: Reinita de Charcos

RANGE. Breeds from Alaska and southern Northwest Territories across Canada to central Labrador and Newfoundland, south to northwestern Washington, and east to central Michigan, northeastern Ohio, southeastern West Virginia, Pennsylvania, New York, and Massachusetts. Winters mostly from southern Mexico to Colombia, Venezuela, Ecuador, Peru, Guyana, Surinam, French Guiana, and the West Indies.

STATUS. Locally common.

HABITAT. Generally inhabits thickets along edges of woodland swamps, ponds, montane woodlands with bogs, pools of standing water, and (rarely) wooded riparian swamps and streams with numerous fallen trees. Prefers woodlands and shrubs around standing water rather than moving streams. Nests in roots of a fallen tree, over or adjacent to water, or on the ground among fallen trees, at the base of living trees, in rotten stumps, and under overhanging banks or cuts. Very high densities in mangrove swamps (Lynch 1989). Uncommon in agricultural habitats (Robbins et al. 1992). Strongly territorial on spring migration at temporary ponds in Texas (Rappole and Warner 1976).

SPECIAL HABITAT REQUIREMENTS. Cool, shady, wet, brushy areas with open pools. Mangrove swamps in winter.

Further Reading. United States and Canada: DeGraaf and Rudis 1986, Farrand 1983c, Godfrey 1966, Griscom and Sprunt 1979, Harrison 1984, Johnsgard 1979, Winker et al. 1992. Mexico and Central America: Edwards 1972, Land 1970, Rappole et al. 1983, Stiles and Skutch 1989. South America: Meyer de Schauensee and Phelps 1978, Rappole et al. 1983. Caribbean: Raffaele 1989, Rappole et al. 1983.

Louisiana Waterthrush
Seiurus motacilla

Canada (quebec): Fauvette Hoche-queue

Costa Rica: Piquigrande, Reinita Acuática

Guatemala: Chipe de Agua Surena, Chipe-Tordo de Luisiana, Pizpita de Río

Mexico: Chipe Suelero Gorjiblanco

Puerto Rico: Pizpita de Río

Venezuela: Reinita de Luisiana

RANGE. Breeds from eastern Nebraska, central Iowa, and east-central Minnesota to central New York and New England, south to eastern Texas, central Louisiana, central Georgia, and the Carolinas. Winters in the West Indies, Mexico, Central America, Colombia, and Venezuela.

STATUS. Uncommon.

HABITAT. Favors extensive bottomland forests with moss-covered logs and rank undergrowth along rapidly moving streams. Typically builds nest on the ground under roots, or under steep banks along streams. Also nests in upturned roots of fallen trees over or near water. Also sometimes inhabits shrub-grown bogs or areas near swamp pools or lake edges. The earliest arriving fall migrant passerine on the northern Gulf Coast. In winter, a forest understory resident (Tuxtla Mountains, Veracruz), occasionally in second growth.

SPECIAL HABITAT REQUIREMENTS. Woodlands with flowing water.

Further Reading. United States and Canada: DeGraaf and Rudis 1986, Eaton 1958, Farrand 1983c, Godfrey 1966, Griscom and Sprunt 1979, Harrison 1984. Mexico and Central America: Edwards 1972, Land 1970, Rappole et al. 1983, Stiles and Skutch 1989. South America: Meyer de Schauensee and Phelps 1978, Rappole et al. 1983. Caribbean: Raffaele 1989, Rappole et al. 1983.

Kentucky Warbler
Oporornis formosus

Canada (Quebec): Fauvette de Kentucky

Costa Rica: Reinita Cachetinegra

Guatemala: Chipe Cachetinegra, Reinita Cachetinegro

Mexico: Chipe Cachetinegro

Puerto Rico: Reinita de Kentucky

Venezuela: Reinita Hermosa

RANGE. Breeds from southeastern Nebraska, southwestern Wisconsin, southern Michigan, central Ohio, southern Pennsylvania, and southeastern New York south to eastern Texas, the Gulf Coast, central Georgia, and South Carolina. Winters from Oaxaca, Veracruz, and the Yucatán Penninsula south to Colombia and Venezuela; occasionally in Virgin Islands.

STATUS. Common but declining in the western part of the breeding range. Populations reduced where woodlands overbrowsed by deer.

HABITAT. Inhabits shrubby woodland borders and the understory of damp or shady deciduous woods, favoring moist, heavily vegetated ravines and bottomlands. Often found near water and at low elevations. Generally nests on or just above the ground among plants at the base of shrubs and trees, or under branches of fallen limbs. Occasionally places nest near or on the ground in shrubs. Commonly parasitized by Brown-headed Cowbirds. In winter prefers shady understory of mature moist forest and tall second growth, and thickets at forest edge and in gaps; a moist tropical forest specialist throughout its Neotropical winter range (Lynch 1989, Mabey and Morton 1992). Exploits young second growth in Panama (Greenberg in Morse 1989:218). Rarely in agricultural habitats (Robbins et al. 1992). Habitually insectivorous and territorial on the wintering grounds (Morton 1980b). A common sedentary winter resident in mature and cloud forest of Tuxtla Mountains, Veracruz (Rappole and Warner 1980). In migration also in low scrub.

SPECIAL HABITAT REQUIREMENTS. Dense understory.

Further Reading. United States and Canada: DeGaris 1936, Farrand 1983c, Gibbs and Faaborg 1990, Godfrey 1966, Griscom and Sprunt 1979, Harrison 1984, Morse 1989. Mexico and Central America: Edwards 1972, Land 1970, Rappole et al. 1983, Stiles and Skutch 1989. South America: Meyer de Schauensee and Phelps 1978, Rappole et al. 1983. Caribbean: Raffaele 1989, Rappole et al. 1983.

Connecticut Warbler
Oporornis agilis

Canada (Quebec): Fauvette à Gorge Grise

Costa Rica: Reinita Ojianillada

Mexico: Chipe Cabecigrís Ojianillado

Puerto Rico and Venezuela: Reinita de Connecticut

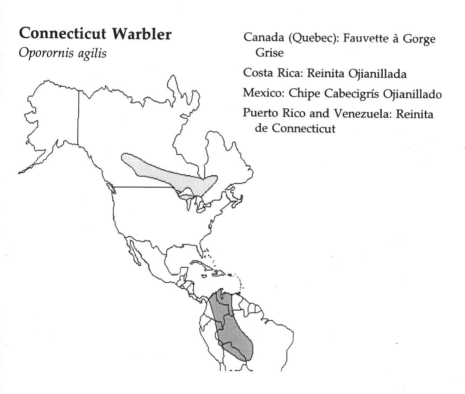

RANGE. Breeds from east-central British Columbia east across Canada to west-central Quebec, and south to southern Manitoba, northern Minnesota, northern Wisconsin, central Michigan, and south-central Ontario. Migrates through (or over) the West Indies, with a few records from Panama. Winters in Colombia, Venezuela, and eastern Brazil. Individuals may occasionally show up in southern Central America.

STATUS. Uncommon and local.

HABITAT. When breeding, generally inhabits cold, damp black spruce and tamarack bogs, and prefers areas with scattered trees and grassy openings. At the extremes of the breeding range, inhabits well-drained ridges and aspen woods. Conceals nest in a mound of moss or beside a clump of dry grass on or near the ground. Usually nests in open forests with widely spaced trees such as aspen and balsam. In winter prefers second growth and brushy forest edges, especially in moist spots.

SPECIAL HABITAT REQUIREMENTS. Breeding: spruce bogs; migration and winter: woodland, thickets.

Further Reading. United States and Canada: Godfrey 1966, Griscom and Sprunt 1979, Harrison 1975, 1984, Walkinshaw and Dyer 1961. Central America: Rappole et al. 1983, Ridgely and Gwynne 1989, Stiles and Skutch 1989. South America: Meyer de Schauensee and Phelps 1978, Rappole et al. 1983, Ridgely and Gwynne 1989. Caribbean: Raffaele 1989, Rappole et al. 1983, Ridgely and Gwynne 1989.

Mourning Warbler
Oporornis philadelphia

Canada (Quebec): Fauvette Triste

Costa Rica: Reinita Enlutada

Guatemala: Chipe Enlutado, Reinita Enlutada

Mexico: Chipe Cabecigrís Filadélfico

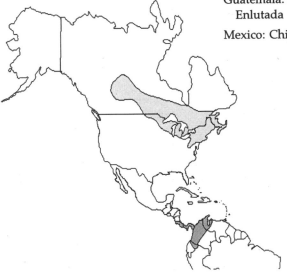

RANGE. Breeds east of the Rocky Mountains across Canada and the northern United States from northeastern British Columbia to Newfoundland, south to North Dakota and central New England through the Appalachian Mountains to Virginia. Winters from southern Nicaragua to Venezuela, Colombia, and Ecuador.

STATUS. Locally common to uncommon. Declining since the early 1980s in the eastern part of the breeding range.

HABITAT. Inhabits shrubby second growth, especially regenerating hardwood clearcuts, dense undergrowth in open woods or shelterwood cuts, shrubby margins of lowland swamps or bogs, and forest clearings or burned areas that have brambles, shrubs, and saplings. May occur in partially open coniferous and deciduous woodlands with herb and shrub understories. Conceals nests in dense herbaceous or shrubby vegetation on or near the ground. Tends to nest in edges along woodland or clearing edges, along logging trails, or at the edges of bogs and marshes. In winter in low thickets in young second growth and overgrown, weedy pastures.

SPECIAL HABITAT REQUIREMENTS. Extensive stands of saplings or dense shrubs.

Further Reading. United States and Canada: Cox 1960, DeGraaf and Rudis 1986, Farrand 1983c, Godfrey 1966, Griscom and Sprunt 1979, Harrison 1984. Mexico and Central America: Edwards 1972, Land 1970, Rappole et al. 1983, Stiles and Skutch 1989. South America: Meyer de Schauensee and Phelps 1978, Rappole et al. 1983.

457

MacGillivray's Warbler
Oporornis tolmiei

Canada (Quebec): Fauvette des Buissons

Costa Rica: Reinita de Tupidero

Guatemala: Chipe de Tupidero, Reinita de MacGillivray, Reinita de Tupidero

Mexico: Chipe Cabecigrís de Tolmie

RANGE. Breeds from southeastern Alaska, southwestern Yukon, northern British Columbia, southern Alberta, northwestern Saskatchewan, and southwestern South Dakota south, primarily in the mountains, to southern California, central Arizona, and southern New Mexico. Winters from central Mexico and southern Baja California to western Panama.

STATUS. Common to uncommon.

HABITAT. Prefers dense, moist, brushy habitat such as riparian willow and alder. Also, early successional stages (dense thickets) of cutover or burned woodlands or other low shrubby habitats. Also inhabits low vegetation such as blackberry, salmonberry, cherry, currant, serviceberry, snowberry, poison oak, ninebark, and spirea. Builds nest about 1 m above the ground, often in dense weedy growth, attached to several stems of plants; sometimes on the ground. In winter prefers second growth, scrubby old fields, and hedgerows.

SPECIAL HABITAT REQUIREMENTS. Low, dense woody vegetation.

Further Reading. United States and Canada: Farrand 1983c, Godfrey 1966, Griscom and Sprunt 1979, Harrison 1984, Morrison 1981, Morse 1989, Terres

1980. Mexico and Central America: Edwards 1972, Land 1970, Ridgely and Gwynne 1989.

Common Yellowthroat
Geothlypis trichas

Canada (Quebec): Fauvette Masquée

Costa Rica: Antifacito Norteño

Guatemala: Antifacito, Chipe Carinegro

Mexico: Mascarita Norteña

Puerto Rico: Reinita Pica Tierra

Venezuela: Reinita Gargantiamarilla

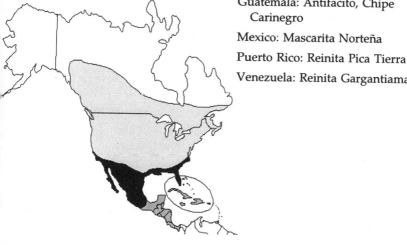

RANGE. Breeds from southeastern Alaska to northern Alberta and Newfoundland south to northern Baja California, in Mexico to Oaxaca and Veracruz, and to southern Texas, the Gulf Coast, and southern Florida. Winters along the Pacific Coast from northern California across southern Arizona, southern New Mexico, southern Texas, the Gulf States, and South Carolina, and along the Atlantic Coast from New Jersey, Virginia, and Delaware to Florida; also throughout the West Indies, Mexico, Central America, and casually to northern Colombia. Populations at the southern limits of range are considered separate species by some authors, and southern range limits are poorly understood.

STATUS. Common but declining in the eastern part of the breeding range; increasing in the western part of the breeding range.

HABITAT. A brush-forager, Common Yellowthroat typically inhabits areas with a mixture of dense, lush herbaceous vegetation with small woody plants (mainly shrubs and small trees), in damp or wet situations: edges of beaver ponds, marshes, riparian thickets, brushy pastures, wet meadows. Occasionally found in dry thickets, shelterbelts, or dense undergrowth in open, commonly hilly, woodlands. Builds a bulky cup nest of grass, leaves, and bark, well hidden on the ground in a grass tussock or similar vegetation. Occasionally locates nest in shrubs or a tangle of briars up to 1 m above the ground. Frequently parasitized by cowbirds. Common in newly regenerated hardwood clearcuts. In winter prefers wet meadows, reedy marshes, and dense, shrubby vegetation in wet areas or adjacent to open water. Also found in scrub thicket and other open habitats on the Yucatán Peninsula, seldom in forest. Common in rice-growing areas and citrus plantations (Robbins et al. 1992).

SPECIAL HABITAT REQUIREMENTS. Dense growth of low vegetation.

Further Reading. United States and Canada: Godfrey 1966, Griscom and Sprunt 1979, Harrison 1984, Hofslund 1959, Low and Mansell 1983, Stewart 1953, Terres 1980. Mexico and Central America: Edwards 1972, Land 1970, Lynch 1992, Ridgely and Gwynne 1989, Stiles and Skutch 1989. South America: Meyer de Schauensee and Phelps 1978, Rappole et al. 1983. Caribbean: Raffaele 1989, Rappole et al. 1983.

Hooded Warbler
Wilsonia citrina

Canada (Quebec): Fauvette à Capuchon

Costa Rica: Reinita Encapuchada

Guatemala: Chipe Careto, Reinita de Capucha

Mexico: Chipe Encapuchado

Puerto Rico and Venezuela: Reinita de Capucha

RANGE. Breeds from southeastern Iowa, northern Illinois, extreme southern Michigan and Ontario, southern New York, and New England south to eastern Texas, the Gulf of Mexico, and northern Florida. Migrates through the Greater Antilles and Central America. Winters from Mexico to Panama. Fairly common in the Virgin Islands; uncommon in Puerto Rico.

STATUS. Common, generally increasing.

HABITAT. Generally inhabits moist, forested regions of mixed hardwoods of beech, maple, hickory, and oak with dense undergrowth. An undergrowth foraging specialist, Hooded Warbler is associated with canopy gaps in mature forests and may benefit from patch cutting (McComb 1985). In the Southeast, also inhabits cypress-gum swamplands. Builds a cuplike nest, usually in a fork of saplings, shrubs, or in herbaceous vegetation, less than 2 m above the ground. In winter generally inhabits low, moist thickets, forest edges, second-growth woodlands, and hedgerows. Much more abundant in moist forest than in dry evergreen woodland on U.S. Virgin Islands (Askins et al. 1992). In Veracruz, Mexico, males outnumbered females 8 to 1 in primary forest, whereas females were slightly more numerous in secondary forest (Rappole and Warner 1980). Regularly occurs in pine plantations and in cacao (Robbins et al. 1992).

SPECIAL HABITAT REQUIREMENTS. Low, dense, deciduous woody vegetation.

Further Reading. United States and Canada: Bent 1953b, DeGraaf and Rudis 1986, Farrand 1983c, Godfrey 1966, Griscom and Sprunt 1979, Harrison 1984, Odum 1931. Mexico and Central America: Edwards 1972, Land 1970, Rappole et al. 1983, Stiles and Skutch 1989. South America: Meyer de Schauensee and Phelps 1978, Rappole et al. 1983. Caribbean: Raffaele 1989, Rappole et al. 1983.

Wilson's Warbler

Wilsonia pusilla

Canada: Fauvette à Calotte Noire

Costa Rica: Reinita Gorrinegra

Guatemala: Chipe Careto, Reinita de Capucha

Mexico: Chipe Coroninegro

RANGE. Breeds from northern Alaska, northern Yukon, northern Ontario, southeastern Labrador, and Newfoundland south to southern California, central Nevada, northern Utah, northern New Mexico, central Ontario, northern New England (northern White Mountains), and Nova Scotia. Winters from southern California and southern Texas to Panama.

STATUS. Common, but declining over much of the breeding range.

HABITAT. Prefers wet clearings in early stages of regeneration. Also inhabits peat or laurel bogs with scattered young or dwarf spruces and tamaracks, and riparian willow and alder thickets. In the West, dense riparian thickets at higher elevations, boggy montane thickets (Dobkin 1992). Generally nests on the ground, sometimes in loose colonies. More abundant in young (40–75 yr) than older (>105 yr) Douglas-fir stands in the Pacific Northwest (Manuwal and Huff 1987). Usually builds nest at the base of a small tree or shrub, often well concealed in a grass hummock. Occasionally, places nest above the ground in low, dense tangles of vegetation. Not normally parasitized by cowbirds. In winter frequents forest understory, openings and edges, second growth, coffee plantations, brushy fields, dooryard trees, and shrubbery (Robbins et al. 1987). In isolated cloud forest in Chiapas, Mexico (where the deforestation rate is among the highest in Latin America), Wilson's Warbler is a forest understory specialist and, together with the canopy specialists Townsend's Warbler and *Auduboni* group of Yellow-rumped Warbler, dominates the autumn avifauna (Vidal-Rodriguez 1992).

SPECIAL HABITAT REQUIREMENTS. Shrub swamps, bog thickets.

Further Reading. United States and Canada: Beal 1907, Bent 1953b, DeGraaf and Rudis 1986, Farrand 1983c, Godfrey 1966, Griscom and Sprunt 1979, Harrison 1984, Morrison 1981, Stewart 1973, Stewart et al. 1977. Mexico and Central America: Edwards 1972, Land 1970, Rappole et al. 1983, Stiles and Skutch 1989.

Canada Warbler
Wilsonia canadensis

Canada (Quebec): Fauvette du Canada

Costa Rica: Reinita Pechirrayada

Guatemala: Chipe Collarejo, Reinita Collareja

Mexico: Chipe de Collar

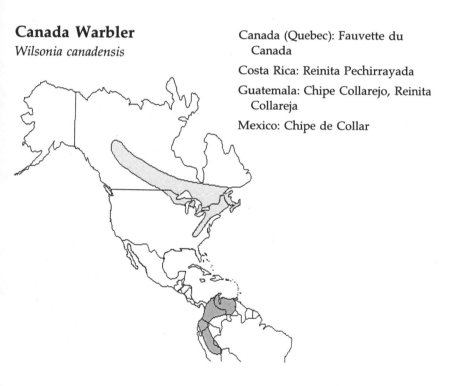

RANGE. Breeds from central Alberta east to southern Quebec and Nova Scotia, south to southern Manitoba, central Minnesota, central Michigan, and through the Appalachian Mountains to northern Georgia. Migrates through Central America. Winters in Venezuela, Colombia, Ecuador, Peru, and Brazil.

STATUS. Locally common. Declining since the early 1980s in the eastern part of the breeding range.

HABITAT. Inhabits a variety of vegetative types from lowlands to uplands and coniferous to deciduous. May be a bird of deciduous-coniferous transition areas (Morse 1989:63). Favors shrubby undergrowth in cool, moist, mature woodlands, streamside thickets, and weedy ravines. Builds nest on or near the ground on mossy logs or stumps, in cavities in banks, among roots of fallen trees, or in mossy hummocks. In the southern Appalachians, occurs in rhododendron thickets. In winter frequents thickets in taller second growth and forest, coffee plantations, semi-open areas, and hedgerows.

SPECIAL HABITAT REQUIREMENTS. Breeding: forest undergrowth, especially deciduous shrubs along streams in coniferous forests, bogs; migration and winter: woodlands, humid thickets.

Further Reading. United States and Canada: Bent 1953b, DeGraaf and Rudis 1986, Farrand 1983c, Godfrey 1966, Griscom and Sprunt 1979, Harrison 1984, Krause 1965. Mexico and Central America: Edwards 1972, Land 1970, Rappole et al. 1983, Stiles and Skutch 1989. South America: Meyer de Schauensee and Phelps 1978, Rappole et al. 1983. Caribbean: Raffaele 1989, Rappole et al. 1983.

Red-faced Warbler
Cardellina rubrifrons

Guatemala: Chipe de Cara Colorada, Reinita Carirroja

Mexico: Chipe Carirrojo

RANGE. Breeds from central Arizona and southwestern New Mexico south into northern Mexico to Chihuahua, Sinaloa, and western Durango. Winters from Sinaloa and Durango south through the highlands of Mexico to the borders of El Salvador and western Honduras.

STATUS. Locally common.

HABITAT. Generally prefers Douglas-fir and Engelmann spruce forests at elevations of 2,100 to 3,000 m, but also inhabits ponderosa pine, oak, aspen, and riparian stands and seems to favor southern exposures. Places nest nearly always on the ground, concealed beneath or beside a sheltering log, rock, sapling, or tuft of grass, usually on a well-drained bank or hillside.

SPECIAL HABITAT REQUIREMENTS. Breeding: montane fir and pine forests; winter: humid montane forest and riparian woodland.

Further Reading. United States and Canada: Griscom and Sprunt 1979, Harrison 1984, Morse 1989, Scott and Gottfried 1983. Mexico and Central America: Edwards 1972, Land 1970, Rappole et al. 1983.

Painted Redstart
Myioborus pictus

Guatemala: Pavito Ocotero, Rey Chipe

Mexico: Pavito Aliblanco

RANGE. Breeds from northwestern and central Arizona, southwestern New Mexico, and western Texas south through the mountains of Central America to Nicaragua; casually in southern California. Winters from northwestern Mexico south through the breeding range.

STATUS. Common. The only one of the 10 New World *Myioborus* species that breeds in the United States.

HABITAT. Mainly inhabits timbered desert mountain canyons, gulches, and rugged slopes in coniferous and deciduous woodlands, generally near water. Especially favors dense thickets and oaks in secluded canyons near streams. Nearly always places nest on the ground under a rock, tree root, or grass tuft that provides overhead shelter, and usually on a sloping bank or rocky canyon wall near water.

SPECIAL HABITAT REQUIREMENTS. Oak-pine forest, pinyon-juniper woodland.

Further Reading. United States and Canada: Bent 1953b, Griscom and Sprunt 1979, Harrison 1984, Marshall and Balda 1974, Oberholser 1974b. Mexico and Central America: Edwards 1972, Land 1970, Rappole et al. 1983.

Yellow-breasted Chat
Icteria virens

Canada (Quebec): Fauvette Polyglotte

Costa Rica: Reinita Grande

Guatemala: Buscabreña, Chipe Grande

Mexico: Chipe Piquigrueso

RANGE. Breeds from southern British Columbia, North Dakota, southern Minnesota, southern Ontario, Vermont, and New Hampshire south to south-central Baja California, the Gulf Coast, north-central Florida, and Mexico. Winters from southern Texas and southern Florida south through Central America and western Panama.

STATUS. Common, but has declined in New England and much of eastern North America because of reversion of old fields to forest. Recent increase in western North America.

HABITAT. Favors extensive areas of shrubby habitat not overtopped by trees. Also, ravine or streamside thickets of vines, briars, small trees, and tall shrubs. Also inhabits forest edges, hedgerows, shrub thickets at stock tanks, overgrown pastures, scrub country, and extensive areas of early successional stages of hardwood forest regeneration. Usually builds nest 1 to 2 m above the ground in dense small bushes, vines, or briars. May sometimes nest in groups or colonies, but maintains separate territory. In winter, territorial in scrub thickets, young second growth, and at woodland edges.

SPECIAL HABITAT REQUIREMENTS. Dense shrublands and shrub swamps with few overtopping young trees.

Further Reading. United States and Canada: Bent 1953b, Dennis 1958, Godfrey 1966, Griscom and Sprunt 1979, Harrison 1984, Petrides 1938, Thompson and Nolan 1973. Mexico and Central America: Edwards 1972, Land 1970, Rappole et al. 1983, Stiles and Skutch 1989.

Olive Warbler
Peucedramus taeniatus

Guatemala: Chipe Aceitunado, Ocoterito

Mexico: Peucedramo

RANGE. Breeds from central and southeastern Arizona, southwestern New Mexico, northern Chihuahua, northern Coahuila, southern Nuevo León, and western Tamaulipas south through the highlands of Mexico and Central America to north-central Nicaragua. Winters throughout the breeding range, except in Arizona and New Mexico, where it withdraws southward.

STATUS. Fairly common.

HABITAT. Generally found near the summits of mountains in the southwestern United States above 2,500 m in mixed pine-fir forests; usually observed near the tops of coniferous trees. Builds a cup-shaped nest, usually placed near the end of a conifer limb, sometimes hidden by pine needles or a cluster of mistletoe; usually high (10 to 25 m) above the ground.

SPECIAL HABITAT REQUIREMENTS. Montane coniferous forest.

Further Reading. United States and Canada: Bent 1953a, Griscom and Sprunt 1979, Harrison 1984. Mexico and Central America: Edwards 1972, Land 1970, Rappole et al. 1983.

Hepatic Tanager
Piranga flava

Costa Rica: Tangara Bermeja

Guatemala: Piranga Rojiza, Quitrique de los Altiplanos

Mexico: Tangara Roja Piquioscura

Venezuela: Cardenal Avispero

RANGE. Breeds from southeastern California and northwestern and central Arizona through New Mexico and western Texas, Nuevo León, and Tamaulipas south through the highlands of Mexico and Central America to central Argentina. Winters from northern Mexico to the southern part of the breeding range; casually in southern California and southern Arizona. Individuals from the United States only occur as far south as Guatemala. Resident from Costa Rica south to Argentina.

STATUS. Fairly common.

HABITAT. Generally favors open coniferous forest, pine and pinyon-juniper woodlands between 1,600 and 2,500 m in elevation, but also inhabits the more montane pine-oak, and riparian woodland. South of Costa Rica, breeds in open, humid forest, scrub, and orchards. Builds a flat, saucer-shaped nest, usually in a fork near the end of a horizontal tree branch, 5 to 15 m above the ground. In winter, migrants found in lowland woodland and forest, second growth, and in clearings with scattered trees.

SPECIAL HABITAT REQUIREMENTS. Open coniferous forest; from Costa Rica south, open humid forest and orchards.

Further Reading. United States and Canada: Bent 1958, Farrand 1983c, Phillips et al. 1964. Mexico and Central America: Edwards 1972, Land 1970, Rappole et al. 1983, Ridgely and Gwynne 1989, Stiles and Skutch 1989. South America: Meyer de Schauensee and Phelps 1978, Rappole et al. 1983.

Summer Tanager
Piranga rubra

Canada (Quebec): Tangara Vermillon

Costa Rica: Tangara Veranera

Guatemala: Piranga Colorada, Piranga Migratoria, Quitrique Colorado

Mexico: Tangara Roja Migratoria

Venezuela: Cardenal Migratorio

RANGE. Breeds from southeastern California and southern Nevada to central Oklahoma, and from southeastern Nebraska to New Jersey south to the Gulf Coast and northern Mexico. Migrates through the West Indies. Winters mainly from Mexico to Bolivia; rare winter visitor in southern temperate areas.

STATUS. Common, but possibly declining since the early 1980s over much of its breeding range.

HABITAT. Generally inhabits dry, open woodlands of oaks, pines, and hickories in the Southeast; but only rich bottomland forests at the northern edge of its range. Inhabits low-elevation willows and cottonwoods, and streamside vegetation in canyons in the Southwest. In winter found in shrubby areas, clearings with scattered trees, second-growth woodland, and forest edges. Builds a flimsy, flat, shallow cup nest on a horizontal limb (often in oaks) 3 to 10 m above the ground. The Summer Tanager is a habitat generalist on the wintering grounds on the Yucatán Peninsula (Lynch 1989). In western Mexico (Sinaloa south to Chiapas) probably restricted to undisturbed tropical deciduous forest, and therefore likely sensitive to deforestation in the lowlands (Hutto 1992).

SPECIAL HABITAT REQUIREMENTS. Deciduous forest, riparian woodland.

Further Reading. United States and Canada: Bent 1958, Fitch and Fitch 1955, Forbush and May 1955, Godfrey 1966, Johnsgard 1979, Potter 1973, Terres 1980. Mexico and Central America: Edwards 1972, Land 1970, Rappole et al. 1983, Stiles and Skutch 1989. South America: Meyer de Schauensee and Phelps 1978, Rappole et al. 1983. Caribbean: Rappole et al. 1983.

Scarlet Tanager
Piranga olivacea

Canada (Quebec): Tangara Écarlate

Costa Rica: Tangara Escarlata

Guatemala: Piranga Escarlata,
Quitrique Rojo

Mexico: Tangara Rojinegra Migratoria

Puerto Rico: Escarlatina

Venezuela: Cardenal Migratorio
Alinegro

RANGE. Breeds from southern Manitoba, western Ontario, southern Quebec, and New Brunswick south to eastern North Dakota, central Nebraska, southern Kansas, eastern Oklahoma, central Arkansas, northern Alabama, and northern Georgia; casually in the western United States. Winters from Panama and Colombia south, east of the Andes, through eastern Ecuador and Peru to northwestern Bolivia in South America.

STATUS. Common.

HABITAT. Generally inhabits mature or nearly mature deciduous and mixed deciduous-coniferous woodlands, roadside shade trees, wooded parks, and large shade trees of suburbs. In the Great Plains states, primarily inhabits mature hardwood forests of river valleys, hillsides, and valleys. Builds a shallow, saucer-shaped nest, usually well out on a horizontal limb of a large tree, usually in a leaf cluster or in a position where nest is shaded from above. A common cowbird host. In winter found in forest or woodlands;

highly vulnerable to tropical deforestation because of fairly restricted winter range (Morton 1992).

SPECIAL HABITAT REQUIREMENTS. Deciduous forest and mature deciduous or mixed woodlands.

Further Reading. United States and Canada: DeGraaf and Rudis 1986, Farrand 1983c, Forbush and May 1955, Godfrey 1966, Johnsgard 1979, Shy 1984, Terres 1980. Mexico and Central America: Edwards 1972, Land 1970, Rappole et al. 1983, Stiles and Skutch 1989. South America: Meyer de Schauensee and Phelps 1978, Rappole et al. 1983. Caribbean: Raffaele 1989, Rappole et al. 1983.

Western Tanager
Piranga ludoviciana

Canada (Quebec): Tangara à Tête Rouge

Costa Rica: Tangara Carirroja

Guatemala: Piranga Cabecirroja, Quitrique de Frente Colorada

Mexico: Tangara Aliblanca Migratoria

RANGE. Breeds from southeastern Alaska, northern British Columbia, southern Northwest Territories, northern Alberta, and central Saskatchewan south to northern Baja California, southern Nevada, southwestern Utah, central and southeastern Arizona, southern New Mexico, and western Texas, and east to eastern Montana, western South Dakota, northwestern Nebraska, central Colorado, and central New Mexico. Winters from Baja California and central Mexico south to Costa Rica, casually to western Panama.

STATUS. Common; increasing.

HABITAT. Generally inhabits relatively open, mature montane coniferous forests and woodlands up to 3,200 m in elevation. Breeds less frequently in mixed forests and in deciduous forests in the mountains, along rivers, or in gulches and canyons at lower elevations. Generally more abundant in old-growth Douglas-fir/ponderosa pine than in younger stands (Hejl and Woods 1991). Builds a shallow, compact, saucer-shaped nest, saddled in a fork of a horizontal branch well out from the trunk. Generally not parasitized by cowbirds. Usually locates nest in a coniferous tree. In winter found in second-growth woodlands and forest edges.

SPECIAL HABITAT REQUIREMENTS. Open coniferous forest and mixed coniferous-deciduous woodland; in Middle America in winter, highland pine forest.

Further Reading. United States and Canada: Beal 1907, Bent 1958, Godfrey 1966, Harrison 1979, Johnsgard 1979. Mexico and Central America: Edwards 1972, Land 1970, Rappole et al. 1983, Ridgely and Gwynne 1989, Stiles and Skutch 1989.

Rose-breasted Grosbeak
Pheucticus ludovicianus

Canada (Quebec): Gros-bec à Poitrine Rose

Costa Rica: Picogrueso Pechirrosado

Guatemala: Piquigrueso Pechirrosado, Realejo de Pecho Rosado

Mexico: Picogrueso Pechirrosa

Puerto Rico: Piquigrueso Rosado

Venezuela: Picogordo Degollado

West Indies: Degollado

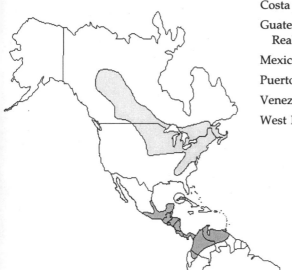

RANGE. Breeds from southern Northwest Territories across southern Canada to Nova Scotia, south to north-central North Dakota and Kansas, central Oklahoma, southern Missouri, central Indiana, and Ohio to central New Jersey and south along the Appalachians to northern Georgia. Migrates through the West Indies, but a rare visitor to Puerto Rico and the Virgin Islands. Winters from central Mexico to Venezuela, Colombia, Ecuador, and Peru; rarely in the southwestern United States.

STATUS. Common, but declining since the early 1980s over much of its breeding range.

HABITAT. Occurs in areas with large trees, openings, and thick shrubs or brush, or second-growth deciduous or mixed woods, borders of swamps and streams, dense growths of small trees (regenerating clearcuts in the sapling stage), and shrubs along edges of woods and pastures. Builds a flimsy nest, usually in a fork of a deciduous tree or shrub, about 3 to 5 m above the ground. Occasionally nests in coniferous trees. Common cowbird host.

SPECIAL HABITAT REQUIREMENTS. Forest edges with dense brush or thick sapling stands.

Further Reading. United States and Canada: Farrand 1983c, Forbush and May 1955, Godfrey 1966, Harrison 1975, McAtee 1908, Terres 1980. Mexico and Central America: Edwards 1972, Land 1970, Rappole et al. 1983, Ridgely and Gwynne 1989, Stiles and Skutch 1989. South America: Meyer de Schauensee and Phelps 1978, Rappole et al. 1983. Caribbean: Bond 1947, Raffaele 1989, Rappole et al. 1983.

Black-headed Grosbeak
Pheucticus melanocephalus

Canada (Quebec): Gros-bec à Tête Noire

Mexico: Picogrueso Pechicafé

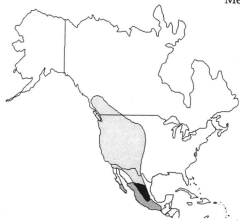

RANGE. Breeds from southern British Columbia east to southern Saskatchewan and central Kansas south to southern Mexico. Winters mainly in western Mexico from southern Baja California and southern Sonora south to Oaxaca. Rare in the southwestern United States during winter.

STATUS. Common; generally increasing.

HABITAT. Primarily inhabits relatively open stands of deciduous forests in uplands or floodplains, but also found in or near orchards, brushy woodlands or chaparral, edges or transitions between grasslands and woodlands, riparian woodland or thickets, and parks or suburbs with many trees. Oc-

475

cupies Douglas-fir stands partially opened by logging in the interior Pacific Northwest (Medin 1985). Builds a bulky, open-cup nest 1 to 4 m above the ground in a fork of a variety of shrubs or small trees. Usually nests (about 80% of time) in deciduous trees and shrubs. Not normally parasitized by cowbirds.

SPECIAL HABITAT REQUIREMENTS. Deciduous forest and woodland, oak-pine and pinyon-juniper woodland.

Further Reading. United States and Canada: Bent 1968a, Farrand 1983c, Godfrey 1966, Hill 1988, Johnsgard 1979, Terres 1980, Weston 1947. Mexico: Edwards 1972, Peterson and Chalif 1973, Rappole et al. 1983.

Blue Grosbeak
Guiraca caerulea

Canada (Quebec): Gros-bec Bleu
Guatemala: Piquigrueso Azul
Mexico: Picogrueso Azul
Puerto Rico: Azulejo
West Indies: Azulejón, Azulejo Real

RANGE. Breeds from southern California, southern Nevada, southern Colorado, Nebraska, south-central North Dakota, southern Ohio, and New Jersey south to the Gulf Coast and central Florida, and through Mexico into Costa Rica. Winters from Mexico south to Panama, and in the Bahamas and Cuba.

STATUS. Fairly common and increasing, but rare and local in the northeastern part of its breeding range.

HABITAT. Inhabits weedy pastures, old fields with saplings, forest edges, streamside thickets, hedgerows, swampy thickets, and willows along irrigation ditches. In the West, inhabits mesquite woods. Builds a compact, rather deep nest 1 to 3 m above the ground in a low tree, shrub, tangle of vines, or briars, typically at the edge of an open area. In winter, found in open shrubby areas with scattered bushes and trees in lowlands. Coastal scrub is the only winter habitat used on the northern Yucatán Peninsula (Lynch 1992).

SPECIAL HABITAT REQUIREMENTS. Open woodlands, thickets, cultivated lands; in winter also in second growth and weedy fields.

Further Reading. United States and Canada: Forbush and May 1955, Godfrey 1966, Johnsgard 1979, Stabler 1959, Terres 1980. Mexico and Central America: Edwards 1972, Land 1970, Rappole et al. 1983, Ridgely and Gwynne 1989, Stiles and Skutch 1989. Caribbean: Bond 1947, Raffaele 1989, Rappole et al. 1983.

Lazuli Bunting
Passerina amoena

Canada (Quebec): Bruant Azuré
Mexico: Colorín Aliblanco

RANGE. Breeds from southern British Columbia to central North Dakota and northeastern South Dakota south to northwestern Baja California, southern Nevada, central Arizona, central New Mexico, and central Texas. Winters from southern Baja California and southern Arizona south to Guerrero and central Veracruz, Mexico.

STATUS. Common to uncommon.

HABITAT. Inhabits a variety of habitats from near sea level to 3,200 m in the Sierra Nevada and 2,600 m in the Rocky Mountains of Colorado. Generally found in diverse habitats with an abundance of shrubs, low trees, and herbaceous vegetation such as riparian thickets and woodlands, brushy draws of prairies and dry brushy hillsides, wooded valleys, aspen woodlands. In much of the arid West, found mainly in riparian vegetation or thickets resulting from irrigation. Builds a coarsely woven cup nest attached to supporting stalks or a fork of a low shrub or vine tangle, usually about 1 m above the ground.

SPECIAL HABITAT REQUIREMENTS. Arid brushy canyons, riparian thickets, chaparral.

Further Reading. United States and Canada: Bent 1968a, Emlen et al. 1975, Grinnell and Miller 1944, Harrison 1979, Johnsgard 1979, Terres 1980. Mexico: Edwards 1972, Peterson and Chalif 1973, Rappole et al. 1983.

Indigo Bunting
Passerina cyanea

Canada (Quebec): Bruant Indigo
Costa Rica: Azulillo Norteño
Guatemala: Azulillo Norteño, Ruicito
Mexico: Colorín Azul
Puerto Rico: Gorrión Azul
Venezuela: Azulillo
West Indies: Azulejo

RANGE. Breeds from southeastern Saskatchewan and northern Minnesota to southern New Brunswick, south to southern New Mexico, central and southeastern Texas, the Gulf Coast, and central Florida; locally in central Colorado, southwestern Utah, central Arizona, and southern California. Winters primarily in southern Mexico, the Greater Antilles (except Hispaniola), and the Virgin Islands south to Panama; also in southern Florida and infrequently in coastal Texas and elsewhere in the southern part of the breeding range.

STATUS. Common but declining over much of its breeding range, especially in the East, as old fields revert to forest.

HABITAT. Generally associated with edges of woods, old burns, open brushy fields, roadside thickets, open deciduous woodlands, and brushy ravines. Tends to be more numerous along streams, and avoids deep woods. Builds a cup nest in a crotch of a bush, shrub, or low tree, in a tangle of berry vines, or in canebrakes. Frequently parasitized by Brown-headed Cowbird. Partial or improvement cuts in the mixed mesophytic forest of the Southeast are beneficial to the Indigo Bunting (McComb et al. 1989). On wintering grounds, irregular flocks inhabit agricultural areas (Rappole and Warner 1980), especially fallow rice fields (Robbins et al. 1992), and grassy roadsides bounded by thickets; seldom in forest.

SPECIAL HABITAT REQUIREMENTS. Forest edges, old fields.

Further Reading. United States and Canada: Bent 1968a, DeGraaf and Rudis 1986, Forbush and May 1955, Godfrey 1966, Harrison 1975. Mexico and Central America: Edwards 1972, Land 1970, Rappole et al. 1983, Stiles and Skutch 1989. South America: Meyer de Schauensee and Phelps 1978, Rappole et al. 1983. Caribbean: Bond 1947, Raffaele 1989, Rappole et al. 1983.

Varied Bunting
Passerina versicolor

Guatemala: Azulillo Morado,
Cuatrocolores Morado

Mexico: Colorín Oscuro

RANGE. Breeds in south-central and southeastern Arizona, southwestern and southeastern New Mexico, and southern Texas south to Mexico and Guatemala. Winters from southern Texas south throughout the breeding range.

STATUS. Local and uncommon.

HABITAT. Inhabits mesquite or thorny shrubs, brushy pastures, dense vegetation with cottonwoods, foothill canyons, and generally hilly and rocky terrain; tends to avoid heavily wooded areas. Builds a compact cup nest 1 to 3 m above the ground in the crotch of a shrub, low tree, or vine tangle.

SPECIAL HABITAT REQUIREMENTS. Thorny shrubs.

Further Reading. United States and Canada: Bent 1968a, Harrison 1979, Oberholser 1974b, Terres 1980. Mexico and Central America: Edwards 1972, Land 1970, Rappole et al. 1983.

Painted Bunting
Passerina ciris

Costa Rica: Azulillo Sietecolores, Siete Colores

Guatemala: Azulillo Pinato, Cuatrocolores Pinato

Mexico: Colorín Sietecolores

West Indies: Arco-iris (male), Verderol (female)

RANGE. Breeds from southeastern New Mexico and southern Missouri south to southern Alabama and into Mexico, also along Atlantic Coast from southeastern North Carolina south to central Florida. Winters from southeastern Texas, central Florida, the Bahamas, and Cuba south through Mexico to Panama.

STATUS. Locally common; declining in the south-central United States.

HABITAT. Inhabits open country with brushy and weedy fields, hedges, edges of woods, roadside shrubs, gullies, thickets along streambanks, shelterbelts, and gardens. Nests in a variety of deciduous shrubs, small trees, and vines. Attaches shallow cup nest to twigs or other supporting vegetation, 1 to 3 m above the ground, in bushes, low trees, or vine tangles. Raises two and sometimes three broods each year; susceptible to parasitism by Brown-headed Cowbirds. Common, but irregularly, singly or in small flocks in overgrown fields, brushy pastures, shrubby areas, and woodland borders in lowland areas in winter.

SPECIAL HABITAT REQUIREMENTS. Open brush areas, riparian thickets; in migration and winter, open grassy, weedy, or scrub habitats.

Further Reading. United States and Canada: Bent 1968a, Farrand 1983c, Harrison 1979, Johnsgard 1979, Terres 1980. Mexico and Central America: Edwards 1972, Land 1970, Rappole et al. 1983, Stiles and Skutch 1989. Caribbean: Bond 1947, Rappole et al. 1983.

Dickcissel
Spiza americana

Canada (Quebec): Dickcissel

Costa Rica: Sabanero Arrocero

Guatemala and Puerto Rico: Sabanero Americano

Mexico: Espiza

Venezuela: Arrocero Americano

RANGE. Breeds from eastern Montana and southern Canada to New York, south to central Colorado, southern Texas, and central Alabama. In the eastern portion of the range, breeds sporadically and irregularly. Winters mostly from Mexico (Michoacán south), primarily along the Pacific slope through Middle America to Colombia, Venezuela, Guyana, Surinam, and French Guiana. Winters locally (in small numbers) in coastal lowlands from southern New England south to Florida and west to southern Texas. Migrates through Baja California and the western Caribbean. Accidental in Puerto Rico.

STATUS. Common but may be declining in the Great Plains; rare and local in the East; overall, populations are declining.

HABITAT. Generally inhabits grasslands having tall grasses, forbs, or shrubs but also fields planted to such crops as alfalfa, clover, and timothy. Builds a bulky cup nest on the ground or attached to forks in shrubs, vines, or low trees. Locates nest in a variety of situations such as marshes, hayfields, abandoned or fallow croplands, roadsides, fencerows, and grasslands. Also frequents abandoned or fallow croplands in breeding season and winter. May experience population declines because of changes in agricultural practices in the grasslands of Venezuela.

SPECIAL HABITAT REQUIREMENTS. Dense herbaceous cover and song perches.

Further Reading. United States and Canada: Fretwell 1986, Godfrey 1966, Gross 1921, Harmeson 1974, Overmire 1962, Taber 1947, Tate and Tate 1982, Zimmerman 1982. Mexico and Central America: Edwards 1972, Land 1970, Rappole et al. 1983, Stiles and Skutch 1989. South America: Fretwell 1986, Meyer de Schauensee and Phelps 1978, Rappole et al. 1983. Caribbean: Raffaele 1989, Rappole et al. 1983.

Green-tailed Towhee
Pipilo chlorurus

Canada (Quebec): Tohi à Queue Verte
Mexico: Rascador Migratorio

RANGE. Breeds from southeastern Washington, southern Idaho, southwestern Montana, northwestern and southeastern Wyoming south through the interior mountains to southern California, southern Nevada, and central Arizona to western Texas. Winters from southern California to western and southern Texas south to central Mexico; casual east of the breeding range in fall and winter.

STATUS. Fairly common.

HABITAT. Generally inhabits relatively arid and brushy foothills with shrubs such as sagebrush, deerbush, snowberry, wild rose, spirea, mountain-mahogany, manzanita, waxberry, and chokecherry; also shrubsteppe, brushy montane slopes from 800 m elevation in California to 3,400 m in Arizona. Tends to breed at higher elevations in the south than in the north. Considered a near-obligate of sagebrush landscapes. Builds a large, loosely constructed and deeply cupped nest on the ground or in low shrubs such as sagebrush, waxberry, and snowberry, usually less than 1 m above the ground.

SPECIAL HABITAT REQUIREMENTS. Arid brush in mountains; lowland habitats in nonbreeding season.

Further Reading. United States and Canada: Bent 1968a, Braun et al. 1976, Harrison 1979, Johnsgard 1979, Morton 1992. Mexico: Edwards 1972, Peterson and Chalif 1973, Rappole et al. 1983.

Rufous-sided Towhee
Pipilo erythrophthalmus

Canada (Quebec): Tohi Commun
Guatemala: Zarcero
Mexico: Rascador Pinto Oscuro

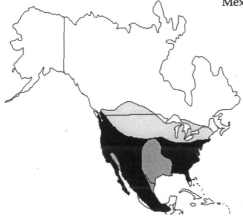

RANGE. Breeds from southern British Columbia to southern Maine south to southern Baja California, Guatemala, northern Oklahoma, eastern Louisiana, and southern Florida. Winters from southern British Columbia, Utah, Colorado, the southern Great Lakes area, and along the Atlantic Coast south throughout the breeding range.

STATUS. Common and widespread, but rapidly declining in parts of the eastern United States because of forest succession; recently increasing in the western United States.

HABITAT. Generally inhabits dense brushy fields and pastures, edges of woods, open woodlands, hedgerows, roadside thickets, and clearings. In the West, inhabits montane shrubby slopes: sagebrush, willows, and chaparral above the deserts, as well as riparian thickets. Generally invades logged sites in western coniferous forests. Builds a bulky nest, usually in a depression in the ground but sometimes up to 1 m above the ground, in low shrubs such as huckleberry, blueberry, coffeeberry, or sagebrush. Conceals and protects nest with overhanging bushes, logs, vines, or a clump of grass. Frequently parasitized by cowbirds.

SPECIAL HABITAT REQUIREMENTS. Dense brushy cover.

Further Reading. United States and Canada: Baumann 1959, Bent 1968a, Davis 1960, Godfrey 1966, Medin 1985. Mexico and Central America: Edwards 1972, Land 1970, Peterson and Chalif 1973, Rappole et al. 1983.

Botteri's Sparrow
Aimophila botterii

Costa Rica: Sabanero Pechianteado

Guatemala: Sabanero Zacaterillo

Mexico: Gorrión de Botteri Común

RANGE. Breeds from southeastern Arizona and extreme southern Texas south to Costa Rica. Winters from northern Mexico south throughout the breeding range. Individuals from the United States winter in Mexico. All individuals occurring south of Mexico are residents.

STATUS. Rare and local.

HABITAT. Only inhabits areas with dense, tall grass, breeding in open grassland and savanna, especially in areas with scattered brush or shrubs. Favors tall-grass habitats with mesquite and catclaw in Arizona; prefers salt-grass with some yucca, pricklypear, and mesquite in the coastal prairies of Texas. Builds nest on the ground among tall grasses, at the base of a tuft of grass, or sometimes under a projecting mat of grass. In Costa Rica found in open, barren, boulder-strewn grasslands.

SPECIAL HABITAT REQUIREMENTS. Open grassland with scattered shrubs or small trees.

Further Reading. United States and Canada: Bent 1968b, Cottam and Knappen 1939, Farrand 1983c, Oberholser 1974b. Mexico and Central America: Edwards 1972, Land 1970, Rappole et al. 1983, Stiles and Skutch 1989.

Cassin's Sparrow
Aimophila cassinii

Mexico: Gorrión de Cassin

RANGE. Breeds from southeastern Arizona, New Mexico, central and northeastern Colorado, southwestern Nebraska, and Kansas south into Mexico and Texas. Singing males may appear sporadically from southern California to South Dakota. Winters from southeastern Arizona and western and south-central Texas into Mexico.

STATUS. Common but has been declining.

HABITAT. Prefers open grassland and shortgrass plains with a few scattered shrubs or small trees. Also frequents mesquite grasslands if the mesquites are small with open areas throughout, but will not usually inhabit areas that are entirely grass unless surrounded by a fence for perching. Occasionally occurs in or near mountainous areas, on grassy slopes with scattered yuccas or small oaks. Favors sandy prairies with scattered sage, yucca, cactus, mesquite, and shinnery oaks in Oklahoma. Apparently can breed where no drinking water is available locally. May nest either on the ground or up to 0.5 m above the ground in low bushes or among tangled branches of cacti. Typically places ground nests at the foot of small shrubby plants, concealed in weeds or placed in a tuft of grass.

SPECIAL HABITAT REQUIREMENTS. Shortgrass prairies with scattered shrubs.

Further Reading. United States and Canada: Bent 1968b, Johnsgard 1979. Mexico and Central America: Edwards 1972, Peterson and Chalif 1973.

Rufous-crowned Sparrow

Aimophila ruficeps

Mexico: Gorrión Bigotudo
Coronirrufo

RANGE. Breeds from central California, southwestern Utah, southeastern Colorado, and central Oklahoma south to Oaxaca and central-western Veracruz, Mexico. Winters from northeastern New Mexico, northern Texas, and south-central Oklahoma south throughout the breeding range.

STATUS. Locally common.

HABITAT. Inhabits dry and desertlike habitats, preferring rocky, brushy, relatively arid hillsides with extensive bare areas. Also in rocky glades on the Great Plains; low ridges and foothills covered with scattered shrubs or trees and grass in Arizona; rocky slopes with large boulders, small cedars, and stunted oaks in Oklahoma; and grassy hillsides with scattered rocks and shrubs, especially sagebrush, and coastal scrub in California. Usually builds nest in a small depression on the ground, often near or under a clump of grass, or at the base of a shrub or small tree. Also locates nest up to 0.5 m or so above ground, wedged among dense vertical growing branches in shrubs and low trees.

SPECIAL HABITAT REQUIREMENTS. Rocky, arid slopes with scattered brush and grass.

Further Reading. United States and Canada: Bent 1968b, Farrand 1983c, Johnsgard 1979, Terres 1980, Verner and Boss 1980. Mexico and Central America: Edwards 1972, Peterson and Chalif 1973, Rappole et al. 1983.

Chipping Sparrow

Spizella passerina

Canada (Quebec): Pinson Familier

Costa Rica and Guatemala: Chimbito
Común

Mexico: Gorrión Coronirrufo
Cejiblanco

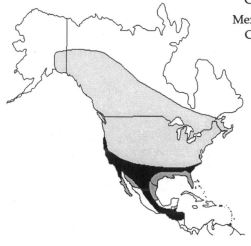

RANGE. Breeds from east-central and southeastern Alaska and central Yukon to northern Manitoba, southern Quebec, and southwestern Newfoundland south to southwestern and east-central California, central and eastern Texas, the Gulf Coast, and northwestern Florida through the highlands of Mexico to Nicaragua. Winters from central California, northern Texas, Tennessee, and Maryland south in Mexico throughout the breeding range. Northern populations winter south to southern Mexico; Central American populations are sedentary.

STATUS. Common but declining in western parts of the breeding range.

HABITAT. Inhabits open coniferous and deciduous woodlands, forest edges and clearings, wooded borders of lakes and rivers, mountain scrub, gardens, residential areas, farms, orchards, and grassland habitats with scattered trees. Prefers habitats with trees surrounded by an open area with only herbaceous vegetation and some open ground for foraging. Occupies successional habitats after logging, burning, and, because of affinity for open stands, older stands of western mixed conifer forest. More abundant in managed (80 yr, thinned) than in unmanaged conifer stands in the interior Pacific Northwest (Mannan and Meslow 1984). In winter, favors weedy fields and dry scrubland. In Costa Rica prefers open, grassy woodlands and savannas. Builds nest less than 8 m, but usually 1 to 3 m, above ground in trees, especially conifers, shrubs, or vines. Generally locates nest near the trunk and top of smaller trees, or lower in the branches and farther from the trunk in larger open-grown trees, usually well concealed.

Further Reading. United States and Canada: Beal and McAtee 1912, Bent 1968b, Buech 1982, DeGraaf and Rudis 1986, Farrand 1983c, Forbush and May 1955, Godfrey 1966, Johnsgard 1979, Mannon and Meslow 1984, Walkinshaw 1944. Mexico and Central America: Edwards 1972, Land 1970, Stiles and Skutch 1989.

Clay-colored Sparrow
Spizella pallida

Canada (Quebec): Pinson des Plaines
Mexico: Gorrión Indefinido Rayado

RANGE. Breeds from eastern British Columbia and west-central and southern Northwest Territories east to central Ontario, and south to eastern Washington, central Montana, eastern Colorado, northern Iowa, central and southeastern Michigan, and southwestern Quebec. Winters from central Texas south to Oaxaca, and rarely to Chiapas, Mexico.

STATUS. Locally common, declining in the western parts of the breeding range.

489

HABITAT. Prefers midwestern mixedgrass prairies with scattered low thickets of shrubs such as wolfberry; will inhabit a variety of dry, uncultivated shrubby habitats, including grasslands with taller shrubs or small trees, brushy hillsides, overgrown clearings and pastures, parklands, brushy woodland edges, burned-over areas, weedy thickets along roads, fencerows, railroad tracks and fields, shelterbelts, regenerating clearcuts, and other early-successional disturbed habitats. Builds nest either on the ground, well hidden in a tuft of grass at the base of a shrub or near a clump of weeds, or up to 1.5 m above ground in a low shrub or small tree. Commonly uses snowberry, rose, serviceberry, and conifers for nesting. Commonly parasitized by cowbirds.

SPECIAL HABITAT REQUIREMENTS. Open brushland.

Further Reading. United States and Canada: Bent 1968b, Buech 1982, Forbush and May 1955, Fox 1961, Hussong 1946, Johnsgard 1979, Knapton 1978, Salt 1966. Mexico and Central America: Edwards 1972, Peterson and Chalif 1973.

Brewer's Sparrow
Spizella breweri

Canada (Quebec): Pinson de Brewer
Mexico: Gorrión de Brewer

RANGE. Breeds from southwestern Yukon and northwestern interior British Columbia to southwestern Saskatchewan south, generally east of the Cascades and coast range, to southern California, central Arizona, central Colorado, and southwestern South Dakota. Winters from southern interior California to central Texas south into Mexico.

STATUS. Once common, now declining over much of its breeding range.

HABITAT. Inhabits open, shrub-dominated habitats; arid sagebrush country in the western United States and scrub balsam-willow habitats in timberline areas of western Canada, as well as bunchgrass prairie with rabbitbrush, dry, brushy mountain meadows, and pinyon-juniper woodlands. Strongly associated with sagebrush steppe communities, where it is the dominant bird species. Builds nest in shrubs, especially sagebrush, almost always located about 1 m or less above the ground. At timberline, locates nest about 15 m above ground in birch trees, well concealed overhead by interlocking branches. Rarely nests directly on the ground. Treatment of sagebrush with herbicide results in fewer nesting birds in the years of spraying (Best 1972) and complete habitat abandonment in subsequent years (Schroeder and Sturges 1975). Uncommon cowbird host (Dobkin 1972).

SPECIAL HABITAT REQUIREMENTS. Brushland and sagebrush; in winter, desert scrub, creosote bush.

Further Reading. United States and Canada: Bent 1968b, Godfrey 1966, Knopf et al. 1990, Peterson and Best 1987, Reynolds 1981, Wiens et al. 1987, 1990. Mexico and Central America: Edwards 1972, Peterson and Chalif 1973.

Field Sparrow
Spizella pusilla

Canada (Quebec): Pinson des Champs
Mexico: Gorrión Indefinido Oriental

RANGE. Breeds from northwestern and southeastern Montana and northern North Dakota to southwestern Quebec and southern New Brunswick south to western Kansas, southern Texas, the Gulf Coast, and southern Georgia. Winters from Kansas to Massachusetts, and south to Mexico, the Gulf Coast, and southern Florida.

STATUS. Fairly common but declining over much of its breeding range.

HABITAT. Occurs in a variety of habitats that provide low grassy areas and shrubs or low trees, including old fields and pastures overgrown with briar thickets or deciduous underbrush, brushy fencerows, cut-over pine forests and burned-over woodlands wherever briars and brush have regenerated, edges of open, unplowed fields, sagebrush flats, forest edges, and other similar habitats. Early in the breeding season, usually builds nest on or near the ground in weed clumps or tufts of grass; later in the season, builds nest as high as 1 m or more above the ground in shrubs or small trees. Locates nest in a wide range of plant species—grape vines, cinquefoil, blackberry bushes, boxelders, small oaks, and hickories.

SPECIAL HABITAT REQUIREMENTS. Abandoned fields or other open areas with low shrubs or trees.

Further Reading. United States and Canada: Bent 1968b, Best 1977, 1978, DeGraaf and Rudis 1986, Farrand 1983c, Forbush and May 1955, Fretwell 1986, Johnsgard 1979. Mexico and Central America: Edwards 1972, Peterson and Chalif 1973.

Black-chinned Sparrow   Mexico: Gorrión Indefinido Oriental
Spizella atrogularis

RANGE. Breeds from south-central California east to southern Nevada and southwestern Utah, south to Arizona, New Mexico, western Texas, and

Mexico. Winters from coastal California, southern Arizona, New Mexico, and Texas, south into Baja California and Oaxaca, Mexico.

STATUS. Uncommon; declining over much of its breeding range.

HABITAT. In desert regions, inhabits tall, dense sagebrush or other brushland areas covered with a variety of plant species. Prefers slopes with rocky outcrops and scattered pinyon or juniper trees. In the far west, inhabits dry chaparral habitat with a variety of shrubs and scrub oak. Generally builds a compact cup nest of dry grasses, often lined with animal hair, typically placed at the base of a shrub or in the lower portions of sage and shrub, occasionally up to 1 m above the ground.

SPECIAL HABITAT REQUIREMENTS. Chaparral and sage habitat with rocky outcrops.

Further Reading. United States and Canada: Phillips et al. 1964. Mexico and Central America: Edwards 1972, Peterson and Chalif 1973, Rappole et al. 1983.

Vesper Sparrow
Pooecetes gramineus

Canada (Quebec): Pinson Vespéral

Guatemala: Sabanero Torito, Semillero Torito

Mexico: Gorrión Zacatero Coliblanco

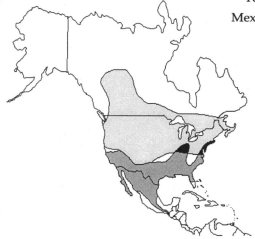

RANGE. Breeds from southern Northwest Territories and central Saskatch-ewan to southern Quebec and Nova Scotia, south to eastern and southern California, central New Mexico, Kansas, and North Carolina. Winters from central California, central Texas, southern Illinois, and Connecticut south to Mexico and Guatemala, the Gulf Coast, and central Florida.

STATUS. Fairly common. Once locally common in the northeastern United States, but populations are declining in eastern North America.

HABITAT. Favors sparsely vegetated dry uplands but also occurs in a variety of habitats throughout its range. In the western United States, inhabits open grasslands and sagebrush flats, pinyon-juniper associations, open meadows and farmlands, and low grassy areas of alpine and subalpine meadows. In the East, inhabits shortgrass meadows, pastures, hayfields, country road-sides, prairie edges, blueberry barrens, edges of potato fields and other tilled croplands, coastal beachgrass, and farther north, forest clearings and burned-over areas. Builds nest in a depression on the ground, frequently near small patches of bare ground, where the vegetation is low and sparse, or at the base of a dirt clod, clump of weeds, or tussock of grass, often well concealed by surrounding live or dead vegetation. Commonly parasitized by cowbirds. In winter, inhabits grasslands.

SPECIAL HABITAT REQUIREMENTS. Open areas with short herbaceous veg-etation and conspicuous song perches.

Further Reading. United States and Canada: Andrle and Carroll 1988, Bent 1968b, Best and Rodenhouse 1984, DeGraaf and Rudis 1986, Farrand 1983c, Forbush and May 1955, Godfrey 1966, Johnsgard 1979, Wray and Whitmore 1979, Zeranski and Baptist 1990:246–247. Mexico and Central America: Ed-wards 1972, Land 1970, Peterson and Chalif 1973, Rappole et al. 1983.

Lark Sparrow
Chondestes grammacus

Canada (Quebec): Pinson à Joues Marron

Guatemala: Sabanero Alondra

Mexico: Gorrión Arlequín

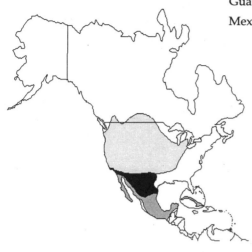

RANGE. Breeds from western Oregon and Washington, north into southern Canada, east through the Great Plains to the Missouri River, and south throughout the Southwest into Mexico. Winters from central California, southern Arizona, and Texas south to Baja California, throughout Mexico, in parts of the Gulf Coast up through South Carolina, and in the Bahamas and Cuba.

STATUS. Fairly common, but declining in much of the eastern part of its breeding range, and may be declining somewhat in the western parts as well.

HABITAT. Generally inhabits open prairies and other open lands. In the spring, frequently found along roadsides with grassy vegetation, but prefers open areas with scattered brush and trees. Also inhabits forest edges, cultivated areas, orchards, fields, and savannas. In the West, at lower elevations: prairie, grasslands with scattered trees, shrubs, sagebrush thickets. Builds nest either on the ground or in low tree or shrub. Sometimes reuses nests of other species (McNair 1984). Breeds in the open but retires to the borders of open woodlands or brushy areas after the young have fledged. Nests in grassland/prairie habitats, where it is sometimes parasitized by cowbirds.

495

SPECIAL HABITAT REQUIREMENTS. Dry fields with scattered shrubs or open forest edges.

Further Reading. United States and Canada: Forbush and May 1955, Godfrey 1966, Newman 1970. Mexico and Central America: Land 1970, Peterson and Chalif 1973, Rappole et al. 1983. Caribbean: Bond 1947, Rappole et al. 1983.

Black-throated Sparrow
Amphispiza billineata

Canada (Quebec): Pinson à Gorge Noire

Mexico: Gorrión Gorjinegro Carirrayado

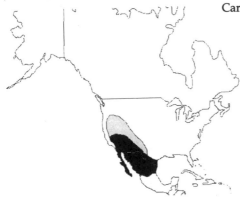

RANGE. Breeds from southeastern Oregon and northern California, east through the Great Basin, south into Baja California and Mexico. Winters from southern Arizona, New Mexico, and Texas, south into Baja California and Jalisco, Mexico.

STATUS. Declining in northern parts of its range.

HABITAT. Inhabits pastureland with thin grass and scattered mesquite, yucca, pricklypear, and cholla cacti. Generally found in dry uplands but extends into the depths of Death Valley, California. Generally conceals nest near the ground in small bushes or a variety of cactus species. Usually locates nest less than 0.5 m above the ground, fastened among forking branches of low shrubs. Builds nest with small twigs and fibers of sage, frequently lining it with the fur of animals found in the area, such as rabbits.

SPECIAL HABITAT REQUIREMENTS. Arid areas and desert scrub with scattered shrubs including cactus, sage, and mesquite.

Further Reading. United States and Canada: Godfrey 1966, Johnsgard 1979. Mexico and Central America: Edwards 1972, Peterson and Chalif 1973.

Lark Bunting
Calamospiza melanocorys

Canada (Quebec): Pinson Noir et Blanc

Mexico: Llanero Alipálido

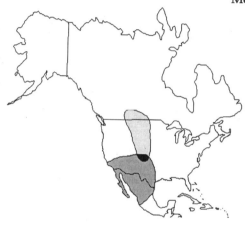

RANGE. Breeds from southern Alberta to southwestern Minnesota, and south, east of the Rockies, to eastern New Mexico, the Texas Panhandle, and northwestern Missouri; also locally or sporadically in southern California and Utah to west-central Texas. Winters from southern California to north-central Texas south into Jalisco, Mexico, and southern Louisiana.

STATUS. Common in the West, but populations fluctuate dramatically; may be declining in the eastern part of its breeding range with the loss or fragmentation of native prairie.

HABITAT. On the western Great Plains, inhabits mixed shortgrass prairie and other areas of predominately low growth, but also areas of taller grasses with scattered shrubs, abandoned cropland, and disturbed grasslands. Also inhabits sagebrush, fenced pastures, cultivated or fallow alfalfa or clover croplands, weedy roadsides, meadows, and areas of relatively barren ground. Builds nest in a depression in the ground, usually well concealed by grasses or other prairie plants, often located near the base of a plant or plant debris. Heavy grazing in shortgrass prairie degrades nesting habitat and can lower population density (Finch et al. 1987).

SPECIAL HABITAT REQUIREMENTS. Open habitats with relatively short, herbaceous vegetation.

Further Reading. United States and Canada: Bent 1968b, Farrand 1983c, Forbush and May 1955, Godfrey 1966, Johnsgard 1979. Mexico and Central America: Edwards 1972, Peterson and Chalif 1973, Rappole et al. 1983.

Savannah Sparrow
Passerculus sandwichensis

Canada (Quebec): Pinson d'Ipswich

Costa Rica: Sabanero Zanjero

Guatemala: Sabanero Zanjero, Semillero Zanjero

Mexico: Gorrión Sabanero Común

RANGE. Breeds from Alaska and northern Yukon to northern Labrador and Newfoundland, south in coastal regions to west-central California, and in the interior to central California, northern New Mexico, Mexico, Nebraska, Kentucky, and New Jersey. Winters from southern British Columbia and southern Nevada to southern Kentucky and, east of the Appalachians, from

Massachusetts south through Mexico to El Salvador, the Bahamas, and Cuba. Resident in coastal southern California.

STATUS. Common throughout the western part of its range, but declining in the eastern United States. Uncommon in southern New England because of loss of grassland habitats.

HABITAT. Inhabits open moist or wet areas with grass or grasslike vegetation. Occurs in hayfields, abandoned or weedy croplands, pastures, coastal and inland marshes, grassy dunes, wet meadow zones of ponds, lakes and streams, prairies, open grasslands, bogs, open moist areas of mountain parks and meadows, and tundra. Places nest in a natural hollow or scratched-out depression in the ground, among thick herbaceous cover, usually well hidden, not only by the dense cover surrounding the nest but also by overhanging vegetation. Winters in grasslands.

SPECIAL HABITAT REQUIREMENTS. Dense grassy or herbaceous vegetation of moderate height.

Further Reading. United States and Canada: Bent 1968b, DeGraaf and Rudis 1986, Forbush and May 1955, Johnsgard 1979, Potter 1972, Tate and Tate 1982, Welsh 1975, Zeranski and Baptist 1990:248. Mexico and Central America: Edwards 1972, Land 1970, Peterson and Chalif 1973, Rappole et al. 1983, Stiles and Skutch 1989. Caribbean: Bond 1947, Rappole et al. 1983.

Grasshopper Sparrow
Ammodramus savannarum

Canada (Quebec): Pinson Sauterelle

Costa Rica: Sabanero Colicorto

Cuba: Chamberguito

Dominican Republic: Tumbarrocio

Guatemala: Sabanero Colicorto,
  Semillero Llanerito

Mexico: Gorrión Saltamonte

Puerto Rico: Chingolo Chicharra,
  Gorrión Chicharra

West Indies: Oiseau Canne

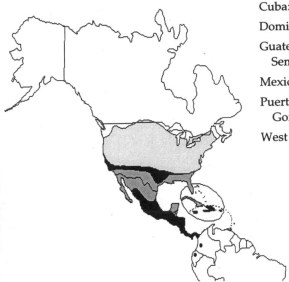

RANGE. Breeds from southern interior British Columbia and southern Alberta to southwestern Quebec and southern Maine south to southern California, central Colorado, northern and south-central Texas, central Georgia, and central North Carolina. There are also local resident populations throughout Central America south to Colombia and Ecuador, and in the West Indies on Jamaica, Hispaniola, and Puerto Rico. Winters from central California (rare) and southern Arizona to Tennessee and North Carolina south to Central America. Birds from North America winter from the southern United States to Costa Rica.

STATUS. Fairly common, but declining in much of its breeding range. Now rare in many northeastern states because of loss of grassland habitats. The Florida subspecies, *floridanus*, is endangered.

HABITAT. Prefers prairies in the West and cultivated grasslands, especially those with orchardgrass, alfalfa, and red clover in the East. Inhabits mixed-grass, shortgrass, and tallgrass prairies (generally >10 ha in size), sage prairies, small grain fields, and weedy fallow fields. Avoids fields where shrubs compose more than a third of the cover, but will occupy grassy habitats with some scattered trees. Builds nest in a slight depression on the ground, usually well hidden at the base of a clump of grass or other vegetation, with vegetation arched over the top. Often nests singly or in small colonies. In Central America found in dry, open savanna grasslands, often with scattered volcanic boulders.

SPECIAL HABITAT REQUIREMENTS. Continuous tall herbaceous cover and conspicuous song perches.

Further Reading. United States and Canada: Andrle and Carroll 1988, Bent 1968b, DeGraaf and Rudis 1986, Forbush and May 1955, Godfrey 1966, Johnsgard 1979, Samson 1980, Smith 1963, Tate and Tate 1980, Zeranski and Baptist 1990:249. Mexico and Central America: Edwards 1972, Land 1970, Peterson and Chalif 1973, Rappole et al. 1983, Stiles and Skutch 1989. South America: Hilty and Brown 1986, Rappole et al. 1983. Caribbean: Bond 1947, Raffaele 1989, Rappole et al. 1983.

Lincoln's Sparrow
Melospiza lincolnii

Canada (Quebec): Pinson de Lincoln

Costa Rica and Guatemala: Sabanero de Lincoln

Mexico: Gorrión de Lincoln

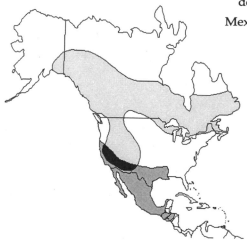

RANGE. Breeds from western central Alaska across most of Canada, south along the Pacific Coast and the Rocky Mountains in southern California and northern New Mexico, and into the northern Great Lakes states and northern

New England. Winters from southern California, southern Arizona, Texas, and New Mexico south throughout Mexico to Costa Rica. Migrates throughout continental North America between its breeding and wintering ranges.

STATUS. Common.

HABITAT. Prefers bogs, wet meadows, and riparian (willow) thickets. Also inhabits hedgerows, fencerows, and the understory of open woodlands, as well as forest edges, clearings, and shrubby areas. Usually places nest on the ground in a shallow depression. Builds a rather frail structure of leaves, moss, and some grasses. Not normally parasitized by cowbirds. In winter prefers brushy fields, either dry or marshy.

SPECIAL HABITAT REQUIREMENTS. Thickets along the edges of fields, waterways, or in wet meadows.

Further Reading. United States and Canada: Finch and Reynolds 1988, Forbush and May 1955, Godfrey 1966, Zeranski and Baptist 1990:252–253. Mexico and Central America: Land 1970, Peterson and Chalif 1973, Rappole et al. 1983, Stiles and Skutch 1989.

Swamp Sparrow
Melospiza georgiana

Canada (Quebec): Pinson des Marais
Mexico: Gorrión Georgiana

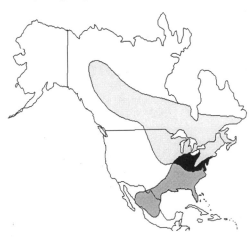

RANGE. Breeds from west-central and southern Northwest Territories and northern Manitoba across to southern Labrador, and south to northeastern and east-central British Columbia, the Dakotas, northern Illinois, and Maryland. Winters from eastern Nebraska through the Great Lakes region to Massachusetts, south to Texas, the Gulf Coast, and Florida, and west across New Mexico to southeastern Arizona, south to Jalisco, Mexico.

STATUS. Common.

HABITAT. Wetlands with shrubs, rank marsh grasses, sedges, and reeds are characteristic habitat. Inhabits brushy wet meadows, sloughs, bogs, swamps, freshwater marshes, swampy shorelines of lakes or streams, and rarely coastal brackish meadows. Avoids heavily wooded wetlands. Often builds nest among cattail stalks, on clumps of bent-over vegetation, on sedge tussocks, or in shrubs, frequently directly over water that may be 0.5 m or more deep. Usually places nest about 0.3 m above ground or water, preferably in areas with mixed vegetation rather than in pure cattails. Parasitized by cowbirds. In winter, frequents springs, seeps, and open brooks that have brushy cover nearby.

SPECIAL HABITAT REQUIREMENTS. Swampy wetlands with rank emergent vegetation.

Further Reading. United States and Canada: Bent 1968c, DeGraaf and Rudis 1986, Farrand 1983c, Forbush and May 1955, Godfrey 1966, Greenberg and Droege 1990, Low and Mansell 1983, Martin et al. 1951. Mexico and Central America: Peterson and Chalif 1973, Rappole et al. 1983.

White-crowned Sparrow
Zonotrichia leucophrys

Canada (Quebec): Pinson à Couronne Blanche

Mexico: Gorrión Gorriblanco

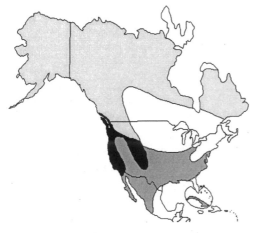

RANGE. Breeds throughout Alaska, the northern regions of Canada surrounding the Arctic Ocean-Hudson Bay region, east to the Atlantic, south through the Yukon and British Columbia, through the Rocky Mountains, west to the Pacific Coast and the Sierra Nevada. Winters throughout most of the United States, except Florida and the northern Great Plains states, and south to Guanajuato, Mexico.

STATUS. Common to abundant, but declining in the western parts of its range.

HABITAT. Frequents valleys, brushy hillsides, roadside vegetation, and cultivated fields. In arctic regions, inhabits open stunted tree growth and brushlands. In the western United States, open coniferous and aspen forests with shrub understory, riparian thickets, montane meadows with shrub thickets. Often builds nest consisting of grassy materials, mosses, and lichens, lined with rootlets or animal hairs, placing it on the ground in a moss or lichen bed, in grassy areas, but sometimes on the lower branches of dense shrubs. Most nests are well concealed and difficult to locate. Not normally parasitized by cowbirds. Uses edge habitats and brush piles during winter.

SPECIAL HABITAT REQUIREMENTS. Thickets, hedgerows, or edge.

Further Reading. United States and Canada: Beal and McAtee 1912, For-bush and May 1955, Godfrey 1966, Petrinovich and Patterson 1983. Mexico and Central America: Peterson and Chalif 1973, Rappole et al. 1983.

Bobolink
Dolichonyx oryzivorus

Canada (Quebec): Goglu

Costa Rica and Venezuela: Tordo Arrocero

Guatemala: Chambergo, Charlatán

Mexico: Tordo Migratorio

West Indies: Chambergo

RANGE. Breeds from southern interior British Columbia across southern Canada and central Ontario south to eastern Oregon, central Colorado, central Illinois, and central New Jersey. Migrates through Central America, the Caribbean, and northern South America. Winters in southern South America (mostly east of the Andes) from Peru and central Brazil south to northern Argentina.

STATUS. Locally common; numbers have declined throughout much of the United States since the early 1900s with losses of grassland and increasing proportions of alfalfa in existing hay fields.

HABITAT. Prefers large open fields of tall grass, grassy hay fields, clover, or grain crops, but also inhabits wet meadows, ungrazed to lightly grazed mixed-grass prairies, and fallow fields. Bobolink abundance is greatest in the largest hay fields with the lowest proportion of alfalfa. Factors in the decline of bobolink populations are: declining area in hay crops, increased use of alfalfa, earlier hay-cropping dates, and earlier rotation of hay fields to other crops (Bollinger and Gavin 1992). Builds nest on the ground, usually in a hollow scraped in the ground or in a natural depression, rarely above ground attached to plant stems. Always locates nest in dense stands of tall vegetation such as hay, clover, or thick growths of weeds. During migration, frequents marshes and grain fields.

SPECIAL HABITAT REQUIREMENT. Large expanses of grassland or old (>8 yr) hay fields with little or no alfalfa.

Further Reading. United States and Canada: Beal 1900, Bent 1958, Bollinger and Gavin 1989, 1992, DeGraaf and Rudis 1986, Forbush and May 1955, Godfrey 1966, Johnsgard 1979, Orians 1985. Mexico and Central America: Edwards 1972, Land 1970, Peterson and Chalif 1973, Rappole et al. 1983, Stiles and Skutch 1989. South America: Meyer de Schauensee and Phelps 1978, Rappole et al. 1983. Caribbean: Bond 1947, Raffaele 1989, Rappole et al. 1983.

Red-winged Blackbird
Agelaius phoeniceus

Canada (Quebec): Carouge à Épaulettes

Costa Rica and Mexico: Tordo Sargento

Guatemala: Tordo Alirrojo, Tordo Capitán

West Indies: Chirriador, Mayito de la Ciénaga, Toti de la Ciénaga

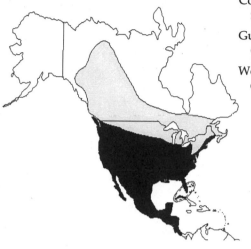

RANGE. Breeds from the southern tip of Alaska, and Yukon down to northern Washington, across the northern part of Canada and the United States, including Idaho, Montana, Wyoming, the Great Lakes states, and New England. Resident in the rest of the United States south throughout Mexico to Costa Rica. Also resident in the Bahamas and Cuba. Northern birds migrate as far south as Mexico.

STATUS. Abundant, but declining in eastern and central regions of the United States.

HABITAT. Breeds in marshes and agricultural areas, usually where there are wetlands; also along the edge of hay fields, old fields, ditches, and pastures. Prefers areas with trees nearby and where open habitat edges, especially marsh/field edges, are highly interspersed. Extremely territorial, defending areas of 30 to 40 square meters. Nests in a deep, narrow cup of grass, reeds, and weed rootlets, usually attached to emergent vegetation (particularly cattails) about 1 m above ground. Also nests in weeds and low brush patches, croplands such as alfalfa and cereal grains, even upland areas of mixed chaparral. Commonly parasitized by cowbirds. Forms large flocks in the winter and moves throughout fields and marshy areas.

SPECIAL HABITAT REQUIREMENTS. Emergent vegetation adjacent to open fields with scattered tall shrubs, trees.

Further Reading. United States and Canada: Albers 1978, Beal 1900, Bent 1958, Case and Hewitt 1963, Godfrey 1966, Low and Mansell 1983, Mott et al. 1972, Orians 1961, 1985, Payne 1969. Mexico and Central America: Land 1970, Peterson and Chalif 1973, Rappole et al. 1983, Stiles and Skutch 1989. South America: Meyer de Schauensee and Phelps 1978, Rappole et al. 1983. Caribbean: Bond 1947, Rappole et al. 1983.

Eastern Meadowlark
Sturnella magna

Canada (Quebec): Sturnelle des Prés

Costa Rica: Carmelo, Zacatera, Zacatero Común

Cuba and Isle of Pines: Sabanero

Guatemala: Peruchio

Mexico: Pradero Tortilla-con-chile

Venezuela: Perdigón

RANGE. Breeds from northern Minnesota and southern Ontario across to southern New Brunswick, south through the eastern United States to Texas, the Gulf Coast, and Florida, west to southwestern South Dakota, central Nebraska, and central Arizona, south throughout Central America and also to South America as far as Colombia in the west and Brazil in the east. Resident on Cuba. Winters from central Arizona, Kansas, central Wisconsin, New England, and Nova Scotia south throughout the breeding range, casually farther north. Northern populations winter as far south as Mexico.

STATUS. Fairly common, although there have been widespread, long-term, and continuing declines in the eastern and central United States because of loss of grassland habitats.

HABITAT. Prefers pastures, but also occurs in other grass-dominated habitats such as hay fields, grassy meadows, tallgrass prairies, open fields of corn, alfalfa, and clover, and weedy orchards. Prefers moist meadows and lowlands at the western edge of its breeding range, where distribution overlaps that of the Western Meadowlark. Builds nest on the ground in a natural depression or scrape, well concealed by a canopy of vegetation bent over the nest, preferably in cover 0.3 to 0.5 m high. During the winter months in Panama, this species is fairly common in savannas and grassy fields in the lowlands of the Pacific slope.

SPECIAL HABITAT REQUIREMENTS. Open grasslands with elevated singing perches such as fences, poles, or lone trees.

Further Reading. United States and Canada: Bent 1958, Godfrey 1966, Johnsgard 1979, Lanyon 1957, Roseberry and Klimstra 1970, Tate and Tate 1982, Zeranski and Baptist 1990:259–260. Mexico and Central America: Land 1970, Peterson and Chalif 1973, Rappole et al. 1983, Ridgely and Gwynne 1989, Stiles and Skutch 1989. South America: Meyer de Schauensee and Phelps 1978, Rappole et al. 1983. Caribbean: Bond 1947, Rappole et al. 1983.

Western Meadowlark
Sturnella neglecta

Canada (Quebec): Sturnelle de l'Ouest
Mexico: Pradero Gorjeador

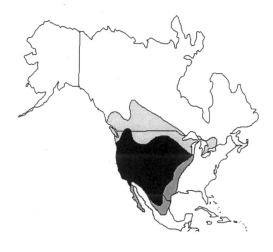

RANGE. Breeds from central British Columbia, north-central Alberta, Saskatchewan, and central Canada, south throughout most of the western United States. Found in most states west of the Mississippi. Resident throughout the Pacific, southern Rocky Mountain, and southern Great Plains states. Winters also in Oklahoma, Texas, and along the coasts of Mexico and Baja California.

STATUS. Common, but consistently declining in much of its breeding range.

HABITAT. Typically inhabits tallgrass and mixedgrass prairie, grasslands, savannas, cultivated fields, hay fields, wet meadows, and pastures, preferring open fields with perch sites such as fences, old logs, or dead trees. Builds nest in a shallow depression on dry ground in open grassland, often in grass or a small grass tuft, sometimes in rocky areas. Usually uses grasses for nest material. An uncommon cowbird host.

SPECIAL HABITAT REQUIREMENTS. Open grasslands.

Further Reading. United States and Canada: Godfrey 1966, Lanyon 1957, Orians 1985. Mexico and Central America: Edwards 1972, Peterson and Chalif 1973, Rappole et al. 1983.

Yellow-headed Blackbird
Xanthocephalus xanthocephalus

Canada (Quebec): Carogue à Tête
Jaune

Mexico: Tordo Cabeciamarillo

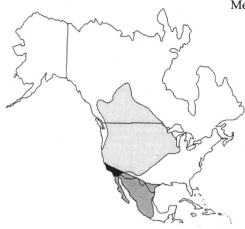

RANGE. Breeds from southern and central Canada, throughout the western United States, west of the Mississippi River. Winters from southern California, Arizona, New Mexico, and Texas south to Guerrero, Mexico.

STATUS. Common; increasing in western North America.

HABITAT. Inhabits freshwater marshes of cattails, bulrushes, and reeds in open country. Generally nests in colonies. Builds a woven basketlike cup nest of marsh vegetation lined with fine grass and attaches it to reeds and cattails up to 1 m above water or sometimes in willows in wet areas. North Dakota populations increased dramatically during 1962–1982 (Besser 1985); depredates sunflower seed crops, though not as destructively as Red-winged Blackbirds (Twedt et al. 1991). Winters in open cultivated fields, pastures, and marshes.

SPECIAL HABITAT REQUIREMENTS. Emergent marshes.

Further Reading. United States and Canada: Beal 1900, Forbush and May 1955, Godfrey 1966, Leonard and Picman 1986, Low and Mansell 1983, Orians 1980, 1985, Willson 1966. Mexico and Central America: Peterson and Chalif 1973, Rappole et al. 1983.

Brewer's Blackbird
Euphagus cyanocephalus

Canada (Quebec): Mainate à Tête Pourprée

Guatemala: Tordo de Brewer

Mexico: Tordo Ojiclaro

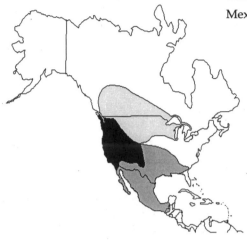

RANGE. Breeds from southwestern and central British Columbia to southern Ontario, south throughout the northern United States; resident in the Pacific Coast, Great Plains, and Rocky Mountain states. Winters from southern British Columbia, east-central Montana, and the northern portions of the Gulf States, south to Baja California, Oaxaca, Mexico, Guatemala, southern Texas, the Gulf Coast, and Florida.

STATUS. Common, but has been declining in western North America.

HABITAT. Prefers to be near water in habitats such as riparian woodlands, aspen groves, parklands, agricultural lands, disturbed grasslands, and marshes; often found near human habitations, especially in the eastern parts of range. Uses bulrushes and pines for roosting and daytime resting places and displays from the tops of pine trees. Nests singly or in loose colonies on the ground or in trees and shrubs 6 to 10 m above the ground. Places the cup-shaped nest usually at or near the end of a branch. Commonly parasitized by cowbirds. In winter, frequents pastures and fields.

SPECIAL HABITAT REQUIREMENTS. Marshes, agricultural areas.

Further Reading. United States and Canada: Beal 1900, Godfrey 1966, Orians 1985, Terres 1980, Williams 1952. Mexico and Central America: Peterson and Chalif 1973, Rappole et al. 1983.

Bronzed Cowbird
Molothrus aeneus

Costa Rica: Vaquero Ojirrojo

Guatemala: Tordo Ojirrojo, Vaquero Ojirrojo

Mexico: Tordo Ojirrojo

RANGE. Resident from extreme southeastern California, southern Arizona, New Mexico, and Texas south through central Mexico to Panama.

STATUS. Locally common.

HABITAT. Inhabits mostly open country with occasional tree patches or large tall shrubs. Prefers humid, hot climate, often in areas where cattle are grazed, and is common in areas of human habitation. Builds no nest; generally lays its eggs in the nests of other birds, preferably those nesting in brush, semi-open to open ranch, farm, and residential areas. In Costa Rica this species frequents open country, especially agricultural land, and often is seen along roadsides.

SPECIAL HABITAT REQUIREMENTS. Open areas with scattered trees or shrubs.

Further Reading. United States and Canada: Godfrey 1966, Oberholser 1974b. Mexico and Central America: Land 1970, Peterson and Chalif 1973, Rappole et al. 1983, Stiles and Skutch 1989.

Brown-headed Cowbird
Molothrus ater

Canada (Quebec): Vacher
Mexico: Tordo Cabecicafé

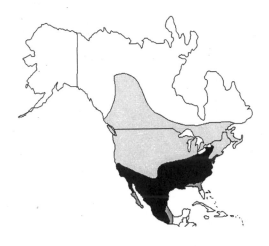

RANGE. Breeds from southeastern Alaska, northern British Columbia, and southern Northwest Territories east to southern Quebec and southern Newfoundland, and south to northern Baja California, Guerrero, Puebla, and Veracruz, Mexico, the Gulf Coast, and central Florida. Winters from northern California, central Arizona, the Great Lakes region, and New England south to the Isthmus of Tehuantepec in Mexico, the Gulf Coast, and southern Florida.

STATUS. Common, but has been declining in eastern North America.

HABITAT. Prefers habitats where low or scattered trees are interspersed with grassland vegetation. Originally occupied open grasslands and avoided unbroken forestlands, but because of agriculture, cattle grazing, and deforestation, occupies a much expanded range. Now found in open coniferous and deciduous woodlands, forest edges, brushy thickets, agricultural land, and suburban areas. Builds no nest; lays its eggs in the nests of more than 100 species of birds, particularly flycatchers, finches, vireos, and warblers.

SPECIAL HABITAT REQUIREMENTS. Habitats with open grassy spaces.

Further Reading. United States and Canada: Beal 1900, Bent 1958, DeGraaf and Rudis 1986, Forbush and May 1955, Godfrey 1966, Johnsgard 1979, Mayfield 1965. Mexico and Central America: Peterson and Chalif 1973, Rappole et al. 1983.

Orchard Oriole
Icterus spurius

Canada (Quebec): Oriole des Vergers

Costa Rica and Mexico: Bolsero Castaño

Guatemala: Bolsero Café, Chorcha Café

Venezuela: Turpial de Huertos

RANGE. Breeds from southeastern Saskatchewan, southern Manitoba, and central Minnesota east to northern Massachusetts, south to Michoacan, Mexico, Texas, the Gulf Coast, and central Florida, and west to eastern Colorado. Migrates through Cuba. Winters from southern Mexico (including the Yucatán Peninsula) to Colombia and Venezuela, casually to southern Texas, rarely in coastal California.

STATUS. Locally common, but showing long-term decline in the central United States.

HABITAT. Prefers orchards and open country with a few scattered trees. Also breeds in residential areas, farmlands, shelterbelts, woodland margins, and lightly wooded river bottoms. At times, it may inhabit marshes and bordering trees. Avoids heavily wooded or dense forests. Suspends semi-pendulous nest, well concealed by dense foliage, from a fork or crotch of a variety of deciduous trees and shrubs 1 to 20 m, typically 3 to 6 m, above the ground. Commonly parasitized by cowbirds. Commonly nests in trees also supporting Eastern Kingbird nests.

SPECIAL HABITAT REQUIREMENTS. Open woodlands or open areas with scattered trees.

515

Further Reading. United States and Canada: Bent 1958, DeGraaf and Rudis 1986, Dennis 1948, Forbush and May 1955, Godfrey 1966, Johnsgard 1979, Tate and Tate 1982. Mexico and Central America: Land 1970, Peterson and Chalif 1973, Rappole et al. 1983, Stiles and Skutch 1989. South America: Meyer de Schauensee and Phelps 1978, Rappole et al. 1983. Caribbean: Bond 1947, Rappole et al. 1983.

Hooded Oriole
Icterus cucullatus

Mexico: Bolsero Cuculado

RANGE. Breeds from northern coastal and central California, southern Nevada, central Arizona, and western Texas south into northern Mexico. Resident in southern Baja California and throughout mainland Mexico. Northern populations winter in Mexico.

STATUS. Common.

HABITAT. Inhabits palm trees, mesquite, dry shrubs, and some deciduous and riparian woodlands; often found around ranches and towns. Usually constructs a large, hanging nest with a variety of grasses, Spanish moss, thin branches, as well as dry vegetables, hair, and other local materials woven together, and suspends it from the limbs of trees or cacti. In winter, common in coastal scrub on the northern Yucatán Peninsula (Lynch 1992).

SPECIAL HABITAT REQUIREMENTS. Riparian woodland, palms, mesquite; in southern parts of breeding range, farms, towns.

Further Reading. United States and Canada: Phillips et al. 1964, Terres 1980. Mexico and Central America: Land 1970, Peterson and Chalif 1973, Rappole et al. 1983.

Northern Oriole
Icterus galbula

Canada (Quebec): Oriole de Baltimore

Costa Rica and Mexico: Bolsero Norteño

Guatemala: Bolsero Norteño, Chorcha Norteña

Puerto Rico: Calandria de Norte

Venezuela: Turpial de Baltimore

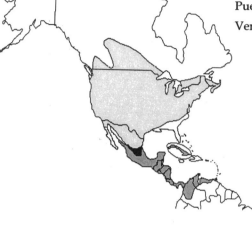

RANGE. Breeds from southern interior British Columbia and central Alberta to central Maine and central Nova Scotia, south to southern Texas, Mexico, the central Gulf States, central North Carolina, and Delaware. Winters from Mexico to Colombia and Venezuela. Also in Cuba, Jamaica, and Puerto Rico.

STATUS. Common, but declining in eastern North America.

HABITAT. In the East, inhabits orchards, deciduous forest edges, wooded river bottoms, upland forests, partially wooded suburban areas, parks, and shelterbelts. In the West, prefers semiarid mesquite groves and deciduous trees bordering streams or irrigation ditches in open country, prairie, or cultivated areas. Usually attaches pendant nest by its rim to the tip of a long drooping branch, 3 to 20 m, but typically 8 to 10 m, above the ground. Most frequently uses large trees, especially elms and cottonwoods growing in the open, but will use a wide variety of deciduous trees throughout its range. In winter this species can be found in cocoa and coffee plantations, semiopen forest canopy, and savanna groves, but will also freely descend to shrub layer to feed along edges and in second growth.

SPECIAL HABITAT REQUIREMENTS. Tall deciduous trees for nesting.

517

Further Reading. United States and Canada: Bent 1958, DeGraaf and Rudis 1986, Forbush and May 1955, Godfrey 1966, Johnsgard 1979. Mexico and Central America: Land 1970, Rappole et al. 1983, Stiles and Skutch 1989. South America: Meyer de Schauensee and Phelps 1978, Rappole et al. 1983. Caribbean: Raffaele 1989, Rappole et al. 1983.

Scott's Oriole
Icterus parisorum

Mexico: Bolsero Parisino

RANGE. Breeds from southern California, Nevada, Utah, western Colorado, central New Mexico, and western Texas south into Mexico. Winters from southern California to southern Mexico.

STATUS. Common.

HABITAT. Prefers pinyon-juniper woodlands of montane semidesert areas, live oak–yucca associations, and sycamores and cottonwoods in canyons. Also uses Joshua-tree habitat. Constructs a cup-shaped nest with fibers of local grasses, weeds, and other vegetative material and suspends it from branches of almost any available tree, including Joshua trees and yucca plants.

SPECIAL HABITAT REQUIREMENTS. Yucca, pinyon-juniper, or oak trees in arid areas.

Further Reading. United States and Canada: Phillips et al. 1964, Terres 1980. Mexico and Central America: Peterson and Chalif 1973, Rappole et al. 1983.

Lesser Goldfinch
Carduelis psaltria

Costa Rica: Jilguero Menor
Mexico: Jilguero Dorsioscuro

RANGE. Resident from southwestern Washington, western Oregon, northern California, northern Utah, northern Colorado, northwestern Oklahoma, and central and southern Texas south to Baja California throughout Mexico and Central America. In Costa Rica, an uncommon, nomadic resident in upper parts of Valle Central and Revenlazon drainage, especially on the slopes of Irazú and Turrialba volcanoes. Also along Pacific slope of Cordillera de Talamanca, 850 to 2,150 m. Range extends south to Venezuela, northern Colombia, northern Peru, and western Cuba.

STATUS. Common. Relentless persecution for the cage-bird trade has greatly reduced numbers in parts of Latin America. In Panama, common locally. Most numerous from Chiriquí to Cocle; in central Panama recorded mainly from Pacific side.

HABITAT. Generally inhabits scattered trees, woodland edge, second growth, open fields, pastures, and human habitation. Often found in drier foothill regions, in the deserts, and up to 2,400 m in elevation and usually near water. Saddles nest on a branch of a shrub or tree. Generally locates nest in dense foliage, 1 to 10 m above the ground. Sometimes nests in loose colonies. In Mexico in rather open pine or pine-oak woodlands, or in other woodland edge or in open country with few or many scattered trees and thickets, weed-grown fields, overgrown vacant lots, parks, gardens, orchards.

SPECIAL HABITAT REQUIREMENTS. Open areas with scattered trees, wood-land edges, second growth, fields; in southern parts of breeding range, farms, towns.

Further Reading. United States and Canada: Coutlee 1968, Edwards 1972, Lindsdale 1957, Verner and Boss 1980.

American Goldfinch
Carduelis tristis

Canada (Quebec): Chardonneret Jaune
Mexico: Jilguero Canario

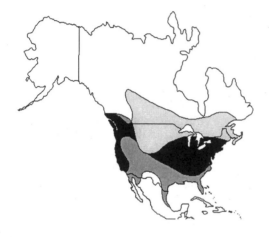

RANGE. Breeds from southern British Columbia and north-central Alberta east to central Ontario and southwestern Newfoundland, and south to California, southern Colorado, northeastern Texas, central Alabama, and South Carolina. Winters from southern British Columbia, the northern United States, southern Ontario, and Nova Scotia south to Veracruz, Mexico, the Gulf Coast, and southern Florida.

STATUS. Common, but has been declining in much of the eastern United States.

HABITAT. Frequents habitats with thistles or cattails, such as open weedy fields, marshes, farmyards, pastures with scattered trees, prairie thickets, shelterbelts, forest edges, and wooded residental areas. In the West, inhabits riparian areas, especially those with willows present along streams, ditches, and ponds, and wooded suburbs. Usually builds nest in a cluster of upright branches or on a horizontal limb of a wide variety of trees or shrubs, typically 1.5 to 5 m above the ground, commonly near water. Delays nesting until there is an abundant supply of seeds, particularly those of composites and thistles, to feed the young. Commonly parasitized by cowbirds.

SPECIAL HABITAT REQUIREMENTS. Open weedy fields and marshes with scattered woody growth for nesting.

Further Reading. United States and Canada: Bent 1968a, Coutlee 1968, DeGraaf and Rudis 1986, Forbush and May 1955, Godfrey 1966, Holcomb 1969, Johnsgard 1979, Middleton 1978, 1979, Stokes 1950. Mexico and Central America: Peterson and Chalif 1973, Rappole et al. 1983.

Evening Grosbeak
Coccothraustes vespertinus

Canada (Quebec): Gros-bec Errant

Mexico: Fringilido Piquigrueso Norteño

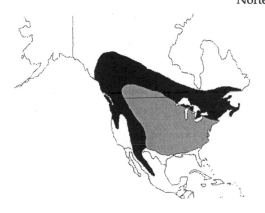

RANGE. Breeds from southwestern and north-central British Columbia, northern Alberta, southern Manitoba, and southern Quebec to Nova Scotia and south, in the mountains, to central California, west-central and eastern Nevada, central and southeastern Arizona, southern New Mexico, the Mexico highlands to Michoacán, Puebla, and west-central Veracruz, and, east of the Rocky Mountains, to Minnesota, Michigan, and Massachusetts. Winters throughout the breeding range and south, sporadically, to southern California, southern Arizona, Oaxaca, the Gulf Coast, and central Florida. Mexican populations are resident.

STATUS. Locally abundant, but recently declining in eastern North America.

HABITAT. Favors coniferous forests (primarily spruce and fir) throughout most of its range, often extending into areas where trees are quite sparse and into mixed forests. Usually places nest on a horizontal limb of a conifer, 5 to 25 m above the ground. Builds a shallow cup, usually in a dense cluster of leaves near the end of a branch. In winter, forms large flocks and may move downslope to oak or pine-oak habitats, parks, and towns.

SPECIAL HABITAT REQUIREMENTS. Conifers.

Further Reading. United States and Canada: Forbush and May 1955, Godfrey 1966, Verner and Boss 1980. Mexico and Central America: Edwards 1972, Peterson and Chalif 1973.

Breeding and Wintering Habitat Use

We attempt here to summarize the major U.S. forest types, grasslands, and deserts used by Neotropical migrant birds for breeding and overwintering. We indicate use of wetland and other open habitats commonly encountered within forests, as well as wooded habitats within deserts and grasslands, and coastal habitats.

The information provided in the tables at the end of the appendix was compiled from many sources. Tables 5–10 are abstracted from DeGraaf et al. 1991, and Table 11 is exerpted from Edwards 1972, Oberholser 1974a, Peterson and Chalif 1973, Phillips et al. 1964, and Rappole et al. 1983. The 20 habitat matrices for the eastern and western forest types are based on the forest cover type groups in Eyre 1980. The habitat matrices presented for the Great Plains are based on the regions described in the Soil Conservation Service's *Land Resource Regions and Major Land Resource Areas of the United States* (1981).

Nonforest habitat types within these broad forest cover types and Great Plains regions, and their use by birds in both the breeding and nonbreeding seasons, were developed from DeGraaf 1978a, 1978b, 1979, 1980; DeGraaf and Rudis 1986; Evans and Kirkman 1981; Hamel et al. 1982; Johnsgard 1979; Thomas 1979; Verner and Boss 1980; and many unpublished reports provided by U.S. Forest Service regional offices and colleagues.

The Nearctic

Eastern Forest Types

EASTERN WHITE PINE–RED PINE–JACK PINE. Eastern white pine (*Pinus strobus*), red pine (*P. resinosa*), or jack pine (*P. banksiana*) compose most of the stocking. Pine forests in the northeastern and Great Lakes states are essentially pure stands, usually on lighter soils. Eastern white pine occurs from the Maritime Provinces across the Great Lakes states to Manitoba, and down the Appalachian Mountains to Georgia. It is shade tolerant and also occurs as a scattered tree in other forest types, and on many soils. Red pine is most extensive in the Great Lakes states and southern Ontario, and also extends east to New England, Quebec, and the Maritime Provinces, where it usually occurs on small outwash areas, rocky slopes, or hilltops. It is shade intolerant and occurs in even-aged stands. Jack pine is mainly found in the Great Lakes states; it characteristically originates after fire and is a short-lived, intolerant pioneer on dry, sandy soils.

RED SPRUCE–BALSAM FIR. Red spruce (*Picea rubens*) or balsam fir (*Abies balsamea*) compose most of the stocking. These species frequently occur to-

525

gether from the Maritime Provinces and adjacent Quebec to northern New England, New York, and the Appalachians. Either may be pure or compose a majority of the stocking. Paper birch (*Betula papyrifera*), quaking aspen (*Populus tremuloides*), bigtooth aspen (*P. grandidentata*), red maple (*Acer rubrum*), eastern white pine, and northern white cedar (*Thuja occidentalis*) are common associates. Red spruce is long-lived and shade tolerant; disturbance creates conditions favorable for establishment of balsam fir.

LONGLEAF PINE–SLASH PINE. Longleaf pine (*Pinus palustris*) or slash pine (*P. elliottii*) compose a majority of the stocking. This type occurs on the Gulf and Atlantic coastal plains from Louisiana to South Carolina, on a range of sites from sandy ridges to poorly drained flatwoods. Excluding fire allows slash pine to become established, and hardwoods and shrubs commonly proliferate. Where longleaf pine stands are treated with prescribed fire, an open understory results.

LOBLOLLY PINE–SHORTLEAF PINE. Loblolly (*Pinus taeda*) and shortleaf (*P. echinata*) pines together compose the majority of the stocking. Loblolly pine predominates except on drier sites. This forest type occurs from Delaware south along the Atlantic coastal plain and Piedmont to Florida and west along the Gulf coastal plain to east Texas. Typically found on moist sites, it spreads to drier sites if fire is controlled. It is succeeded by upland oaks.

OAK-PINE. Upland oaks and pines (usually loblolly or shortleaf) each compose 25% of the stocking. They generally occur from east Texas to Georgia on upland sites on the Gulf coastal plain and Piedmont, and north in smaller areas through the Appalachians to include table mountain pine (*P. pungens*), oak–Virginia pine (*P. virginiana*), and oak–pitch pine (*P. resinosa*).

OAK-HICKORY. Upland oaks and hickories (*Carya* spp.) compose most of the stocking, and pines constitute less than 25% of the stocking. Oak-hickory forests occur across a wide geographic range from Texas, Missouri, and Iowa to southern New England, with many oak and other hardwood species involved under various physiographic conditions.

OAK-GUM-CYPRESS. In bottomland forests of the lower Mississippi River Valley and its major tributaries from the Ohio River south, tupelo (*Nyssa* spp.), blackgum (*N. sylvatica*), sweetgum (*Liquidambar styraciflua*), oak (*Quercus* spp.), or bald cypress (*Taxodium distichum*), singly or in combination, compose most of the stocking; pines contribute less than 25% of the stocking.

ELM-ASH-COTTONWOOD. Elm (*Ulmus* spp.), ash (*Fraxinus* spp.), cottonwood (*Populus deltoides*), or red maple compose most of the stocking in these forests. Common associates in river bottoms (especially in the Missouri River drainage) are sycamore (*Platanus* spp.) and willow (*Salix* spp.). On uplands, including those in the Great Lakes states, western New York, and southern New England, common associates are red maple and American beech (*Fagus grandifolia*).

NORTHERN HARDWOODS. Sugar (*Acer saccharum*) or red maple, American beech, or yellow birch (*Betula alleghaniensis*), singly or in combination, compose most of the stocking. This forest type varies geographically in its composition. It extends from the Maritime Provinces through Wisconsin and south through the central Appalachians. Sugar maple is characteristic of the type. Beech is absent in much of its western extent and on wetter sites in the East, where red maple and yellow birch also become common. Balsam fir and red spruce are common associates in the Northeast, and aspen is common throughout. Northern red oak (*Quercus rubra*), white ash (*Fraxinus americana*), eastern white pine, paper birch, and eastern hemlock (*Tsuga canadensis*) are commonly associated in the central and southern parts of the range, where this forest type is often called mixed woods.

ASPEN-BIRCH. Quaking and bigtooth aspens or paper birch compose a majority of the stocking. The aspens and paper birch are transcontinental in distribution. All are pioneer types that establish themselves after fire and clearcutting. The aspens are unique in that almost all stands regenerate from root suckers. This forest type is short-lived and is succeeded on dry sites by red pine, red maple, or oaks; on intermediate sites by white pine; on moist fertile sites by northern hardwoods; and on the wettest sites by balsam fir. Paper birch is succeeded by spruce-fir in the northern parts of its range and by northern hardwoods and eastern hemlock on well-drained, fertile sites elsewhere.

Western Forest Types

DOUGLAS-FIR. Douglas-fir (*Pseudotsuga menziesii*) composes most of the stocking. Common associates are western hemlock (*Tsuga heterophylla*), western redcedar (*Thuja plicata*), true firs (*Abies* spp.), redwood (*Sequoia sempervirens*), ponderosa pine (*Pinus ponderosa*), and larch (*Larix* spp.). This forest type predominates in the Pacific Northwest but also occurs (decreasing southward) throughout the Rocky Mountains south to northern New Mexico.

HEMLOCK–SITKA SPRUCE. Western hemlock and/or Sitka spruce (*Picea sitchensis*) compose most of the stocking. Common associates include Douglas-fir, silver fir (*Abies amabilis*), and western redcedar. This type composes the coastal forests of Washington and Oregon.

REDWOOD. Redwood composes most of the stocking. It is restricted to the California fog belt, extending from southernmost Oregon south along the Pacific Coast to the Santa Lucia Mountains and inland to the reaches of coastal fogs. Common associates are Douglas-fir, grand fir (*Abies grandis*), and tanoak (*Lithocarpus densiflorus*).

PONDEROSA PINE. Ponderosa pine composes most of the stocking. Common associates in the western part of the range (California, Oregon) include Jeffrey (*P. jeffreyi*) and sugar (*P. lambertiana*) pines; to the north, Douglas-fir and incense-cedar (*Libocedrus decurrens*); to the east, limber (*P. flexillis*), Arizona (*P. ponderosa* var. *arizonica*), and Chihuahua (*P. leiophylla* var. *chihuahuana*) pines; and throughout, white fir (*Abies concolor*). This forest type is generally distributed to the west, north, and east of the Great Basin and the deserts of the Southwest.

WESTERN WHITE PINE–LARCH. Western white pine (*Pinus monticola*) composes most of the stocking. The type attains its best development in northern Idaho and northwestern Montana. Common associates include western redcedar, larch, white fir, Douglas-fir, lodgepole pine (*P. contorta*), and Engelmann spruce (*Picea engelmannii*). Such admixtures produce the "mixed conifer" type, as it is known in the West. Western larch (*Larix occidentalis*) composes most of the stocking in some areas between the Columbia River in eastern Washington and the west slopes of the Rocky Mountains in Montana. Common associates are Douglas-fir, grand fir, western redcedar, and western white pine.

LODGEPOLE PINE. Lodgepole pine composes most of the stocking; this mid-elevation type occurs to 3,500 m in the Rocky Mountains, to 3,700 m in California, and to 1,950 m in Oregon and Washington. Best development is on moist, sandy, or gravelly loam. Common associates are subalpine fir (*Abies lasiocarpa*), western white pine, Engelmann spruce, aspen, and larch.

FIR-SPRUCE. The true firs, Engelmann spruce, or Colorado blue spruce (*Picea pungens*) compose most of the stocking. Common associates are lodgepole pine and, at high elevations, mountain hemlock (*Tsuga mertensiana*).

ASPEN-HARDWOODS. Quaking aspen or red alder (*Alnus rubra*) compose a majority of the stocking. Quaking aspen is the most common and extensive hardwood in the western United States. It occurs primarily at middle elevations on a variety of sites in the Rocky Mountain cordillera, where it is usually succeeded by interior Douglas-fir. Quaking aspen is usually the first tree to dominate burns and other disturbed areas, where it produces even-aged stands. Where conifer seed sources are absent, aspen may exist as a virtual climax, where it vegetatively reproduces repeatedly, developing into all-aged stands.

All western aspen communities have an herbaceous understory, commonly forbs, but sometimes grasses and sedges. In the northern portion of this forest type's range in the West, willows, common bearberry, and buffaloberry are common understory shrubs. Farther south, snowberry, choke-cherry, and western serviceberry are more common.

Red alder is essentially coastal and the most important hardwood of the Pacific Northwest; best growth is on moist, rich, loamy bottomlands.

CHAPARRAL. Chaparral consists of heavily branched, dwarfed trees or shrubs, commonly evergreens, whose canopy at maturity covers at least 50% of the ground. Common constituent plants include oaks (*Quercus* spp.), mountain-mahogany (*Cercocarpus* spp.), silktassel (*Garrya* spp.), ceanothus (*Ceanothus* spp.), manzanita (*Arctostaphylos* spp.), and chamise (*Adlenostoma* spp.).

PINYON-JUNIPER. Pinyon pines (primarily pinyon [*P. edulis*], Mexican pinyon [*P. cembroides*], singleleaf pinyon [*P. monophylla*]) and junipers (primarily Utah [*Juniperus osteosperma*], alligator [*J. deppeana*], and one-seed [*J. monosperma*]) compose most of the stocking. This forest type is widely distributed throughout the semiarid West, usually on dry, shallow, rocky soils of mesas, benches, and canyon walls.

Eastern Open, Wetland, Plains, Deserts, and Other Nonforest Habitats

FIELD, GLADE, ORCHARD. Primarily grasses, hay fields, abandoned agricultural land, and fruit orchards with grassy ground cover.

PASTURE, WET OR SEDGE MEADOW. Agricultural lands that are too wet, steep, or rocky for crops; meadows dominated by grasses or sedges (*Carex* spp.) with soils that are saturated or seasonally flooded.

FRESH MARSH, POND. Palustrine and lacustrine wetlands, permanently flooded, containing emergents such as cattails (*Typha* spp.), bulrushes (*Scirpus* spp.), rushes (*Juncus* spp.), and floating-leaved plants such as spatterdock (*Nuphar* spp.) and water lily (*Nymphaea* spp.).

WOODED SWAMP, BOG, SHRUB SWAMP. Palustrine, forested wetlands, either needle-leaved evergreen or broad-leaved deciduous; dominant plants are Atlantic white-cedar (*Chamaecyparis thyoides*), black spruce (*Picea mariana*), or red maple. Wooded swamps are seasonally or permanently flooded; bogs are permanently flooded.

LAKE, STREAM, RIVER. Stratified lacustrine wetlands; permanently flowing watercourses of any width.

SAND PINE, SCRUB OAK. Southeastern and southern woodlands on droughty, infertile, coarse-textured or sandy soils that support any of the scrub oaks (turkey [*Q. laevis*], bluejack [*Q. incana*], blackjack [*Q. marilandica*], or post [*Q. stellata* var. *margaretta*]) or sand pine (*Pinus clausa*).

POCOSINS. Bay-swamp and pond pine (*Pinus serotina*) woodlands with boggy soils in which broadleaf evergreens predominate: blackgum (*Nyssa sylvatica* var. *biflora*), redbay (*Persea borbonia*), sweetbay magnolia (*Magnolia virginiana*), loblolly bay (*Gordonia lasianthus*), and some coniferous associates of pond pine, especially Atlantic white-cedar.

EVERGLADES, MANGROVES. Palustrine wetlands in southern Florida that are semi-permanently flooded, dominated by saw grass (*Cladium jamaicense*); estuarine, intertidal wetlands dominated by mangrove (*Rhizophora* spp.).

ALPINE TUNDRA, KRUMMHOLZ. Elevated slopes above timberline characterized by low, shrubby, slow-growing woody plants and a ground cover of boreal lichens, sedges, and grasses; the transition zone from subalpine forest to alpine tundra is characterized by dwarfed, wind-sheared trees.

Great Plains Habitats

GULF PRAIRIES AND MARSHES. Moderate to tall dense open grasslands dominated by seacoast bluestem (*Andropogon littoralis*) and coastal sacahuista (*Spartina spartinae*).

EAST TEXAS PRAIRIES, CROSS TIMBERS, PINEYWOODS, AND POST OAK SAVANNA. Open grassy savanna to dense brushland occurring in north-central Texas; pine-hardwood forest and grazing lands of east Texas; midgrass prairie dominated by little bluestem (*Andropogon scoparius*) and shin oak (*Quercus mohriana*).

SOUTH TEXAS SHRUB-GRASSLAND. Vegetation ranges from desert grass-shrub vegetation in west Texas to mixed oak savanna in the eastern Edwards Plateau region to open grassland on the Rio Grande plain.

SOUTHERN PLAINS. Open to moderately dense short grasslands occurring from southeastern Colorado and central Oklahoma south through eastern New Mexico and Texas Panhandle. Natural vegetation is characterized by grama (*Bouteloua* spp.) and buffalograss (*Buchloe* spp.).

CENTRAL PLAINS. Grasslands ranging from short grasses in the West to tall grasses in the East. Includes the region from southeastern Wyoming and northeastern Colorado east through Indiana.

NORTHERN PLAINS. Plains region from north-central Montana and northwestern Minnesota south to southeastern Wyoming and northwestern Iowa. Area supports grassland vegetation dominated by western wheatgrass (*Agropyron smithii*) and needlegrass (*Stipa*) in the West and by little bluestem in the East.

WETLAND AND RIPARIAN HABITATS. All Great Plains wetland and riparian habitats, including marshes, ponds, lakes, streams, stock ponds, and woodlands associated with wetlands.

SHELTERBELTS AND WOODLOTS. Planted bands of trees that serve as windbreaks to protect fields or farmsteads; other wooded areas surrounded by agricultural lands.

PINK-OAK, BRUSHY WOODLAND, BADLANDS-JUNIPER. Includes dry woodlands of pines and oaks with or without a brushy understory. Badlands-juniper region is largely devoid of vegetation but may have scattered junipers on suitable sites.

Southwestern and Western Nonforest Habitats

GREAT BASIN SHRUBSTEPPE. Open to dense stands of shrubs and low trees, including big sagebrush (*Artemisia tridentata*), saltbush (*Atriplex confertifolia*), greasewood (*Sarcobatus vermiculatus*), or creosote bush (*Larrea divaricata*).

SONORAN DESERT SCRUB. Open to dense vegetation of shrubs, low trees, and succulents dominated by paloverde (*Cercidium microphyllum*), pricklypear (*Opuntia* spp.), and giant saguaro (*Cereus giganteus*).

CHIHUAHUAN DESERT SCRUB. Open stands of creosote bush and large succulents (Texas longhorn [*Ferocactus pringlei*], blue barrel [*Echinocactus platyaconthus*]) in southern New Mexico and southwest Texas.

MOJAVE DESERT SCRUB. Located between the Great Basin desert scrub and the Sonoran Desert scrub, it is intermediate between them, sharing plant species of both but containing the endemic arboreal leaf succulent Joshua tree (*Yucca brevifolia*).

DESERT RIPARIAN DECIDUOUS WOODLAND, MARSH. Woodlands, especially of cottonwoods, that occur where desert streams provide sufficient moisture for a narrow band of trees and shrubs along the margins.

ANNUAL GRASSLANDS, FARMS. Grasslands dominated by wild oat (*Avena* spp.), ripgut brome (*B. rigidus*), soft chess (*B. mollis*), bur clover (*Medicago hispida*), and filaree (*Erodium* spp.) with less than 5% woody cover.

RIVER, RIPARIAN WOODLAND, SUBALPINE MARSH. Occurs at elevations where stream conditions provide sufficient permanent moisture for emergent plants or for a narrow band of deciduous trees and shrubs; at low elevation characterized by cottonwood and sycamore, at midelevation by white alder (*Alnus rhombifolia*) and bigleaf maple (*Acer macrophyllum*), and at high elevation by willow.

MOUNTAIN AND ALPINE MEADOWS. Sedges (*Carex* spp.) and grasslike plants (*Heleocharis* spp., *Scirpus* spp.) above treeline.

RELICT CONIFER FOREST, MADREAN EVERGREEN WOODLAND. Warm-temperate forests and woodlands in the Southwest. Relict conifer forests consist of small populations of Arizona cypress (*Cupressus arizonica*) and

531

closed-cone pines (durango pine [*P. durangensis*], Apache pine [*P. engelman-nii*], Chihuahua pine [*P. leiophylla* var. *chihuahuana*], bishop pine (*P. muricata*), and knobcone pine (*P. attenuata*), restricted to canyons and suitable slopes along drainages. Madrean evergreen woodland is composed primarily of evergreen oaks but includes madrean pines.

Arctic and Coastal Habitats

DRY TUNDRA.   Treeless high-arctic landscape characterized by discontinuous vegetation of various herbs and prostrate shrubs on mesic sites, with bare mineral soil a dominant component; sometimes referred to as steppe-desert.

MOIST AND WET TUNDRA.   Treeless low-arctic landscape with continuous mats of vegetation in all but the most exposed sites, fairly high floristic diversity, and erect shrubs in riparian situations; sometimes referred to as a bog and meadow landscape. Moist tundra: saturated but without standing water; wet tundra: containing standing water.

TUNDRA (THAW) LAKES.   Shallow water bodies formed and drained cyclically as polygonization occurs in the meadows of former lake basins; the low-centered polygons coalesce to form ponds, which further coalesce to form lakes. Draining of these lakes allows the cycle to repeat.

OCEAN CLIFFS AND ROCKY SHORES.   Rocky headlands, sheer rock faces, and high-energy intertidal substrates composed of bedrock, boulders, or stones fronting the open ocean.

OCEAN BEACHES, BARS, AND FLATS (EASTERN AND WESTERN COASTS). Sloping or level unconsolidated sand, gravel, or cobble shores, mostly devoid of vegetation and exposed to wind and waves of the open ocean or coastal bays.

BAYS AND ESTUARIES (EASTERN AND WESTERN COASTS).   Coastal water bodies semi-enclosed by land but with periodic or continuous access to the ocean; salinity wide-ranging because of evaporation (in saline lagoons) and mixing of seawater with runoff from land.

ISLANDS AND SPOIL DEPOSITS (EASTERN AND WESTERN COASTS).   Isolated offshore or estuarine land masses, either naturally occurring or formed by deposition of dredged material, often with rocky shores or sand or gravel beaches.

The Neotropics

Shrubsteppe

ALPINE.   A region of grasses, sedges, mosses, and low, stunted shrubs located above timberline. Wetter alpine areas in the northern Andes are referred to as *páramo*, whereas drier alpine grasslands of the central and southern Andes are called *puna*.

DESERT. Dry regions characterized by sparse growths of cacti, low xerophytic shrubs, and ephemeral, moisture-sensitive herbaceous plants.

GRASSLAND. Open, treeless areas covered with herbaceous vegetation. The pampas of eastern Argentina, Uruguay, and southern Brazil and the campos of central Brazil are examples, as are treeless pastures in the Caribbean.

SAVANNA. A grassland with scattered trees. The cerrado of central Brazil, the lowland Caribbean pine savanna of northeastern Nicaragua, the campina of Amazonia, and the mesquite (*Prosopis* spp.)/acacia (*Acacia* spp.) pastures of the Caribbean are examples.

SCRUB. Low, dense growths of woody, often thorny vegetation, exemplified by the coppice of many Caribbean islands, the matorral of Chile, the chaco of southern Bolivia, Paraguay, and northern Argentina, and the caatinga of northeastern Brazil.

Forest

CONIFEROUS. Closed canopy growths of trees dominated by gymnosperms such as spruce, fir, and pine, generally found at higher elevations. Here generally implies unburned forests, commonly with broadleaf understory including shrubs in the Melastometaceae and Rubiaceae. Note that burned and especially burned and grazed conifer forests are poor habitat for migrant birds, except for the Palm Warbler.

MIXED CONIFEROUS/BROADLEAF. Forests dominated by pine or other conifers mixed with temperate tree species such as oak, sweetgum, and maple, generally found at midelevations of Middle America.

DECIDUOUS BROADLEAF. Closed-canopy forests that lose their leaves seasonally; characteristic of drier sites (e.g., limestone outcrops) or regions.

BROADLEAF EVERGREEN. Forests located in frost-free regions of moderate to abundant rainfall throughout the year. There are several different kinds depending on the amounts of rainfall, quality of soil, and drainage.

GALLERY. Broadleaf forest located on the floodplains of rivers in generally drier regions.

Aquatic

FRESHWATER WETLANDS. Seasonal or perennially wet areas of open, nonsaline water and/or marsh.

BRACKISH WETLANDS. Mangroves and marshes on sites subject to seasonal, tidal, or perennial flooding by weakly to moderately saline water, generally along the coast.

SEASHORE. Tidal flats, beach, littoral, and dune habitats subject to exposure to seawater.

COASTAL MARINE. Close offshore ocean environments.

Table 5. Neotropical migrants breeding in eastern forest types

Species	White–red–jack pine	Spruce–fir	Longleaf–slash pine	Loblolly–shortleaf pine	Oak–pine	Oak–hickory	Oak–gum–cypress	Elm–ash–cottonwood	Maple–beech–birch	Aspen–birch
Double-crested Cormorant							X			
Anhinga							X			
American Bittern							X			
Least Bittern							X			
Great Blue Heron							X	X		
Great Egret							X			
Snowy Egret							X			
Little Blue Heron							X			
Cattle Egret							X			
Green Heron							X	X		
Black-crowned Night-Heron							X			
Yellow-crowned Night-Heron							X			
Glossy Ibis							X			
White-faced Ibis							X			
Wood Duck	X	X					X	X	X	
Gadwall								X		
Ring-necked Duck	X	X							X	
Hooded Merganser	X	X				X	X	X	X	X
Black Vulture			X	X	X	X	X			
Turkey Vulture	X		X	X	X	X	X	X	X	
Osprey							X			
American Swallow-tailed Kite			X		X		X			
Mississippi Kite				X	X		X			
Sharp-shinned Hawk	X	X	X	X	X	X	X	X	X	X
Cooper's Hawk	X	X	X	X	X	X	X	X	X	X
Red-shouldered Hawk	X		X	X		X	X	X	X	X
Broad-winged Hawk	X	X	X	X	X	X			X	X
Red-tailed Hawk	X	X	X	X	X	X		X	X	X
American Kestrel	X	X	X	X	X	X		X	X	X
Merlin	X	X	X	X						
Peregrine Falcon	X	X	X	X	X	X		X	X	X
Yellow Rail		X								
King Rail							X			
Virginia Rail								X		
Sora							X			

Appendix A

Table 5—*cont.*

Species	White–red–jack pine	Spruce–fir	Longleaf–slash pine	Loblolly–shortleaf pine	Oak–pine	Oak–hickory	Oak–gum–cypress	Elm–ash–cottonwood	Maple–beech–birch	Aspen–birch
Purple Gallinule							X			
Common Moorhen							X	X		
Solitary Sandpiper								X		
Spotted Sandpiper								X		
Upland Sandpiper					X	X	X			
Common Snipe							X	X		
Mourning Dove	X	X	X	X	X	X	X		X	X
Black-billed Cuckoo	X	X			X	X			X	X
Yellow-billed Cuckoo	X		X	X	X	X	X	X	X	X
Burrowing Owl			X	X	X	X				
Long-eared Owl	X	X			X	X	X			
Common Nighthawk	X	X	X	X	X	X	X	X	X	X
Chuck-will's-widow			X	X	X	X				
Whip-poor-will	X	X	X	X	X	X	X	X	X	X
Chimney Swift	X	X	X	X	X	X	X	X	X	X
Ruby-throated Hummingbird	X	X	X	X	X	X	X	X	X	X
Belted Kingfisher	X	X			X	X	X	X	X	X
Yellow-bellied Sapsucker	X	X	X	X	X	X	X	X	X	X
Olive-sided Flycatcher	X	X							X	X
Eastern Wood-Pewee	X	X	X	X	X	X	X	X	X	X
Yellow-bellied Flycatcher	X	X							X	
Acadian Flycatcher					X	X	X	X		
Alder Flycatcher	X	X						X	X	X
Willow Flycatcher						X	X	X	X	X
Least Flycatcher	X	X				X		X	X	X
Eastern Phoebe	X	X	X	X	X	X	X	X	X	X
Great Crested Flycatcher	X	X	X	X	X	X	X	X	X	
Eastern Kingbird	X		X	X	X	X		X	X	X
Purple Martin	X	X	X	X	X	X	X	X	X	X
Tree Swallow	X	X	X	X			X	X	X	X
Northern Rough-winged Swallow	X	X	X	X	X		X	X	X	X
Bank Swallow	X	X			X	X		X	X	X
Cliff Swallow	X	X		X	X		X	X	X	X

536

Table 5—*cont.*

Species	White–red–jack pine	Spruce–fir	Longleaf–slash pine	Loblolly–shortleaf pine	Oak–pine	Oak–hickory	Oak–gum–cypress	Elm–ash–cottonwood	Maple–beech–birch	Aspen–birch
Barn Swallow	X	X	X	X	X	X	X	X	X	X
Bewick's Wren					X	X				
House Wren	X	X	X	X	X	X	X	X	X	X
Marsh Wren							X	X		
Golden-crowned Kinglet	X	X	X	X	X	X	X	X	X	X
Ruby-crowned Kinglet	X	X	X	X	X		X	X	X	
Blue-gray Gnatcatcher		X	X	X	X	X	X			
Eastern Bluebird	X		X	X	X	X		X	X	X
Veery	X	X				X		X	X	X
Gray-cheeked Thrush		X								
Swainson's Thrush	X	X							X	X
Hermit Thrush	X	X	X		X		X	X	X	X
Wood Thrush	X	X		X	X	X	X	X	X	X
American Robin	X	X	X	X	X	X	X	X	X	X
Gray Catbird	X	X	X	X	X	X	X	X	X	X
American Pipit							X			
Cedar Waxwing	X	X	X	X	X	X	X	X	X	X
Loggerhead Shrike	X		X	X	X	X	X			X
White-eyed Vireo				X	X	X	X	X	X	X
Solitary Vireo	X	X	X	X	X		X		X	X
Yellow-throated Vireo			X		X		X	X	X	
Warbling Vireo						X	X	X	X	
Philadelphia Vireo									X	X
Red-eyed Vireo	X	X		X	X	X	X	X	X	X
Bachman's Warbler							X			
Blue-winged Warbler						X			X	X
Golden-winged Warbler					X	X			X	X
Tennessee Warbler	X	X							X	X
Orange-crowned Warbler		X					X			
Nashville Warbler	X	X							X	X
Northern Parula	X	X		X	X	X	X	X	X	X
Yellow Warbler	X	X				X	X	X	X	X
Chestnut-sided Warbler	X	X			X	X			X	X
Magnolia Warbler	X	X							X	

Table 5—*cont.*

Species	White–red–jack pine	Spruce–fir	Longleaf–slash pine	Loblolly–shortleaf pine	Oak–pine	Oak–hickory	Oak–gum–cypress	Elm–ash–cottonwood	Maple–beech–birch	Aspen–birch
Cape May Warbler	X	X								
Black-throated Blue Warbler	X	X				X			X	X
Yellow-rumped Warbler	X	X	X	X	X	X	X	X	X	X
Black-throated Green Warbler	X	X		X	X		X		X	
Blackburnian Warbler		X				X			X	
Yellow-throated Warbler	X		X	X	X	X	X	X		
Pine Warbler	X		X	X	X				X	
Prairie Warbler	X		X	X	X	X	X			X
Palm Warbler	X	X	X	X						
Bay-breasted Warbler	X	X								
Blackpoll Warbler		X								
Cerulean Warbler					X	X		X		
Black-and-white Warbler	X	X	X	X	X	X	X	X	X	X
American Redstart	X	X				X	X	X	X	X
Prothonotary Warbler							X	X		
Worm-eating Warbler				X	X	X	X			
Swainson's Warbler	X						X			
Ovenbird	X	X			X	X	X	X	X	X
Northern Waterthrush	X	X					X	X	X	
Louisiana Waterthrush					X	X	X	X		
Kentucky Warbler					X	X	X	X		
Mourning Warbler	X	X							X	X
Common Yellowthroat	X	X	X	X	X	X	X	X	X	X
Hooded Warbler					X	X	X	X		
Wilson's Warbler		X								
Canada Warbler	X	X							X	X	X
Yellow-breasted Chat	X		X	X	X	X	X	X		
Summer Tanager			X	X	X	X	X			
Scarlet Tanager	X					X	X	X	X	X
Rose-breasted Grosbeak	X	X				X			X	X
Blue Grosbeak					X	X	X	X		
Indigo Bunting	X			X	X	X	X	X	X	X
Painted Bunting				X		X	X			

Table 5—*cont.*

Species	White–red–jack pine	Spruce–fir	Longleaf–slash pine	Loblolly–shortleaf pine	Oak–pine	Oak–hickory	Oak–gum–cypress	Elm–ash–cottonwood	Maple–beech–birch	Aspen–birch
Dickcissel				X	X	X				
Rufous-sided Towhee	X		X	X	X	X	X	X	X	X
Chipping Sparrow	X	X	X	X	X	X			X	X
Lark Sparrow				X		X				
Lincoln's Sparrow							X			
Swamp Sparrow		X								
White-crowned Sparrow				X	X	X				
Red-winged Blackbird							X	X		
Brewer's Blackbird				X	X	X				
Brown-headed Cowbird	X	X	X	X	X	X	X	X	X	X
Orchard Oriole					X	X	X	X		
Northern Oriole						X	X	X	X	X
American Goldfinch										X
Evening Grosbeak	X	X			X	X		X	X	

Appendix A

Table 6. Neotropical migrants breeding in eastern nonforest habitats

Species	Field, glade, orchard	Pasture, wet or sedge meadows	Fresh marsh, pond	Wooded swamp, bog, shrub swamp	Lake, stream, river	Sand pine-scrub oak	Pocosins	Everglades, mangroves, tropical hardwoods	Alpine tundra, krummholtz
Pied-billed Grebe			X	X	X			X	
Brown Pelican								X	
Double-crested Cormorant					X			X	
Neotropic Cormorant					X			X	
Anhinga			X		X			X	
American Bittern		X	X		X			X	
Least Bittern		X	X					X	
Great Blue Heron		X	X		X		X	X	
Great Egret		X	X				X	X	
Snowy Egret		X	X				X	X	
Little Blue Heron			X				X	X	
Tricolored Heron			X		X		X	X	
Reddish Egret			X		X		X	X	
Cattle Egret	X	X	X				X	X	
Green Heron		X	X	X	X		X	X	
Black-crowned Night-Heron			X	X	X		X	X	
Yellow-crowned Night-Heron			X	X	X		X	X	
White Ibis			X					X	
Glossy Ibis			X				X	X	
Roseate Spoonbill								X	
Wood Stork			X		X			X	
Fulvous Whistling-Duck			X					X	
Wood Duck			X	X	X		X	X	
Green-winged Teal		X	X					X	
Mallard	X	X	X		X				
Northern Pintail			X						
Blue-winged Teal			X					X	
Northern Shoveler			X					X	
Gadwall			X	X					
American Wigeon		X	X		X				
Redhead			X		X				
Ring-necked Duck			X	X	X				
Lesser Scaup			X		X				
Hooded Merganser			X	X	X				
Red-breasted Merganser			X		X				

Table 6—*cont.*

Species	Field, glade, orchard	Pasture, wet or sedge meadows	Fresh marsh, pond	Wooded swamp, bog, shrub swamp	Lake, stream, river	Sand pine-scrub oak	Pocosins	Everglades, mangroves, tropical hardwoods	Alpine tundra, krummholtz
Ruddy Duck			X		X				
Black Vulture	X	X				X	X	X	
Turkey Vulture	X	X				X	X	X	
Osprey			X		X		X	X	
American Swallow-tailed Kite			X					X	
Mississippi Kite						X	X		
Northern Harrier		X	X				X	X	
Sharp-shinned Hawk				X				X	
Cooper's Hawk	X	X							
Red-shouldered Hawk				X			X	X	
Broad-winged Hawk								X	
Red-tailed Hawk	X	X				X	X	X	
American Kestrel	X	X				X	X	X	
Merlin	X	X	X				X	X	
Peregrine Falcon	X	X	X		X		X	X	X
Yellow Rail		X	X					X	
Black Rail			X				X	X	
King Rail			X					X	
Virginia Rail			X					X	
Sora			X					X	
Purple Gallinule			X					X	
Common Moorhen				X		X			X
American Coot			X		X			X	
Sandhill Crane		X	X					X	
Piping Plover					X				
Killdeer		X	X		X			X	
Greater Yellowlegs		X							
Lesser Yellowlegs		X							
Solitary Sandpiper			X	X	X				
Spotted Sandpiper					X			X	
Upland Sandpiper		X							
Eskimo Curlew	X	X	X						
Whimbrel	X		X		X			X	
Common Snipe		X	X	X				X	
Laughing Gull	X								

Table 6—*cont.*

Species	Field, glade, orchard	Pasture, wet or sedge meadows	Fresh marsh, pond	Wooded swamp, bog, shrub swamp	Lake, stream, river	Sand pine-scrub oak	Pocosins	Everglades, mangroves, tropical hardwoods	Alpine tundra, krummholtz
Gull-billed Tern	X				X				
Common Tern					X				
Black Tern			X		X				
White-crowned Pigeon								X	
Mourning Dove	X	X				X	X	X	
Yellow-billed Cuckoo						X	X	X	
Mangrove Cuckoo								X	
Burrowing Owl						X			
Long-eared Owl	X	X	X						
Short-eared Owl	X	X	X					X	
Common Nighthawk	X	X	X		X	X		X	
Chuck-will's-widow						X		X	
Whip-poor-will	X							X	
Chimney Swift	X	X	X		X				
Ruby-throated Hummingbird	X	X						X	
Belted Kingfisher			X		X			X	
Yellow-bellied Sapsucker	X					X	X	X	
Acadian Flycatcher				X			X		
Alder Flycatcher				X					
Willow Flycatcher	X								
Least Flycatcher	X				X				
Eastern Phoebe	X	X		X		X	X	X	
Great Crested Flycatcher						X	X	X	
Western Kingbird								X	
Eastern Kingbird	X	X	X			X		X	
Gray Kingbird								X	
Scissor-tailed Flycatcher								X	
Horned Lark	X	X							
Purple Martin	X	X	X		X		X	X	
Tree Swallow	X	X	X	X	X			X	
Northern Rough-winged Swallow	X	X	X		X			X	
Bank Swallow	X	X	X		X				
Cliff Swallow	X	X	X		X				
Barn Swallow	X	X	X		X				

Table 6—*cont.*

Species	Field, glade, orchard	Pasture, wet or sedge meadows	Fresh marsh, pond	Wooded swamp, bog, shrub swamp	Lake, stream, river	Sand pine-scrub oak	Pocosins	Everglades, mangroves, tropical hardwoods	Alpine tundra, krummholtz
Bewick's Wren	X	X							
House Wren	X	X		X		X	X	X	
Sedge Wren		X						X	
Marsh Wren		X	X					X	
Golden-crowned Kinglet				X		X	X		X
Ruby-crowned Kinglet				X		X	X	X	X
Blue-gray Gnatcatcher				X		X		X	
Eastern Bluebird	X	X				X		X	
Veery				X					
Gray-cheeked Thrush									X
Hermit Thrush				X			X	X	
Wood Thrush				X			X		
American Robin	X	X		X		X	X	X	
Gray Catbird	X	X		X		X	X	X	
American Pipit	X	X			X				
Cedar Waxwing	X	X		X	X	X		X	
Loggerhead Shrike	X	X				X		X	
White-eyed Vireo	X			X		X	X	X	
Solitary Vireo				X		X	X	X	
Yellow-throated Vireo				X	X		X	X	
Warbling Vireo	X	X	X	X	X				
Red-eyed Vireo	X	X		X			X		
Black-whiskered Vireo								X	
Bachman's Warbler							X		
Blue-winged Warbler	X	X							
Golden-winged Warbler	X	X							
Orange-crowned Warbler						X	X	X	
Nashville Warbler				X					
Northern Parula				X	X		X	X	
Yellow Warbler	X	X	X	X	X			X	
Chestnut-sided Warbler		X		X					
Magnolia Warbler									X
Cape May Warbler								X	X
Black-throated Blue Warbler								X	
Yellow-rumped Warbler				X		X	X	X	

Appendix A

Table 6—*cont.*

Species	Field, glade, orchard	Pasture, wet or sedge meadows	Fresh marsh, pond	Wooded swamp, bog, shrub swamp	Lake, stream, river	Sand pine-scrub oak	Pocosins	Everglades, mangroves, tropical hardwoods	Alpine tundra, krummholtz
Black-throated Green Warbler							X	X	
Yellow-throated Warbler						X	X	X	
Pine Warbler						X	X	X	
Prairie Warbler	X					X	X	X	
Palm Warbler		X		X		X	X	X	
Bay-breasted Warbler	X								
Blackpoll Warbler									X
Black-and-white Warbler				X		X	X	X	
American Redstart	X			X				X	
Prothonotary Warbler				X	X		X		
Worm-eating Warbler							X		
Swainson's Warbler							X		
Ovenbird								X	
Northern Waterthrush				X			X	X	
Louisiana Waterthrush				X	X				
Kentucky Warbler				X	X				
Common Yellowthroat	X	X	X	X	X	X	X	X	
Hooded Warbler				X			X		
Wilson's Warbler				X					
Canada Warbler				X					
Yellow-breasted Chat	X	X	X	X		X	X	X	
Summer Tanager						X	X	X	
Rose-breasted Grosbeak	X								
Blue Grosbeak	X	X				X	X		
Indigo Bunting	X	X				X	X	X	
Painted Bunting						X	X	X	
Dickcissel								X	
Rufous-sided Towhee	X					X		X	
Chipping Sparrow	X	X				X		X	
Field Sparrow	X					X			
Vesper Sparrow	X	X				X			
Lark Sparrow	X								
Savannah Sparrow	X	X	X			X	X	X	
Grasshopper Sparrow	X					X		X	
Lincoln's Sparrow				X	X				

Table 6—*cont.*

Species	Field, glade, orchard	Pasture, wet or sedge meadows	Fresh marsh, pond	Wooded swamp, bog, shrub swamp	Lake, stream, river	Sand pine-scrub oak	Pocosins	Everglades, mangroves, tropical hardwoods	Alpine tundra, krummholtz
Swamp Sparrow		X	X	X			X	X	
White-crowned Sparrow	X					X			
Bobolink	X	X							
Red-winged Blackbird	X	X	X	X	X		X	X	
Eastern Meadowlark	X					X		X	
Western Meadowlark	X								
Brewer's Blackbird	X								
Brown-headed Cowbird	X	X	X	X		X	X		
Orchard Oriole	X						X		
Northern Oriole	X	X		X	X			X	
American Goldfinch	X	X	X	X		X	X	X	

Appendix A

Table 7. Neotropical migrants breeding in Great Plains habitats

Species	Gulf prairies and marshes	East Texas prairies, cross timber, pineywoods, and post oak savanna	South Texas shrub-grassland	Southern plains	Central plains	Northern plains	Wetland and riparian habitats	Shelterbelts and woodlots	Pine–oak, brushy woodland, badlands-juniper
Pied-billed Grebe	X						X		
Eared Grebe	X						X		
Western Grebe	X						X		
American White Pelican	X						X		
Double-crested Cormorant	X						X		
Anhinga	X						X		
American Bittern	X						X		
Least Bittern	X						X		
Great Blue Heron	X						X		
Great Egret	X						X		
Snowy Egret	X						X		
Little Blue Heron	X						X		
Cattle Egret	X	X	X	X	X		X		
Green Heron	X								
Black-crowned Night-Heron	X						X		
Yellow-crowned Night-Heron	X						X		
Black-bellied Whistling-Duck	X								
Greater White-fronted Goose	X						X		
Snow Goose	X						X		
Wood Duck	X	X					X		
Green-winged Teal	X	X	X	X	X	X	X		
Mallard	X	X	X	X	X	X	X		
Northern Pintail	X	X	X	X	X	X	X		
Blue-winged Teal	X						X		
Cinnamon Teal	X				X	X	X		
Northern Shoveler	X						X		
Gadwall	X						X		
American Wigeon	X						X		
Canvasback	X						X		
Redhead	X						X		

Table 7—*cont.*

Species	Gulf prairies and marshes	East Texas prairies, cross timber, pineywoods, and post oak savanna	South Texas shrub-grassland	Southern plains	Central plains	Northern plains	Wetland and riparian habitats	Shelterbelts and woodlots	Pine-oak, brushy woodland, badlands-juniper
Ring-necked Duck	X						X		
Lesser Scaup	X	X		X		X	X		
Hooded Merganser	X						X		
Red-breasted Merganser	X						X		
Ruddy Duck	X				X	X	X		
Black Vulture	X	X	X	X					
Turkey Vulture	X	X	X	X	X	X	X	X	X
Osprey							X		
Mississippi Kite	X	X	X	X			X		
Northern Harrier	X	X	X	X	X	X	X	X	X
Sharp-shinned Hawk	X	X	X	X	X	X	X	X	X
Cooper's Hawk	X	X	X	X	X	X	X	X	X
Red-shouldered Hawk	X	X	X				X		
Broad-winged Hawk	X	X					X	X	X
Swainson's Hawk	X	X	X	X	X	X	X	X	X
Zone-tailed Hawk			X						
Red-tailed Hawk	X	X	X	X	X	X	X	X	X
Ferruginous Hawk	X	X	X	X	X	X			X
American Kestrel	X	X	X	X	X	X	X	X	X
Merlin	X	X	X			X	X	X	X
Peregrine Falcon	X	X							X
Prairie Falcon	X	X				X			X
Yellow Rail	X								
Black Rail	X						X		
King Rail	X						X		
Virginia Rail	X						X		
Sora	X						X		
Purple Gallinule	X						X		
Common Moorhen	X						X		
American Coot	X			X	X	X	X		
Sandhill Crane	X						X		

Appendix A

Table 7—*cont.*

Species	Gulf prairies and marshes	East Texas prairies, cross timber, pineywoods, and post oak savanna	South Texas shrub-grassland	Southern plains	Central plains	Northern plains	Wetland and riparian habitats	Shelterbelts and woodlots	Pine–oak, brushy woodland, badlands-juniper
Whooping Crane	X						X		
Black-bellied Plover	X						X		
Snowy Plover	X						X		
Piping Plover	X						X		
Killdeer	X		X	X	X	X	X		
Mountain Plover	X			X	X	X			
Black-necked Stilt	X						X		
American Avocet	X			X	X	X	X		
Greater Yellowlegs	X						X		
Lesser Yellowlegs	X						X		
Willet	X				X	X	X		
Spotted Sandpiper	X	X	X	X	X	X	X		
Upland Sandpiper				X	X	X	X		
Long-billed Curlew	X			X	X	X	X		
Marbled Godwit	X					X	X		
Ruddy Turnstone	X								
Red Knot	X								
Sanderling	X								
Western Sandpiper	X						X		
Least Sandpiper	X						X		
Short-billed Dowitcher	X						X		
Long-billed Dowitcher	X								
Common Snipe	X						X		
Wilson's Phalarope	X			X	X	X	X		
Franklin's Gull					X	X	X		
Bonaparte's Gull	X						X		
Ring-billed Gull	X				X	X	X		
California Gull						X	X		
Herring Gull	X						X		
Caspian Tern	X						X		
Common Tern							X		

Table 7—_cont._

Species	Gulf prairies and marshes	East Texas prairies, cross timber, pineywoods, and post oak savanna	South Texas shrub-grassland	Southern plains	Central plains	Northern plains	Wetland and riparian habitats	Shelterbelts and woodlots	Pine-oak, brushy woodland, badlands-juniper
Forster's Tern	X						X		
Least Tern				X	X				
Black Tern							X		
White-winged Dove	X		X						
Mourning Dove		X	X	X	X	X	X	X	X
Black-billed Cuckoo						X	X	X	X
Yellow-billed Cuckoo							X	X	
Burrowing Owl			X	X	X	X			
Long-eared Owl							X	X	X
Short-eared Owl	X	X	X	X	X	X	X		
Common Nighthawk	X	X	X	X	X	X	X	X	X
Common Poorwill			X	X	X	X			
Chuck-will's-widow	X	X					X	X	
Whip-poor-will	X						X	X	
Chimney Swift	X	X	X	X	X	X	X	X	
Buff-bellied Hummingbird			X						
Ruby-throated Hummingbird	X	X	X	X	X	X	X	X	
Black-chinned Hummingbird		X	X	X					
Broad-tailed Hummingbird			X				X	X	
Belted Kingfisher	X	X	X				X		
Yellow-bellied Sapsucker	X	X	X				X	X	
Olive-sided Flycatcher							X		
Western Wood-Pewee			X	X			X	X	
Eastern Wood-Pewee		X	X				X	X	
Acadian Flycatcher	X	X	X				X		
Willow Flycatcher					X	X	X	X	X
Least Flycatcher							X	X	
Eastern Phoebe	X	X	X				X	X	
Say's Phoebe	X	X	X	X	X	X	X	X	X
Vermilion Flycatcher	X	X	X				X		
Ash-throated Flycatcher		X	X	X			X		

Table 7—*cont.*

Species	Gulf prairies and marshes	East Texas prairies, cross timber, pineywoods, and post oak savanna	South Texas shrub-grassland	Southern plains	Central plains	Northern plains	Wetland and riparian habitats	Shelterbelts and woodlots	Pine–oak, brushy woodland, badlands-juniper
Great Crested Flycatcher	X	X	X	X	X	X	X	X	X
Brown-crested Flycatcher			X						
Cassin's Kingbird			X	X	X		X	X	X
Western Kingbird	X	X	X	X	X	X	X	X	X
Eastern Kingbird	X	X	X	X	X	X	X	X	X
Scissor-tailed Flycatcher	X	X	X	X	X		X	X	
Rose-throated Becard			X						
Horned Lark	X		X	X	X	X			
Purple Martin	X	X	X		X		X	X	
Tree Swallow	X				X		X	X	
Violet-green Swallow						X	X		
Northern Rough-winged Swallow	X		X	X	X	X	X		
Bank Swallow	X			X	X	X	X		
Cliff Swallow		X	X	X	X	X	X	X	
Barn Swallow	X	X	X	X	X	X	X	X	X
Bewick's Wren		X	X	X	X		X	X	
House Wren	X	X	X				X	X	X
Sedge Wren	X				X	X	X		
Marsh Wren	X		X		X	X	X		
Golden-crowned Kinglet		X	X				X	X	X
Ruby-crowned Kinglet	X	X	X				X	X	X
Blue-gray Gnatcatcher	X	X	X				X	X	X
Eastern Bluebird	X	X	X		X		X	X	
Veery	X					X	X	X	
Swainson's Thrush	X		X						
Hermit Thrush	X	X	X				X	X	
Wood Thrush		X					X		
American Robin	X	X	X	X	X	X	X	X	X
Gray Catbird	X	X	X		X	X	X	X	X
Sage Thrasher			X						

Table 7—*cont.*

Species	Gulf prairies and marshes	East Texas prairies, cross timber, pineywoods, and post oak savanna	South Texas shrub-grassland	Southern plains	Central plains	Northern plains	Wetland and riparian habitats	Shelterbelts and woodlots	Pine–oak, brushy woodland, badlands-juniper
American Pipit	X	X	X	X			X		
Sprague's Pipit	X	X	X	X			X	X	
Cedar Waxwing	X	X	X	X	X	X	X	X	X
Loggerhead Shrike	X	X	X	X	X	X	X	X	X
White-eyed Vireo	X	X	X				X	X	
Bell's Vireo		X	X	X			X		X
Black-capped Vireo		X	X	X					
Solitary Vireo	X	X	X.	X			X		
Yellow-throated Vireo	X	X	X				X	X	
Warbling Vireo	X	X	X				X	X	X
Red-eyed Vireo	X	X	X		X		X	X	X
Yellow-green Vireo			X						
Blue-winged Warbler					X		X	X	
Orange-crowned Warbler	X	X	X				X		X
Northern Parula	X	X	X				X		
Yellow Warbler		X	X	X			X	X	
Yellow-rumped Warbler	X	X	X	X	X	X	X	X	X
Golden-cheeked Warbler		X	X						
Yellow-throated Warbler	X	X	X				X		
Cerulean Warbler		X					X		
Black-and-white Warbler	X	X	X				X	X	X
American Redstart		X			X	X	X	X	
Prothonotary Warbler	X	X					X		
Worm-eating Warbler		X					X		
Swainson's Warbler		X					X		
Ovenbird	X						X		X
Northern Waterthrush							X		
Louisiana Waterthrush	X	X					X		
Kentucky Warbler		X					X	X	
Common Yellowthroat	X	X	X	X	X	X	X	X	X
Hooded Warbler		X					X		

Appendix A

Table 7—*cont.*

Species	Gulf prairies and marshes	East Texas prairies, cross timber, pineywoods, and post oak savanna	South Texas shrub-grassland	Southern plains	Central plains	Northern plains	Wetland and riparian habitats	Shelterbelts and woodlots	Pine–oak, brushy woodland, badlands-juniper
Wilson's Warbler	X					X			
Yellow-breasted Chat	X	X	X	X		X	X	X	
Summer Tanager	X	X	X				X	X	
Scarlet Tanager							X	X	
Rose-breasted Grosbeak	X						X	X	
Black-headed Grosbeak							X	X	X
Blue Grosbeak		X	X	X	X		X	X	X
Lazuli Bunting				X	X	X	X	X	X
Indigo Bunting	X	X	X	X	X		X	X	X
Painted Bunting	X	X	X	X					
Dickcissel	X	X	X	X	X	X			
Green-tailed Towhee			X	X	X			X	
Rufous-sided Towhee	X	X	X	X	X	X	X	X	X
Cassin's Sparrow	X		X	X	X				
Rufous-crowned Sparrow		X	X	X					
Chipping Sparrow	X	X	X	X	X	X	X	X	X
Clay-colored Sparrow	X		X			X	X	X	
Brewer's Sparrow		X	X	X	X	X			
Field Sparrow	X	X	X	X	X	X	X	X	X
Vesper Sparrow	X	X	X	X	X	X			X
Lark Sparrow	X	X	X	X	X	X	X	X	X
Black-throated Sparrow			X	X					
Lark Bunting		X	X	X	X	X	X	X	X
Savannah Sparrow	X	X	X	X	X	X	X		
Grasshopper Sparrow		X	X	X	X	X			X
Lincoln's Sparrow		X	X			X	X		
Swamp Sparrow	X	X	X	X			X	X	
White-crowned Sparrow	X	X	X	X		X	X	X	X
Bobolink					X	X	X		
Red-winged Blackbird	X	X	X	X	X	X	X	X	
Eastern Meadowlark	X	X	X	X	X	X	X		

Table 7—*cont.*

Species	Gulf prairies and marshes	East Texas prairies, cross timber, pineywoods, and post oak savanna	South Texas shrub-grassland	Southern plains	Central plains	Northern plains	Wetland and riparian habitats	Shelterbelts and woodlots	Pine-oak, brushy woodland, badlands-juniper
Western Meadowlark	X	X	X	X	X	X	X	X	X
Yellow-headed Blackbird	X				X	X	X		X
Brewer's Blackbird	X	X	X	X	X	X	X		X
Brown-headed Cowbird	X	X	X	X	X	X	X	X	X
Orchard Oriole	X	X	X				X	X	
Northern Oriole	X	X	X	X	X	X	X	X	X
Lesser Goldfinch			X	X					
American Goldfinch	X	X	X	X	X	X	X	X	X
Evening Grosbeak					X	X	X	X	X

Table 8. Neotropical migrants breeding in western forest types

Species	Douglas-fir	Hemlock–sitka spruce	Redwood	Ponderosa pine	Larch/white pine	Lodgepole pine	Fir–spruce	Aspen/hardwoods	Chaparral	Pinyon–juniper
Wood Duck		X	X	X	X		X	X		
Hooded Merganser	X	X			X					
Turkey Vulture	X	X	X	X	X	X	X	X	X	X
Osprey	X	X	X	X	X	X	X	X	X	X
Sharp-shinned Hawk	X	X	X	X	X	X	X	X	X	X
Cooper's Hawk	X	X	X	X	X	X	X	X	X	X
Red-shouldered Hawk	X			X				X		
Swainson's Hawk				X				X	X	X
Red-tailed Hawk	X	X	X	X	X	X	X	X	X	X
American Kestrel	X	X	X	X	X	X	X	X	X	X
Merlin	X	X	X	X	X	X	X	X	X	X
Peregrine Falcon	X	X	X	X	X	X	X	X	X	X
Prairie Falcon				X				X	X	X
Band-tailed Pigeon	X	X	X	X		X	X	X	X	X
Mourning Dove	X	X	X	X		X	X	X	X	X
Flammulated Owl	X			X	X	X	X	X		
Long-eared Owl	X	X	X	X	X	X	X	X	X	X
Short-eared Owl				X				X	X	X
Common Nighthawk	X	X		X			X	X	X	X
Common Poorwill	X			X			X	X	X	X
Whip-poor-will				X						
Black Swift	X	X					X	X		
Vaux's Swift	X	X	X	X	X	X	X	X		
White-throated Swift	X	X	X	X	X	X	X	X		X
White-eared Hummingbird				X	X	X	X	X		
Lucifer Hummingbird									X	X
Black-chinned Hummingbird	X	X	X	X	X		X	X	X	
Calliope Hummingbird	X			X	X	X	X	X		
Broad-tailed Hummingbird	X		X	X	X	X	X	X	X	
Rufous Hummingbird	X	X	X	X	X	X	X	X	X	X
Allen's Hummingbird	X	X	X	X		X	X	X	X	
Belted Kingfisher	X		X	X	X			X		X
Yellow-bellied Sapsucker	X	X	X		X	X	X	X	X	
Williamson's Sapsucker	X		X	X	X	X	X	X		

Table 8—*cont.*

Species	Douglas-fir	Hemlock–sitka spruce	Redwood	Ponderosa pine	Larch/white pine	Lodgepole pine	Fir–spruce	Aspen/hardwoods	Chaparral	Pinyon–juniper
Olive-sided Flycatcher	X	X	X	X	X	X	X	X		
Western Wood-Pewee	X	X	X	X	X	X	X	X	X	X
Willow Flycatcher	X	X		X	X			X		X
Hammond's Flycatcher	X	X	X	X	X		X	X		
Dusky Flycatcher	X			X	X	X	X		X	X
Gray Flycatcher				X					X	X
"Western" Flycatcher	X	X	X	X	X	X	X	X		X
Say's Phoebe				X				X	X	X
Ash-throated Flycatcher				X				X	X	X
Brown-crested Flycatcher				X					X	X
Cassin's Kingbird				X					X	X
Western Kingbird				X					X	X
Rose-throated Becard									X	X
Purple Martin	X	X	X	X		X	X	X		
Tree Swallow	X	X	X	X	X	X	X	X	X	X
Violet-green Swallow	X	X	X	X	X	X	X	X	X	X
Northern Rough-winged Swallow	X						X	X	X	X
Cliff Swallow	X	X	X	X			X	X	X	X
Barn Swallow	X	X	X	X			X	X	X	X
Bewick's Wren	X		X	X			X	X	X	X
Rock Wren	X			X	X	X	X	X	X	X
House Wren	X	X	X	X	X	X	X	X	X	X
Golden-crowned Kinglet	X	X	X	X	X	X	X	X		
Ruby-crowned Kinglet	X	X	X	X	X	X	X	X	X	X
Blue-gray Gnatcatcher	X		X	X				X	X	X
Mountain Bluebird	X			X	X	X	X	X	X	X
Western Bluebird	X		X	X	X	X	X	X	X	X
Townsend's Solitaire	X	X	X	X	X	X	X	X	X	X
Swainson's Thrush	X	X	X	X	X	X	X	X	X	X
Hermit Thrush	X	X	X	X	X	X	X	X	X	X
American Robin	X	X	X	X	X	X	X	X	X	X
Gray Catbird				X				X		X
Cedar Waxwing	X	X	X	X			X	X	X	X
Phainopepla									X	X
Loggerhead Shrike	X			X				X	X	X

Table 8—*cont.*

Species	Douglas–fir	Hemlock–sitka spruce	Redwood	Ponderosa pine	Larch/white pine	Lodgepole pine	Fir–spruce	Aspen/hardwoods	Chaparral	Pinyon–juniper
Solitary Vireo	X	X	X	X	X	X	X	X	X	X
Warbling Vireo	X	X	X	X		X	X	X	X	
Red-eyed Vireo	X			X				X		
Orange-crowned Warbler	X	X	X	X			X	X	X	X
Nashville Warbler	X	X	X		X			X	X	
Virginia's Warbler				X					X	X
Yellow Warbler	X	X	X	X	X	X	X	X	X	X
Yellow-rumped Warbler	X	X	X	X	X	X	X	X	X	X
Black-throated Gray Warbler				X	X			X	X	X
Townsend's Warbler	X	X	X	X	X	X	X	X		
Hermit Warbler	X	X	X	X			X	X	X	
Grace's Warbler	X			X						
Northern Waterthrush				X				X		
MacGillivray's Warbler	X	X	X	X	X	X	X	X	X	
Common Yellowthroat	X	X	X	X	X	X	X	X		
Wilson's Warbler	X	X	X	X	X	X	X	X		
Red-faced Warbler				X			X	X		
Painted Redstart				X				X		X
Yellow-breasted Chat				X	X			X	X	
Hepatic Tanager				X				X		
Summer Tanager				X				X		
Western Tanager	X	X	X	X	X	X	X	X	X	X
Black-headed Grosbeak	X	X	X	X	X		X	X	X	X
Lazuli Bunting	X		X	X			X	X	X	X
Indigo Bunting				X						X
Green-tailed Towhee				X				X	X	X
Rufous-sided Towhee	X	X	X	X	X	X	X	X	X	X
Rufous-crowned Sparrow				X				X	X	
Chipping Sparrow	X	X	X	X	X	X	X	X	X	X
Lincoln's Sparrow	X	X	X			X	X	X		
White-crowned Sparrow	X	X	X	X		X	X	X	X	X
Brewer's Blackbird	X			X			X	X	X	X
Northern Oriole	X	X	X	X	X		X	X	X	X

Table 8—*cont.*

Species	Douglas-fir	Hemlock–sitka spruce	Redwood	Ponderosa pine	Larch/white pine	Lodgepole pine	Fir–spruce	Aspen/hardwoods	Chaparral	Pinyon–juniper
Scott's Oriole										X
Lesser Goldfinch			X	X				X	X	X
American Goldfinch	X	X	X	X				X	X	X
Evening Grosbeak	X	X	X	X	X	X	X	X	X	X

Appendix A

Table 9. Neotropical migrants breeding in southwestern and western nonforest habitats

Species	Great Basin shrubsteppe	Sonoran Desert scrub	Chihuahuan Desert scrub	Mojave Desert scrub	Desert riparian and canyon woodlands, marshes	Annual grasslands, farms	River, riparian woodlands, subalpine marshes	Mountain and alpine meadows	Relict conifer forests, madrean evergreen woodlands
Pied-billed Grebe					X		X		
Eared Grebe					X		X		
Western Grebe					X		X		
Double-crested Cormorant					X		X		
American Bittern					X		X		
Least Bittern					X				
Great Blue Heron					X	X	X	X	
Great Egret					X		X		
Snowy Egret					X		X		
Cattle Egret					X	X			
Green Heron					X		X		
Black-crowned Night-Heron					X		X	X	
White-faced Ibis					X	X	X		
Black-bellied Whistling-Duck					X				
Greater White-fronted Goose						X	X		
Snow Goose					X	X	X		
Wood Duck					X		X		
Green-winged Teal					X	X	X	X	
Mallard					X	X	X	X	
Northern Pintail					X	X	X	X	
Blue-winged Teal					X	X	X	X	
Cinnamon Teal					X	X	X	X	
Northern Shoveler					X	X	X	X	
Gadwall					X	X	X	X	
American Wigeon					X	X	X	X	
Canvasback					X		X		
Redhead					X		X		
Ring-necked Duck					X		X		
Lesser Scaup					X	X	X		
Hooded Merganser							X		
Red-breasted Merganser							X		

Table 9—*cont.*

Species	Great Basin shrubsteppe	Sonoran Desert scrub	Chihuahuan Desert scrub	Mojave Desert scrub	Desert riparian and canyon woodlands, marshes	Annual grasslands, farms	River, riparian woodlands, subalpine marshes	Mountain and alpine meadows	Relict conifer forests, madrean evergreen woodlands
Ruddy Duck					X		X	X	
Black Vulture		X	X						
Turkey Vulture	X	X	X	X	X	X	X	X	
Osprey					X		X		
Northern Harrier	X				X	X	X	X	
Sharp-shinned Hawk	X	X			X		X	X	
Cooper's Hawk	X						X	X	X
Common Black-Hawk					X				
Gray Hawk					X				
Red-shouldered Hawk						X	X		
Swainson's Hawk	X	X	X	X		X	X	X	
Zone-tailed Hawk					X	X			
Red-tailed Hawk	X	X	X	X	X	X	X	X	
American Kestrel	X	X	X	X	X	X	X	X	X
Merlin	X	X	X	X	X	X	X		
Peregrine Falcon	X	X	X	X	X	X	X		X
Prairie Falcon	X	X	X	X	X	X	X	X	
Virginia Rail					X		X	X	
Sora					X		X	X	
Common Moorhen					X				
American Coot					X	X	X	X	
Sandhill Crane					X	X	X		
Whooping Crane					X				
Killdeer	X	X	X	X	X	X	X	X	
Mountain Plover						X		X	
Black-necked Stilt					X		X		
American Avocet					X	X	X	X	
Greater Yellowlegs					X		X		
Willet					X		X		
Spotted Sandpiper					X	X	X	X	
Upland Sandpiper						X		X	

Appendix A

Table 9—cont.

Species	Great Basin shrubsteppe	Sonoran Desert scrub	Chihuahuan Desert scrub	Mojave Desert scrub	Desert riparian and canyon woodlands, marshes	Annual grasslands, farms	River, riparian woodlands, subalpine marshes	Mountain and alpine meadows	Relict conifer forests, madrean evergreen woodlands
Whimbrel						X	X		
Long-billed Curlew					X	X	X	X	
Marbled Godwit						X	X	X	
Least Sandpiper					X	X	X		
Long-billed Dowitcher					X		X		
Common Snipe					X	X	X	X	
Wilson's Phalarope					X		X		
Franklin's Gull					X	X	X	X	
Ring-billed Gull					X	X	X		
California Gull					X	X	X		
Herring Gull					X		X		
Gull-billed Tern						X			
Caspian Tern							X		
Forster's Tern					X	X	X	X	
Black Tern					X	X	X	X	
Band-tailed Pigeon	X	X			X	X	X		
White-winged Dove		X	X	X	X				
Mourning Dove	X	X	X	X	X	X	X	X	X
Black-billed Cuckoo							X		
Yellow-billed Cuckoo	X	X	X	X	X		X		
Flammulated Owl	X				X		X		
Elf Owl		X	X		X				
Burrowing Owl	X		X	X		X		X	
Long-eared Owl	X	X	X	X	X	X	X	X	
Short-eared Owl	X	X	X	X	X	X	X	X	
Lesser Nighthawk		X	X	X		X			
Common Nighthawk	X	X	X	X	X	X	X	X	
Common Poorwill	X	X	X	X	X	X	X		
Whip-poor-will		X			X				
Black Swift				X			X	X	
Vaux's Swift							X	X	

Table 9—*cont.*

Species	Great Basin shrubsteppe	Sonoran Desert scrub	Chihuahuan Desert scrub	Mojave Desert scrub	Desert riparian and canyon woodlands, marshes	Annual grasslands, farms	River, riparian woodlands, subalpine marshes	Mountain and alpine meadows	Relict conifer forests, madrean evergreen woodlands
White-throated Swift	X	X	X	X	X		X	X	
Broad-billed Hummingbird		X	X		X				
Violet-crowned Hummingbird		X							
Blue-throated Hummingbird		X	X		X				
Magnificent Hummingbird		X			X				
Lucifer Hummingbird		X	X			X			
Black-chinned Hummingbird	X	X		X	X		X	X	X
Costa's Hummingbird		X		X					
Calliope Hummingbird	X						X	X	
Broad-tailed Hummingbird	X	X					X	X	
Rufous Hummingbird	X	X					X	X	X
Allen's Hummingbird		X					X	X	
Elegant Trogon					X		X		
Belted Kingfisher					X		X		
Williamson's Sapsucker	X				X		X		
Northern Beardless-Tyrannulet		X			X				
Olive-sided Flycatcher							X		X
Greater Pewee		X			X				
Western Wood-Pewee					X		X		
Willow Flycatcher	X				X		X	X	
Hammond's Flycatcher							X		
Dusky Flycatcher	X						X	X	
Gray Flycatcher	X								
"Western" Flycatcher							X	X	
Buff-breasted Flycatcher		X			X				
Say's Phoebe	X	X	X	X		X	X		
Vermilion Flycatcher		X	X		X				
Dusky-capped Flycatcher		X			X				
Ash-throated Flycatcher		X	X	X	X		X		X

Appendix A

Table 9—*cont.*

Species	Great Basin shrubsteppe	Sonoran Desert scrub	Chihuahuan Desert scrub	Mojave Desert scrub	Desert riparian and canyon woodlands, marshes	Annual grasslands, farms	River, riparian woodlands, subalpine marshes	Mountain and alpine meadows	Relict conifer forests, madrean evergreen woodlands
Brown-crested Flycatcher		X			X		X		
Sulphur-bellied Flycatcher		X			X				
Cassin's Kingbird	X	X	X	X	X	X			
Thick-billed Kingbird		X	X		X				
Western Kingbird	X	X	X	X	X	X	X		
Eastern Kingbird	X		X			X	X		
Rose-throated Becard		X	X						
Horned Lark	X	X	X	X			X	X	X
Purple Martin		X			X	X	X	X	
Tree Swallow					X	X	X	X	
Violet-green Swallow	X				X	X	X	X	
Northern Rough-winged Swallow	X				X	X	X		
Bank Swallow	X	X			X	X	X		
Cliff Swallow						X	X	X	X
Barn Swallow	X	X	X	X	X	X	X		X
Rock Wren	X	X	X	X					
Bewick's Wren		X	X	X			X	X	X
House Wren	X	X	X		X	X	X	X	
Marsh Wren	X				X		X	X	
Ruby-crowned Kinglet					X		X		
Blue-gray Gnatcatcher	X	X	X		X		X		
Mountain Bluebird	X	X	X	X			X	X	X
Western Bluebird	X				X	X	X	X	
Townsend's Solitaire	X				X		X	X	
Veery							X		
Swainson's Thrush					X		X		
Hermit Thrush						X	X	X	
American Robin	X	X	X	X	X	X	X	X	
Gray Catbird							X		
Sage Thrasher	X	X		X		X	X		

Table 9—*cont.*

Species	Great Basin shrubsteppe	Sonoran Desert scrub	Chihuahuan Desert scrub	Mojave Desert scrub	Desert riparian and canyon woodlands, marshes	Annual grasslands, farms	River, riparian woodlands, subalpine marshes	Mountain and alpine meadows	Relict conifer forests, madrean evergreen woodlands
Bendire's Thrasher		X	X		X	X			
Water Pipit	X					X	X	X	
Cedar Waxwing	X	X	X	X	X	X	X		
Phainopepla		X	X		X				
Loggerhead Shrike	X	X	X	X	X	X	X	X	
Bell's Vireo				X	X	X			
Gray Vireo	X								
Solitary Vireo	X				X		X		X
Warbling Vireo					X		X		
Orange-crowned Warbler	X				X		X		
Nashville Warbler							X		
Virginia's Warbler	X				X		X		
Lucy's Warbler		X	X	X	X		X		
Yellow Warbler	X				X		X	X	
Yellow-rumped Warbler	X				X		X		
Black-throated Gray Warbler	X	X		X	X		X		X
Townsend's Warbler							X		
Hermit Warbler							X		
Grace's Warbler							X		
Ovenbird							X		
MacGillivray's Warbler	X						X	X	
Common Yellowthroat					X		X	X	
Wilson's Warbler					X		X	X	
Red-faced Warbler					X		X		
Painted Redstart					X		X		
Yellow-breasted Chat	X				X		X		
Olive Warbler							X		
Hepatic Tanager							X		
Summer Tanager					X		X		
Western Tanager	X				X		X		
Rose-breasted Grosbeak							X		

Table 9—*cont.*

Species	Great Basin shrubsteppe	Sonoran Desert scrub	Chihuahuan Desert scrub	Mojave Desert scrub	Desert riparian and canyon woodlands, marshes	Annual grasslands, farms	River, riparian woodlands, subalpine marshes	Mountain and alpine meadows	Relict conifer forests, madrean evergreen woodlands
Black-headed Grosbeak	X				X		X		X
Blue Grosbeak					X	X	X		
Lazuli Bunting	X	X			X	X	X	X	
Indigo Bunting					X		X		
Varied Bunting		X	X		X				
Green-tailed Towhee	X	X			X		X		
Rufous-sided Towhee	X	X			X		X		X
Botteri's Sparrow		X							
Cassin's Sparrow		X	X			X			
Rufous-crowned Sparrow	X	X	X				X		
Chipping Sparrow	X	X	X	X	X	X	X	X	
Clay-colored Sparrow		X	X			X			
Black-chinned Sparrow	X	X	X						X
Vesper Sparrow	X	X	X	X	X	X	X	X	
Lark Sparrow	X	X				X	X		
Black-throated Sparrow	X	X	X	X	X				
Savannah Sparrow	X	X	X		X	X	X	X	
Grasshopper Sparrow		X	X	X	X	X	X		
Lincoln's Sparrow					X		X	X	
Swamp Sparrow							X	X	
White-crowned Sparrow	X	X	X	X	X	X	X	X	
Bobolink							X		
Red-winged Blackbird					X	X	X	X	
Eastern Meadowlark		X	X						
Western Meadowlark	X	X	X	X	X	X	X	X	
Yellow-headed Blackbird					X		X		
Brewer's Blackbird					X	X	X	X	
Bronzed Cowbird		X			X				
Brown-headed Cowbird	X	X	X	X	X	X	X	X	
Hooded Oriole		X			X				
Northern Oriole					X	X	X		

Table 9—*cont.*

Species	Great Basin shrubsteppe	Sonoran Desert scrub	Chihuahuan Desert scrub	Mojave Desert scrub	Desert riparian and canyon woodlands, marshes	Annual grasslands, farms	River, riparian woodlands, subalpine marshes	Mountain and alpine meadows	Relict conifer forests, madrean evergreen woodlands
Scott's Oriole		X	X	X					X
Lesser Goldfinch	X	X	X	X	X	X			
American Goldfinch	X				X	X	X	X	
Evening Grosbeak	X				X		X		

Table 10. Neotropical migrants breeding in arctic and coastal habitats

Species	Dry tundra	Moist tundra	Tundra lakes	Oceanic cliffs & rocky shorelines	Eastern ocean beaches	Eastern bays & estuaries	Eastern islands & spoils	Western ocean beaches	Western bays & estuaries	Western islands & spoils
Pied-billed Grebe						X			X	
American White Pelican									X	X
Brown Pelican						X	X		X	X
Double-crested Cormorant				X		X	X		X	X
Neotropic Cormorant						X	X			
American Bittern						X	X		X	X
Least Bittern						X	X		X	X
Great Blue Heron				X		X	X		X	X
Great Egret						X	X		X	X
Snowy Egret						X	X		X	X
Little Blue Heron						X	X			
Tricolored Heron						X	X		X	X
Reddish Egret						X	X		X	X
Cattle Egret						X	X		X	X
Green Heron						X	X		X	X
Black-crowned Night-Heron						X	X		X	X
Yellow-crowned Night-Heron						X	X			
White Ibis						X	X			
Glossy Ibis						X	X			
White-faced Ibis						X	X		X	X
Roseate Spoonbill						X	X			
Wood Stork						X			X	X
Greater White-fronted Goose		X	X			X			X	
Snow Goose	X		X					X		X
Green-winged Teal		X	X							
Mallard	X	X	X			X	X		X	X
Northern Pintail	X	X	X							
Blue-winged Teal		X	X							
Northern Shoveler		X	X							

Table 10—_cont._

Species	Dry tundra	Moist tundra	Tundra lakes	Oceanic cliffs & rocky shorelines	Eastern ocean beaches	Eastern bays & estuaries	Eastern islands & spoils	Western ocean beaches	Western bays & estuaries	Western islands & spoils
Gadwall						X	X		X	X
American Wigeon	X	X	X							
Canvasback		X	X							
Redhead			X							
Lesser Scaup		X	X			X	X		X	X
Red-breasted Merganser						X	X		X	X
Ruddy Duck			X							
Osprey						X	X		X	X
Peregrine Falcon	X	X	X	X	X	X	X	X	X	X
Black Rail						X			X	
King Rail						X				
Virginia Rail						X			X	
Sora						X			X	
Sandhill Crane		X	X							
Whooping Crane		X	X							
Black-bellied Plover	X	X				X			X	
American Golden-Plover	X	X				X			X	
Snowy Plover					X	X		X	X	
Wilson's Plover					X	X		X	X	
Semipalmated Plover	X		X			X			X	
Piping Plover					X					
American Oystercatcher					X	X	X			
Greater Yellowlegs	X	X	X			X			X	
Lesser Yellowlegs	X	X	X			X			X	
Willet					X	X	X			
Wandering Tattler	X	X								
Eskimo Curlew	X	X								
Whimbrel	X									
Hudsonian Godwit		X				X			X	
Ruddy Turnstone	X	X								
Surfbird	X			X						
Red Knot	X					X			X	

Table 10—*cont.*

Species	Dry tundra	Moist tundra	Tundra lakes	Oceanic cliffs & rocky shorelines	Eastern ocean beaches	Eastern bays & estuaries	Eastern islands & spoils	Western ocean beaches	Western bays & estuaries	Western islands & spoils
					Eastern			Western		
Sanderling	X	X	X				X			X
Semipalmated Sandpiper	X	X	X		X	X		X	X	
Western Sandpiper	X	X	X							
Least Sandpiper	X	X	X							
White-rumped Sandpiper	X	X	X			X			X	
Baird's Sandpiper	X	X	X							
Pectoral Sandpiper	X									
Dunlin		X	X			X			X	
Stilt Sandpiper	X	X	X							
Buff-breasted Sandpiper	X									
Short-billed Dowitcher		X	X							
Long-billed Dowitcher	X	X	X							
Common Snipe		X	X							
Red-necked Phalarope		X	X							
Red Phalarope	X	X	X				X			X
Laughing Gull					X	X	X	X	X	X
Ring-billed Gull					X	X	X	X	X	X
Herring Gull			X	X	X	X	X	X	X	X
Western Gull				X				X	X	X
Glaucous-winged Gull				X				X	X	
Sabine's Gull	X	X	X		X	X	X	X	X	X
Gull-billed Tern					X	X	X	X	X	X
Caspian Tern					X	X	X	X	X	X
Royal Tern						X	X			
Elegant Tern								X	X	X
Sandwich Tern						X	X			
Roseate Tern					X	X	X	X	X	X
Common Tern					X	X	X	X	X	X
Forster's Tern						X	X		X	X
Least Tern					X	X	X	X	X	X
Sooty Tern				X	X	X	X			

Table 10—*cont.*

Species	Dry tundra	Moist tundra	Tundra lakes	Oceanic cliffs & rocky shorelines	Eastern — Eastern ocean beaches	Eastern bays & estuaries	Eastern islands & spoils	Western — Western ocean beaches	Western bays & estuaries	Western islands & spoils
Black Skimmer					X	X	X			
White-crowned Pigeon							X			
Short-eared Owl	X									
Common Nighthawk					X			X		
Black Swift				X						
White-throated Swift				X						
Belted Kingfisher				X		X			X	
Horned Lark	X				X			X		
Sedge Wren						X				
Marsh Wren						X			X	
American Pipit	X	X								
Prairie Warbler							X			
Savannah Sparrow	X	X			X	X		X	X	
Red-winged Blackbird						X			X	

Table 11. Wintering habitats used by Neotropical migrants

Species	Shrubsteppe					Forest					Aquatic			
	Alpine	Desert	Grassland	Savanna	Scrub	Coniferous	Coniferous-broadleaf	Deciduous-broadleaf	Broadleaf evergreen	Gallery	Fresh wetlands	Brackish wetlands	Seashore	Coastal marine
Pied-billed Grebe											X	X		X
Eared Grebe											X	X		X
Western Grebe											X	X		X
American White Pelican											X	X		X
Brown Pelican												X		X
Double-creasted Cormorant											X	X		X
Neotropic Cormorant											X	X		
Anhinga											X	X		
American Bittern											X	X		
Least Bittern											X	X		
Great Blue Heron											X	X		
Great Egret											X	X		
Snowy Egret											X	X		
Little Blue Heron				X							X	X		
Tricolored Heron											X	X		
Reddish Egret													X	
Cattle Egret			X	X							X	X		
Green Heron											X	X		
Black-crowned Night-Heron											X	X		
Yellow-crowned Night-Heron											X	X	X	

Species	C1	C2	C3	C4	C5	C6	C7	C8	C9	C10	C11	C12
White Ibis	X	X										
Glossy Ibis	X	X										
White-faced Ibis	X	X										
Roseate Spoonbill	X											
Wood Stork	X	X										
Fulvous Whistling-Duck	X	X										
Black-bellied Whistling-Duck	X	X										
Greater White-fronted Goose	X	X										
Snow Goose	X	X										
Wood Duck	X	X										
Green-winged Teal	X	X										
Mallard	X	X										
Northern Pintail	X	X										
Blue-winged Teal	X	X										
Cinnamon Teal	X	X										
Northern Shoveler	X	X										
Gadwall	X	X										
American Wigeon	X	X										
Canvasback	X	X										
Redhead	X	X										
Ring-necked Duck	X	X										
Lesser Scaup	X	X										
Hooded Merganser	X	X										
Red-breasted Merganser	X											
Ruddy Duck	X	X										
Black Vulture											X	X
Turkey Vulture								X		X	X	X
Osprey					X		X	X				
American Swallow-tailed Kite										X		
Mississippi Kite			X	X			X		X		X	X
Northern Harrier			X	X					X	X		
Sharp-shinned Hawk	X	X				X						

Table 11—*cont.*

Species	Alpine	Desert	Grassland	Savanna	Scrub	Coniferous	Coniferous-broadleaf	Deciduous-broadleaf	Broadleaf evergreen	Gallery	Fresh wetlands	Brackish wetlands	Seashore	Coastal marine
				Shrubsteppe				Forest				Aquatic		
Cooper's Hawk						×	×		×					
Common Black-Hawk									×			×		
Gray Hawk				×	×			×		×				
Red-shouldered Hawk				×	×			×		×				
Broad-winged Hawk								×	×	×	×			
Swainson's Hawk			×	×	×									
Zone-tailed Hawk				×	×		×	×		×				
Red-tailed Hawk	×				×	×	×	×						
Ferruginous Hawk		×	×	×	×									
American Kestrel			×	×	×									
Merlin						×	×					×		
Peregrine Falcon											×	×		
Prairie Falcon		×	×	×										
Yellow Rail											×	×		
Black Rail											×	×		
King Rail											×			
Virginia Rail											×			
Sora											×	×		
Purple Gallinule											×	×		
Common Moorhen											×	×		

Species						
American Coot		X			X	
Sandhill Crane	X	X			X	X
Whooping Crane	X	X	X		X	X
Black-bellied Plover	X			X		X
American Golden-Plover	X			X	X	
Snowy Plover			X			X
Wilson's Plover						X
Semipalmated Plover						X
Piping Plover						X
Killdeer	X	X		X	X	
Mountain Plover						X
American Oystercatcher			X	X	X	X
Black-necked Stilt			X	X	X	X
American Avocet			X	X	X	X
Greater Yellowlegs			X	X	X	X
Lesser Yellowlegs			X		X	
Solitary Sandpiper					X	X
Willet					X	X
Wandering Tattler	X	X		X	X	
Spotted Sandpiper	X	X	X	X	X	X
Upland Sandpiper					X	
Eskimo Curlew					X	
Whimbrel	X		X	X	X	X
Long-billed Curlew	X	X			X	X
Hudsonian Godwit					X	X
Marbled Godwit	X		X	X	X	X
Ruddy Turnstone						X
Surfbird						X
Red Knot						X
Sanderling						X
Semipalmated Sandpiper	X		X	X	X	X
Western Sandpiper					X	X

Table 11—*cont.*

Species	Shrubsteppe					Forest					Aquatic			
	Alpine	Desert	Grassland	Savanna	Scrub	Coniferous	Coniferous-broadleaf	Deciduous-broadleaf	Broadleaf evergreen	Gallery	Fresh wetlands	Brackish wetlands	Seashore	Coastal marine
Least Sandpiper			X								X	X	X	
White-rumped Sandpiper			X								X	X	X	
Baird's Sandpiper			X								X		X	
Pectoral Sandpiper			X								X		X	
Dunlin											X	X	X	
Stilt Sandpiper											X			
Buff-breasted Sandpiper			X	X							X			
Short-billed Dowitcher											X	X	X	
Long-billed Dowitcher											X	X	X	
Common Snipe			X								X			
Wilson's Phalarope											X	X	X	
Red-necked Phalarope												X	X	X
Red Phalarope												X	X	X
Laughing Gull												X	X	X
Franklin's Gull			X								X	X	X	X
Bonaparte's Gull											X	X	X	X
Ring-billed Gull											X	X	X	X
California Gull												X	X	X
Herring Gull												X	X	X
Western Gull												X		X

Glaucous-winged Gull											X	X	X
Sabine's Gull												X	X
Gull-billed Tern										X	X	X	
Caspian Tern											X	X	X
Royal Tern										X	X	X	X
Elegant Tern											X	X	X
Sandwich Tern											X	X	X
Roseate Tern											X	X	
Common Tern											X	X	
Forster's Tern										X	X	X	
Least Tern										X	X	X	X
Sooty Tern												X	X
Black Tern										X	X	X	
Black Skimmer											X	X	
White-crowned Pigeon				X			X	X			X		
Red-billed Pigeon				X			X						
Band-tailed Pigeon				X			X	X					
White-winged Dove	X	X	X	X			X						
Mourning Dove		X	X	X									
Black-billed Cuckoo			X	X			X						
Yellow-billed Cuckoo				X			X		X				
Mangrove Cuckoo				X					X		X		
Flammulated Owl					X	X	X	X					
Elf Owl	X												
Burrowing Owl		X	X	X									
Long-eared Owl					X	X	X						
Short-eared Owl		X	X							X			
Lesser Nighthawk	X	X	X	X			X						
Common Nighthawk		X	X										
Common Poorwill	X		X	X									
Chuck-will's-widow							X	X	X				
Whip-poor-will					X	X	X	X					

Table 11—*cont.*

Species	Shrubsteppe					Forest					Aquatic			
	Alpine	Desert	Grassland	Savanna	Scrub	Coniferous	Coniferous-broadleaf	Deciduous-broadleaf	Broadleaf evergreen	Gallery	Fresh wetlands	Brackish wetlands	Seashore	Coastal marine
Black Swift	X	X	X	X	X									
Chimney Swift		X	X	X				X	X	X				
Vaux's Swift			X					X	X					
White-throated Swift		X		X	X		X							
Broad-billed Hummingbird		X					X							
White-eared Hummingbird				X		X	X							
Buff-bellied Hummingbird				X	X		X							
Violet-crowned Hummingbird		X			X	X	X	X						
Blue-throated Hummingbird		X			X	X	X							
Magnificent Hummingbird						X	X							
Lucifer Hummingbird		X			X	X	X							
Ruby-throated Hummingbird				X	X			X	X					
Black-chinned Hummingbird				X	X			X						
Costa's Hummingbird		X			X	X	X							
Calliope Hummingbird						X	X							
Broad-tailed Hummingbird	X			X		X								
Rufous Hummingbird					X	X	X							
Allen's Hummingbird					X									
Elegant Trogon					X		X	X		X	X			
Belted Kingfisher											X	X		

Yellow-bellied Sapsucker					X	X	X	X	X	
Williamson's Sapsucker					X	X				
Northern Beardless-Tyrannulet	X		X	X						
Olive-sided Flycatcher							X	X	X	
Greater Pewee					X	X	X			
Western Wood-Pewee					X	X		X		
Eastern Wood-Pewee				X	X	X	X	X	X	
Yellow-bellied Flycatcher							X	X	X	
Acadian Flycatcher								X	X	
Alder Flycatcher				X			X		X	X
Willow Flycatcher				X			X		X	X
White-throated Flycatcher		X	X	X						
Least Flycatcher			X	X			X		X	
Hammond's Flycatcher					X	X	X		X	
Dusky Flycatcher				X					X	
Gray Flycatcher	X		X						X	
Pine Flycatcher					X	X	X			
"Western" Flycatcher					X	X	X	X	X	
Buff-breasted Flycatcher					X	X	X			
Eastern Phoebe			X	X					X	
Say's Phoebe	X	X	X	X						
Vermilion Flycatcher	X		X	X						
Dusky-capped Flycatcher							X	X	X	
Ash-throated Flycatcher	X	X	X	X			X		X	
Great Crested Flycatcher							X	X	X	
Brown-crested Flycatcher				X			X		X	
Sulphur-bellied Flycatcher							X	X	X	
Tropical Kingbird			X	X						
Cassin's Kingbird		X	X							
Thick-billed Kingbird			X	X						
Western Kingbird			X							
Eastern Kingbird			X				X	X	X	

Table 11—*cont.*

Species	Alpine	Desert	Grassland	Savanna	Scrub	Coniferous	Coniferous-broadleaf	Deciduous-broadleaf	Broadleaf evergreen	Gallery	Fresh wetlands	Brackish wetlands	Seashore	Coastal marine
	Shrubsteppe					**Forest**					**Aquatic**			
Gray Kingbird				X	X							X		
Scissor-tailed Flycatcher			X	X	X					X		X		
Rose-throated Becard				X	X									
Horned Lark	X	X	X[a]										X	
Purple Martin			X	X							X	X		
Gray-breasted Martin			X	X							X	X		
Tree Swallow											X	X		
Violet-green Swallow	X		X	X		X	X	X						
Northern Rough-winged Swallow			X	X				X	X		X	X		
Bank Swallow			X	X						X	X			
Cliff Swallow			X	X							X			
Cave Swallow			X	X							X			
Barn Swallow			X	X							X			
Rock Wren		X			X									
Bewick's Wren		X			X									
House Wren					X		X	X	X	X				
Sedge Wren			X								X			
Marsh Wren											X	X		
Golden-crowned Kinglet					X	X	X	X		X				
Ruby-crowned Kinglet						X	X	X						

Blue-gray Gnatcatcher					X	X			X	X		
Eastern Bluebird				X		X	X	X				
Mountain Bluebird	X					X	X					
Western Bluebird				X	X	X	X					
Townsend's Solitaire						X	X					
Veery									X			
Gray-cheeked Thrush									X			
Swainson's Thrush									X			
Hermit Thrush						X	X	X		X		
Wood Thrush									X	X		
American Robin						X	X	X				
Gray Catbird					X				X	X		
Sage Thrasher		X										
Bendire's Thrasher		X										
American Pipit		X	X								X	
Sprague's Pipit			X								X	
Cedar Waxwing								X	X	X		
Phainopepla		X			X	X	X	X				
Loggerhead Shrike				X	X							
White-eyed Vireo					X			X	X	X		
Bell's Vireo					X					X		
Black-capped Vireo					X							
Gray Vireo					X							
Solitary Vireo						X	X	X				
Yellow-throated Vireo								X	X	X		
Warbling Vireo						X	X	X				
Philadelphia Vireo								X	X	X		
Red-eyed Vireo								X	X	X		
Yellow-green Vireo								X	X	X		
Black-whiskered Vireo								X	X	X		X
Bachman's Warbler											X	
Blue-winged Warbler								X		X		X

Table 11—*cont.*

Species	Shrubsteppe					Forest					Aquatic			
	Alpine	Desert	Grassland	Savanna	Scrub	Coniferous	Coniferous-broadleaf	Deciduous-broadleaf	Broadleaf evergreen	Gallery	Fresh wetlands	Brackish wetlands	Seashore	Coastal marine
Golden-winged Warbler					X				X	X				
Tennessee Warbler					X	X		X	X	X				
Orange-crowned Warbler					X	X		X	X	X				
Nashville Warbler				X			X							
Virginia's Warbler						X								
Colima Warbler							X							
Lucy's Warbler				X	X					X				
Northern Parula				X	X			X	X	X		X		
Yellow Warbler				X	X					X		X		
Chestnut-sided Warbler					X	X			X	X				
Magnolia Warbler					X	X		X	X	X				
Cape May Warbler					X	X		X	X	X				
Black-throated Blue Warbler					X	X[b]	X		X					
Yellow-rumped Warbler			X	X	X	X	X				X			
Black-throated Gray Warbler						X	X	X						
Townsend's Warbler						X	X	X	X					
Hermit Warbler						X	X							
Black-throated Green Warbler					X	X			X	X				
Golden-cheeked Warbler						X	X		X					
Blackburnian Warbler								X	X	X				

Species	1	2	3	4	5	6	7	8	9	10
Yellow-throated Warbler	X			X				X		
Grace's Warbler				X	X					
Pine Warbler					X					
Kirtland's Warbler			X							
Prairie Warbler	X		X							
Palm Warbler	X		X		X^c			X		
Bay-breasted Warbler				X	X					
Blackpoll Warbler			X	X						
Cerulean Warbler				X						
Black-and-white Warbler	X		X	X	X					
American Redstart	X		X	X	X					
Prothonotary Warbler			X	X						
Worm-eating Warbler			X	X						
Swainson's Warbler			X	X						
Ovenbird				X	X					
Northern Waterthrush		X		X						
Louisiana Waterthrush			X	X						
Kentucky Warbler			X	X						
Connecticut Warbler				X						
Mourning Warbler				X						
MacGillivray's Warbler		X		X						
Common Yellowthroat	X	X		X						
Hooded Warbler			X	X	X					
Wilson's Warbler			X	X	X					
Canada Warbler			X	X						
Red-faced Warbler				X	X					
Painted Redstart				X	X					
Yellow-breasted Chat	X			X	X					
Olive Warbler				X						
Hepatic Tanager			X	X	X					
Summer Tanager				X	X					
Scarlet Tanager				X	X					

Table 11—*cont.*

Species	Alpine	Desert	Grassland	Savanna	Scrub	Coniferous	Coniferous-broadleaf	Deciduous-broadleaf	Broadleaf evergreen	Gallery	Fresh wetlands	Brackish wetlands	Seashore	Coastal marine
Western Tanager						X	X			X				
Rose-breasted Grosbeak						X	X	X	X	X				
Black-headed Grosbeak						X	X	X	X	X				
Blue Grosbeak				X	X									
Lazuli Bunting			X	X										
Indigo Bunting				X	X									
Varied Bunting					X									
Painted Bunting				X	X					X				
Dickcissel			X	X										
Green-tailed Towhee				X			X	X						
Rufous-sided Towhee				X	X	X								
Botteri's Sparrow			X	X										
Cassin's Sparrow			X	X	X									
Rufous-crowned Sparrow				X	X		X							
Chipping Sparrow				X		X	X							
Clay-colored Sparrow			X	X										
Brewer's Sparrow		X												
Field Sparrow			X	X	X									
Black-chinned Sparrow				X	X									
Vesper Sparrow			X											

582

	1	2	3	4	5	6	7	8
Lark Sparrow				X				
Black-throated Sparrow	X		X	X	X			
Lark Bunting			X	X				X
Savannah Sparrow			X	X				
Grasshopper Sparrow			X	X				X
Lincoln's Sparrow				X	X	X		
Swamp Sparrow				X				
White-crowned Sparrow			X	X				
Bobolink			X	X				X
Red-winged Blackbird				X				
Eastern Meadowlark			X	X				
Western Meadowlark			X	X				
Yellow-headed Blackbird			X	X				X
Brewer's Blackbird		X	X	X				
Bronzed Cowbird			X	X	X			
Brown-headed Cowbird			X	X	X			
Orchard Oriole			X	X				X
Hooded Oriole		X	X	X				
Northern Oriole	X	X	X	X	X			
Scott's Oriole		X	X	X				
Lesser Goldfinch			X	X	X	X		
American Goldfinch			X	X	X	X		
Evening Grosbeak				X	X	X		

Notes: a = grazed; b = montane; c = if burned.

Population Changes by Physiographic Region

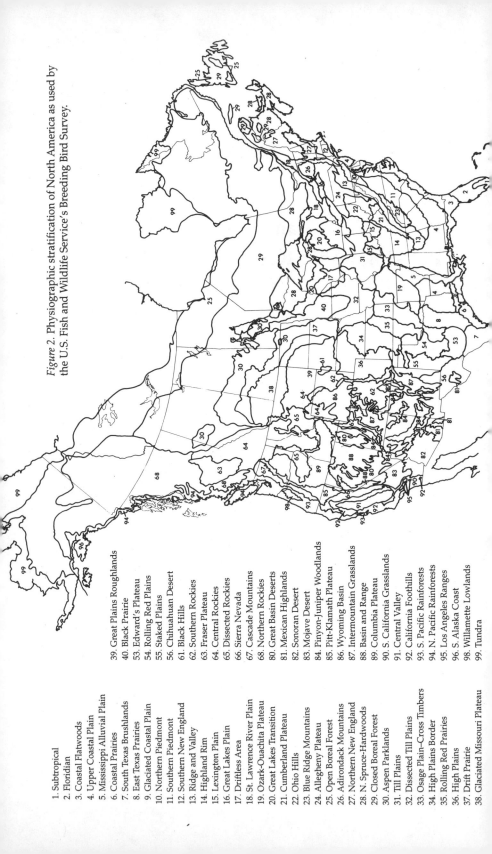

Figure 2. Physiographic stratification of North America as used by the U.S. Fish and Wildlife Service's Breeding Bird Survey.

1. Subtropical
2. Floridian
3. Coastal Flatwoods
4. Upper Coastal Plain
5. Mississippi Alluvial Plain
6. Coastal Prairies
7. South Texas Brushlands
8. East Texas Prairies
9. Glaciated Coastal Plain
10. Northern Piedmont
11. Southern Piedmont
12. Southern New England
13. Ridge and Valley
14. Highland Rim
15. Lexington Plain
16. Great Lakes Plain
17. Driftless Area
18. St. Lawrence River Plain
19. Ozark-Ouachita Plateau
20. Great Lakes Transition
21. Cumberland Plateau
22. Ohio Hills
23. Blue Ridge Mountains
24. Allegheny Plateau
25. Open Boreal Forest
26. Adirondack Mountains
27. Northern New England
28. N. Spruce-Hardwoods
29. Closed Boreal Forest
30. Aspen Parklands
31. Till Plains
32. Dissected Till Plains
33. Osage Plain–Cross Timbers
34. High Plains Border
35. Rolling Red Prairies
36. High Plains
37. Drift Prairie
38. Glaciated Missouri Plateau

39. Great Plains Roughlands
40. Black Prairie
53. Edward's Plateau
54. Rolling Red Plains
55. Staked Plains
56. Chihuahuan Desert
61. Black Hills
62. Southern Rockies
63. Fraser Plateau
64. Central Rockies
65. Dissected Rockies
66. Sierra Nevada
67. Cascade Mountains
68. Northern Rockies
80. Great Basin Deserts
81. Mexican Highlands
82. Sonoran Desert
83. Mojave Desert
84. Pinyon-Juniper Woodlands
85. Pitt-Klamath Plateau
86. Wyoming Basin
87. Intermountain Grasslands
88. Basin and Range
89. Columbia Plateau
90. S. California Grasslands
91. Central Valley
92. California Foothills
93. S. Pacific Rainforests
94. N. Pacific Rainforests
95. Los Angeles Ranges
96. S. Alaska Coast
97. Willamette Lowlands
98. Tundra
99. Tundra

This appendix shows population changes of Neotropical migrants over the long term (1966–1994) and short term (1980–1994) first by physiographic region, then for eastern, central, and western North America north of Mexico, and last for continental North America north of Mexico, excluding Alaska. Species included have shown a statistically significant ($P < 0.05$ or $P < 0.01$) change as detected by the U.S. Fish and Wildlife Service's Breeding Bird Survey (I = increase, D = decrease). Population trends within physiographic regions are given only for species reported from at least 14 BBS routes. Trends within eastern, central, and western regions of the United States and Canada are given for species reported from at least 25 routes, and trends for the continent are given for species reported from at least 50 routes. Trends should be interpreted with caution: species vary in detectability and not all habitats are surveyed adequately. Trends for land birds, for example, are more reliable than those for water or marsh birds.

	Species	Long term (1966–1994)	Short term (1980–1994)
Floridian	Anhinga		D
	Little Blue Heron		D
	White Ibis		D
	Turkey Vulture	D	
	Osprey	I	
	Red-tailed Hawk	D	
	Sandhill Crane	I	
	Killdeer	D	D
	Mourning Dove	I	I
	Common Nighthawk	D	
	Chimney Swift		I
	Eastern Kingbird	D	
	Eastern Bluebird	D	
	White-eyed Vireo	I	I
	Northern Parula		I
	Common Yellowthroat		I
	Rufous-sided Towhee	D	D
	Red-winged Blackbird	D	D
	Eastern Meadowlark	D	D
Coastal Flatwoods	Great Blue Heron	I	I
	Great Egret		I
	Wood Duck	I	I
	Turkey Vulture	I	
	Osprey	I	
	Broad-winged Hawk		I
	Killdeer		I

587

Appendix B

Species	Long term (1966–1994)	Short term (1980–1994)
Laughing Gull	I	I
Yellow-billed Cuckoo		D
Common Nighthawk	D	
Northern Rough-winged Swallow		D
Eastern Bluebird		I
Wood Thrush		D
American Robin		I
Loggerhead Shrike	D	
Yellow-throated Vireo	I	
Red-eyed Vireo	D	D
Prothonotary Warbler		I
Swainson's Warbler	I	D
Common Yellowthroat	D	
Hooded Warbler		D
Blue Grosbeak	I	I
Painted Bunting	D	
Rufous-sided Towhee	D	
Bachman's Sparrow	D	
Eastern Meadowlark	D	D
Brown-headed Cowbird	I	D
Upper Coastal Plain Great Blue Heron	I	I
Cattle Egret	I	
Black-crowned Night-Heron	D	
Canada Goose	I	I
Wood Duck	I	
American Black Duck		D
Mallard	I	I
Black Vulture	I	I
Turkey Vulture	I	I
Osprey	I	I
Cooper's Hawk		I
Red-shouldered Hawk	I	I
Red-tailed Hawk	I	I
Herring Gull	I	
Mourning Dove	D	D
Black-billed Cuckoo	D	D
Yellow-billed Cuckoo	D	D
Common Nighthawk	D	
Chuck-will's-widow	D	D
Eastern Wood-Pewee	D	
Acadian Flycatcher	I	
Eastern Phoebe	I	I
Eastern Kingbird	D	
Barn Swallow	I	I

Population Changes by Physiographic Region

	Species	Long term (1966–1994)	Short term (1980–1994)
	Bewick's Wren	D	
	Marsh Wren	D	D
	Eastern Bluebird	I	I
	Wood Thrush	D	
	American Robin	I	I
	Cedar Waxwing	I	I
	Loggerhead Shrike	D	
	White-eyed Vireo		D
	Yellow Warbler		D
	Pine Warbler	I	I
	Prairie Warbler	D	
	Prothonotary Warbler	D	D
	Common Yellowthroat	D	
	Hooded Warbler		I
	Yellow-breasted Chat	I	I
	Summer Tanager		D
	Scarlet Tanager	D	
	Blue Grosbeak	I	I
	Rufous-sided Towhee	D	
	Bachman's Sparrow		D
	Chipping Sparrow	D	
	Field Sparrow	D	
	Vesper Sparrow	D	D
	Grasshopper Sparrow	D	
	Red-winged Blackbird	D	D
	Eastern Meadowlark	D	D
	Orchard Oriole	D	
	Northern (Baltimore) Oriole	D	D
	American Goldfinch		I
Mississippi Alluvial Plain	Great Blue Heron	I	I
	Cattle Egret	I	
	Mourning Dove	D	
	Yellow-billed Cuckoo	D	
	Belted Kingfisher	I	
	Eastern Wood-Pewee	D	
	Eastern Kingbird	D	
	Northern Rough-winged Swallow	I	I
	Barn Swallow	I	I
	Blue-gray Gnatcatcher	D	
	American Robin	I	I
	Gray Catbird	D	D
	White-eyed Vireo	D	
	Common Yellowthroat	D	
	Yellow-breasted Chat	D	

	Species	Long term (1966–1994)	Short term (1980–1994)
	Indigo Bunting	D	
	Painted Bunting	D	D
	Dickcissel		I
	Field Sparrow	D	
	Red-winged Blackbird	I	
	Orchard Oriole	D	
	Northern (Baltimore) Oriole	D	D
Coastal Prairies	Common Nighthawk	D	
	Chimney Swift	I	
	Dickcissel		D
	Eastern Meadowlark	D	D
	Brown-headed Cowbird	D	
South Texas Brushlands	Turkey Vulture		I
	Lesser Nighthawk	I	
	Brown-crested Flycatcher		I
	Western Kingbird	I	
	Scissor-tailed Flycatcher	I	I
	Painted Bunting	D	
	Cassin's Sparrow	D	
	Black-throated Sparrow	D	
	Red-winged Blackbird	I	
	Eastern Meadowlark		D
	Brown-headed Cowbird	D	D
	Orchard Oriole	D	D
East Texas Prairies	Great Blue Heron	I	
	Little Blue Heron		D
	Turkey Vulture		D
	Red-tailed Hawk	I	
	Killdeer	D	D
	Mourning Dove	D	D
	Yellow-billed Cuckoo	D	D
	Common Nighthawk		D
	Chimney Swift	D	D
	Eastern Phoebe	I	
	Great Crested Flycatcher		D
	Eastern Kingbird		D
	Scissor-tailed Flycatcher	D	
	Eastern Bluebird	I	
	American Robin	I	
	Loggerhead Shrike	D	D
	Painted Bunting	D	
	Lark Sparrow	D	D
	Red-winged Blackbird	D	D

Species	Long term (1966–1994)	Short term (1980–1994)
Eastern Meadowlark		D
Brown-headed Cowbird	D	D
Orchard Oriole	D	

Northern Piedmont

Species	Long term (1966–1994)	Short term (1980–1994)
Great Blue Heron	I	I
Green Heron		D
Wood Duck	I	
Black Vulture	I	
Red-shouldered Hawk	I	
Red-tailed Hawk	I	
Yellow-billed Cuckoo		D
Eastern Phoebe	I	I
Great Crested Flycatcher	D	
Tree Swallow		I
Northern Rough-winged Swallow	I	
Horned Lark	D	
Purple Martin	D	D
Barn Swallow	D	D
House Wren		I
Eastern Bluebird	I	I
American Robin	I	I
Gray Catbird	I	I
Cedar Waxwing	I	
American Redstart	D	
Common Yellowthroat		D
Indigo Bunting	D	
Rufous-sided Towhee	D	
Field Sparrow	D	D
Vesper Sparrow	D	D
Grasshopper Sparrow	D	D
Red-winged Blackbird	D	D
Eastern Meadowlark	D	D
Northern (Baltimore) Oriole	D	
American Goldfinch		D

Southern Piedmont

Species	Long term (1966–1994)	Short term (1980–1994)
Great Blue Heron	I	I
Red-shouldered Hawk	I	
Red-tailed Hawk	I	
Killdeer		I
Eastern Wood-Pewee		D
Eastern Phoebe	I	I
Eastern Kingbird	D	
Purple Martin	I	I
Barn Swallow	D	D
House Wren		I
Blue-gray Gnatcatcher	I	I
Eastern Bluebird	I	I

Appendix B

	Species	Long term (1966–1994)	Short term (1980–1994)
	Wood Thrush	D	
	Gray Catbird	D	D
	Cedar Waxwing	I	
	Loggerhead Shrike	D	
	White-eyed Vireo	I	
	Yellow-throated Vireo	I	I
	Red-eyed Vireo	I	I
	Northern Parula	I	I
	Yellow Warbler	D	
	Pine Warbler	I	I
	Black-and-white Warbler	I	
	Common Yellowthroat	D	I
	Blue Grosbeak		D
	Indigo Bunting	I	
	Rufous-sided Towhee	D	
	Chipping Sparrow		I
	Field Sparrow	D	D
	Red-winged Blackbird	D	D
	Eastern Meadowlark	D	D
	American Goldfinch		I
Southern New England	Mallard		I
	Turkey Vulture	I	I
	Broad-winged Hawk	D	D
	Red-tailed Hawk	I	I
	American Kestrel	D	D
	Killdeer	D	
	Herring Gull		D
	Mourning Dove	I	
	Black-billed Cuckoo	D	D
	Yellow-billed Cuckoo	D	D
	Whip-poor-will		D
	Eastern Wood-Pewee		D
	Least Flycatcher	D	
	Eastern Phoebe		I
	Great Crested Flycatcher	D	D
	Eastern Kingbird	D	D
	Northern Rough-winged Swallow	I	I
	House Wren	D	D
	Blue-gray Gnatcatcher	I	
	Eastern Bluebird		I
	Hermit Thrush		I
	Wood Thrush	D	
	Gray Catbird	I	
	Cedar Waxwing	I	
	Yellow-throated Vireo		D
	Red-eyed Vireo		D

Population Changes by Physiographic Region

	Species	Long term (1966–1994)	Short term (1980–1994)
	Blue-winged Warbler	D	D
	Yellow Warbler	I	
	Yellow-rumped (Myrtle) Warbler	I	
	Black-throated Green Warbler	I	I
	Pine Warbler	I	I
	Prairie Warbler	D	D
	Black-and-white Warbler		D
	Common Yellowthroat	D	D
	Scarlet Tanager	D	D
	Rose-breasted Grosbeak	D	D
	Indigo Bunting	D	D
	Rufous-sided Towhee	D	D
	Chipping Sparrow	I	I
	Field Sparrow	D	D
	Red-winged Blackbird	D	D
	Eastern Meadowlark	D	D
	Northern (Baltimore) Oriole	D	D
	American Goldfinch		I
Ridge and Valley	Great Blue Heron	I	I
	Wood Duck	I	I
	Mallard	I	
	Turkey Vulture		I
	Sharp-shinned Hawk	I	
	Red-shouldered Hawk	I	
	Red-tailed Hawk	I	I
	Mourning Dove	I	
	Yellow-billed Cuckoo	D	D
	Whip-poor-will	D	
	Belted Kingfisher	D	
	Eastern Wood-Pewee	D	D
	Eastern Phoebe		I
	Horned Lark	D	
	Tree Swallow	I	I
	Bank Swallow		D
	Cliff Swallow	I	I
	Barn Swallow	I	D
	House Wren	D	
	Eastern Bluebird	I	I
	Wood Thrush	D	D
	Cedar Waxwing	I	
	White-eyed Vireo	D	D
	Solitary Vireo		I
	Yellow-throated Vireo	D	
	Red-eyed Vireo		I

	Species	Long term (1966–1994)	Short term (1980–1994)
	Blue-winged Warbler	D	
	Yellow-throated Warbler	D	
	Prairie Warbler	D	D
	Black-and-white Warbler	D	
	Prothonotary Warbler	D	
	Ovenbird	I	I
	Kentucky Warbler	D	D
	Yellow-breasted Chat	D	
	Summer Tanager	D	D
	Blue Grosbeak	D	
	Indigo Bunting	D	D
	Rufous-sided Towhee	D	D
	Chipping Sparrow	D	
	Field Sparrow	D	D
	Grasshopper Sparrow	D	D
	Swamp Sparrow	I	
	Bobolink		D
	Red-winged Blackbird	D	D
	Eastern Meadowlark	D	D
	Brown-headed Cowbird	D	D
	Orchard Oriole	D	D
	Northern (Baltimore) Oriole	D	D
Highland Rim	Great Blue Heron	I	I
	Green Heron	D	D
	Cooper's Hawk	I	I
	Broad-winged Hawk		D
	Red-tailed Hawk	I	I
	Killdeer	I	
	Yellow-billed Cuckoo	D	D
	Chimney Swift	D	D
	Belted Kingfisher	D	
	Eastern Phoebe		I
	Horned Lark	D	
	Purple Martin	I	I
	Northern Rough-winged Swallow	I	
	Barn Swallow	D	D
	Bewick's Wren	D	
	House Wren	I	I
	Eastern Bluebird	I	I
	Wood Thrush		D
	American Robin	I	I
	Gray Catbird	D	D
	Cedar Waxwing	I	I
	Loggerhead Shrike	D	

Species	Long term (1966–1994)	Short term (1980–1994)
White-eyed Vireo	D	D
Red-eyed Vireo		I
Northern Parula		I
Yellow Warbler		D
Yellow-throated Warbler	I	I
Pine Warbler	I	I
Prairie Warbler	D	
Cerulean Warbler	D	
American Redstart	D	D
Common Yellowthroat		D
Yellow-breasted Chat	D	
Scarlet Tanager	I	
Blue Grosbeak	I	I
Indigo Bunting	D	D
Dickcissel	D	
Rufous-sided Towhee	D	
Chipping Sparrow		I
Field Sparrow	D	D
Grasshopper Sparrow	D	
Red-winged Blackbird		D
Eastern Meadowlark	D	D
Brown-headed Cowbird		I
Orchard Oriole	D	D
American Goldfinch		I
Lexington Plain		
Great Blue Heron	I	I
Green Heron	D	
Wood Duck	I	
Mallard	I	I
Red-tailed Hawk	I	
American Kestrel		D
Killdeer	I	
Yellow-billed Cuckoo	D	
Chimney Swift	D	D
Eastern Wood-Pewee	D	
Eastern Phoebe		I
Eastern Kingbird		D
Horned Lark		I
Barn Swallow		D
House Wren	I	I
Eastern Bluebird		I
American Robin	I	I
Warbling Vireo	I	
Red-eyed Vireo	I	
Common Yellowthroat		D
Yellow-breasted Chat	D	
Scarlet Tanager	I	

Appendix B

	Species	Long term (1966–1994)	Short term (1980–1994)
	Blue Grosbeak	I	I
	Indigo Bunting	D	D
	Rufous-sided Towhee	D	
	Field Sparrow	D	D
	Grasshopper Sparrow	D	
	Red-winged Blackbird	D	D
	Eastern Meadowlark	D	D
	Orchard Oriole	I	D
	Northern (Baltimore) Oriole	I	
	American Goldfinch		I
Great Lakes Plain	American Bittern	D	
	Great Blue Heron	I	
	Wood Duck	I	I
	Mallard	I	
	Blue-winged Teal	D	D
	Turkey Vulture	I	
	Cooper's Hawk	I	
	Sandhill Crane	I	I
	Killdeer	I	
	Spotted Sandpiper	D	
	Common Snipe		I
	Ring-billed Gull	I	
	Black Tern	D	
	Mourning Dove	I	
	Yellow-billed Cuckoo	D	D
	Belted Kingfisher		D
	Eastern Wood-Pewee	I	
	Eastern Phoebe		I
	Eastern Kingbird	D	D
	Purple Martin	D	D
	Tree Swallow	I	
	Barn Swallow	I	
	House Wren		I
	Marsh Wren		D
	Eastern Bluebird	I	I
	American Robin	I	I
	Gray Catbird	I	I
	Cedar Waxwing	I	
	Yellow-throated Vireo	I	I
	Red-eyed Vireo	I	I
	Yellow Warbler	I	
	Common Yellowthroat	I	
	Yellow-breasted Chat	D	D
	Scarlet Tanager	I	I
	Indigo Bunting	I	D
	Dickcissel	D	

	Species	Long term (1966–1994)	Short term (1980–1994)
	Chipping Sparrow	I	I
	Field Sparrow	D	D
	Vesper Sparrow	D	D
	Savannah Sparrow	D	
	Grasshopper Sparrow	D	D
	Henslow's Sparrow	D	D
	Bobolink	D	D
	Red-winged Blackbird	D	D
	Eastern Meadowlark	D	D
	Western Meadowlark	D	D
	Brown-headed Cowbird	D	
	Orchard Oriole		D
	Northern (Baltimore) Oriole		D
	American Goldfinch	I	I
Driftless Area	Mallard	I	I
	Turkey Vulture	I	I
	Red-tailed Hawk	I	
	Killdeer	I	I
	Belted Kingfisher	D	D
	Eastern Phoebe		I
	Horned Lark		I
	Tree Swallow	I	
	House Wren	I	I
	Sedge Wren		I
	Blue-gray Gnatcatcher	I	I
	Eastern Bluebird	I	I
	American Robin	I	I
	Cedar Waxwing	I	I
	Yellow Warbler		I
	American Redstart	I	
	Ovenbird		I
	Scarlet Tanager		I
	Rose-breasted Grosbeak	I	
	Chipping Sparrow		I
	Field Sparrow	D	
	Vesper Sparrow	D	
	Savannah Sparrow		I
	Grasshopper Sparrow	D	
	Bobolink	D	D
	Red-winged Blackbird		D
	Western Meadowlark	D	D
	Brown-headed Cowbird	D	
St. Lawrence River Plain	Great Blue Heron	I	
	Mallard	I	I
	American Kestrel		D

Appendix B

	Species	Long term (1966–1994)	Short term (1980–1994)
	Common Moorhen	D	
	Killdeer	D	D
	Ring-billed Gull	I	I
	Mourning Dove	I	I
	Chimney Swift	D	
	Ruby-throated Hummingbird	I	
	Eastern Wood-Pewee		D
	Eastern Phoebe		I
	Horned Lark	D	D
	Purple Martin		D
	Tree Swallow	I	
	Barn Swallow		D
	House Wren	I	
	Ruby-crowned Kinglet		D
	Eastern Bluebird		I
	Hermit Thrush		I
	Gray Catbird	D	
	Cedar Waxwing	I	
	Yellow-throated Vireo		I
	Warbling Vireo	I	I
	Red-eyed Vireo		I
	Nashville Warbler	I	
	Yellow Warbler	I	
	Chestnut-sided Warbler	I	I
	Magnolia Warbler	I	I
	Black-throated Blue Warbler	I	
	Yellow-rumped (Myrtle) Warbler	I	
	Blackburnian Warbler		I
	Black-and-white Warbler	I	I
	Ovenbird	I	I
	Northern Waterthrush		I
	Mourning Warbler	I	
	Rufous-sided Towhee		I
	Vesper Sparrow	D	
	Savannah Sparrow	D	
	Swamp Sparrow	I	I
	Bobolink		D
	Red-winged Blackbird		D
	Eastern Meadowlark	D	D
	Brown-headed Cowbird	D	D
	Northern (Baltimore) Oriole		D
Ozark-Ouachita Plateau	Great Blue Heron	I	
	Green Heron		D

	Species	Long term (1966–1994)	Short term (1980–1994)
	Wood Duck	I	I
	Common Nighthawk	D	
	Acadian Flycatcher	D	
	Eastern Phoebe	I	
	Horned Lark		D
	Barn Swallow		D
	Eastern Bluebird		I
	Wood Thrush	D	
	American Robin	I	
	Gray Catbird	D	
	Loggerhead Shrike	D	D
	Bell's Vireo	D	D
	Blue-winged Warbler		I
	Prairie Warbler	D	
	Ovenbird	D	
	Rufous-sided Towhee	D	
	Field Sparrow	D	
	Lark Sparrow	D	
	Red-winged Blackbird		D
	Eastern Meadowlark		D
	Orchard Oriole	D	
	Northern (Baltimore) Oriole		I
Great Lakes Transition	American Bittern	D	
	Wood Duck	I	I
	Turkey Vulture		I
	Red-tailed Hawk	I	I
	Sandhill Crane	I	I
	Yellow-billed Cuckoo	D	
	Least Flycatcher	D	
	Eastern Phoebe		I
	Eastern Kingbird	D	D
	Purple Martin "	D	D
	Tree Swallow	I	
	House Wren		I
	Sedge Wren	I	
	Eastern Bluebird		I
	Veery	D	D
	Warbling Vireo	D	
	Red-eyed Vireo	I	I
	Yellow Warbler		I
	Chestnut-sided Warbler	I	
	Yellow-rumped (Myrtle) Warbler	I	
	American Redstart		I
	Ovenbird	I	I
	Common Yellowthroat	I	

	Species	Long term (1966–1994)	Short term (1980–1994)
	Scarlet Tanager	I	
	Dickcissel	D	
	Rufous-sided Towhee	D	
	Field Sparrow	D	
	Vesper Sparrow	D	D
	Savannah Sparrow	D	
	Grasshopper Sparrow	D	
	Henslow's Sparrow	D	
	Swamp Sparrow	I	
	Bobolink		D
	Red-winged Blackbird		D
	Eastern Meadowlark	D	D
	Western Meadowlark	D	D
	Brown-headed Cowbird	D	
	Northern (Baltimore) Oriole		D
Cumberland Plateau	Red-shouldered Hawk	I	
	Chimney Swift	I	
	Mourning Dove	I	I
	Ruby-throated Hummingbird	D	
	Belted Kingfisher		D
	Eastern Wood-Pewee	D	D
	Acadian Flycatcher		D
	Great Crested Flycatcher	D	D
	Purple Martin		D
	Northern Rough-winged Swallow	D	
	Eastern Bluebird		I
	Wood Thrush		D
	American Robin	I	I
	Gray Catbird	D	D
	Cedar Waxwing	I	I
	Northern Parula		D
	Yellow Warbler		D
	Prairie Warbler	D	
	Cerulean Warbler	D	
	Black-and-white Warbler	D	D
	American Redstart	D	
	Kentucky Warbler		D
	Common Yellowthroat	D	
	Yellow-breasted Chat	D	
	Summer Tanager	D	D
	Scarlet Tanager	I	I
	Indigo Bunting	D	
	Rufous-sided Towhee	D	
	Field Sparrow	D	D

Species	Long term (1966–1994)	Short term (1980–1994)
Red-winged Blackbird		D
Eastern Meadowlark	D	D

Ohio Hills

Species	Long term (1966–1994)	Short term (1980–1994)
Great Blue Heron	I	I
Cooper's Hawk		I
Red-tailed Hawk	I	
Killdeer	I	I
Mourning Dove	I	I
Yellow-billed Cuckoo	D	D
Ruby-throated Hummingbird		I
Belted Kingfisher	D	
Eastern Wood-Pewee	D	D
Great Crested Flycatcher	D	
Horned Lark	D	
Purple Martin	D	
Northern Rough-winged Swallow		I
Barn Swallow		D
House Wren	I	I
Eastern Bluebird	D	
Wood Thrush	I	
American Robin	I	I
Gray Catbird		I
White-eyed Vireo	I	I
Yellow-throated Vireo		I
Warbling Vireo	D	
Blue-winged Warbler		D
Golden-winged Warbler	D	
Yellow Warbler		I
Yellow-throated Warbler	I	I
Prairie Warbler	D	D
Cerulean Warbler	D	
Black-and-white Warbler	D	
Ovenbird		I
Yellow-breasted Chat	D	
Rose-breasted Grosbeak	I	I
Indigo Bunting	D	D
Rufous-sided Towhee	D	
Field Sparrow	D	D
Vesper Sparrow	D	
Grasshopper Sparrow	D	D
Bobolink	I	
Red-winged Blackbird	D	D
Eastern Meadowlark	D	
Brown-headed Cowbird	D	
American Goldfinch	D	D

	Species	Long term (1966–1994)	Short term (1980–1994)
Blue Ridge Mountains	Eastern Wood-Pewee	D	
	Acadian Flycatcher	D	
	Barn Swallow	D	D
	Wood Thrush	D	
	Gray Catbird	D	D
	Black-and-white Warbler	D	
	Yellow-breasted Chat	D	
	Rufous-sided Towhee	D	
	Chipping Sparrow	D	D
	Field Sparrow	D	D
	Eastern Meadowlark	D	D
	Brown-headed Cowbird	D	D
Allegheny Plateau	Great Blue Heron	I	
	Wood Duck		I
	Mallard	I	I
	Turkey Vulture	I	
	Red-tailed Hawk	I	
	Killdeer	D	D
	Spotted Sandpiper	D	
	Mourning Dove	I	I
	Yellow-billed Cuckoo		D
	Ruby-throated Hummingbird	I	
	Belted Kingfisher	D	
	Eastern Wood-Pewee	D	D
	Acadian Flycatcher	I	
	Eastern Phoebe		I
	Great Crested Flycatcher		D
	Eastern Kingbird	D	D
	Horned Lark	D	
	Purple Martin	D	D
	Tree Swallow	I	I
	Barn Swallow		D
	Hermit Thrush	I	I
	Wood Thrush	D	D
	American Robin	D	
	White-eyed Vireo		D
	Solitary Vireo	I	I
	Warbling Vireo	I	
	Red-eyed Vireo	I	I
	Golden-winged Warbler	D	
	Nashville Warbler	D	
	Chestnut-sided Warbler		I
	Black-throated Blue Warbler	I	I
	Yellow-rumped (Myrtle) Warbler	I	I

Species	Long term (1966–1994)	Short term (1980–1994)
Cerulean Warbler	D	
Black-and-white Warbler	D	
Ovenbird	I	I
Hooded Warbler	I	I
Canada Warbler	D	D
Yellow-breasted Chat	D	
Rose-breasted Grosbeak		D
Rufous-sided Towhee	D	D
Chipping Sparrow	D	D
Field Sparrow	D	D
Vesper Sparrow	D	
Savannah Sparrow	D	D
Grasshopper Sparrow	D	
Henslow's Sparrow	D	D
Red-winged Blackbird	D	D
Eastern Meadowlark	D	D
Brown-headed Cowbird	D	D
Northern (Baltimore) Oriole		D
American Goldfinch	D	D
Adirondack Mountains Mourning Dove	I	I
Chimney Swift	D	D
Ruby-throated Hummingbird		I
Olive-sided Flycatcher	D	D
Eastern Wood-Pewee	D	
Least Flycatcher	D	D
Eastern Phoebe		I
Great Crested Flycatcher	D	D
Eastern Kingbird	D	D
Bank Swallow	D	D
Barn Swallow	D	
House Wren		D
Veery	D	D
Hermit Thrush	I	I
Wood Thrush	D	D
Gray Catbird	D	D
Cedar Waxwing		D
Solitary Vireo	I	I
Red-eyed Vireo	I	I
Chestnut-sided Warbler	D	D
Magnolia Warbler	I	
Yellow-rumped (Myrtle) Warbler	I	

	Species	Long term (1966–1994)	Short term (1980–1994)
	American Redstart	D	D
	Common Yellowthroat	D	D
	Canada Warbler	D	D
	Scarlet Tanager	D	D
	Rose-breasted Grosbeak	D	D
	Rufous-sided Towhee	D	D
	Field Sparrow	D	D
	Savannah Sparrow	D	
	Red-winged Blackbird	D	D
	Brown-headed Cowbird	D	D
	Northern (Baltimore) Oriole	D	D
	American Goldfinch	D	
	Evening Grosbeak	I	I
Northern New England	Great Blue Heron	I	
	Mallard	I	I
	Turkey Vulture	I	
	Red-tailed Hawk	I	
	Common Snipe		I
	Mourning Dove	I	I
	Black-billed Cuckoo		D
	Chimney Swift		D
	Least Flycatcher	D	
	Eastern Phoebe		I
	Great Crested Flycatcher	I	I
	Eastern Kingbird		D
	Barn Swallow		D
	Eastern Bluebird	I	I
	Veery	D	D
	Hermit Thrush		I
	Wood Thrush		D
	Gray Catbird	D	D
	Cedar Waxwing	I	
	Solitary Vireo	I	I
	Red-eyed Vireo		D
	Nashville Warbler		D
	Northern Parula		I
	Chestnut-sided Warbler	D	D
	Magnolia Warbler		D
	Yellow-rumped (Myrtle) Warbler		I
	Blackburnian Warbler	D	
	Pine Warbler	I	I
	Ovenbird		I
	Common Yellowthroat	D	D
	Rufous-sided Towhee	D	D

	Species	Long term (1966–1994)	Short term (1980–1994)
	Chipping Sparrow		I
	Field Sparrow	D	
	Vesper Sparrow		D
	Savannah Sparrow		I
	Red-winged Blackbird	D	D
	Eastern Meadowlark	D	D
	Northern (Baltimore) Oriole		D
	American Goldfinch		D
	Evening Grosbeak	I	
Northern Spruce-Hardwoods	Great Blue Heron	I	
	Wood Duck	I	
	Ring-necked Duck		I
	Hooded Merganser		I
	Red-breasted Merganser	D	
	Turkey Vulture	I	I
	Osprey	I	I
	Broad-winged Hawk		D
	Sandhill Crane	I	I
	Killdeer	D	D
	Spotted Sandpiper		D
	Upland Sandpiper	I	I
	Common Snipe	D	D
	Ring-billed Gull	I	
	Common Tern	D	
	Mourning Dove	I	I
	Black-billed Cuckoo	D	D
	Whip-poor-will		D
	Chimney Swift	D	D
	Belted Kingfisher	D	D
	Olive-sided Flycatcher	D	D
	Eastern Wood-Pewee	D	D
	Least Flycatcher	D	D
	Eastern Phoebe	I	I
	Eastern Kingbird		D
	Horned Lark	D	D
	Purple Martin		D
	Tree Swallow		D
	Bank Swallow		D
	Cliff Swallow		D
	Barn Swallow	D	D
	Ruby-crowned Kinglet	D	
	Eastern Bluebird		I
	Veery	D	D
	Swainson's Thrush	D	D
	Hermit Thrush	I	I

	Species	Long term (1966–1994)	Short term (1980–1994)
	Wood Thrush	D	D
	Gray Catbird	D	D
	Cedar Waxwing	I	
	Solitary Vireo	I	I
	Yellow-throated Vireo		I
	Philadelphia Vireo		I
	Red-eyed Vireo	I	I
	Northern Parula		D
	Yellow Warbler	I	
	Magnolia Warbler	I	
	Yellow-rumped (Myrtle) Warbler	I	
	Blackburnian Warbler		I
	Pine Warbler	I	I
	Bay-breasted Warbler	D	D
	American Redstart		D
	Ovenbird		D
	Northern Waterthrush		D
	Mourning Warbler		D
	Common Yellowthroat	D	D
	Canada Warbler		D
	Scarlet Tanager		D
	Rose-breasted Grosbeak	D	D
	Rufous-sided Towhee	D	
	Chipping Sparrow		D
	Vesper Sparrow	D	
	Savannah Sparrow	D	D
	Lincoln's Sparrow	I	D
	Bobolink	D	D
	Red-winged Blackbird	D	D
	Eastern Meadowlark	D	D
	Brown-headed Cowbird	D	D
	Northern (Baltimore) Oriole	D	D
	American Goldfinch	D	
	Evening Grosbeak	D	D
Closed Boreal Forest	Broad-winged Hawk	I	
	Spotted Sandpiper	D	
	Herring Gull		D
	Belted Kingfisher	D	
	Yellow-bellied Flycatcher	I	
	Bank Swallow	D	D
	Ruby-crowned Kinglet		I
	Veery	I	
	Swainson's Thrush		D
	American Robin	I	
	Cedar Waxwing	I	

Species	Long term (1966–1994)	Short term (1980–1994)
Red-eyed Vireo		I
Tennessee Warbler		D
Yellow Warbler		D
Chestnut-sided Warbler	D	D
Bay-breasted Warbler		D
Black-and-white Warbler		D
Ovenbird		D
Mourning Warbler	D	D
Common Yellowthroat	D	
Rose-breasted Grosbeak		D
Chipping Sparrow	I	
Brown-headed Cowbird	D	

Aspen Parklands

Species	Long term (1966–1994)	Short term (1980–1994)
Pied-billed Grebe	D	
American Bittern	D	D
Northern Pintail	I	
Gadwall	I	
Canvasback		D
Redhead	I	
Northern Harrier	D	D
Red-tailed Hawk	I	I
American Coot		D
Killdeer	D	D
Lesser Yellowlegs	D	D
Mourning Dove	I	
Short-eared Owl	D	
Common Nighthawk	D	D
Belted Kingfisher	D	
Olive-sided Flycatcher		D
Eastern Wood-Pewee		I
Least Flycatcher	I	
Great Crested Flycatcher	I	
Horned Lark	D	D
Tree Swallow		I
Barn Swallow		D
House Wren	I	I
Sedge Wren		I
Marsh Wren	I	
Ruby-crowned Kinglet	I	
Hermit Thrush		D
American Robin	I	I
Gray Catbird		D
Warbling Vireo	I	
Tennessee Warbler		D
Yellow Warbler	I	
American Redstart	I	
Ovenbird		D
Northern Waterthrush	I	

Appendix B

	Species	Long term (1966–1994)	Short term (1980–1994)
	Mourning Warbler		D
	Clay-colored Sparrow	D	
	Vesper Sparrow		I
	Savannah Sparrow	I	I
	Grasshopper Sparrow		D
	Lincoln's Sparrow	I	
	Red-winged Blackbird		D
	Western Meadowlark	D	D
	Yellow-headed Blackbird	I	
	Northern (Baltimore) Oriole	I	
Till Plains	Great Blue Heron	I	
	Wood Duck		I
	Red-tailed Hawk	I	I
	American Kestrel	I	I
	Killdeer	I	I
	Mourning Dove		I
	Common Nighthawk	D	
	Yellow-billed Cuckoo	D	D
	Whip-poor-will	D	D
	Chimney Swift	D	D
	Belted Kingfisher	I	
	Acadian Flycatcher		D
	Eastern Phoebe	I	I
	Great Crested Flycatcher	D	
	Eastern Kingbird	D	D
	Purple Martin	D	
	Barn Swallow		D
	House Wren	I	I
	Eastern Bluebird	I	I
	Wood Thrush		D
	American Robin	I	I
	Cedar Waxwing	I	I
	Loggerhead Shrike	D	
	White-eyed Vireo		D
	Red-eyed Vireo	D	
	Northern Parula	I	I
	Yellow Warbler	I	I
	Yellow-throated Warbler		I
	American Redstart	D	
	Common Yellowthroat		D
	Yellow-breasted Chat	D	D
	Rose-breasted Grosbeak	I	
	Indigo Bunting		D
	Dickcissel	D	
	Chipping Sparrow	I	I
	Field Sparrow	D	D

	Species	Long term (1966–1994)	Short term (1980–1994)
	Vesper Sparrow	D	D
	Savannah Sparrow		D
	Grasshopper Sparrow	D	
	Bobolink	D	D
	Red-winged Blackbird	D	
	Eastern Meadowlark	D	I
	Brown-headed Cowbird		I
Dissected Till Plains	Great Blue Heron	I	I
	Green Heron		I
	Wood Duck	I	
	Mallard	I	I
	Red-tailed Hawk		I
	American Kestrel		I
	Killdeer	I	I
	Mourning Dove	D	
	Chimney Swift		D
	Great Crested Flycatcher		D
	Eastern Kingbird	D	D
	Horned Lark	D	D
	Purple Martin	D	D
	Tree Swallow	I	
	Cliff Swallow	I	I
	Barn Swallow		D
	House Wren	I	
	Eastern Bluebird		I
	American Robin	I	I
	Cedar Waxwing	I	I
	Loggerhead Shrike	D	
	Common Yellowthroat	D	D
	Indigo Bunting		D
	Dickcissel	D	
	Chipping Sparrow	I	I
	Field Sparrow	D	D
	Vesper Sparrow	I	
	Savannah Sparrow		D
	Bobolink	D	D
	Red-winged Blackbird	D	D
	Eastern Meadowlark	D	
	Orchard Oriole	D	
	American Goldfinch	D	D
Osage Plain–Cross Timbers	Great Egret	I	
	Little Blue Heron		D
	Cattle Egret	I	
	Green Heron	D	
	Red-tailed Hawk	I	I
	Killdeer		D

Appendix B

	Species	Long term (1966–1994)	Short term (1980–1994)
	Upland Sandpiper		D
	Eastern Phoebe		I
	Great Crested Flycatcher	D	
	Eastern Kingbird	D	D
	Purple Martin	I	
	Cliff Swallow	I	
	Bewick's Wren		I
	Eastern Bluebird		I
	American Robin	I	
	Loggerhead Shrike	D	D
	Bell's Vireo	D	
	Yellow-breasted Chat	D	
	Indigo Bunting		I
	Painted Bunting		I
	Field Sparrow	D	
	Lark Sparrow	D	
	Eastern Meadowlark	D	D
	Brown-headed Cowbird	D	
	Orchard Oriole	D	D
	Northern (Baltimore) Oriole		D
High Plains Border	Great Blue Heron		D
	Turkey Vulture	I	
	Northern Harrier	D	
	Swainson's Hawk	D	D
	Killdeer	I	
	Upland Sandpiper	I	
	Mourning Dove		D
	Yellow-billed Cuckoo		D
	Burrowing Owl		D
	Common Nighthawk	I	
	Western Kingbird		D
	Scissor-tailed Flycatcher	D	D
	Northern Rough-winged Swallow	I	
	Cliff Swallow		D
	Barn Swallow		D
	House Wren	I	I
	American Robin	I	I
	Warbling Vireo	I	
	Dickcissel		I
	Lark Bunting	D	
Rolling Red Prairies	Swainson's Hawk		D
	American Kestrel		D
	Killdeer		D
	Mourning Dove	D	D

610

Species	Long term (1966–1994)	Short term (1980–1994)
Yellow-billed Cuckoo		D
Eastern Phoebe	I	
Great Crested Flycatcher		D
Western Kingbird	D	D
Eastern Kingbird	D	D
Scissor-tailed Flycatcher	D	
Barn Swallow		D
Eastern Bluebird		I
American Robin	I	
Loggerhead Shrike	D	D
Bell's Vireo	D	
Field Sparrow	D	
Lark Sparrow	D	D
Eastern Meadowlark	D	
Brown-headed Cowbird	D	
Orchard Oriole	D	
Northern (Baltimore) Oriole	D	D
High Plains Great Blue Heron	I	
Northern Harrier	D	
Ferruginous Hawk		I
American Avocet		D
Long-billed Curlew	D	D
Mourning Dove	I	
Barn Swallow	I	
House Wren	I	I
American Robin	I	I
Loggerhead Shrike	I	
Dickcissel		I
Cassin's Sparrow		I
Brewer's Sparrow	D	
Grasshopper Sparrow		D
Western Meadowlark	D	.
Yellow-headed Blackbird		I
Orchard Oriole		I
Drift Prairie Pied-billed Grebe		I
Eared Grebe	I	
American Bittern	D	
Northern Pintail	D	
Gadwall		I
American Wigeon		I
Ruddy Duck		I
Northern Harrier	D	
Red-tailed Hawk		I
American Kestrel	I	
Sora		I

Species	Long term (1966–1994)	Short term (1980–1994)
American Coot		I
Killdeer		D
Willet	D	
Upland Sandpiper	I	
Marbled Godwit	D	D
Ring-billed Gull		I
Mourning Dove	I	
Common Nighthawk		I
Least Flycatcher	I	I
Western Kingbird	I	I
Eastern Kingbird	I	I
Tree Swallow		I
Northern Rough-winged Swallow		D
Cliff Swallow	I	I
Barn Swallow	I	
House Wren	I	I
Sedge Wren		I
Marsh Wren		I
American Robin	I	I
Sprague's Pipit	D	D
Loggerhead Shrike		I
Warbling Vireo	I	
Yellow Warbler		I
Dickcissel	I	I
Chipping Sparrow	I	I
Clay-colored Sparrow	D	
Vesper Sparrow	I	I
Red-winged Blackbird	D	
Orchard Oriole	I	
Northern (Baltimore) Oriole	I	
American Goldfinch	I	I
Glaciated Missouri Plateau Great Blue Heron		I
Mallard	I	
Northern Pintail	D	
Gadwall		I
Red-tailed Hawk	I	
Ferruginous Hawk	I	
Sora		I
Killdeer	D	D
Upland Sandpiper	I	I
Wilson's Phalarope		D
Ring-billed Gull	I	
Black Tern		D
Mourning Dove	I	I
Common Nighthawk	I	I

Species	Long term (1966–1994)	Short term (1980–1994)
Western Kingbird	I	I
Eastern Kingbird	I	I
Tree Swallow	I	
House Wren	I	I
American Robin	I	I
Loggerhead Shrike	I	I
Warbling Vireo	I	
Vesper Sparrow	I	I
Lark Bunting		I
Savannah Sparrow		I
Red-winged Blackbird	D	
Western Meadowlark		I
Northern (Baltimore) Oriole	D	

Great Plains Roughlands

Species	Long term (1966–1994)	Short term (1980–1994)
Green-winged Teal	I	
Gadwall	I	
American Wigeon	I	
Swainson's Hawk	I	I
Red-tailed Hawk	I	I
Ferruginous Hawk	I	I
American Kestrel	I	
Killdeer	D	D
Upland Sandpiper		D
Common Snipe		D
Mourning Dove	I	
Black-billed Cuckoo	D	D
Common Nighthawk		D
Belted Kingfisher	D	D
Least Flycatcher		I
Western Kingbird	I	I
Eastern Kingbird	I	I
Tree Swallow		I
Barn Swallow	I	
Rock Wren		D
House Wren	I	I
American Robin	I	
Black-headed Grosbeak	I	
Dickcissel	D	
Brewer's Sparrow	D	
Field Sparrow		I
Lark Bunting		I
Grasshopper Sparrow	D	
Bobolink	D	D
Red-winged Blackbird	D	D
Brown-headed Cowbird	I	I
Northern (Baltimore) Oriole		D

	Species	Long term (1966–1994)	Short term (1980–1994)
Black Prairie	Pied-billed Grebe		D
	Red-tailed Hawk	I	
	American Coot		D
	Ring-billed Gull	I	I
	Mourning Dove		D
	Black-billed Cuckoo	D	
	Belted Kingfisher		D
	Eastern Phoebe	I	
	Western Kingbird	I	I
	Eastern Kingbird	I	
	Purple Martin		D
	Cliff Swallow	I	
	Barn Swallow		D
	Marsh Wren	D	D
	Eastern Bluebird	I	I
	American Robin	I	
	Cedar Waxwing	I	
	Common Yellowthroat		D
	Indigo Bunting	D	D
	Grasshopper Sparrow	D	
	Bobolink		D
	Red-winged Blackbird		D
	Western Meadowlark	D	D
	Yellow-headed Blackbird		D
	Orchard Oriole	I	
Edward's Plateau	Killdeer		D
	Mourning Dove	D	D
	Yellow-billed Cuckoo		D
	Common Nighthawk		D
	Chimney Swift		D
	Eastern Phoebe		D
	Ash-throated Flycatcher	I	
	Barn Swallow	I	
	Eastern Bluebird	I	D
	Bell's Vireo	D	
	Painted Bunting	D	
	Dickcissel	D	
	Cassin's Sparrow	D	D
	Rufous-crowned Sparrow	D	D
	Lark Sparrow	D	D
	Grasshopper Sparrow	D	D
	Red-winged Blackbird		D
	Eastern Meadowlark		D
	Orchard Oriole	D	D
	Lesser Goldfinch	D	
Chihuahuan Desert	Swainson's Hawk		I
	Red-tailed Hawk	I	

	Species	Long term (1966–1994)	Short term (1980–1994)
	Mourning Dove	D	
	Ash-throated Flycatcher		I
	Horned Lark		D
	Cliff Swallow		I
	Loggerhead Shrike	D	
	Western Meadowlark	I	
	Scott's Oriole		I
Southern Rockies	Red-tailed Hawk	I	
	Western Wood-Pewee	D	
	Violet-green Swallow	D	D
	Cliff Swallow		I
	Orange-crowned Warbler		I
	Yellow Warbler		I
	MacGillivray's Warbler	D	
Fraser Plateau	Olive-sided Flycatcher	D	
	Hammond's Flycatcher	D	
	Swainson's Thrush	D	
	Wilson's Warbler	D	
Central Rockies	Pied-billed Grebe		D
	Green-winged Teal	I	I
	Red-tailed Hawk	I	
	Sora	I	
	Killdeer		D
	Common Snipe		D
	Mourning Dove	D	
	Rufous Hummingbird		I
	Barn Swallow		D
	House Wren		I
	Solitary Vireo	I	
	Warbling Vireo	I	I
	Red-eyed Vireo	D	D
	Orange-crowned Warbler		I
	Nashville Warbler	I	
	Yellow Warbler	D	
	Western Tanager		I
	Black-headed Grosbeak	I	I
	Lazuli Bunting		I
	Rufous-sided Towhee	I	I
	Clay-colored Sparrow		I
	Vesper Sparrow		I
	Lincoln's Sparrow	I	
	Bobolink		D
	Brown-headed Cowbird		D
	Northern (Bullock's) Oriole	I	
	Evening Grosbeak	I	I

	Species	Long term (1966–1994)	Short term (1980–1994)
Dissected Rockies	Mallard	I	
	Red-tailed Hawk	I	
	American Kestrel		D
	Sora		D
	Sandhill Crane	I	I
	Killdeer	D	D
	Spotted Sandpiper		D
	Long-billed Curlew		I
	Eastern Kingbird	I	I
	Horned Lark	D	
	Tree Swallow	I	
	Cliff Swallow	I	
	House Wren	I	I
	Swainson's Thrush	D	
	American Robin		D
	Western Tanager	I	I
	Brewer's Sparrow	D	D
	Red-winged Blackbird	D	
	Brewer's Blackbird	D	D
	Evening Grosbeak		D
Sierra Nevada	Band-tailed Pigeon	D	D
	Olive-sided Flycatcher	D	D
	Western Wood-Pewee	D	D
	Golden-crowned Kinglet	D	
	Townsend's Solitaire	D	D
	American Robin	D	
	Solitary Vireo	I	I
	Orange-crowned Warbler		D
	Chipping Sparrow	D	D
	Brewer's Blackbird		D
	Brown-headed Cowbird	D	D
Cascade Mountains	Rufous Hummingbird		D
	Olive-sided Flycatcher	D	
	Hammond's Flycatcher		I
	Tree Swallow		I
	Golden-crowned Kinglet		D
	American Robin	I	
	Solitary Vireo		I
	Warbling Vireo	I	I
	Hermit Warbler		I
	Black-headed Grosbeak		I
	Lazuli Bunting		I
Northern Rocky Mountains	Ruby-crowned Kinglet	I	
	American Robin	I	I
	Yellow-rumped (Myrtle) Warbler	I	I

Population Changes by Physiographic Region

	Species	Long term (1966–1994)	Short term (1980–1994)
Great Basin Deserts	Horned Lark	D	
Sonoran Desert	Ash-throated Flycatcher	I	
	Western Kingbird	I	
	Northern Rough-winged Swallow		D
	Cliff Swallow	I	I
Mojave Desert	Red-tailed Hawk		D
	Mourning Dove		D
	Say's Phoebe		D
Pinyon-Juniper Woodlands	American Kestrel		D
	Common Nighthawk		D
	Black-chinned Hummingbird		I
	Gray Flycatcher	I	
	Horned Lark	D	
	Bewick's Wren	I	
	Solitary Vireo		I
	Virginia's Warbler		I
	Yellow-rumped (Audubon's) Warbler		D
	Blue Grosbeak	I	
	Chipping Sparrow	D	
Pitt-Klamath Plateau	Mallard	I	I
	Cinnamon Teal		I
	Red-tailed Hawk	I	I
	Killdeer	D	D
	Common Snipe	D	
	Ring-billed Gull		I
	Cliff Swallow	I	I
	Barn Swallow	D	D
	Townsend's Solitaire	I	
	Hermit Thrush	I	
	Solitary Vireo	I	
	Lazuli Bunting		I
	Rufous-sided Towhee	I	I
	Chipping Sparrow	D	
	Brewer's Blackbird		I
Wyoming Basin	Red-tailed Hawk	I	I
	Wilson's Phalarope	D	D
	Vesper Sparrow		I
Intermountain Grasslands	Red-tailed Hawk	I	I
	Mourning Dove	D	D
	Western Wood-Pewee		I

Appendix B

	Species	Long term (1966–1994)	Short term (1980–1994)
	Western Kingbird		D
	Horned Lark		D
	Bewick's Wren	I	I
	Sage Thrasher		D
	Loggerhead Shrike	D	
	Brewer's Sparrow		D
	Cinnamon Teal		D
Basin and Range	Killdeer	D	D
	Loggerhead Shrike	D	D
	Vesper Sparrow		D
	Savannah Sparrow	D	
	Brewer's Blackbird	D	D
Columbia Plateau	Red-tailed Hawk	I	I
	Ferruginous Hawk	I	
	Killdeer	D	D
	Long-billed Curlew	I	
	Common Snipe	D	
	Mourning Dove	I	
	Western Wood-Pewee	I	I
	Gray Flycatcher		I
	Violet-green Swallow		I
	House Wren	I	
	Black-headed Grosbeak	I	I
	Lazuli Bunting	I	
	Chipping Sparrow	D	D
	Brewer's Sparrow	D	D
	Lark Sparrow		D
	Western Meadowlark		D
	Brewer's Blackbird		D
	Brown-headed Cowbird		I
Southern California Grasslands	Red-tailed Hawk	I	
	American Kestrel		D
	Red-winged Blackbird		D
Central Valley	Mallard		I
	Red-tailed Hawk	I	
	American Kestrel	D	
	Killdeer	D	
	Mourning Dove	D	D
	Western Kingbird	I	I
	Horned Lark	D	D
California Foothills	Mallard	I	I
	Turkey Vulture		I
	Cooper's Hawk		D
	Red-shouldered Hawk	I	

Species	Long term (1966–1994)	Short term (1980–1994)
American Kestrel	D	D
American Coot		D
Killdeer	D	D
Mourning Dove	D	
Black-chinned Hummingbird		D
Belted Kingfisher	D	D
Olive-sided Flycatcher	D	
Western Wood-Pewee	D	D
Northern Rough-winged Swallow	D	
Bewick's Wren	D	D
Western Bluebird	D	
American Robin		D
Warbling Vireo		D
Yellow Warbler	D	D
Chipping Sparrow	D	D
Black-chinned Sparrow	D	
Western Meadowlark	D	
Brewer's Blackbird	D	D

Southern Pacific Rainforests

Species	Long term (1966–1994)	Short term (1980–1994)
American Kestrel	D	
Killdeer	D	D
Band-tailed Pigeon	D	D
Mourning Dove	D	D
Common Nighthawk	D	
Rufous Hummingbird	D	
Allen's Hummingbird		D
Olive-sided Flycatcher	D	D
Western Wood-Pewee	D	
Western Kingbird		D
Barn Swallow	D	D
Bewick's Wren	D	
Golden-crowned Kinglet	D	
Western Bluebird	D	D
Warbling Vireo		D
Orange-crowned Warbler	D	D
MacGillivray's Warbler	D	
Common Yellowthroat	I	I
Yellow-rumped (Audubon's) Warbler		D
Wilson's Warbler		D
Western Tanager	D	
Lazuli Bunting	D	
Chipping Sparrow	D	D
Lark Sparrow	D	
White-crowned Sparrow	D	D
Western Meadowlark	D	D

	Species	Long term (1966–1994)	Short term (1980–1994)
	Brown-headed Cowbird	D	
	Northern (Bullock's) Oriole	D	
	Lesser Goldfinch	D	
	American Goldfinch	D	
	Evening Grosbeak		D
Northern Pacific Rainforests	Band-tailed Pigeon	D	
	Rufous Hummingbird	D	
	Olive-sided Flycatcher	D	
	Tree Swallow		D
	Barn Swallow	D	D
	Bewick's Wren		D
	Yellow Warbler	D	
	Rufous-sided Towhee		I
	Brown-headed Cowbird		D
	American Goldfinch	D	
Eastern Region—	Double-crested Cormorant	I	
U.S. and Canada, all states	Anhinga		D
and provinces east of the	Great Blue Heron	I	I
Mississippi River	Little Blue Heron	D	
	Black-crowned Night-Heron	D	
	Wood Duck	I	I
	Mallard	I	
	Ring-necked Duck		I
	Hooded Merganser		I
	Turkey Vulture	I	I
	Osprey	I	I
	Sharp-shinned Hawk	I	
	Cooper's Hawk	I	I
	Red-tailed Hawk	I	I
	American Coot	D	D
	Sandhill Crane	I	I
	Spotted Sandpiper		D
	Laughing Gull	I	
	Ring-billed Gull	I	
	Herring Gull	D	
	Common Tern	D	
	Forster's Tern	I	
	Least Tern		D
	Black Tern	D	
	Mourning Dove	I	I
	Black-billed Cuckoo		D
	Yellow-billed Cuckoo	D	D
	Common Nighthawk	D	D
	Chuck-will's-widow	D	

Species	Long term (1966–1994)	Short term (1980–1994)
Whip-poor-will	D	D
Chimney Swift	D	D
Ruby-throated Hummingbird	I	I
Belted Kingfisher	D	D
Olive-sided Flycatcher	D	D
Eastern Wood-Pewee	D	D
Yellow-bellied Flycatcher	I	
Acadian Flycatcher	I	
Least Flycatcher	D	D
Eastern Phoebe	I	I
Eastern Kingbird	D	D
Purple Martin		D
Tree Swallow		D
Bank Swallow		D
Barn Swallow	D	D
Bewick's Wren	D	
House Wren	I	I
Sedge Wren	I	
Marsh Wren		D
Ruby-crowned Kinglet	D	I
Blue-gray Gnatcatcher	I	I
Eastern Bluebird	I	I
Veery	D	D
Swainson's Thrush		D
Hermit Thrush	I	I
Wood Thrush	D	D
American Robin	I	I
Cedar Waxwing	I	
Loggerhead Shrike	D	
Solitary Vireo	I	I
Yellow-throated Vireo	I	I
Red-eyed Vireo	I	I
Golden-winged Warbler	D	
Yellow Warbler	I	
Magnolia Warbler	I	
Yellow-rumped (Myrtle) Warbler	I	
Pine Warbler	I	I
Prairie Warbler	D	
Bay-breasted Warbler		D
Blackpoll Warbler		D
Cerulean Warbler	D	
Prothonotary Warbler		D
Ovenbird	I	
Mourning Warbler		D
Common Yellowthroat	D	D
Hooded Warbler	I	

	Species	Long term (1966–1994)	Short term (1980–1994)
	Canada Warbler		D
	Yellow-breasted Chat	D	
	Rose-breasted Grosbeak		D
	Blue Grosbeak	I	
	Indigo Bunting	D	D
	Painted Bunting	D	
	Dickcissel	D	
	Rufous-sided Towhee	D	D
	Field Sparrow	D	D
	Vesper Sparrow	D	D
	Savannah Sparrow	D	D
	Grasshopper Sparrow	D	
	Bobolink	D	D
	Red-winged Blackbird	D	D
	Eastern Meadowlark	D	D
	Western Meadowlark	D	D
	Brown-headed Cowbird	D	D
	Orchard Oriole	D	
	Northern (Baltimore) Oriole	D	D
	American Goldfinch	D	
	Evening Grosbeak		D
Central Region—	Eared Grebe		D
U.S. and Canada, all states	Double-crested Cormorant	I	
and provinces between the	Great Blue Heron	I	
Rocky Mountains and the	Snowy Egret		I
Mississippi River	Cattle Egret		D
	Green Heron		D
	White Ibis	I	
	Green-winged Teal		I
	Mallard	I	
	Northern Pintail	D	
	Gadwall	I	I
	Northern Harrier	D	
	Red-tailed Hawk	I	I
	Ferruginous Hawk	I	I
	American Kestrel	I	
	Sora		I
	Killdeer		D
	Upland Sandpiper	I	
	Long-billed Curlew	D	D
	Franklin's Gull	D	
	Ring-billed Gull	I	I
	California Gull	I	
	Mourning Dove		D
	Black Tern	D	
	Yellow-billed Cuckoo	D	D
	Common Poorwill	I	

Species	Long term (1966–1994)	Short term (1980–1994)
Chimney Swift		D
Eastern Wood-Pewee	D	
Acadian Flycatcher	D	
Vermilion Flycatcher	D	
Western Kingbird	I	
Horned Lark		D
Barn Swallow	I	
Rock Wren	D	D
Bewick's Wren		I
House Wren	I	I
Marsh Wren		I
Eastern Bluebird	I	I
Veery	D	
American Robin	I	I
Cedar Waxwing	I	
Loggerhead Shrike	D	
White-eyed Vireo	D	
Bell's Vireo	D	
Pine Warbler		I
Prairie Warbler	D	
Prothonotary Warbler	D	
Ovenbird	D	
Kentucky Warbler	D	
Common Yellowthroat	D	D
Black-headed Grosbeak	I	I
Painted Bunting	D	
Dickcissel	D	I
Cassin's Sparrow	D	
Rufous-crowned Sparrow	D	
Chipping Sparrow		I
Brewer's Sparrow	D	
Field Sparrow	D	
Lark Sparrow	D	D
Black-throated Sparrow	D	
Grasshopper Sparrow	D	
Bobolink	D	D
Red-winged Blackbird	D	D
Eastern Meadowlark	D	D
Orchard Oriole	D	
Northern (Baltimore) Oriole		D
Lesser Goldfinch	D	
Western Region—		
Pied-billed Grebe	D	
Eared Grebe	I	
Double-crested Cormorant	I	
American Bittern	D	

**Western Region—
U.S. and Canada, Rocky Mountains to the Pacific, excluding Alaska**

623

Appendix B

Species	Long term (1966–1994)	Short term (1980–1994)
Great Egret	I	I
Snowy Egret	I	
Green Heron	I	
White-faced Ibis	I	I
Wood Duck	I	
Northern Pintail	D	D
Gadwall	I	I
Canvasback		D
Redhead	I	
Ring-necked Duck	I	
Turkey Vulture		I
Osprey	I	I
Sharp-shinned Hawk	I	
Red-shouldered Hawk	I	
Swainson's Hawk	I	
Red-tailed Hawk	I	I
Ferruginous Hawk	I	
American Kestrel		D
Prairie Falcon	D	
American Coot		D
Killdeer	D	D
Lesser Yellowlegs	D	D
Long-billed Curlew	I	
Ring-billed Gull		I
Band-tailed Pigeon	D	D
Mourning Dove	D	D
Burrowing Owl	I	I
Short-eared Owl	D	
Lesser Nighthawk	I	
Black Swift	D	
Broad-tailed Hummingbird		I
Allen's Hummingbird		D
Rufous Hummingbird	D	
Olive-sided Flycatcher	D	D
Western Wood-Pewee	D	
Least Flycatcher	I	
Gray Flycatcher	I	I
Ash-throated Flycatcher	I	
Horned Lark	D	D
Cliff Swallow	I	
Barn Swallow		D
Bewick's Wren		D
House Wren	I	I
Marsh Wren	I	
Golden-crowned Kinglet	D	
Sprague's Pipit	D	D

Species	Long term (1966–1994)	Short term (1980–1994)
Loggerhead Shrike	D	
Solitary Vireo	I	I
Warbling Vireo	I	I
Red-eyed Vireo		D
Tennessee Warbler		D
Virginia's Warbler	I	
Ovenbird		D
Common Yellowthroat	I	
Yellow-breasted Chat		I
Western Tanager		I
Rose-breasted Grosbeak	I	
Blue Grosbeak	I	I
Rufous-sided Towhee		I
Rufous-crowned Sparrow	I	
Chipping Sparrow	D	
Clay-colored Sparrow	D	
Brewer's Sparrow	D	D
Black-chinned Sparrow	D	
Vesper Sparrow		I
Savannah Sparrow	I	
Grasshopper Sparrow	D	D
Lincoln's Sparrow	I	
White-crowned Sparrow	D	
Bobolink		D
Western Meadowlark	D	D
Yellow-headed Blackbird	I	
Brewer's Blackbird	D	D
Northern (Baltimore) Oriole	I	
Lesser Goldfinch	D	
Continental—		
all U.S. and Canada,		
excluding Alaska Eared Grebe	I	
Double-crested Cormorant	I	
Anhinga		D
American Bittern	D	
Great Blue Heron	I	I
Great Egret		I
Snowy Egret		I
Little Blue Heron		D
Green Heron		D
Black-crowned Night-Heron	I	
Wood Duck	I	I
Mallard	I	I
Northern Pintail	D	
Blue-winged Teal		D
Gadwall	I	I
Canvasback	D	

Appendix B

Species	Long term (1966–1994)	Short term (1980–1994)
Redhead	I	I
Ringed-neck Duck	I	I
Hooded Merganser		I
Black Vulture	I	
Turkey Vulture		I
Osprey	I	I
Cooper's Hawk	I	
Broad-winged Hawk		D
Swainson's Hawk	I	I
Red-tailed Hawk	I	I
Ferruginous Hawk	I	I
Merlin	I	
Sora		I
American Coot		D
Sandhill Crane	I	I
Killdeer	D	D
Upland Sandpiper	I	
Long-billed Curlew		D
Common Snipe		D
Laughing Gull	I	
Franklin's Gull		I
Ring-billed Gull	I	I
Herring Gull	D	D
Forster's Tern	I	I
Band-tailed Pigeon	D	D
Mourning Dove		D
Black-billed Cuckoo		D
Yellow-billed Cuckoo	D	D
Lesser Nighthawk	I	I
Common Nighthawk		D
Chuck-will's Widow	D	
Chimney Swift	D	D
Ruby-throated Hummingbird		I
Broad-tailed Hummingbird		I
Rufous Hummingbird	D	
Belted Kingfisher	D	D
Olive-sided Flycatcher	D	D
Western Wood-Pewee	D	D
Eastern Wood-Pewee	D	D
Yellow-bellied Flycatcher	I	
Least Flycatcher	D	
Gray Flycatcher	I	I
Black Phoebe	I	D
Eastern Phoebe	I	I
Ash-throated Flycatcher	I	

Species	Long term (1966–1994)	Short term (1980–1994)
Western Kingbird	I	
Horned Lark	D	D
Purple Martin		D
Tree Swallow	I	
Cliff Swallow	I	
Barn Swallow		D
Rock Wren		D
House Wren	I	I
Sedge Wren		I
Marsh Wren	I	I
Ruby-crowned Kinglet		I
Blue-gray Gnatcatcher	I	I
Eastern Bluebird	I	I
Townsend's Solitaire		D
Veery	D	D
Hermit Thrush	I	I
Wood Thrush	D	D
American Robin	I	I
Gray Catbird	D	
Sprague's Pipit	D	
Cedar Waxwing	I	
Loggerhead Shrike	D	
Bell's Vireo	D	
Solitary Vireo	I	I
Yellow-throated Vireo	I	I
Warbling Vireo	I	I
Philadelphia Vireo		I
Red-eyed Vireo	I	I
Golden-winged Warbler	D	
Tennessee Warbler		D
Orange-crowned Warbler		I
Yellow Warbler	I	
Magnolia Warbler	I	
Yellow-rumped (Myrtle) Warbler	I	
Pine Warbler	I	I
Prairie Warbler	D	
Bay-breasted Warbler		D
Blackpoll Warbler		D
Cerulean Warbler	D	
Prothonotary Warbler	D	D
Ovenbird	I	
Kentucky Warbler		D
Mourning Warbler		D
Common Yellowthroat	D	D
Hooded Warbler		I
Wilson's Warbler		D

Appendix B

Species	Long term (1966–1994)	Short term (1980–1994)
Canada Warbler		D
Yellow-breasted Chat		I
Western Tanager		I
Rose-breasted Grosbeak		D
Black-headed Grosbeak		I
Blue Grosbeak	I	I
Indigo Bunting	D	D
Painted Bunting	D	
Dickcissel	D	
Rufous-sided Towhee	D	
Cassin's Sparrow	D	I
Clay-colored Sparrow	D	
Brewer's Sparrow	D	D
Field Sparrow	D	D
Vesper Sparrow	D	
Lark Sparrow	D	D
Black-throated Sparrow	D	
Savannah Sparrow	D	
Grasshopper Sparrow	D	D
Lincoln's Sparrow	I	
White-crowned Sparrow	D	
Bobolink	D	D
Red-winged Blackbird	D	D
Eastern Meadowlark	D	D
Western Meadowlark	D	
Yellow-headed Blackbird	I	
Brewer's Blackbird	D	D
Brown-headed Cowbird	D	
Orchard Oriole	D	D
Northern (Baltimore) Oriole		D
Lesser Goldfinch	D	
American Goldfinch	D	
Evening Grosbeak		D

Literature Cited

Albers, P. H. 1978. Habitat selection by breeding Red-winged Blackbirds. Wilson Bull. 90:619–634.

Aldrich, E. C. 1945. Nesting of the Allen Hummingbird. Condor 47:137–148.

Allen, A. A. 1933. The Bank Swallows' story. Bird Lore 35:116–125.

Allen, J. N. 1980. The ecology and behavior of the Long-billed Curlew in southeastern Washington. Wildl. Monogr. 73:1–67.

Allen, R. P. 1952. The Whooping Crane. Nat. Audubon Soc. Res. Rep. 3. National Audubon Society, New York. 246 pp.

Allen, R. W., and M. M. Nice. 1952. A study of the breeding biology of the Purple Martin (*Progne subis*). Am. Midl. Nat. 47:606–665.

Alverson, W. S., D. M. Waller, and S. L. Solheim. 1988. Forests too deer: edge effects in northern Wisconsin. Conserv. Biol. 2:348–358.

Ambuel, B., and S. A. Temple. 1982. Songbird populations in southern Wisconsin forests: 1954 and 1979. J. Field Ornithol. 53:149–158.

Ambuel, B., and S. A. Temple. 1983. Area-dependent changes in the bird communities and vegetation of southern Wisconsin forests. Ecology 64:1057–1068.

American Ornithologists' Union. 1983. Check-list of North American birds. 6th ed. American Ornithologists' Union, Washington, D.C. 877 pp.

American Ornithologists' Union. 1985. Thirty-fifth supplement to the American Ornithologists' Union, *Check-list of North American birds*. Auk 102:680–686.

American Ornithologists' Union. 1987. Thirty-sixth supplement to the American Ornithologists' Union, *Check-list of North American birds*. Auk 104:591–596.

American Ornithologists' Union. 1989. Thirty-seventh supplement to the American Ornithologists' Union, *Check-list of North American birds*. Auk 106:591–596.

American Ornithologists' Union. 1991. Thirty-eighth supplement to the American Ornithologists' Union, *Check-list of North American birds*. Auk 108:750–754.

Anderson, B. W., and S. A. Laymon. 1989. Creating habitat for the Yellow-billed Cuckoo (*Coccyzus americana*). Pages 468–472 *in* Proceedings of the California Riparian Systems Conference. Gen. Tech. Rep. PSW-110. U.S. Dept. Agric., Forest Serv. Pacific Southwest Forest and Range Exp. Stn., Berkeley, Calif.

Anderson, J. M. 1977. Yellow Rail (*Coturnicops noveboracensis*). Pages 66–70 *in* G. C. Sanderson, ed., Management of migratory shore and upland game birds in North America. International Association of Fish and Wildlife Agencies, Washington, D.C.

Literature Cited

Andrle, R. F., and J. H. Carroll. 1988. The atlas of breeding birds in New York State. Cornell University Press, Ithaca, N.Y.

Aney, W. C. 1984. The effects of patch size on bird communities of remnant old-growth pine stands in western Montana. M.S. thesis, University of Montana, Missoula. 98 pp.

Armstrong, W. H. 1958. Nesting and food habits of the Long-eared Owl in Michigan. Publ. Mus. Mich. State Univ. Biol. Ser. 1:61–96.

Arrenhius, O. 1921. Species and area. J. Ecology 9:95–99.

Askins, R. A., D. N. Ewert, and R. L. Norton. 1992. Abundance of wintering migrants in fragmented and continuous forests in the U.S. Virgin Islands. Pages 197–206 in J. M. Hagan III and D. W. Johnston, eds., Ecology and conservation of Neotropical migrant landbirds. Smithsonian Institution Press, Washington, D.C. 609 pp.

Askins, R. A., J. F. Lynch, and R. Greenberg. 1990. Population declines in migratory birds in eastern North America. Curr. Ornithol. 7:1–57.

Askins, R. A., and M. J. Philbrick. 1987. Effects of changes in regional forest abundance on the decline and recovery of a forest bird community. Wilson Bull. 99:7–21.

Askins, R. A., M. J. Philbrick, and D. S. Sugeno. 1987. Relationship between the regional abundance of forest and the composition of forest bird communities. Biol. Conserv. 39:129–152.

Austin, G. R. 1964. The world of the Red-tailed Hawk. J. B. Lippincott, Philadelphia and New York. 128 pp.

Bailey, A. M., and R. J. Niedrach. 1965. Birds of Colorado, vol. 2. Denver Museum of Natural History, Denver, Colo. 895 pp.

Baird, K. J., and J. P. Rieger. 1989. A restoration design for Least Bell's Vireo habitat in San Diego County. Pages 462–467 in Proceedings of the California Riparian Systems Conference. Gen. Tech. Rep. PSW-110. U.S. Dept. Agric., Forest Serv. Pacific Southwest Forest and Range Exp. Stn., Berkeley, Calif.

Baird, T. H. 1990. Changes in breeding bird populations between 1930 and 1985 in the Quaker Run Valley of Allegheny State Park, New York. N.Y. State Mus. Bull. 477.

Baker, B. W. 1944. Nesting of the American Redstart. Wilson Bull. 56:83–90.

Balda, R. P., B. C. McKnight, and C. D. Johnson. 1975. Flammulated Owl migration in the southwestern United States. Wilson Bull. 87:520–533.

Baldwin, D. H., G. V. Burger, and F. H. Kortright. 1964. Cousins by the dozens. Pages 15–22 in J. P. Linduska and A. L. Nelson, eds., Waterfowl tomorrow. U.S. Dept. of the Interior, Fish and Wildl. Serv., Washington, D.C.

Baldwin, P. H., and W. F. Hunter. 1963. Nesting and nest visitors of the Vaux's Swift in Montana. Auk 80:81–85.

Baldwin, P. H., and N. K. Zaczkowski. 1963. Breeding biology of the Vaux's Swift. Condor 65:400–406.

Balgooyen, T. G. 1976. Behavior and ecology of the American Kestrel (*Falco sparverius* L.) in the Sierra Nevada of California. Univ. Calif. Publ. Zool. 103:1–83.

Baltosser, W. H. 1989. Nectar availability and habitat selection by hummingbirds in Guadalupe Canyon. Wilson Bull. 191:559–578.

Barlow, J. C. 1962. Natural history of the Bell's Vireo, *Vireo bellii* Audubon. Univ. Kans. Mus. Nat. Hist. Publ. 12:241–296.

Bartgis, R. 1992. Loggerhead Shrike (*Lanius ludovicianus*). Pages 281–297 in K. J. Schneider and D. M. Pence, eds., Migratory nongame birds of management concern in the Northeast. U.S. Fish and Wildl. Serv., Region 5, Newton Corner, Mass.

Bateman, H. A., Jr. 1977. King Rail (*Rallus elegans*). Pages 93–104 *in* G. C. Sanderson, ed., Management of migratory shore and upland game birds in North America. International Association of Fish and Wildlife Agencies, Washington, D.C.

Baumann, S. A. 1959. The breeding cycle of the Rufous-sided Towhee (*Pipilo erythrophthalmus* L.) in central California. Wasmann J. Biol. 17:161–220.

Beal, F. E. L. 1900. Food of the Bobolink, blackbirds, and grackles. Biol. Surv. Bull. 13. U.S. Dept. of Agric., Washington, D.C. 77 pp.

Beal, F. E. L. 1904. Some common birds in their relation to agriculture. Farmers' Bull. 54. U.S. Dept. of Agric., Washington, D.C. 48 pp.

Beal, F. E. L. 1907. Birds of California in their relation to the fruit industry. Biol. Surv. Bull. 30, part 1. U.S. Dept. of Agric., Washington, D.C. 122 pp.

Beal, F. E. L. 1911. Food of the woodpeckers of the United States. Biol. Surv. Bull. 37. U.S. Dept. of Agric., Washington, D.C. 64 pp.

Beal, F. E. L. 1912. Food of our more important flycatchers. Biol. Surv. Bull. 44. U.S. Dept. of Agric., Washington, D.C. 67 pp.

Beal, F. E. L. 1915a. Food of the robins and bluebirds of the United States. Bull. 171. U.S. Dept. of Agric., Washington, D.C. 31 pp.

Beal, F. E. L. 1915b. Food habits of the thrushes of the United States. Bull. 280. U.S. Dept. of Agric., Washington, D.C. 23 pp.

Beal, F. E. L. 1918. Food habits of the swallows, a family of valuable native birds. Bull. 619. U.S. Dept. of Agric., Washington, D.C. 28 pp.

Beal, F. E. L., and W. L. McAtee. 1912. Food of some well-known birds of forest, farm, and garden. Farmers' Bull. 506. U.S. Dept. of Agric., Washington, D.C.

Beason, R. C., and E. C. Franks. 1974. Breeding behavior of the Horned Lark. Auk 91: 65–74.

Beaver, D. L., and P. H. Baldwin. 1975. Ecological overlap and the problem of competition and sympatry in the Western and Hammond's flycatchers. Condor 77:1–13.

Bechard, M. J., R. L. Knight, D. G. Smith, and R. E. Fitzner. 1990. Nest sites of sympatric hawks (*Buteo* spp.) in Washington. J. Field Ornithol. 61:159–170.

Becker, D. M., and C. H. Sieg. 1987. Home range and habitat utilization of breeding male Merlins, *Falco columbarius*, in southeastern Montana. Can. Field-Nat. 101:398–403.

Bednarz, J. C. 1988. Swainson's Hawk. Pages 87–96 *in* R. L. Glinski, B. G. Pendleton, M. B. Moss, M. N. LeFranc, Jr., B. A. Millsap, and S. W. Hoffman, eds., Proceedings of the Southwest Raptor Management Symposium and Workshop. Sci. and Tech. Ser. 11. Natl. Wildl. Fed., Washington, D.C. 395 pp.

Bednarz, J. C., and J. J. Dinsmore. 1982. Nest-sites and habitat of Red-shouldered and Red-tailed Hawks in Iowa. Wilson Bull. 94:31–45.

Bednarz, J. C., D. Klem, Jr., L. J. Goodrich, and S. E. Senner. 1990. Migration counts of raptors at Hawks Mountain, Pennsylvania, as indicators of population trends, 1934–1986. Auk 107:96–109.

Bellrose, F. C. 1976. Ducks, geese, and swans of North America. Stackpole Books, Harrisburg, Penn. 544 pp.

Bennett, L. J. 1938. The Blue-winged Teal, its ecology and management. Collegiate Press, Ames, Iowa. 144 pp.

Literature Cited

Bennett, S. E. 1980. Interspecific competition and the niche of the American Redstart (*Setophaga ruticilla*) in wintering and breeding communities. Pages 319–335 *in* A. Keast and E. S. Morton, eds., Migrant birds in the Neotropics. Smithsonian Institution Press, Washington, D.C.

Bent, A. C. 1921. Life Histories of North American gulls and terns. U.S. Natl. Mus. Bull. 113. 345 pp. (Reprinted 1963 by Dover Publishers, New York.)

Bent, A. C. 1927. Life histories of North American shore birds. Part 1. U.S. Natl. Mus. Bull. 142, Washington, D.C. 420 pp. (Reprinted 1962 by Dover Publishers, New York.)

Bent, A. C. 1929. Life histories of North American shore birds. Part 2. U.S. Natl. Mus. Bull. 146. 412 pp. (Reprinted 1962 by Dover Publishers, New York.)

Bent, A. C. 1932. Life histories of North American gallinaceous birds. U.S. Natl. Mus. Bull. 162. 490 pp. (Reprinted 1963 by Dover Publishers, New York.)

Bent, A. C. 1939. Life histories of North American woodpeckers. U.S. Natl. Mus. Bull. 174. 334 pp. (Reprinted 1964 by Dover Publishers, New York.)

Bent, A. C. 1940a. Life histories of North American cuckoos, goatsuckers, hummingbirds, and their allies, part 1. U.S. Natl. Mus. Bull. 176, pp. 1–244. (Reprinted 1964 by Dover Publishers, New York.)

Bent, A. C. 1940b. Life histories of North American cuckoos, goatsuckers, hummingbirds, and their allies, part 2. U.S. Natl. Mus. Bull. 176, pp. 245–506. (Reprinted 1964 by Dover Publishers, New York.)

Bent, A. C. 1942. Life histories of North American flycatchers, larks, swallows, and their allies. U.S. Natl. Mus. Bull. 179. 555 pp. (Reprinted 1963 by Dover Publishers, New York.)

Bent, A. C. 1948. Life histories of North American nuthatches, wrens, thrashers, and their allies. U.S. Natl. Mus. Bull. 195. 475 pp. (Reprinted 1964 by Dover Publishers, New York.)

Bent, A. C. 1949. Life histories of North American thrushes, kinglets, and their allies. U.S. Natl. Mus. Bull. 196. 454 pp. (Reprinted 1964 by Dover Publishers, New York.)

Bent, A. C. 1950. Life histories of North American wagtails, shrikes, vireos, and their allies. U.S. Natl. Mus. Bull. 197. 411 pp. (Reprinted 1965 by Dover Publishers, New York.)

Bent, A. C. 1953a. Life histories of North American wood warblers, part 1. U.S. Natl. Mus. Bull. 203, pp. 1–367. (Reprinted 1965 by Dover Publishers, New York.)

Bent, A. C. 1953b. Life histories of North American wood warblers, part 2. U.S. Natl. Mus. Bull. 203, pp. 367–734. (Reprinted 1965 by Dover Publishers, New York.)

Bent, A. C. 1958. Life histories of North American blackbirds, orioles, tanagers, and allies. U.S. Natl. Mus. Bull. 211. 549 pp. (Reprinted 1965 by Dover Publishers, New York.)

Bent, A. C. 1968a. Life histories of North American cardinals, grosbeaks, buntings, towhees, finches, sparrows, and allies, part 1. U.S. Natl. Mus. Bull. 237, pp. 1–602. (Reprinted 1968 by Dover Publishers, New York.)

Bent, A. C. 1968b. Life histories of North American cardinals, grosbeaks, buntings, towhees, finches, sparrows, and allies, part 2. U.S. Natl. Mus. Bull. 237, pp. 603–1,248. (Reprinted 1968 by Dover Publishers, New York.)

Bent, A. C. 1968c. Life histories of North American cardinals, grosbeaks, buntings, towhees, finches, sparrows, and allies, part 3. U.S. Natl. Mus. Bull. 237, pp. 1,249–1,889. (Reprinted 1968 by Dover Publishers, New York.)

Berger, A. J. 1951. Nesting density of Virginia and Sora Rails in Michigan. Condor 53: 202.

Bergman, R. D., P. Swain, and W. W. Weller. 1970. A comparative study of nesting Forster's and Black Terns. Wilson Bull. 82:435–444.

Bertin, R. I. 1977. Breeding habitats of the Wood Thrush and Veery. Auk 79:303–311.

Besser, J. F. 1985. Changes in breeding blackbird numbers in North Dakota from 1967 to 1981–82. Prairie Nat. 17:133–142.

Best, L. B. 1972. First year effects of sagebrush control on two sparrows. J. Wildl. Manage. 36:534–544.

Best, L. B. 1977. Nesting biology of the Field Sparrow. Auk 94:308–319.

Best, L. B. 1978. Field Sparrow reproductive success and nesting ecology. Auk 95:9–22.

Best, L. B., and N. L. Rodenhouse. 1984. Territory preference of Vesper Sparrows in cropland. Wilson Bull. 96:72–82.

Beyer, L. K. 1938. Nest life of the Bank Swallow. Wilson Bull. 50:122–137.

Birch, T. W., and E. H. Wharton. 1982. Land-use change in Ohio, 1952–1979. Resource Bull. NE-70. U.S. Dept. Agric., Forest Serv., Northeastern Forest Exp. Stn., Broomall, Penn.

Bird, D. M., ed. 1983. Biology and management of Bald Eagles and Ospreys. Harpell Press, Ste. Anne de Bellevue, Quebec.

Blair, C. L., and F. Schitoskey, Jr. 1982. Breeding biology and diet of the Ferruginous Hawk in South Dakota. Wilson Bull. 94:46–54.

Blake, E. R. 1949. The nest of the Colima Warbler in Texas. Wilson Bull. 61:65–67.

Blake, E. R. 1953. Birds of Mexico. University of Chicago Press, Chicago. 644 pp.

Blake, J. G., and B. A. Loiselle. 1992. Habitat use by Neotropical migrants at La Selva Biological Station and Braulio Carrillo National Park, Costa Rica. Pages 257–272 in J. M. Hagan III and D. W. Johnston, eds., Ecology and conservation of Neotropical migrant landbirds. Smithsonian Institution Press, Washington, D.C. 609 pp.

Blake, J. G., G. E. Niemi, and J. A. Hanowski. 1992. Drought and annual variation in bird populations: effects of migratory strategy and breeding habitat. Pages 419–430 in J. M. Hagan III and D. W. Johnston, eds., Ecology and conservation of Neotropical migrant landbirds. Smithsonian Institution Press, Washington, D.C. 609 pp.

Blake, J. G., F. G. Stiles, and B. A. Loiselle. 1990. Birds of LaSelva Biological Station: habitat use, trophic composition, and migrants. Pages 161–182 in A. H. Gentry, ed., Four Neotropical rainforests. Yale University Press, New Haven, Conn.

Blancher, P. J., and R. J. Robertson. 1987. Effect of food supply on the breeding biology of Western Kingbirds. Ecology 68:723–732.

Bolen, E. G. 1967a. The ecology of the Black-bellied Tree Duck in southern Texas. Utah State University, Logan. 133 pp.

Bolen, E. G. 1967b. Nesting boxes of Black-bellied Tree Ducks. J. Wildl. Manage. 31: 794–797.

Bolen, E. G., and B. J. Forsyth. 1967. Foods of the Black-bellied Tree Duck in south Texas. Wilson Bull. 79:43–49.

Bollinger, E. K., and T. A. Gavin. 1989. The effects of site quality on breeding-site fidelity in Bobolinks. Auk 106:584–594.

Bollinger, E. K., and T. A. Gavin. 1992. Eastern Bobolink populations: ecology and conservation in an agricultural landscape. Pages 497–506 in J. M. Hagan III and D.

W. Johnston, eds., Ecology and conservation of Neotropical migrant landbirds. Smithsonian Institution Press, Washington, D.C. 609 pp.

Bond, J. 1947. Field guide to the birds of the West Indies. MacMillan, New York. 257 pp.

Bosakowski, T. 1986. Short-eared Owl winter roosting strategies. Am. Birds 40:237–240.

Brackbill, H. 1943. A nesting study of the Wood Thrush. Wilson Bull. 55:73–87.

Braun, C. E., M. F. Baker, R. L. Eng, J. S. Gashwiler, and M. H. Schroeder. 1976. Wilson Ornithological Society conservation committee report on effects of alteration of sagebrush communities on the associated avifauna. Wilson Bull. 88:165–171.

Breckenridge, W. J. 1956. Measurements of the habitat niche of the Least Flycatcher. Wilson Bull. 68:47–51.

Brewster, W. 1885. The Oystercatcher (Haematopus pallidus) in Massachusetts. Auk 2:384.

Brewster, W. 1891. Notes on Bachman's Warbler (Helminthophila bachmanii). Auk 8:149–157.

Brewster, W. 1938. The birds of the Lake Umbagog region of Maine, part 4. Compiled by Ludlow Griscom. Bull. Mus. Comp. Zool. 66:525–620.

Briggs, S. A., and J. H. Criswell. 1979. Gradual silencing of spring in Washington. Atl. Nat. 32:19–26.

Brigham, R. M., and M. R. Barclay. 1992. Lunar influence on foraging and nesting activity of Common Poorwills (Phalaenoptilus nuttallii). Auk 109:315–320.

Briskie, J. V., and S. G. Sealy. 1989. Determination of clutch size in the Least Flycatcher. Auk 106:269–278.

Brittingham, M. C., and S. A. Temple. 1983. Have cowbirds caused forest songbirds to decline? BioScience 33:31–35.

Brooks, B. L., and S. A. Temple. 1990. Dynamics of a Loggerhead Shrike population in Minnesota. Wilson Bull. 102:441–450.

Brooks, R. P., and W. J. Davis. 1987. Habitat selection by breeding Belted Kingfishers (Ceryle alcyon). Am. Midl. Nat. 117:63–70.

Brooks, R. T., and T. W. Birch. 1988. Changes in New England forests and forest owners: implications for wildlife habitat resources and management. Trans. N. Am. Wildl. Nat. Res. Conf. 53:78–87.

Brown, D. E. 1977. White-winged Dove (Zenaida asiatica). Pages 247–272 in G. C. Sanderson, ed., Management of migratory shore and upland game birds in North America. International Association of Fish and Wildlife Agencies, Washington, D.C.

Brown, L., and D. Amadon. 1968. Eagles, hawks, and falcons of the world, vol. 1. McGraw-Hill, New York. 414 pp.

Buech, R. R. 1982. Nesting ecology and cowbird parasitism of Clay-colored, Chipping, and Field Sparrows in a Christmas tree plantation. J. Field Ornithol. 53:363–369.

Buechner, H. K., and J. H. Buechner. 1970. A symposium of the Smithsonian Institution on the Avifauna of northern Latin America. Smithson. Contrib. Zool. 26:1–112.

Bull, E. L., and H. D. Cooper. 1991. Vaux's Swift nests in hollow trees. West. Birds 22:85–91.

Bull, E. L., A. L. Wright, and M. G. Henjum. 1989. Nesting and diet of Long-eared Owls in conifer forests, Oregon. Condor 91:908–912.

Bull, J., and J. Farrand. 1977. The Audubon Society field guide to North American birds (eastern region). Knopf, New York. 775 pp.

Burger, J., and F. Lesser. 1978. Determinants of colony site Selection in Common Terns (*Sterna hirundo*). Colon. Waterbirds 1:118–127.

Burger, J., and L. M. Miller. 1977. Colony and nest site selection in White-faced and Glossy Ibises. Auk 94:664–676.

Burger, J., and J. Shisler. 1978. Nest site selection and competitive interactions of Herring and Laughing Gulls in New Jersey. Auk 95:252–266.

Burleigh, T. D. 1972. Birds of Idaho. Caxton Printers, Caldwell, Idaho. 467 pp.

Busby, D. G., and S. G. Sealy. 1979. Feeding ecology of a population of nesting Yellow Warblers. Can. J. Zool. 57:1670–1681.

Butcher, G. S., W. A. Niering, W. J. Barry, and R. H. Goodwin. 1981. Equilibrium biogeography and the size of nature preserves: an avian case study. Oecologia 49: 29–37.

Butts, K. O. 1973. Life history and habitat requirements of Burrowing Owls in western Oklahoma. M.S. thesis, Oklahoma State University, Stillwater. 188 pp.

Cade, T. J. 1960. Ecology of the peregrine and gyrfalcon populations in Alaska. Univ. Calif. Publ. Zool. 63:151–290.

Cade, T. J. 1982. The falcons of the world. Cornell University Press, Ithaca, N.Y.

Cade, T. J., J. H. Enderson, C. G. Thelander, and C. M. White, eds. 1988. Peregrine Falcon populations: their management and recovery. The Peregrine Fund, Boise, Idaho.

Cain, S. A., and G. M. Castro. 1959. Manual of vegetation analysis. Harper & Row, New York.

Cairns, W. E. 1982. Biology and behavior of breeding Piping Plovers. Wilson Bull. 94: 531–545.

Calder, W. A. 1971. Temperature relationships and nesting of the Calliope Hummingbird. Condor 73:314–321.

Calder, W. A., and E. G. Jones. 1989. Implications of recapture data for migration of the Rufous Hummingbird (*Selasphorus rufus*) in the Rocky Mountains. Auk 106:488–489.

Calder, W. A., N. M. Waser, S. M. Hiebert, D. W. Inouye, and S. Miller. 1983. Site-fidelity, longevity, and population dynamics of Broad-tailed Hummingbirds: a ten-year study. Oecologia 56:359–364.

Carson, R. 1962. Silent spring. Houghton Mifflin, New York.

Carter, J. W. 1992. Upland Sandpiper (*Bartramia longicauda*). Pages 235–251 *in* K. J. Schneider and D. M. Pence, eds., Migratory nongame birds of management concern in the Northeast. U.S. Fish and Wildl. Serv., Region 5, Newton Corner, Mass.

Case, N. A., and O. H. Hewitt. 1963. Nesting and productivity of the Red-winged Blackbird in relation to habitat. Living Bird 2:7–20.

Chapin, E. A. 1925. Food habits of the vireos. Bull. 1355. U.S. Dept. of Agric., Washington, D.C. 43 pp.

Chapman, L. B. 1955. Studies of a Tree Swallow colony. Bird Banding 26:45–70.

Clapp, R. B., D. Morgan-Jacobs, R. C. Banks. 1983. Marine birds of the southeastern United States and Gulf of Mexico. Part 3: Charadriiformes. FWS/OBS-83/30. U.S. Dept. of the Interior, Fish and Wildl. Serv., Div. Biol. Serv., Washington, D.C. 853 pp.

Clark, R. J. 1975. A field study of the Short-eared Owl, *Asio flammeus* (Pontoppidan), in North America. Wildl. Monogr. 47:1–67.

Literature Cited

Coleman, J. S., and J. D. Fraser. 1989. Habitat use and home ranges of Black and Turkey Vultures. J. Wildl. Manage. 53:782–792.

Combellack, C. R. B. 1954. A nesting of Violet-green Swallows. Auk 71:435–442.

Confer, J. L. 1992. Golden-winged Warbler (*Vermivora chrysoptera*). Pages 369–383 *in* K. J. Schneider and D. M. Pence, eds., Migratory nongame birds of management concern in the Northeast. U.S. Fish and Wildl. Serv., Region 5, Newton Corner, Mass.

Confer, J. L., and K. Knapp. 1981. Golden-winged Warblers and Blue-winged Warblers: the relative success of a habitat specialist and a generalist. Auk 98:108–114.

Cooch, F. G. 1964. Snows and blues. *In* J. P. Linduska and A. L. Nelson, eds., Waterfowl tomorrow. U.S. Dept. of the Interior, Fish and Wildl. Serv., Washington, D.C. 770 pp.

Cornwell, G. W. 1963. Observations on the breeding biology and behavior of a nesting population of Belted Kingfishers. Condor 65:426–431.

Cottam, C. 1939. Food habits of North American diving ducks. Tech. Bull. 643. U.S. Dept. of Agric., Washington, D.C. 140 pp.

Cottam, C., and W. C. Glazner. 1959. Late nesting of water birds in south Texas. Trans. N. Am. Wildl. Conf. 24:382–394.

Cottam, C., and H. C. Hanson. 1938. Food habits of some arctic birds and mammals. Field Mus. Nat. Hist., Zool. Ser. 20:405–426.

Cottam, C., and P. Knappen. 1939. Food of some uncommon North American birds. Auk 56:138–169.

Cottam, C., and J. B. Trefethen, eds. 1968. Whitewings. D. Van Nostrand, Princeton, N.J. 348 pp.

Coutlee, E. L. 1968. Comparative breeding behavior of Lesser and Lawrence's Goldfinches. Condor 70:228–242.

Cox, G. W. 1960. A life history of the Mourning Warbler. Wilson Bull. 72:5–28.

Craig, E. H., T. H. Craig, and L. R. Powers. 1988. Activity patterns and home-range use of nesting Long-eared Owls. Wilson Bull. 100:204–213.

Craig, G. 1986. Peregrine Falcon. Pages 807–824 *in* R. L. Di Silvestro, ed., Audubon wildlife report 1986. National Audubon Society, New York.

Cramp, S., and K. E. L. Simmons. 1980. Handbook of the birds of Europe, the Middle East, and North Africa. Vol. 2. Hawks to bustards. Oxford University Press, Oxford. 695 pp.

Cramp, S., and K. E. L. Simmons. 1983. Handbook of the birds of Europe, the Middle East, and North Africa. Vol. 3. Waders to gulls. Oxford University Press, Oxford. 913 pp.

Crawford, R. D. 1977. Polygynous breeding of Short-billed Marsh Wrens. Auk 94:359–362.

Cripps, B. J., Jr. 1966. The nesting cycle of the Chestnut-sided Warbler. Raven 37:43–48.

Criswell, J. H. 1975. Breeding bird population studies, 1975. Atl. Nat. 30:175–176.

Crockett, A. B., and H. H. Hadow. 1975. Nest site selection by Williamson and Red-naped Sapsuckers. Condor 77:365–368.

Curtis, J. T. 1956. The vegetation of Wisconsin. University of Wisconsin Press, Madison.

Darley, J. A., D. M. Scott, and N. K. Taylor. 1977. Effects of age, sex and breeding success on site fidelity of Gray Catbirds. Bird Banding 48:145–151.

Darlington, P. J., Jr. 1957. Zoogeography: the geographic distribution of animals. Wiley, New York.

Darveau, M., J. L. DesGranges, and G. Gauthier. 1992. Habitat use by three breeding insectivorous birds in declining maple forests. Condor 94:72–82.

Davis, D. E. 1954. The breeding biology of Hammond's Flycatcher. Auk 71:164–171.

Davis, J. 1960. Nesting behavior of the Rufous-sided Towhee in coastal California. Condor 62:434–456.

Davis, J., G. F. Fisler, and B. S. Davis. 1963. The breeding biology of the Western Flycatcher. Condor 65:337–382.

Davison, V. E., and E. G. Sullivan. 1963. Mourning Doves' selection of foods. J. Wildl. Manage. 27:373–383.

de Calesta, D. S. 1994. Effect of white-tailed deer on songbirds within managed forests in Pennsylvania. J. Wildl. Manage. 58:711–718.

DeGaris, C. F. 1936. Notes on six nests of the Kentucky Warbler (Oporornis formosus). Auk 53:418–428.

DeGraaf, R. M., tech. coord. 1978a. Management of southern forests for nongame birds. Gen. Tech. Rep. SE-14. U.S. Dept. of Agric., Forest Serv., Southeastern Forest Exp. Stn., Asheville, N.C.

DeGraaf, R. M., tech. coord. 1978b. Nongame bird habitat management in the coniferous forests of the western United States. Gen. Tech. Rep. PNW-64. U.S. Dept. of Agric., Forest Serv., Pacific Northwest Forest and Range Exp. Stn., Portland, Oreg.

DeGraaf, R. M., tech. coord. 1979. Management of north-central and northeastern forests for nongame birds. Gen. Tech. Rep. NC-51. U.S. Dept. of Agric., Forest Serv., North Central Forest Exp. Stn., St. Paul, Minn.

DeGraaf, R. M., tech. coord. 1980. Management of western forests and grasslands for nongame birds. Gen. Tech. Rep. INT-81. U.S. Dept. of Agric., Forest Serv., Intermountain Forest and Range Exp. Stn., Ogden, Utah.

DeGraaf, R. M., W. M. Healy, and R. T. Brooks. 1991. Effects of thinning and deer browsing on breeding birds in New England oak woodlands. For. Ecol. Manage. 41:179–191.

DeGraaf, R. M., and D. D. Rudis. 1986. New England wildlife: habitat, natural history, and distribution. Gen. Tech. Rep. NE-108. U.S. Dept. Agric., Forest Serv., Northeastern Forest Exp. Stn., Broomall, Penn. 491 pp.

DeGraaf, R. M., V. E. Scott, R. H. Hamre, L. Ernst, and S. H. Anderson. 1991. Forest and rangeland birds of the United States: natural history and habitat use. Agric. Handb. 688. U.S. Dept. of Agric. Forest Serv., Washington, D.C.

Deis, R. 1981. Again silent spring. Defenders 56:6–10.

DeKirline, L. 1948. Least Flycatcher. Audubon 50:149–153.

DellaSala, D. A., and D. L. Rabe. 1987. Response of Least Flycatchers, Empidonax minimus, to forest disturbances. Biol. Conserv. 41:291–299.

Dennis, J. V. 1948. Observations on the Orchard Oriole in the lower Mississippi Delta. Bird Banding 19:12–20.

Dennis, J. V. 1958. Some aspects of the breeding ecology of the Yellow-breasted Chat (Icteria virens). Bird Banding 29:169–183.

Devitt, O. E. 1939. The Yellow Rail breeding in Ontario. Auk 56:238–243.

Dirzo, R., and M. C. Garcia. 1992. Rates of deforestation in Los Tuxtlas, a Neotropical area in southeast Mexico. Conserv. Biol. 6:84–90.

Dobkin, D. S. 1992. Neotropical migrant landbirds in the northern Rockies and Great Plains. Publ. R1-93-34. U.S. Dept. Agric., Forest Serv. Northern Region, Missoula, Mont.

Literature Cited

Dobkin, D. S., and B. A. Wilcox. 1986. Analysis of natural forest fragments: riparian birds in the Toiyabe Mountains, Nevada. Pages 293–299 *in* J. Verner, M. L. Morrison, and C. J. Ralph, eds., Wildlife 2000: modeling habitat relationships of terrestrial vertebrates. University of Wisconsin Press, Madison.

Droege, S. 1991. Unpubl. summary of Breeding Bird Survey data, 1966–1989, in letter to Neotropical migrant workshop participants. U.S. Fish and Wild. Serv., Office of Migratory Bird Management, Laurel, Md.

Drury, W. H., Jr. 1961. The breeding biology of shorebirds on Bylot Island, Northwest Territories, Canada. Auk 78:176–219.

Dunkle, S. W. 1977. Swainson's Hawks on the Laramie Plains, Wyoming. Auk 94:65–71.

Dzubin, A., H. W. Miller, and G. V. Schildman. 1964. White-fronts. Pages 135–143 *in* J. P. Linduska and A. L. Nelson, eds., Waterfowl tomorrow. U.S. Dept. of the Interior, Fish and Wildl. Serv., Washington, D.C.

Eaton, S. W. 1958. A life history of the Louisiana Waterthrush. Wilson Bull. 70:211–236.

Edwards, E. P. 1972. Field guide to the birds of Mexico. Ernest P. Edwards, Sweet Briar, Va. 300 pp.

Emlen, J. T. 1973. Territorial aggression in wintering warblers at Bahama agave blossoms. Wilson Bull. 85:71–74.

Emlen, S. T., J. D. Rising, and W. L. Thompson. 1975. A behavioral and morphological study of sympatry in the Indigo and Lazuli Buntings of the Great Plains. Wilson Bull. 87:145–177.

Enderson, J. H. 1964. A study of the Prairie Falcon in the central Rocky Mountain Region. Auk 81:332–352.

Errington, P. L., and L. J. Bennett. 1935. Food habits of Burrowing Owls in northwestern Iowa. Wilson Bull. 47:125–128.

Erskine, A. J. 1977. Birds in boreal Canada. Can. Wildl. Serv. Rep. Ser. 41:1–73.

Erskine, A. J. 1984. A preliminary catalogue of bird census plot studies in Canada, part 5. Can. Wildl. Serv. Prog. Notes 144:1–34.

Erskine, A. J., B. T. Collins, E. Hayakawa, and C. Downes. 1992. The cooperative Breeding Bird Survey in Canada, 1989–1991. Can. Wildl. Serv. Prog. Notes 199:1–14.

Evans, D. L. 1982. Status reports on twelve raptors. Spec. Sci. Rep. Wildl. 238. U.S. Dept. of the Interior, Fish and Wildl. Serv., Washington, D.C. 68 pp.

Evans, D. L., and E. K. Bartels. 1981. Impacts of surface mining on Canvasbacks. U.S. Fish and Wildl. Serv., Northern Prairie Wildlife Research Center. Jamestown, N.D. 34 pp.

Evans, K. E., and R. A. Kirkman. 1981. Guide to the bird habitats of the Ozark Plateau. Gen. Tech. Rep. NC-68. U.S. Dept. Agric., Forest Serv., North Central Forest Exp. Stn., St. Paul, Minn.

Eyer, L. E. 1963. Observations on Golden-winged Warblers at Itasca State Park, Minnesota. Jack-Pine Warbler 41:96–109.

Eyre, F. H., ed. 1980. Forest cover types of the United States and Canada. Society of American Foresters, Washington, D.C.

Faaborg, J. 1976. Habitat selection and territorial behavior of the small grebes of North Dakota. Wilson Bull. 88:390–399.

Faaborg, J., and W. J. Arendt. 1992. Long-term declines of winter resident warblers in a Puerto Rican dry forest: which species are in trouble? Pages 57–63 *in* J. M. Hagan

III and D. W. Johnston, eds., Ecology and conservation of Neotropical migrant landbirds. Smithsonian Institution Press, Washington, D.C. 609 pp.

Farrand, J., Jr., ed. 1983a. The Audubon Society master guide to birding: 1. Loons to sandpipers. Knopf, New York. 447 pp.

Farrand, J., Jr., ed. 1983b. The Audubon Society master guide to birding: 2. Gulls to dippers. Knopf, New York. 398 pp.

Farrand, J., Jr., ed. 1983c. The Audubon Society master guide to birding: 3. Old World warblers to sparrows. Knopf, New York. 399 pp.

Ficken, M. S., and R. W. Ficken. 1968. Territorial relationships of Blue-winged Warblers, Golden-winged Warblers, and their hybrids. Wilson Bull. 80:442–451.

Finch, D. M. 1990. Effects of predation and competitor interference on nesting success of House Wrens and Tree Swallows. Condor 92:674–687.

Finch, D. M. 1991. Population ecology, habitat requirements, and conservation of Neotropical migratory birds. Gen. Tech. Rep. RM-205. U.S. Dept. Agric., Forest Serv., Rocky Mountain Forest and Range Exp. Stn., Fort Collins, Colo.

Finch, D. M., S. H. Anderson, and W. A. Hubert. 1987. Habitat suitability index models: Lark Bunting. FWS/OBS-82/10.137. U.S. Dept. of the Interior, Fish and Wildl. Serv., Washington, D.C.

Finch, D. M., and T. Martin. 1991. Research working group of the Neotropical migratory bird program: workplans and reports, 18 October 1991. U.S. Dept. Agric., Forest Serv., Rocky Mountain Forest and Range Exp. Stn., Laramie, Wyo.

Finch, D. M., and R. T. Reynolds. 1988. Bird response to understory variation and conifer succession in aspen forests. Pages 87–96 in J. Emerick, S. Q. Foster, L. Hayden-Wing, J. Hodgson, J. W. Monarch, A. Smith, O. Thorne II, and J. Todd, eds., Issues and technology in the management of impacted western wildlife. Symp. Proc. of the Thorne Ecological Institute, Boulder, Colo.

Fischer, R. B. 1958. The breeding biology of the Chimney Swift, Chaetura pelagica (Linnaeus). N.Y. State Mus. Sci. Serv. Bull. 368:1–141.

Fitch, F. W., Jr. 1950. Life history and ecology of the Scissor-tailed Flycatcher, Muscivora forficata. Auk 67:145–168.

Fitch, H. S. 1974. Observations of the food and nesting of the Broad-winged Hawk (Buteo platypterus) in northeastern Kansas. Condor 76:331–360.

Fitch, H. S., and V. R. Fitch. 1955. Observations on the Summer Tanager in northeastern Kansas. Wilson Bull. 67:45–54.

Fitch, H. S., F. Swenson, and D. F. Tillotson. 1946. Behavior and food habits of the Red-tailed Hawk. Condor 48:205–237.

Fjeldsa, J. 1977. The coot and the moorhen. Biological Monogr., Av-media, Copenhagen. 56 pp.

Fogarty, M. J., and K. A. Arnold. 1977. Common Snipe (Capella gallinago delicata). Pages 189–209 in G. C. Sanderson, ed., Management of migratory shore and upland game birds in North America. International Association of Fish and Wildlife Agencies, Washington, D.C.

Fogden, M. P. L. 1972. The seasonality and population dynamics of equatorial forest birds in Sarawak. Ibis 114:307–343.

Forbush, E. W., and J. B. May. 1955. A natural history of American birds of eastern and central North America. Revised ed. Houghton Mifflin, Boston. 554 pp.

Fox, G. A. 1961. A contribution to the life history of the Clay-colored Sparrow. Auk 78:220–224.

Literature Cited

Fox, G. A. 1964. Notes on the western race of the Pigeon Hawk. Blue Jay 22:140–147.

Frakes, R. A., and R. E. Johnson. 1982. Niche convergence in *Empidonax* flycatchers. Condor 84:286–291.

Fredrickson, L. H. 1970. Breeding biology of American Coots in Iowa. Wilson Bull. 82:445–457.

Fredrickson, L. H. 1971. Common gallinule breeding biology and development. Auk 88:914–919.

Freedman, B., C. Beauchamp, I. A. McLaren, and S. I. Tingley. 1981. Forestry management practices and populations of breeding birds in a hardwood forest in Nova Scotia. Can. Field-Nat. 95:307–311.

Freemark, K., and B. Collins. 1992. Landscape ecology of birds breeding in temperate forest fragments. Pages 443–454 *in* J. M. Hagan III and D. W. Johnston, eds., Ecology and conservation of Neotropical migrant landbirds. Smithsonian Institution Press, Washington, D.C. 609 pp.

Freer, V. M. 1979. Factors affecting site tenacity in New York Bank Swallows. Bird Banding 50:349–357.

Fretwell, S. D. 1986. Distribution and abundance of the Dickcissel. Curr. Ornithol. 4: 211–242.

Gaddis, P. K. 1983. Composition and behavior of mixed-species flocks of forest birds in north-central Florida. Fla. Field Nat. 11:25–34.

Galli, A. E., C. F. Leck, and R. T. T. Forman. 1976. Avian distribution patterns in forest islands of different sizes in central New Jersey. Auk 93:356–364.

Gambona, G. J. 1977. Predation on Rufous Hummingbird by Wied's Flycatcher. Auk 94:157–158.

Gass, C. L. 1979. Territory regulation, tenure, and migration in Rufous Hummingbirds. Can. J. Zool. 57:914–923.

Gauthreaux, S. A., Jr. 1992. The use of weather radar to monitor long-term patterns of trans-Gulf migration in spring. Pages 96–100 *in* J. M. Hagan III and D. W. Johnston, eds., Ecology and conservation of Neotropical migrant landbirds. Smithsonian Institution Press, Washington, D.C. 609 pp.

Gawlik, D. E., and K. L. Bildstein. 1990. Reproductive success and nesting habitat of Loggerhead Shrikes in north-central South Carolina. Wilson Bull. 102:37–48.

Gibb, J. 1956. Food, feeding habits, and territory of the Rock Pipit, *Anthus spinoletta*. Ibis 98:506–530.

Gibbs, J. P., and J. Faaborg. 1990. Estimating the viability of Ovenbird and Kentucky Warbler populations in forest fragments. Conserv. Biol. 4:193–196.

Gibson, F. 1971. The breeding biology of the American Avocet (*Recurvirostra americana*) in central Oregon. Condor 73:444–454.

Gill, F. B. 1980. Historical aspects of hybridization between Blue-winged and Golden-winged Warblers. Auk 97:1–18.

Gilmer, D. S., P. M. Konrad, and R. E. Stewart. 1983. Nesting ecology of Red-tailed Hawks and Great Horned Owls in central North Dakota and their interactions with other large raptors. Prairie Nat. 15:133–143.

Gleason, H. A. 1922. On the relationship between species and area. Ecology 3:158–162.

Glinski, R. L. 1988. Gray Hawk. Pages 83–86 *in* R. L. Glinski, B. G. Pendleton, M. B. Moss, M. N. LeFranc, Jr., B. A. Millsap, and S. W. Hoffman, eds., Proceedings of the Southwest Raptor Management Symposium and Workshop. Sci. and Tech. Ser. 11. Natl. Wildl. Fed., Washington, D.C. 395 pp.

Glinski, R. L., and A. L. Gennaro. 1988. Mississippi Kite. Pages 54–56 *in* R. L. Glinski, B. G. Pendleton, M. B. Moss, M. N. LeFranc, Jr., B. A. Millsap, and S. W. Hoffman, eds., Proceedings of the Southwest Raptor Management Symposium and Workshop. Sci. and Tech. Ser. 11. Natl. Wildl. Fed., Washington, D.C. 395 pp.

Glover, F. A. 1953. Nesting ecology of the Pied-billed Grebe in northwestern Iowa. Wilson Bull. 65:32–39.

Godfrey, W. E. 1966. The birds of Canada. Bryant Press, Toronto. 428 pp.

Goldwasser, S., D. Gaines, and S. R. Wilbur. 1980. The Least Bell's Vireo in California: a de facto endangered race. Am. Birds 34:742–745.

Graber, J. W. 1961. Distribution, habitat requirements, and life history of the Black-capped Vireo (*Vireo atricapilla*). Ecol. Monogr. 31:313–336.

Graber, R., and J. Graber. 1951. Nesting of the Parula Warbler in Michigan. Wilson Bull. 63:75–83.

Graham, D. S. 1988. Responses of five host species to cowbird parasitism. Condor 90:588–591.

Graul, W. D. 1975. Breeding biology of the Mountain Plover. Wilson Bull. 87:6–31.

Greaves, J. M. 1989. Maintaining site integrity for breeding Least Bell's Vireos. Pages 293–298 *in* Proceedings of the California Riparian Systems Conference. Gen. Tech. Rep. PSW-110. U.S. Dept. Agric., Forest Serv., Pacific Southwest Forest and Range Exp. Stn., Berkeley, Calif.

Green, G. A., and R. G. Anthony. 1989. Nesting success and habitat relationships of Burrowing Owls in the Columbia Basin, Oregon. Condor 91:347–354.

Greenberg, R. 1984. The winter exploitation systems of Bay-breasted and Chestnut-sided Warblers in Panama. Univ. Calif. Publ. Zool. 116:1–107.

Greenberg, R., and S. Droege. 1990. Adaptations to tidal marshes in breeding populations of the Swamp Sparrow. Condor 92:393–404.

Greenhalgh, C. M. 1952. Food habits of the California Gull in Utah. Condor 54:302–308.

Grice, D., and J. P. Rogers. 1965. The Wood Duck in Massachusetts. Final report federal aid in wildlife restoration project W-19-R. Mass. Div. of Fisheries and Game, Westboro. 96 pp.

Grinnell, J., and A. H. Miller. 1944. The distribution of birds of California. Pac. Coast Avifauna 27:1–608.

Griscom, L., and A. Sprunt, Jr. 1979. The warblers of America. Doubleday, Garden City, N.Y. 302 pp.

Gross, A. O. 1921. The Dickcissel (*Spiza americana*) of the Illinois prairies. Auk 38:1–26, 163–184.

Grzybowsky, J. A., R. B. Clapp, and J. T. Marshall, Jr. 1986. History and current population status of the Black-capped Vireo in Oklahoma. Am. Birds 40:1151–1161.

Guay, J. W. 1968. The breeding biology of Franklin's Gull (*Larus pipixcan*). Ph.D. dissertation, University of Alberta, Edmonton. 119 pp.

Guinan, D. M., and S. G. Sealy. 1989. Foraging-substrate use by House Wrens nesting in natural cavities in a riparian habitat. Can. J. Zool. 67:61–67.

Gullion, G. W. 1954. The reproductive cycle of American Coots in California. Auk 71:366–412.

Literature Cited

Haas, C. A., and S. A. Sloane. 1989. Low return rates of migratory Loggerhead Shrikes: winter mortality or low site fidelity? Wilson Bull. 101:458–460.

Hagan, J. M., III. 1993. Decline of the Rufous-sided Towhee in the eastern United States. Auk 110:863–874.

Hagan, J. M., III, and D. W. Johnston, eds. 1992. Ecology and conservation of Neotropical migrant landbirds. Smithsonian Institution Press, Washington, D.C. 609 pp.

Hagan, J. M., III, T. L. Lloyd-Evans, J. L. Atwood, and D. S. Wood. 1992. Long-term changes in migratory landbirds in the northeastern United States. Pages 115–130 *in* J. M. Hagan III and D. W. Johnston, eds., Ecology and conservation of Neotropical migrant landbirds. Smithsonian Institution Press, Washington, D.C. 609 pp.

Hagan, J. M., III, and J. R. Walters. 1990. Foraging behavior, reproductive success, and colonial nesting in Ospreys. Auk 107:505–521.

Hagar, J. A. 1966. Nesting of the Hudsonian Godwit at Churchill, Manitoba. Living Bird 5:5–43.

Hall, G. A. 1984a. A long-term population study in an Appalachian spruce forest. Wilson Bull. 96:228–240.

Hall, G. A. 1984b. Population decline of Neotropical migrants in an Appalachian forest. Am. Birds 38:14–18.

Hall, R. S., R. L. Glinski, D. H. Ellis, J. M. Ramakka, and D. L. Base. 1988. Ferruginous Hawk. Pages 111–118 *in* R. L. Glinski, B. G. Pendleton, M. B. Moss, M. N. LeFranc, Jr., B. A. Millsap, and S. W. Hoffman, eds., Proceedings of the Southwest Raptor Management Symposium and Workshop. Sci. and Tech. Ser. 11. Natl. Wildl. Fed., Washington, D.C. 395 pp.

Hamel, P. B. 1986. Bachman's Warbler: a species in peril. Smithsonian Institution Press, Washington, D.C. 109 pp.

Hamel, P. B. 1992. Cerulean Warbler (*Dendroica cerulea*). Pages 385–400 *in* K. J. Schneider and D. M. Pence, eds., Migratory nongame birds of management concern in the Northeast. U.S. Fish and Wildl. Serv., Region 5, Newton Corner, Mass.

Hamel, P. B., H. E. LeGrand, Jr., M. R. Lennartz, and S. A. Gauthreaux, Jr. 1982. Bird-habitat relationships on southeastern forest lands. Gen. Tech. Rep. SE-22. U.S. Dept. of Agric., Forest Serv., Southeastern Forest Exp. Stn., Asheville, N.C.

Hann, H. W. 1937. Life history of the Ovenbird in southern Michigan. Wilson Bull. 49:145–237.

Hanson, H. C., and C. W. Kossack. 1963. The Mourning Dove in Illinois. Illinois Dept. Conserv., Tech. Bull. 2, Southern Illinois University, Carbondale. 133 pp.

Harding, K. C. 1931. Nesting habits of the Black-throated Blue Warbler. Auk 48:512–522.

Harmeson, J. P. 1974. Breeding ecology of the Dickcissel. Auk 91:348–359.

Harrison, H. H. 1975. A field guide to birds' nests in the United States east of the Mississippi River. Houghton Mifflin, Boston. 257 pp.

Harrison, H. H. 1979. A field guide to western birds' nests. Houghton Mifflin, Boston. 279 pp.

Harrison, H. H. 1984. Wood warblers' world. Simon and Schuster, New York. 335 pp.

Hartshorne, J. M. 1962. Behavior of the Eastern Bluebird at the nest. Living Bird 1:131–149.

Haug, E. A., and L. W. Oliphant. 1990. Movements, activity patterns, and habitat use of Burrowing Owls in Saskatchewan. J. Wildl. Manage. 54:27–35.

Heintzelman, D. S. 1979. Hawks and owls of North America. Universe Books, New York. 195 pp.

Hejl, S. J., and R. E. Woods. 1991. Bird assemblages in old-growth and rotation-aged Douglas-fir/ponderosa pine stands in the northern Rocky Mountains: a preliminary assessment. Pages 93–100 in D. M. Baumgartner and J. E. Lotan, eds., symposium proceedings, Interior Douglas-fir: the species and its management. Washington State University, Pullman.

Hendricks, B. J., and J. P. Rieger. 1989. Description of nesting habitat for Least Bell's Vireo in San Diego County. Pages 285–292 in Proceedings of the California Riparian Systems Conference. Gen. Tech. Rep. PSW-110. U.S. Dept. Agric., Forest Serv., Pacific Southwest Forest and Range Exp. Stn., Berkeley, Calif.

Henny, C. J., F. P. Ward, K. E. Riddle, and R. M. Prouty. 1982. Migratory Peregrine Falcons accumulate pesticides in Latin America during winter. Can. Field-Nat. 96: 333–338.

Herklots, G. A. C. 1961. The birds of Trinidad and Tobago. Collins Clear Type Press, London. 287 pp.

Herlugson, C. J. 1981a. Food of adult and nestling Western and Mountain Bluebirds. Murrelet 63:59–65.

Herlugson, C. J. 1981b. Nest site selection of Mountain Bluebirds. Condor 83:252–255.

Herrick, F. H. 1910. Life and behavior of the Cuckoo. J. Exp. Zool. 9:169–233.

Hespendeide, H. A. 1964. Competition and the genus Tyrannus. Wilson Bull. 76:265–281.

Hespendeide, H. A. 1971. Flycatcher habitat selection in the eastern deciduous forest. Auk 88:61–74.

Higgins, K. F., and L. Kirsch. 1975. Some aspects of the breeding biology of the Upland Sandpiper in North Dakota. Wilson Bull. 87:96–102.

Hill, G. E. 1988. Age, plumage brightness, territory quality, and reproductive success in the Black-headed Grosbeak. Condor 90:379–388.

Hilty, S. L., and W. L. Brown. 1986. A guide to the birds of Columbia. Princeton University Press, Princeton, N.J. 836 pp.

Hofslund, P. B. 1959. A life history study of the Yellowthroat, Geothlypis trichas. Proc. Minn. Acad. Sci. 27:144–174.

Hohn, E. O. 1967. Observations on the breeding biology of Wilson's Phalarope Steganopus tricolor in central Alberta. Auk 84:220–244.

Holcomb, L. C. 1969. Breeding biology of the American Goldfinch in Ohio. Bird Banding 40:26–44.

Holcomb, L. C. 1972. Traill's Flycatcher breeding biology. Nebr. Bird Rev. 40:50–68.

Holliman, D. C. 1977. Purple Gallinule, Porphyrula martinica. Pages 105–109 in G. C. Sanderson, ed., Management of migratory shore and upland game birds in North America. International Association of Fish and Wildlife Agencies, Washington, D.C.

Holmes, R. T. 1971. Density, habitat, and the mating system of the Western Sandpiper, Calidris mauri. Oecologia 7:191–208.

Holmes, R. T., and S. K. Robinson. 1988. Spatial patterns, foraging tactics, and diets of ground-foraging birds in a northern hardwoods forest. Wilson Bull. 100:377–394.

Holmes, R. T., and T. W. Sherry. 1988. Assessing population trends of New Hampshire forest birds: local vs. regional patterns. Auk 105:756–768.

Holmes, R. T., and T. W. Sherry. 1992. Site fidelity of migratory warblers in temperate breeding and Neotropical wintering areas: implications for population dynamics, habitat selection, and conservation. Pages 563–575 in J. M. Hagan III and D. W. Johnston, eds., Ecology and conservation of Neotropical migrant landbirds. Smithsonian Institution Press, Washington, D.C. 609 pp.

Holmes, R. T., T. W. Sherry, and L. Reitsma. 1989. Population structure, territoriality, and overwinter survival of two migrant warbler species in Jamaica. Condor 91:545–561.

Holmes, R. T., T. W. Sherry, and F. W. Sturges. 1986. Bird community dynamics in a temperate deciduous forest: long-term trends at Hubbard Brook. Ecol. Monogr. 56: 201–220.

Hooper, R. G., and P. B. Hamel. 1977. Nesting habitat of Bachman's Warbler: a review. Wilson Bull. 89:373–379.

Horak, G. J. 1970. A comparative study of the foods of the Sora and Virginia Rail. Wilson Bull. 82:206–213.

Howe, F. P., and L. D. Flake. 1989. Nesting ecology of Mourning Doves in a cold desert ecosystem. Wilson Bull. 101:467–472.

Howe, J. 1983. The vanishing birds of Veracruz. Defenders 58:18–28.

Howe, M. A., P. H. Geissler, and B. A. Harrington. 1989. Population trends of North American shorebirds based on the International Shorebird Survey. Biol. Conserv. 49:185–199.

Howell, T. R. 1952. Natural history and differentiation in the Yellow-bellied Sapsucker. Condor 54:237–282.

Howell, T. R. 1971. An ecological study of the birds of the lowland pine savanna and adjacent rain forest in northeastern Nicaragua. Living Bird 10:185–242.

Howie, R. R., and R. Ritcey. 1987. Distribution, habitat selection, and densities of Flammulated Owls in British Columbia. Pages 249–254 in R. W. Nero, R. J. Clark, R. J. Knapton, and R. H. Hamre, eds., Biology and conservation of northern forest owls. Gen. Tech. Rep. RM-142. U.S. Dept. Agric., Forest Serv., Rocky Mountain Forest and Range Exp. Stn., Ft. Collins, Colo.

Hubbard, J. P. 1965. Summer birds of the Mogollon Mountains, New Mexico. Condor 67:404–415.

Humphrey, R. C. 1988. Ecology and range expansion of American Oystercatchers in Massachusetts. M.S. thesis, University of Massachusetts, Amherst. 73 pp.

Hunter, W. C. 1990. Handbook for nongame bird management and monitoring in the Southeast Region. U.S. Dept. of the Interior, Fish and Wildl. Serv., Southeast Region, Atlanta, Ga. 198 pp.

Hunter, W. F., and P. H. Baldwin. 1962. Nesting of the Black Swift in Montana. Wilson Bull. 74:409–416.

Hurley, R. J., and E. C. Franks. 1976. Changes in the breeding ranges of two grassland birds. Auk 92:108–115.

Hussell, D. J. T., M. H. Mather, and P. H. Sinclair. 1992. Trends in numbers of tropical and temperate wintering migrant landbirds in migration at Long Point, Ontario, 1961–1988. Pages 101–114 in J. M. Hagan III and D. W. Johnston, eds., Ecology and conservation of Neotropical migrant landbirds. Smithsonian Institution Press, Washington, D.C. 609 pp.

Hussell, D. J. T., and G. W. Page. 1976. Observations on the breeding biology of Black-bellied Plovers on Devon Island, N.W.T., Canada. Wilson Bull. 88:632–653.

Hussong, C. 1946. The Clay-colored Sparrow. Passenger Pigeon 8:3–7.

Hutto, R. L. 1988. Is tropical deforestation responsible for the reported declines in Neotropical migrant populations? Am. Birds 42:375–379.

Hutto, R. L. 1992. Habitat distributions of migratory landbird species in western Mexico. Pages 221–239 *in* J. M. Hagan III and D. W. Johnston, eds., Ecology and conservation of Neotropical migrant landbirds. Smithsonian Institution Press, Washington, D.C. 609 pp.

James, F. C., D. A. Wiedenfeld, and C. E. McCulloch. 1992. Trends in breeding populations of warblers: declines in the southern highlands and increases in the lowlands. Pages 43–56 *in* J. M. Hagan III and D. W. Johnston, eds., Ecology and conservation of Neotropical migrant landbirds. Smithsonian Institution Press, Washington, D.C. 609 pp.

James, R. D. 1976. Foraging behavior and habitat selection of three species of vireos in southern Ontario. Wilson Bull. 88:62–75.

Jarvis, W. L., and W. E. Southern. 1976. Food habits of Ring-billed Gulls breeding in the Great Lakes region. Wilson Bull. 88:621–631.

Jeffrey, R. G. 1977. Band-tailed Pigeon, *Columba fasciata*. Pages 211–245 *in* G. C. Sanderson, ed., Management of migratory shore and upland game birds in North America. International Association of Fish and Wildlife Agencies, Washington, D.C.

Jehl, J. R., Jr. 1973. Breeding biology and systematic relationships of the Stilt Sandpiper. Wilson Bull. 85:115–147.

Johnsgard, P. A. 1975a. North American game birds of upland and shoreline. University of Nebraska Press, Lincoln. 183 pp.

Johnsgard, P. A. 1975b. Waterfowl of North America. Indiana University Press, Bloomington. 575 pp.

Johnsgard, P. A. 1979. Birds of the Great Plains. University of Nebraska Press, Lincoln. 539 pp.

Johnsgard, P. A. 1983. The hummingbirds of North America. Smithsonian Institution Press, Washington, D.C. 303 pp.

Johnsgard, P. A. 1988. North American owls. Smithsonian Institution Press, Washington, D.C. 295 pp.

Johnsgard, P. A. 1990. Hawks, eagles, and falcons of North America. Smithsonian Institution Press, Washington, D.C.

Johnson, A. W. 1965. The birds of Chile and adjacent regions of Argentina, Bolivia, and Peru. Platt Establecimientos Graficos S.A., Buenos Aires. 398 pp.

Johnson, E. J., and L. B. Best. 1982. Factors affecting feeding and brooding of Gray Catbird nestlings. Auk 99:148–156.

Johnson, N.K. 1976. Breeding distribution of Nashville and Virginia's Warblers. Auk 93:219–230.

Johnson, N.K., and J. A. Marten. 1988. Evolutionary genetics of flycatchers. II. Differentiation in the *Empidonax difficilis* complex. Auk 105:177–191.

Johnston, D. W., and J. M. Hagan III. 1992. An analysis of long-term breeding bird censuses from eastern deciduous forests. Pages 75–84 *in* J. M. Hagan III and D. W. Johnston, eds., Ecology and conservation of Neotropical migrant landbirds. Smithsonian Institution Press, Washington, D.C. 609 pp.

Johnston, D. W., and D. L. Winings. 1987. Natural history of Plummers Island, Maryland. XXVII. The declines of forest birds on Plummers Island, Maryland and vicinity. Proc. Biol. Soc. Wash. 100:762–768.

Jones, E. T. 1986. The passerine decline. N. Am. Bird Bander 11:74–75.

Jones, J. C. 1940. Food habits of the American Coot with notes on distribution. Wildl. Res. Bull. 2. U.S. Dept. of the Interior, Bureau of Biological Survey, Washington, D.C. 52 pp.

Jones, S. 1979. Habitat management series for unique or endangered species. Rep. 17. The accipiters—Goshawk, Cooper's Hawk, Sharp-shinned Hawk. Tech. Note 335. U.S. Dept. of the Interior, Bureau of Land Management, Washington, D.C. 51 pp.

Joyner, D. E. 1969. A survey of the ecology and behavior of the Ruddy Duck (*Oxyura jamaicensis*) in northern Utah. M.S. thesis, University of Utah, Salt Lake City. 83 pp.

Karalus, K. E., and A. W. Eckert. 1974. The owls of North America. Doubleday, Garden City, N.Y. 278 pp.

Karr, J. R. 1976. On the relative abundance of migrants from the North Temperate Zone in tropical habitats. Wilson Bull. 88:433–458.

Keast, A. 1980. Migratory Parulidae: what can species co-occurrence in the north reveal about ecological plasticity and wintering patterns? Pages 457–476 *in* A. Keast and E. S. Morton, eds., Migrant birds in the Neotropics. Smithsonian Institution Press, Washington, D.C.

Keast, A., and E. S. Morton. 1980. Migrant birds in the Neotropics. Smithsonian Institution Press, Washington, D.C.

Keeler, J. E. 1977. Mourning Dove, *Zenaida macroura*. Pages 275–298 *in* G. C. Sanderson, ed., Management of migratory shore and upland game birds in North America. International Association Fish and Wildlife Agencies, Washington, D.C.

Keller, M. E., and S. H. Anderson. 1992. Avian use of habitat configurations created by forest cutting in southeastern Wyoming. Condor 94:55–65.

Kendeigh, S. C. 1941. Territorial and mating behavior of the House Wren. Ill. Biol. Monogr. 18:1–120.

Kendeigh, S. C. 1947. Bird population studies in the coniferous forest biome during a spruce budworm outbreak. Ontario Dept. of Lands and Forests, Biol. Bull. 1:1–100.

Kendeigh, S. C. 1982. Bird populations in east-central Illinois: fluctuations, variations, and development over a half century. Ill. Biol. Monogr. 52:1–136.

Kiel, W. H., Jr. 1955. Nesting studies of the coot in southwestern Manitoba. J. Wildl. Manage. 19:189–198.

Kilham, L. 1971. Reproductive behavior of Yellow-bellied Sapsuckers. I. Preference for nesting in *Fomes*-infected aspens and nest hole interactions with flying squirrels, raccoons, and other animals. Wilson Bull. 83:159–171.

King, J. R. 1955. Notes on the life history of Traill's Flycatcher, *Empidonax traillii*, in southeastern Washington. Auk 72:148–173.

Kitchen, D. W., and G. S. Hunt. 1969. Brood habitat of the Hooded Merganser. J. Wildl. Manage. 33:605–609.

Knapton, R. W. 1978. Breeding ecology of the Clay-colored Sparrow. Living Bird 17: 137–158.

Knopf, F. L. 1979. Spatial and temporal aspects of colonial nesting of White Pelicans. Condor 81:353–363.

Knopf, F. L., and J. L. Kennedy. 1981. Differential predation by two species of piscivorous birds. Wilson Bull. 93:554–556.

Knopf, F. L., and J. A. Sedgwick. 1987. Latent population responses of summer birds to a catastrophic, climatological event. Condor 89:869–873.

Knopf, F. L., J. A. Sedgwick, and D. B. Inkley. 1990. Regional correspondence among shrubsteppe bird habitats. Condor 92:45–53.

Knorr, O. A. 1961. The geographical and ecological distribution of the Black Swift in Colorado. Wilson Bull. 73:155–170.

Knowles, C. J., C. J. Stoner, and S. P. Gieb. 1982. Selective use of black-tailed prairie dog towns by Mountain Plovers. Condor 84:71–74.

Knowles, E. H. M. 1942. Nesting habits of the Spotted Sandpiper. Auk 59:583–584.

Knupp, D. M., R. B. Owen, Jr., and J. B. Dimond. 1977. Reproductive biology of American Robins in northern Maine. Auk 94:80–85.

Kondla, N. G. 1973. Nesting of the Black Swift at Johnston's Canyon, Alberta. Can. Field-Nat. 87:64–65.

Konrad, P. M., and D. S. Gilmer. 1984. Observations on the nesting ecology of Burrowing Owls in central North Dakota. Prairie Nat. 16:129–130.

Kramer, E. 1948. Oystercatcher breeding in New Jersey. Auk 65:460.

Krapu, G. L. 1974. Feeding ecology of pintail hens during reproduction. Auk 91:278–290.

Krause, H. 1965. Nesting of a pair of Canada Warblers. Living Bird 4:5–11.

Krauth, S. 1972. The breeding biology of the Common Gallinule. M.S. thesis, University of Wisconsin, Oshkosh. 74 pp.

Kridelbaugh, A. 1983. Nesting ecology of the Loggerhead Shrike in central Missouri. Wilson Bull. 95:303–308.

Kroll, J. C. 1980. Habitat requirements of the Golden-cheeked Warbler: management implications. J. Range Manage. 33:60–65.

Kroodsma, D. E. 1989. Two North American song populations of the Marsh Wren reach distributional limits in the central Great Plains. Condor 91:332–340.

Lack, D. 1956. A review of the genera and nesting habits of swifts. Auk 73:1–32.

Land, H. C. 1970. Birds of Guatemala. Livingston Publishing Co., Wynnewood, Penn. 381 pp.

Lanyon, W. E. 1957. The comparative biology of the meadowlarks, *Sturnella*, in Wisconsin. Publ. Nuttall Ornithol. Club 1:1–67.

Lawrence, L. de K. 1948. Comparative study of the nesting behavior of Chestnut-sided and Nashville Warblers. Auk 65:204–219.

Lawrence, L. de K. 1953a. Nesting life and behavior of the Red-eyed Vireo. Can. Field-Nat. 67:47–77.

Lawrence, L. de K. 1953b. Notes on the nesting of the Blackburnian Warbler. Wilson Bull. 65:135–144.

Lawrence, L. de K. 1967. A comparative life-history study of four species of woodpeckers. Ornithol. Monogr. 5:1–156.

Laymon, S. A., and M. D. Halterman. 1987. Can the western subspecies of the Yellow-billed Cuckoo be saved from extinction? West. Birds 18:19–25.

Laymon, S. A., and M. D. Halterman. 1989. A proposed habitat management plan for Yellow-billed Cuckoos in California. Pages 272–277 in Proceedings of the California Riparian Systems Conference. Gen. Tech. Rep. PSW-110. U.S. Dept. Agric., Forest Serv., Pacific Southwest Forest and Range Exp. Stn., Berkeley, Calif.

Lea, R. B. 1942. A study of the nesting habits of the Cedar Waxwing. Wilson Bull. 54:225–237.

Leck, C. F., and F. L. Cantor. 1979. Seasonality, clutch size, and hatching success in the Cedar Waxwing. Auk 96:196–198.

Literature Cited

Leck, C. F., B. G. Murray, Jr., and J. Swinebroad. 1981. Changes in breeding bird populations at Hutcheson Memorial Forest since 1958. William L. Hutcheson Memorial Forest Bull. 6:8–14.

Leck, C. F., B. G. Murray, Jr., and J. Swinebroad. 1988. Long-term changes in the breeding bird populations of a New Jersey forest. Biol. Conserv. 46:145–157.

Lederer, J. R. 1977. Winter feeding territories in the Townsend's Solitaire. Bird Banding 48:11–18.

LeGrand, H. E., and K. J. Schneider. 1992. Bachman's Sparrow (*Aimophila aestivalis*). Pages 299–313 *in* K. J. Schneider and D. M. Pence, eds., Migratory nongame birds of management concern in the Northeast. U.S. Fish and Wildl. Serv., Region 5, Newton Corner, Mass.

Lemieux, L. 1959. The breeding biology of the Greater Snow Goose on Bylot Island, Northwest Territories. Can. Field-Nat. 73:117–128.

Lemon, R. E., D. M. Weary, and K. J. Norris. 1992. Male morphology and behavior correlate with reproductive success in the American Redstart (*Setophaga ruticilla*). Behav. Ecol. Sociobiol. 29:399–404.

Leonard, M. L., and J. Picman. 1986. Why are nesting Marsh Wrens and Yellow-headed Blackbirds spatially segregated? Auk 103:135–140.

Leonard, M. L., and J. Picman. 1987. Nesting mortality and habitat selection by Marsh Wrens. Auk 104:491–495.

Levenson, H., and J. R. Koplin. 1984. Effects of human activity on the productivity of nesting Ospreys. J. Wildl. Manage. 48:1374–1377.

Levy, S. H. 1959. Thick-billed Kingbird in the United States. Auk 76:92.

Lewis, J. C. 1977. Sandhill Crane, *Grus canadensis*. Pages 5–43 *in* G. C. Sanderson, ed., Management of migratory shore and upland game birds in North America. International Association of Fish and Wildlife Agencies, Washington, D.C.

Ligon, J. D. 1968. The biology of the Elf Owl, *Micrathene whitneyi*. Misc. Publ. 136. University of Michigan Museum of Zoology, Ann Arbor.

Ligon, J. D. 1971. Some factors influencing numbers of the Red-cockaded Woodpecker. Pages 30–43 *in* The ecology and management of the Red-cockaded Woodpecker: symposium proceedings. U.S. Dept. of the Interior, Bureau of Sport Fisheries and Wildlife, Folkston, Ga. (Published in cooperation with Tall Timbers Research Station, Tallahassee, Fla.)

Ligon, J. S. 1961. New Mexico birds and where to find them. University of New Mexico Press, Albuquerque. 360 pp.

Lincer, J. L., and J. A. Sherburne. 1974. Organochlorines in kestrel prey: a north-south dichotomy. J. Wildl. Manage. 38:427–434.

Lindvall, M. L., and J. B. Low. 1982. Nesting ecology and production of Western Grebes at Bear River Migratory Bird Refuge, Utah. Condor 84:66–70.

Lingle, G. R., and N. F. Sloan. 1980. Food habits of White Pelicans during 1976 and 1977 at Chase Lake National Wildlife Refuge, North Dakota. Wilson Bull. 92:123–125.

Linsdale, J. M. 1957. Goldfinches on the Hastings Natural History Reservation. Am. Mid. Nat. 57:1–119.

Litwin, T. S. 1986. Factors affecting avian diversity in a northeastern woodlot. Ph.D. dissertation, Cornell University, Ithaca, N.Y.

Litwin, T. S., and C. R. Smith. 1992. Factors influencing the decline of Neotropical migrants in a northeastern forest fragment: isolation, fragmentation, or mosaic effects? Pages 483–496 *in* J. M. Hagan III and D. W. Johnston, eds., Ecology and conservation of Neotropical migrant landbirds. Smithsonian Institution Press, Washington, D.C. 609 pp.

Loftin, H. 1963a. Notes on autumn bird migrants in Panama. Carib. J. Sci. 3:63–68.

Loftin, H. 1963b. Some repeats and returns of North American migrant birds in Panama. Bird Banding 34:219–221.

Loftin, H. 1967. Florida-banded Snowy Egret recovered in Panama. Fla. Nat. 40:30.

Loftin, H. 1977. Returns and recoveries of banded North American birds in Panama and the tropics. Bird Banding 48:253–258.

Loftin, H., G. I. Child, and S. Bongiorno. 1967. Returns in 1965–1966 of North American migrant birds banded in Panama. Bird Banding 38:151–152.

Loftin, H., D. T. Rogers, Jr., and D. L. Hicks. 1966. Repeats, returns, and recoveries of North American migrant birds banded in Panama. Bird Banding 37:35–44.

Lokemoen, J. T. 1966. Breeding ecology of the Redhead duck in western Montana. J. Wildl. Manage. 30:668–681.

Lopez Ornat, A., and R. Greenberg. 1990. Sexual segregation by habitat in migratory warblers in Quintana Roo, Mexico. Auk 107:539–543.

Lovejoy, T. E., III. 1983. Tropical deforestation and North American migrant birds. Bird Conserv. 1:126–128.

Low, G., and W. Mansell. 1983. North American marsh birds. Harper & Row, New York. 192 pp.

Ludwig, J. P. 1965. Biology and structure of the Caspian Tern (*Hydroprogne caspia*) population of the Great Lakes from 1896–1964. Bird Banding 36:217–233.

Lunk, W. A. 1962. The Rough-winged Swallow, *Stelgidopteryx ruficollis* (Vieillot): a study based on its breeding biology in Michigan. Publ. Nuttall Ornithol. Club 4:1–155.

Lynch, J. F. 1989. Distribution of overwintering Nearctic migrants in the Yucatán Peninsula. I: general patterns of occurrence. Condor 91:515–544.

Lynch, J. F. 1992. Distribution of overwintering Nearctic migrants in the Yucatán Peninsula. II: Use of native and human-modified vegetation. Pages 178–196 *in* J. M. Hagan III and D. W. Johnston, eds., Ecology and conservation of Neotropical migrant landbirds. Smithsonian Institution Press, Washington, D.C. 609 pp.

Lynch, J. F., E. S. Morton, and M. E. Van der Voort. 1985. Habitat segregation between the sexes of wintering Hooded Warblers (*Wilsonia citrina*). Auk 102:714–721.

Lynch, J. F., and D. F. Whigham. 1984. Effects of forest fragmentation on breeding bird communities in Maryland, USA. Biol. Conserv. 28:287–324.

Lynch, J. F., and R. F. Whitcomb. 1978. Effects of the insularization of the eastern deciduous forest on avifaunal diversity and turn over. Pages 461–489 *in* A. Marmelstein, ed., Classification, inventory, and analysis of fish and wildlife habitat. Obs-78/76. U.S. Dept. of the Interior, Fish and Wildl. Serv., Washington, D.C.

Mabey, S. E., and E. S. Morton. 1992. Demography and territorial behavior of wintering Kentucky Warblers in Panama. Pages 329–336 *in* J. M. Hagan III and D. W. Johnston, eds., Ecology and conservation of Neotropical migrant landbirds. Smithsonian Institution Press, Washington, D.C. 609 pp.

MacArthur, R. H., and E. O. Wilson. 1967. The theory of island biogeography. Princeton University Press, Princeton, N.J.

MacKenzie, D. I., and S. G. Sealy. 1981. Nest site selection in Eastern and Western Kingbirds: a multivariate approach. Condor 83:310–321.

MacKenzie, J. P. S. 1977. Birds in peril. Houghton Mifflin, Boston. 191 pp.

Mannan, R. W., and E. C. Meslow. 1984. Bird populations and vegetation characteristics in managed and old-growth forests, northeastern Oregon. J. Wildl. Manage. 48:1219–1238.

Manuwal, D. A. 1970. Notes on the territoriality of Hammond's Flycatcher in western Montana. Condor 72:364–365.

Manuwal, D. A., and M. H. Huff. 1987. Spring and winter bird populations in a Douglas-fir sere. J. Wildl. Manage. 51:586–595.

Marks, J. S. 1986. Nest site characteristics and reproductive success of Long-eared Owls in southwestern Idaho. Wilson Bull. 98:547–560.

Marshall, J. R. 1988. Birds lost from a giant sequoia forest during fifty years. Condor 90:359–372.

Marshall, J., and R. P. Balda. 1974. The breeding ecology of the Painted Redstart. Condor 76:89–101.

Marti, C. D. 1974. Feeding ecology of four sympatric owls. Condor 76:45–61.

Marti, C. D. 1976. A review of prey selection by the Long-eared Owl. Condor 78:331–336.

Martin, A. C., H. S. Zim, and A. L. Nelson. 1951. American wildlife and plants. McGraw-Hill, New York. 500 pp.

Martin, R. F. 1974. Syntopic culvert nesting of Cave and Barn Swallows in Texas. Auk 91:776–782.

Martin, T. E., and J. R. Karr. 1986. Temporal dynamics of Neotropical birds with special reference to frugivores in second-growth woods. Wilson Bull. 98:38–60.

Martin, T. E., and J. J. Roper. 1988. Nest predation and nest site selection of a western population of the Hermit Thrush. Condor 90:51–57.

Marvil, R. E., and A. Cruz. 1989. Impact of Brown-headed Cowbird parasitism on the reproductive success of the Solitary Vireo. Auk 106:476–480.

Matray, P. F. 1974. Broad-winged Hawk nesting and ecology. Auk 91:307–324.

Mayfield, H. 1960. The Kirtland's Warbler. Bull. 40. Cranbrook Institute of Sciences, Bloomfield Hills, Mich. 242 pp.

Mayfield, H. 1965. The Brown-headed Cowbird, with old and new hosts. Living Bird 4:13–27.

Mayhew, W. W. 1958. The biology of the Cliff Swallow in California. Condor 60:7–37.

Mayr, E. 1985. Nearctic region, Neotropical region. Pages 379–382 in B. Campbell and E. Lack, eds., A dictionary for birds. British Trust for Ornithology, Buteo Books, Vermillion, S.D.

McAtee, W. L. 1908. Food habits of the grosbeaks. Bull. 32. U.S. Dept. of Agric., Bureau of Biological Survey, Washington, D.C. 92 pp.

McAtee, W. L. 1935. Food habits of common hawks. Circ. 370. U.S. Dept. of Agric., Bureau of Biological Survey, Washington, D.C. 36 pp.

McAtee, W. L., and F. E. L. Beal. 1912. Some common game, aquatic, and rapacious birds in relation to man. Farmers' Bull. 497. U.S. Dept. of Agric., Washington, D.C. 30 pp.

McCallum, D. A., and F. R. Gehlbach. 1988. Nest-site preferences of Flammulated Owls in western New Mexico. Condor 90:653–661.

McComb, W. C. 1985. Habitat associations of birds and mammals in an Appalachian forest. Proc. Annu. Conf. Southeast. Assoc. Fish Wildl. Agencies 39:420–429.

McComb, W. C., P. L. Groetsch, G. E. Jacoby, and G. A. McPeek. 1989. Response of forest birds to an improvement cut in Kentucky. Proc. Annu. Conf. Southeast. Assoc. Fish Wildl. Agencies 43:313–325.

McGilvrey, F. B., compiler. 1968. A guide to Wood Duck production habitat requirements. Res. Publ. 60. U.S. Dept. of the Interior, Bureau of Sport Fisheries and Wildlife, Washington, D.C. 32 pp.

McNair, D. B. 1984. Reuse of other species' nests by Lark Sparrows. Southwest. Nat. 29:506–509.

McNicholl, M. K. 1971. The breeding biology and ecology of Forster's Tern (*Sterna forsteri*) at Delta, Manitoba. M.S. thesis, University of Manitoba, Winnipeg. 652 pp.

McNulty, F. 1966. The Whooping Crane. Dutton, New York. 190 pp.

McPherson, J. M. 1987. A field study of winter fruit preferences of Cedar Waxwings. Condor 89:293–306.

McShea, W. J., and J. H. Rappole. 1992. Impact of white-tailed deer on forest understory birds in northern Virginia. *In* P. Stangel, ed., Effects of land use on migratory birds. Soc. Conserv. Biol., Washington, D.C.

Meanley, B. 1965. Early-fall food and habitat of the Sora in the Patuxent River Marsh, Maryland. Chesapeake Sci. 6:235–237.

Meanley, B. 1966. Some observations on habitats of the Swainson's Warbler. Living Bird 5:151–165.

Meanley, B. 1969. Natural history of the King Rail. N. Am. Fauna 67:1–108.

Meanley, B. 1971. Natural history of the Swainson's Warbler. N. Am. Fauna 69:1–90.

Meanley, B., and A. G. Meanley. 1958. Nesting habitat of the Black-bellied Tree Duck in Texas. Wilson Bull. 70:94–95.

Meanley, B., and R. T. Mitchell. 1958. Food habits of Bachman's Warbler. Atl. Nat. 13:236–238.

Medin, D. E. 1985. Breeding bird responses to diameter-cut logging in west-central Idaho. Res. Paper Int. 355. U.S. Dept. Agric., For. Serv., Intermountain Forest and Range Exp. Stn., Ogden, Utah. 12 pp.

Medin, D. E., and G. D. Booth. 1989. Responses of birds and small mammals to single-tree selection logging in Idaho. Res. Paper Int. 408. U.S. Dept. Agric., For. Serv., Intermountain Forest and Range Exp. Stn., Ogden, Utah.

Mendall, H. L. 1937. Nesting of the Bay-breasted Warbler. Auk 54:429–439.

Mendall, H. L. 1958. The Ring-necked Duck in the northeast. Sec. Ser. 73. Orono: University of Maine Studies: 1–317.

Metz, K. J. 1991. The enigma of multiple nest building by male Marsh Wrens. Auk 108:170–173.

Meyer de Schauensee, R. 1966. The species of birds of South America and their distribution. Livingston Publishing Co., Wynnewood, Penn.

Meyer de Schauensee, R. 1982. A guide to the birds of South America. 2d ed. Acad. Nat. Sci., Philadelphia, Penn.

Meyer de Schauensee, R., and W. H. Phelps. 1978. A guide to the birds of Venezuela. Princeton University Press, Princeton, N.J. 424 pp.

Middleton, A. L. A. 1978. The annual cycle of the American Goldfinch. Condor 80:401–406.

Middleton, A. L. A. 1979. Influence of age and habitat on reproduction by the American Goldfinch. Ecology 60:418–432.

Miller, A. H. 1931. Systematic revision and natural history of the American shrikes (*Lanius*). Univ. Calif. Publ. Zool. 38:11–242.

Miller, E. V. 1941. Behavior of the Bewick Wren. Condor 43:81–99.

Miller, J. H., and M. T. Green. 1987. Distribution, status, and origin of Water Pipits breeding in California. Condor 89:788–797.

Miller, R. S., and R. W. Nero. 1983. Hummingbird-sapsucker associations in northern climates. Can. J. Zool. 61:1540–1546.

Mock, P. J. 1991. Daily allocation of time and energy of Western Bluebirds feeding nestlings. Condor 93:598–611.

Monroe, B. L. 1968. A distributional survey of the birds of Honduras. Allen Press, Lawrence, Kans. 458 pp.

Moore, R. T. 1939. The Arizona Broad-billed Hummingbird. Auk 56:313–319.

Morel, G., and F. Bourlière. 1962. Ecological relations of the sedentary and migratory avifauna in a Sahel savannah of lower Senegal (in French). Terre et Vie 4: 371–393.

Morris, R. F., W. F. Cheshire, C. A. Miller, and D. G. Mott. 1958. The numerical response of avian and mammalian predators during a gradation of the spruce budworm. Ecology 39:487–494.

Morrison, M. L. 1980. Seasonal aspects of the predatory behavior of Loggerhead Shrikes. Condor 82:296–300.

Morrison, M. L. 1981a. Population trends of the Loggerhead Shrike in the United States. Am. Birds 35:754–757.

Morrison, M. L. 1981b. The structure of western warbler assemblages: analysis of foraging and habitat selection in Oregon. Auk 98:578–588.

Morse, D. H. 1967. Foraging relationships of Brown-headed Nuthatches and Pine Warblers. Ecology 48:94–103.

Morse, D. H. 1970. Ecological aspects of some mixed-species foraging flocks of birds. Ecol. Monogr. 40:119–168.

Morse, D. H. 1971. The insectivorous bird as an adaptive strategy. Annu. Rev. Ecol. Syst. 2:177–200.

Morse, D. H. 1978. Populations of Bay-breasted and Cape May Warblers during an outbreak of the spruce budworm. Wilson Bull. 90:404–413.

Morse, D. H. 1989. American warblers: an ecological and behavioral perspective. Harvard University Press, Cambridge, Mass. 406 pp.

Morse, T. E., J. L. Jakabosky, and V. P. McCrow. 1969. Some aspects of the breeding biology of the Hooded Merganser. J. Wildl. Manage. 33:596–604.

Morton, E. S. 1979. Effective pollination of *Erythrina fusca* by the Orchard Oriole (*Icterus spurius*): coevolved behavioral manipulation? Ann. Missouri Bot. Garden 66: 482–489.

Morton, E. S. 1980a. Our migrant birds: can we continue to take them for granted? Atl. Nat. 33:36–40.

Morton, E. S. 1980b. Adaptations to seasonal changes by migrant land birds in the Panama Canal Zone. Pages 437–453 *in* A. Keast and E. S. Morton, eds., Migrant birds in the Neotropics. Smithsonian Institution Press, Washington, D.C.

Morton, E. S. 1992. What do we know about the future of migrant landbirds? Pages 579–589 *in* J. M. Hagan III and D. W. Johnston, eds., Ecology and conservation of Neotropical migrant landbirds. Smithsonian Institution Press, Washington, D.C. 609 pp.

Morton, E. S., and R. Greenberg. 1989. The outlook for migratory songbirds: "future shock" for birders. Am. Birds 43:178–183.

Morton, E. S., J. F. Lynch, K. Young, and P. Mehlhop. 1987. Do male Hooded Warblers exclude females from nonbreeding territories in tropical forest? Auk 104:133–135.

Mott, D. F., R. R. West, J. W. DeGrazio, and J. L. Guarino. 1972. Foods of the Red-winged Blackbird in Brown County, South Dakota. J. Wildl. Manage. 36:983–987.

Mousley, H. 1931. A study of the home life of the Alder Flycatcher (Empidonax traillii traillii). Auk 48:547–552.

Mousley, H. 1934a. A study of the home life of the Northern Crested Flycatcher (Myiarchus crinitus boreus). Auk 51:207–216.

Mousley, H. 1934b. A study of the home life of the Short-billed Marsh Wren (Cistothorus stellaris). Auk 51:439–445.

Mumford, R. E. 1964. The breeding biology of the Acadian Flycatcher. Misc. Publ. 125. University of Michigan Museum of Zoology, Ann Arbor. 50 pp.

Neff, J. A. 1940. Notes on nesting and other habits of the Western White-winged Dove in Arizona. J. Wildl. Manage. 4:279–290.

Neff, J. A. 1947. Habits, food, and economic status of the Band-tailed Pigeon. N. Am. Fauna 58. Dept. Interior, U.S. Fish and Wildl. Serv., Washington, D.C. 76 pp.

Newman, G. A. 1970. Cowbird parasitism and nesting success of Lark Sparrows in southern Oklahoma. Wilson Bull. 82:304–309.

Newton, I. 1979. Population ecology of raptors. Buteo Books, Vermillion, S.D.

Nickell, W. P. 1965. Habitats, territory, and nesting of the Catbird. Am. Midl. Nat. 73: 433–478.

Nickell, W. P. 1968. Return of northern migrants to tropical winter quarters and banded birds recovered in the United States. Bird Banding 39:107–116.

Nisbet, I. C. T. 1970. Autumn migration of the Blackpoll Warbler: evidence for a long flight provided by regional survey. Bird Banding 41:207–240.

Nisbet, I. C. T., W. H. Drury, and J. Baird. 1963. Weight loss during migration. Part 1: Deposition and consumption of fat by the Blackpoll Warbler (Dendroica striata). Bird Banding 34:107–138.

Nolan, V., Jr. 1978. The ecology and behavior of the Prairie Warbler, Dendroica discolor. Ornithol. Monogr. 26:1–595.

Nolan, V., Jr., and D. P. Wooldridge. 1962. Food habits and feeding behavior of the White-eyed Vireo. Wilson Bull. 74:68–73.

Noon, B. R. 1981. The distribution of an avian guild along a temperate elevational gradient: the importance and expression of competition. Ecol. Monogr. 51:105–124.

Norwine, J., and R. Bingham. 1985. Frequency and severity of droughts in south Texas: 1900–1983. Pages 1–17 in R. D. Brown, ed., Livestock and wildlife management during drought. Caesar Kleberg Wildl. Res. Inst., Texas A&I University, Kingsville, Tex.

Novak, P. G. 1992. Black Tern (Chlidonias niger). Pages 149–170 in K. J. Schneider and D. M. Pence, eds., Migratory nongame birds of management concern in the Northeast. U.S. Fish and Wildl. Serv., Region 5, Newton Corner, Mass.

Novakowski, N. S. 1966. Whooping Crane population dynamics on the nesting grounds. Wood Buffalo National Park, Northwest Territories, Canada. Rep. Ser. 1. Can. Wildl. Serv., Ottawa. 19 pp.

Oberholser, H. C. 1974a. The bird life of Texas, vol. 1. University of Texas Press, Austin. 530 pp.

Oberholser, H. C. 1974b. The bird life of Texas, vol. 2. University of Texas Press, Austin. 539 pp.

Odom, R. R. 1977. Sora, *Porzana carolina*. Pages 57–65 *in* G. C. Sanderson, ed., Management of migratory shore and upland game birds in North America. International Association of Fish and Wildlife Agencies, Washington, D.C.

Odum, E. P. 1931. Notes on the nesting habits of the Hooded Warbler. Wilson Bull. 43:316–317.

Ohlendorf, H. M. 1974. Competitive relationships among kingbirds, *Tyrannus*, in Trans-Pecos Texas. Wilson Bull. 86:357–373.

Ohlendorf, H. M. 1976. Comparative breeding ecology of phoebes in Trans-Pecos Texas. Wilson Bull. 88:255–271.

Oliver, W. W. 1970. The feeding patterns of sapsuckers on ponderosa pine in northeastern California. Condor 72:241.

Olrog, C. C. 1984. Argentina birds: a new field guide (in Spanish). Adm. Parques Nac., Buenos Aires, Argentina.

Orians, G. H. 1961. The ecology of blackbird (*Agelaius*) social systems. Ecol. Monogr. 31:285–312.

Orians, G. H. 1980. Some adaptations of marsh-nesting blackbirds. Monogr. Pop. Biol. 14. Princeton University Press, Princeton, N.J.

Orians, G. H. 1985. Blackbirds of the Americas. University of Washington Press, Seattle.

Ornat, A. L., and R. Greenberg. 1990. Sexual segregation by habitat in migratory warblers in Quintana Roo, Mexico. Auk 107:539–543.

Osborne, D. R., and A. T. Peterson. 1984. Decline of the Upland Sandpiper (*Bartramia longicauda*) in Ohio: an endangered species. Ohio J. Sci. 84(1):8–10.

Ouellet, H. 1993. Bicknell's Thrush. Wilson Bull. 105:545–572.

Overmire, T. G. 1962. Nesting of the Dickcissel in Oklahoma. Auk 79:115–116.

Packard, F. M. 1945. Birds of Rocky Mountain National Park. Auk 62:371–394.

Page, G. W., and L. E. Stenzel. 1981. The breeding status of the Snowy Plover in California. West. Birds 12:1–40.

Palmer, R. S. 1941. A behavior study of the Common Tern. Proc. Boston Soc. Nat. Hist. 42:1–119.

Palmer, R. S. 1962. Handbook of North American birds. Vol. 1. Loons through flamingos. Yale University Press, New Haven, Conn. 567 pp.

Palmer, R. S. 1967. Species accounts. *In* G. D. Stout, ed., The shorebirds of North America. Viking Press, New York. 270 pp.

Palmer, R. D. 1976a. Handbook of North American birds. Vol. 2. Yale University Press, New Haven, Conn. 521 pp.

Palmer, R. D. 1976b. Handbook of North American birds. Vol. 3. Yale University Press, New Haven, Conn. 560 pp.

Parker, J. W., and J. C. Ogden. 1979. The recent history and status of the Mississippi Kite. Am. Birds 33:119–129.

Parmelee, D. F. 1970. Breeding behavior of the Sanderling in the Canadian high Arctic. Living Bird 9:97–146.

Parmelee, D. F., D. W. Greiner, and W. D. Graul. 1968. Summer schedule and breeding biology of the White-rumped Sandpiper in the central Canadian Arctic. Wilson Bull. 80:5–29.

Pasquier, R., and E. S. Morton. 1982. Why birds take winter vacations. Smithsonian 13(7):169–187.

Payne, R. B. 1969. Breeding seasons and reproductive physiology of Tricolored Blackbirds and Red-winged Blackbirds. Univ. Calif. Publ. Zool. 90:1–115.

Pearson, T. G. 1936. Birds of America. Garden City Books, Garden City, N.Y. 289 pp.

Peeters, H. J. 1962. Nuptial behavior of the Band-tailed Pigeon in the San Francisco Bay area. Condor 64:445–470.

Pennington, T. D., and J. Sarukhan, 1968. Arboles tropicales de Mexico. Inst. Nac. Investig. Forestales, Mexico. 413 pp.

Petersen, K. L., and L. B. Best. 1987. Effects of prescribed burning on nongame birds in a sagebrush community. Wildl. Soc. Bull. 15:317–329.

Petersen, K. L., and L. B. Best. 1991. Nest-site selection by Sage Thrashers in southeastern Idaho. Great Basin Nat. 51:261–266.

Peterson, A. J. 1955. The breeding cycle in the Bank Swallow. Wilson Bull. 67:235–286.

Peterson, J. M. C., and C. Fichtel. 1992. Olive-sided Flycatcher (*Contopus borealis*). Pages 149–170 *in* K. J. Schneider and D. M. Pence, eds., Migratory nongame birds of management concern in the Northeast. U.S. Fish and Wildl. Serv., Region 5, Newton Corner, Mass.

Peterson, R. T., and E. L. Chalif. 1973. A field guide to Mexican birds. Houghton Mifflin, Boston. 298 pp.

Peterson, R. T., and E. L. Chalif. 1989. Aves de México: Guía de campo. Translated by M. A. Ramos and I. Castillo. Editorial Diana, World Wildlife Fund, Mexico.

Petit, D. R., L. J. Petit, and K. G. Smith. 1992. Habitat associations of migratory birds overwintering in Belize, Central America. Pages 247–256 *in* J. M. Hagan III and D. W. Johnston, eds., Ecology and conservation of Neotropical migrant landbirds. Smithsonian Institution Press, Washington, D.C. 609 pp.

Petrides, G. A. 1938. A life history study of the Yellow-breasted Chat. Wilson Bull. 50:184–189.

Petrinovich, L., and T. L. Patterson. 1983. The White-crowned Sparrow: reproductive success (1975–1980). Auk 100:811–825.

Phillips, A. R., J. Marshall, and G. Monson. 1964. The birds of Arizona. University of Arizona Press, Tucson. 220 pp.

Phillips, A. R. 1949. Nesting of the Rose-throated Becard in Arizona. Condor 51:137–139.

Pickwell, G. B. 1931. The prairie Horned Lark. Missouri Acad. of Sci., St. Louis. 153 pp.

Picman, J., and A. K. Picman. 1980. Destruction of nests by the Short-billed Marsh Wren. Condor 82:176–179.

Pitelka, F. A. 1940. Breeding behavior of the Black-throated Green Warbler. Wilson Bull. 52:2–18.

Pitelka, F. A. 1959. Numbers, breeding schedule, and territoriality in Pectoral Sandpipers of northern Alaska. Condor 61:233–264.

Place, A. R., and E. W. Stiles. 1992. Living off the wax of the land: bayberries and Yellow-rumped Warblers. Auk 109:334–345.

Platt, J. B. 1976. Sharp-shinned Hawk nesting and nest site selection in Utah. Condor 78:102–103.

Poddar, S., and R. J. Lederer. 1982. Juniper berries as an exclusive winter forage for Townsend's Solitaires. Am. Midl. Nat. 108:34–40.

Porter, D. K., M. A. Strong, J. B. Giezentanner, and R. A. Ryder. 1975. Nest ecology, productivity, and growth of the Loggerhead Shrike on the shortgrass prairie. Southwest Nat. 19:429–436.

Portnoy, J. W., and W. E. Dodge. 1979. Red-shouldered Hawk nesting ecology and behavior. Wilson Bull. 91:104–117.

Poston, H. J. 1974. Home range and breeding biology of the Shoveler. Rep. Ser. 25. Can. Wildl. Serv., Ottawa. 49 pp.

Potter, E. F. 1973. Breeding behavior of the Summer Tanager. Chat 37:35–39.

Potter, P. E. 1972. Territorial behavior in Savannah Sparrows in southeastern Michigan. Wilson Bull. 84:48–59.

Pough, R. H. 1951. Audubon water bird guide. Doubleday, Garden City, N.Y. 352 pp.

Powell, G. V. N., R. Bjork, and M. L. Avila H. In Press. Studies of the migration of the Resplendent Quetzal. ICBP Tech Publ.

Powell, G. V. N., and J. H. Rappole. 1986. The Hooded Warbler. In W. Fosburgh, ed., Audubon wildlife report, vol. 3. National Audubon Society, Washington, D.C.

Powell, G. V. N., J. H. Rappole, and S. A. Sader. 1992. Neotropical migrant landbird use of lowland Atlantic habitats in Costa Rica: a test of remote sensing for identification of habitat. Pages 287–298 in J. M. Hagan III and D. W. Johnston, eds., Ecology and conservation of Neotropical migrant landbirds. Smithsonian Institution Press, Washington, D.C. 609 pp.

Power, H. W. 1966. Biology of the Mountain Bluebird in Montana. Condor 68:351–371.

Power, H. W. 1980. The foraging behavior of Mountain Bluebirds with emphasis on sexual foraging differences. Ornithol. Monogr. 28.

Preble, N. A. 1957. Nesting habits of the Yellow-billed Cuckoo. Am. Midl. Nat. 57:474–482.

Prior, K. A. 1990. Turkey Vulture food habits in southern Ontario. Wilson Bull. 102:706–710.

Pulich, W. M. 1976. The Golden-cheeked Warbler. Texas Parks and Wildl. Dept., Austin. 172 pp.

Pulliam, H. R. 1988. Sources, sinks, and population regulation. Am. Nat. 132:652–661.

Putnam, L. S. 1949. The life history of the Cedar Waxwing. Wilson Bull. 61:141–182.

Raffaele, H. A. 1989. A guide to the birds of Puerto Rico and the Virgin Islands. Princeton University Press, Princeton, N.J. 254 pp.

Rappole, J. H. 1974. Migrants and space: the wintering ground as a limiting factor for migrant populations. Bull. Tex. Ornithol. Soc. 7:2–4.

Rappole, J. H., W. J. McShea, and J. H. Vega-Rivera. 1993. Estimation of species and numbers in upland avian breeding communities. J. Field Ornithol. 64:55–70.

Rappole, J. H., and E. S. Morton. 1985. Effects of habitat alteration on a tropical avian forest community. Pages 1013–1021 in P. A. Buckley, E. S. Morton, R. S. Ridgely, and F. G. Buckley, eds., Neotropical ornithology. Ornithol. Monogr. 36.

Rappole, J. H., E. S. Morton, T. E. Lovejoy III, and J. S. Ruos. 1983. Nearctic avian migrants in the Neotropics. U.S. Dept. of the Interior, Fish and Wildl. Serv., Washington, D.C.

Rappole, J. H., E. S. Morton, and M. A. Ramos. 1992. Density, philopatry, and population estimates for songbird migrants wintering in Veracruz. Pages 337–344 in J. M. Hagan III and D. W. Johnston, eds., Ecology and conservation of Neotropical migrant landbirds. Smithsonian Institution Press, Washington, D.C. 609 pp.

Rappole, J. H., M. A. Ramos, and K. Winker. 1989. Wintering Wood Thrush mortality in southern Veracruz. Auk 106:402–410.

Rappole, J. H., M. A. Ramos, K. Winker, R. J. Oehlenschlager, and D. W. Warner. In Press. Nearctic avian migrants of the Tuxtla Mountains and neighboring lowlands. *In* R. Dirzo and R. Vogt, eds., Natural history of the Tuxtla Mountains of Veracruz. University of Mexico, Mexico City.

Rappole, J. H., and A. R. Tipton. 1992. Evolution of avian migration in the Neotropics. Ornitol. Neotrop. 3:45–55.

Rappole, J. H., and G. Waggerman. 1986. Calling males as an index of density for breeding White-winged Doves. Wildl. Soc. Bull. 14:151–155.

Rappole, J. H., and D. W. Warner. 1976. Relationships between behavior, physiology, and weather in avian transients at migration stopover sites. Oecologia 26:193–212.

Rappole, J. H., and D. W. Warner. 1980. Ecological aspects of avian migrant behavior in Veracruz, Mexico. Pages 353–393 *in* A. Keast and E. S. Morton, eds., Migrant birds in the Neotropics. Smithsonian Institution Press, Washington, D.C.

Rasmussen, D. I. 1941. Biotic communities of the Kaibab Plateau, Arizona. Ecol. Monogr. 11:229–275.

Raynor, G. S. 1941. The nesting habits of the Whip-poor-will. Bird Banding 12:98–104.

Reitsma, L. R., R. T. Holmes, and T. W. Sherry. 1990. Effects of removal of red squirrels, *Tamiasciurus hudsonicus*, and eastern chipmunks, *Tamias striatus*, on nest predation in a northern hardwood forest: an artificial nest experiment. Oikos 57:375–380.

Remsen, J. V., Jr. 1986. Was Bachman's Warbler a bamboo specialist? Auk 103:216–219.

Rendell, W. B., and R. J. Robertson. 1989. Nest-site characteristics, reproductive success, and cavity availability for Tree Swallows breeding in natural cavities. Condor 91:875–885.

Rendell, W. B., and R. J. Robertson. 1990. Influence of forest edge on nest-site selection by Tree Swallows. Wilson Bull. 102:634–644.

Restani, M. 1991. Resource partitioning among three *Buteo* species in the Centennial Valley, Montana. Condor 93:1007–1010.

Reynolds, R. T., and B. D. Linkhart. 1987. The nesting biology of Flammulated Owls in Colorado. Pages 239–248 *in* R. W. Nero, R. J. Clark, R. J. Knapton, and R. H. Hamre, eds., Biology and conservation of northern forest owls. Gen. Tech. Rep. RM-142. U.S. Dept. Agric., Forest Serv., Rocky Mountain Forest and Range Exp. Stn., Ft. Collins, Colo.

Reynolds, R. T., and E. C. Meslow, 1984. Partitioning of food and niche characteristics of coexisting *Accipiter* during breeding. Auk 101:761–779.

Reynolds, R. T., E. C. Meslow, and H. M. Wight. 1982. Nesting habitat of coexisting *Accipiter* in Oregon. J. Wildl. Manage. 46:124–138.

Reynolds, T. D. 1981. Nesting of the Sage Thrasher, Sage Sparrow, and Brewer's Sparrow in southeastern Idaho. Condor 83:61–64.

Reynolds, T. D., and T. D. Rich. 1978. Reproductive biology of the Sage Thrasher (*Oreoscoptes montanus*) on the Snake River Plain in south-central Idaho. Auk 95:580–582.

Rich, T. D. 1978. Nest placement in Sage Thrashers. Wilson Bull. 90:303.

Rich, T. D. 1986. Habitat and nest site selection by Burrowing Owls in the sagebrush steppe of Idaho. J. Wildl. Manage. 50:548–555.

Richards, A. 1988. Shorebirds. Gallery Books, New York. 224 pp.

Literature Cited

Ricklefs, R. E. 1972. Latitudinal variation in breeding productivity of the Rough-winged Swallow. Auk 89:826–936.

Ridgely, R. S., and J. A. Gwynne. 1989. A guide to the birds of Panama. Princeton University Press, Princeton, N.J. 534 pp.

Ridgely, R. S., and G. Tudor. 1989. The birds of South America. Vol. 1. The Oscine passerines. University of Texas Press, Austin.

Risebrough, R. W., and D. B. Peakall. 1988. The relative importance of the several organochlorines in the decline of Peregrine Falcon populations. Pages 449–462 in T. J. Cade, J. H. Enderson, C. G. Thelander, and C. M. White, eds., Peregrine Falcon populations. Peregrine Fund, Boise, Idaho.

Risebrough, R. W., et al. 1989. Investigations of the decline of Swainson's Hawk populations in California. J. Raptor Res. 23:63–71.

Robbins, C. S. 1979. Effect of forest fragmentation on bird populations. Pages 198–212 in R. M. DeGraaf and K. E. Evans, eds., Proceedings of the Workshop Management on North Central and Northeastern Forests for Nongame Birds. GTR NC-51. U.S. Dept. Agric., Forest Serv., St. Paul, Minn.

Robbins, C. S. 1980. Effect of forest fragmentation on breeding bird populations in the Piedmont of the mid-Atlantic region. Atl. Nat. 33:31–36.

Robbins, C. S., B. Bruun, and H. S. Zim. 1983. Birds of North America. Golden Press, New York. 360 pp.

Robbins, C. S., D. Bystrak, and P. H. Geissler. 1986. The Breeding Bird Survey: its first 15 years, 1965–1979. U.S. Dept. Interior, Fish and Wildl. Serv. Publ. 157, Washington, D.C.

Robbins, C. S., D. K. Dawson, and B. A. Dowell. 1989a. Habitat area requirements of breeding forest birds of the Middle Atlantic states. Wildl. Monogr. 103.

Robbins, C. S., B. A. Dowell, D. K. Dawson, J. Colon, F. Espinoza, J. Rodriguez, R. Sutton, and T. Vargas. 1987. Comparison of Neotropical winter bird populations in isolated patches versus extensive forest. Acta Oecol. Oecol. Gen. 8:285–292.

Robbins, C. S., J. W. Fitzpatrick, and P. B. Hamel. 1992. A warbler in trouble: *Dendroica cerulea*. Pages 549–562 in J. M. Hagan III and D. W. Johnston, eds., Ecology and conservation of Neotropical migrant landbirds. Smithsonian Institution Press, Washington, D.C. 609 pp.

Robbins, C. S., J. R. Sauer, R. S. Greenberg, and S. Droege. 1989b. Population declines in North American birds that migrate to the Neotropics. Proc. Natl. Acad. Sci. 86:7658–7662.

Robinson, J. G., and K. H. Redford, eds. 1991. Neotropical wildlife conservation. University of Chicago Press, Chicago. 520 pp.

Robinson, S. K. 1981. Ecological relations and social interactions of Philadelphia and Red-eyed Vireos. Condor 83:16–26.

Robinson, S. K. 1992. Population dynamics of breeding Neotropical migrants in a fragmented Illinois landscape. Pages 408–418 in J. M. Hagan III and D. W. Johnston, eds., Ecology and conservation of Neotropical migrant landbirds. Smithsonian Institution Press, Washington, D.C. 609 pp.

Root, R. B. 1967. The niche exploitation pattern of the Blue-gray Gnatcatcher. Ecol. Monogr. 37:317–350.

Root, R. B. 1969. The behavior and reproductive success of the Blue-gray Gnatcatcher. Condor 71:16–31.

Roseberry, J. L., and W. D. Klimstra. 1970. The nesting ecology and reproductive performance of the Eastern Meadowlark. Wilson Bull. 82:243–267.

Rosenfield, R. N. 1984. Nesting biology of Broad-winged Hawks in Wisconsin. Raptor Res. 18:6–9.

Rotenberry, J. T., and J. A. Wiens. 1989. Reproductive biology of shrubsteppe passerine birds: geographical and temporal variation in clutch size, brood size, and fledging success. Condor 91:1–14.

Rusch, D. H., and P. D. Doerr. 1972. Broad-winged Hawk nesting and food habits. Auk 89:139–145.

Russell, H. N., Jr., and A. M. Woodbury. 1941. Nesting of the Gray Flycatcher. Auk 58:28–37.

Rustad, O. A. 1972. An Eastern Bluebird nesting study in south-central Minnesota. Loon 44:80–84.

Ryan, M. R., and R. B. Renken. 1987. Habitat use by breeding Willets in the northern Great Plains. Wilson Bull. 99:175–189.

Ryan, M. R., R. B. Renken, and J. J. Dinsmore. 1984. Marbled Godwit habitat selection in the northern prairie region. J. Wildl. Manage. 48:1206–1218.

Ryder, J. P. 1967. The breeding biology of Ross' Goose in the Perry River region, Northwest Territories. Rep. Ser. 3. Can. Wildl. Serv., Ottawa. 56 pp.

Sader, S. A., and A. T. Joyce. 1988. Deforestation rates and trends in Costa Rica, 1940 to 1983. Biotropica 20:11–19.

Sakai, H. F., and B. R. Noon. 1991. Nest-site characteristics of Hammond's and Pacific-slope Flycatchers in northwestern California. Condor 93:563–574.

Salomonson, M. G., and R. P. Balda. 1977. Winter territoriality of Townsend's Solitares (*Myadestes townsendi*) in a piñon–juniper–ponderosa pine ecotone. Condor 79:148–161.

Salt, W. R. 1966. A nesting study of *Spizella pallida*. Auk 83:274–281.

Samson, F. B. 1980. Island biogeography and the conservation of prairie birds. Proc. N. Am. Prairie Conf. 7:293–305.

Samuel, D. E. 1971. The breeding biology of Barn and Cliff Swallows in West Virginia. Wilson Bull. 83:284–301.

Sanders, S. D., and M. A. Flett. 1989. Montane riparian habitat and Willow Flycatchers: threats to a sensitive environment and species. Pages 262–266 in Proceedings of the California Riparian Systems Conference. Gen. Tech. Rep. PSW-110. U.S. Dept. Agric., Forest Serv., Pacific Southwest Forest and Range Exp. Stn., Berkeley, Calif.

Sauer, J. R., and S. Droege. 1990. Recent population trends of the Eastern Bluebird. Wilson Bull. 102:239–252.

Sauer, J. R., and S. Droege. 1992. Geographic patterns in population trends of Neotropical migrants in North America. Pages 26–42 in J. M. Hagan III and D. W. Johnston, eds., Ecology and conservation of Neotropical migrant landbirds. Smithsonian Institution Press, Washington, D.C. 609 pp.

Schell, J. H., R. L. Glinski, and H. A. Snyder. 1988. Common Black-Hawk. Pages 65–70 in R. L. Glinski, B. G. Pendleton, M. B. Moss, M. N. LeFranc, Jr., B. A. Millsap, and S. W. Hoffman, eds., Proceedings of the Southwest Raptor Management Symposium and Workshop. Sci. and Tech. Ser. 11. Natl. Wildl. Fed., Washington, D.C. 395 pp.

Schmutz, J. K. 1984. Ferruginous and Swainson's Hawk abundance and distribution in relation to land use in southeastern Alberta. J. Wildl. Manage. 48:1180–1187.

Literature Cited

Schnell, J. H. 1979. Habitat management series for unique or endangered species. Rep. 18. Black Hawk, *Buteogallus anthracinus*. Tech. Note 329. U.S. Dept. of the Interior, Bureau of Land Management, Washington, D.C. 25 pp.

Schrantz, F. G. 1943. Nest life of the Eastern Yellow Warbler. Auk 60:367–387.

Schroeder, M. H., and D. L. Sturges. 1975. The effect on the Brewer's Sparrow of spraying big sagebrush. J. Range Manage. 28:294–297.

Schwartz, P. 1963. Orientation experiments with Northern Waterthrushes wintering in Venezuela. Proc. Ornithol. Congr. 13:481–484.

Schwartz, P. 1964. The Northern Waterthrush in Venezuela. Living Bird 3:169–184.

Scott, V. E., and G. L. Crouch. 1988. Summer birds and mammals of aspen-conifer forests in west-central Colorado. Res. Pap. RM-280. U.S. Dept. of Agric., Forest Serv., Fort Collins, Colo.

Scott, V. E., K. E. Evans, D. R. Patton, and C. P. Stone. 1977. Cavity-nesting birds of North American forests. Agric. Handb. 511. U.S. Dept. of Agric., Washington, D.C. 112 pp.

Scott, V. E., and G. J. Gottfried. 1983. Bird response to timber harvest in a mixed conifer forest in Arizona. Res. Pap. RM-245. U.S. Dept. of Agric., Forest Serv., Fort Collins, Colo. 8 pp.

Sealy, S. G. 1974. Ecological segregation of Swainson's and Hermit Thrushes on Langara Island, British Columbia. Condor 76:350–351.

Sealy, S. G. 1978a. Clutch size and nest placement of the Pied-billed Grebe in Manitoba. Wilson Bull. 90:301–302.

Sealy, S. G. 1978b. Possible influence of food on egg-laying and clutch size in the Black-billed Cuckoo. Condor 80:103–104.

Sedgewick, J. A., and F. L. Knopf. 1989. Regionwide polygyny in Willow Flycatchers. Condor 91:473–475.

Selander, R. K., and J. K. Baker. 1957. The Cave Swallow in Texas. Condor 59:345–363.

Serrao, J. 1985. Decline of forest songbirds. Records of N.J. Birds 11:5–9.

Serrentino, P. 1992. Northern Harrier (*Circus cyaneus*). Pages 89–118 *in* K. J. Schneider and D. M. Pence, eds., Migratory nongame birds of management concern in the Northeast. U.S. Fish and Wildl. Serv., Region 5, Newton Corner, Mass.

Sharp, B. 1985. Avifaunal changes in central Oregon since 1899. West. Birds 16:63–70.

Sherry, T. W., and R. T. Holmes. 1988. Habitat selection by breeding American Redstarts in response to a dominant competitor, the Least Flycatcher. Auk 105:350–364.

Sherry, T. W., and R. T. Holmes. 1992. Population fluctuations in a long-distance Neotropical migrant: demographic evidence for the importance of breeding season events in the American Redstart. Pages 431–442 *in* J. M. Hagan III and D. W. Johnston, eds., Ecology and conservation of Neotropical migrant landbirds. Smithsonian Institution Press, Washington, D.C. 609 pp.

Shields, W. M. 1984. Factors affecting nest and site fidelity in Adirondack Barn Swallows (*Hirundo rustica*). Auk 101:780–789.

Shy, E. 1984. Habitat shift and geographical variation in North American tanagers (Thraupinae: *Piranga*). Oecologia 63:281–285.

Sick, H. 1971. Blackpoll Warbler on winter quarters in Rio de Janeiro, Brazil. Wilson Bull. 83:198–200.

Sick, H. 1985. Brazilian Ornithology (in Portuguese). Editoria Univ. Brasília, Brasília, Brazil.

Siegfried, W. R. 1976. Breeding biology and parasitism in the Ruddy Duck. Wilson Bull. 88:566–574.

Simms, E. 1985. British Warblers. Collins, London.

Skaggs, R. W., D. H. Ellis, W. G. Hunt, and T. H. Johnson. 1988. Peregrine Falcon. Pages 127–136 in R. L. Glinski, B. G. Pendleton, M. B. Moss, M. N. LeFranc, Jr., B. A. Millsap, and S. W. Hoffman, eds., Proceedings of the Southwest Raptor Management Symposium and Workshop. Sci. and Tech. Ser. 11. Natl. Wildl. Fed., Washington, D.C. 395 pp.

Skutch, A. F. 1976. Parent birds and their young. University of Texas Press, Austin.

Slud, P. 1964. The birds of Costa Rica. Bull. Am. Mus. Nat. Hist. 128:1–430.

Smith, D. G., J. R. Murphy, and N. D. Woffindin. 1981. Relationships between jackrabbit abundance and Ferruginous Hawk reproduction. Condor 83:52–56.

Smith, D. G., C. R. Wilson, and H. H. Frost. 1972. The biology of the American Kestrel in central Utah. Southwest. Nat. 17:73–83.

Smith, N. G. 1980. Hawk and vulture migrations in the Neotropics. Pages 51–65 in A. Keast and E. S. Morton, eds., Migrant birds in the Neotropics. Smithsonian Institution Press, Washington, D.C.

Smith, R. L. 1963. Some ecological notes on the Grasshopper Sparrow. Wilson Bull. 75:159–165.

Smith, T. M., and H. H. Shugart. 1987. Territory size variation in the Ovenbird: the role of habitat structure. Ecology 68:695–704.

Smith, W. P. 1934. Observations of the nesting habits of the Black-and-white Warbler. Bird Banding 5:31–36.

Snapp, B. D. 1976. Colonial breeding in the Barn Swallow (*Hirundo rustica*) and its adaptive significance. Condor 78:471–480.

Snow, C. 1974a. Ferruginous Hawk, *Buteo regalis*. Habitat management series for unique or endangered species. Tech. Note 255. U.S. Dept. of the Interior, Bureau of Land Management, Washington, D.C. 23 pp.

Snow, C. 1974b. Prairie Falcon, *Falco mexicanus*. Habitat management series for unique or endangered species. Tech. Note 240. U.S. Dept. of the Interior, Bureau of Land Management, Washington, D.C. 18 pp.

Snyder, H. A., and R. L. Glinski. 1988. Zone-tailed Hawk. Pages 105–110 in R. L. Glinski, B. G. Pendleton, M. B. Moss, M. N. LeFranc, Jr., B. A. Millsap, and S. W. Hoffman, eds., Proceedings of the Southwest Raptor Management Symposium and Workshop. Sci. and Tech. Ser. 11. Natl. Wildl. Fed., Washington, D.C. 395 pp.

Soil Conservation Service. 1981. Land resource regions and major land resource areas of the United States. Agric. Handb. 296. U.S. Dept. of Agric., Washington, D.C.

Speiser, R., and T. Bosakowski. 1988. Nest site preferences of Red-tailed Hawks in the highlands of southeastern New York and northern New Jersey. J. Field Ornithol. 59:361–368.

Spencer, O. R. 1943. Nesting habits of the Black-billed Cuckoo. Wilson Bull. 55:11–22.

Sperry, C. C. 1940. Food habits of a group of shorebirds: woodcock, snipe, knot, and dowitcher. Wildl. Res. Bull. 1. U.S. Dept. of the Interior, Biological Survey, Washington, D.C. 37 pp.

Sprunt, A., Jr. 1955. North American birds of prey. Harper, New York. 227 pp.

Stabler, R. M. 1959. Nesting of the Blue Grosbeak in Colorado. Condor 61:46–48.

Staicer, C. A. 1992. Social behavior of the Northern Parula, Cape May Warbler, and Prairie Warbler wintering in second-growth forest in southwestern Puerto Rico.

Literature Cited

Pages 308–320 *in* J. M. Hagan III and D. W. Johnston, eds., Ecology and conservation of Neotropical migrant landbirds. Smithsonian Institution Press, Washington, D.C. 609 pp.

Steidl, R. J., and C. R. Griffin. 1991. Growth and brood reduction of mid-Atlantic Coast Ospreys. Auk 108:363–370.

Steidl, R. J., C. R. Griffin, L. J. Niles, and K. E. Clark. 1991. Reproductive success and eggshell thinning of a reestablished Peregrine Falcon population. J. Wildl. Manage. 55:294–299.

Stein, R. C. 1958. The behavioral, ecological and morphological characteristics of two populations of the Alder Flycatcher, *Empidonax traillii* (Audubon). N.Y. State Mus. Sci. Serv. Bull. 371:1–63.

Steinhart, P. 1984. Trouble in the tropics. Natl. Wildl. (Dec.–Jan.):16–20.

Stenger, J. 1958. Food habits and available food of Ovenbirds in relation to territory size. Auk 75:335–346.

Stenzel, L. E., H. R. Huber, and G. W. Page. 1976. Feeding behavior and diet of the Long-billed Curlew and Willet. Wilson Bull. 88:314–332.

Stevenson, H. M. 1972. The recent history of Bachman's Warbler. Wilson Bull. 84:344–347.

Stewart, P. A. 1987. Decline in numbers of wood warblers in spring and autumn migrations through Ohio. N. Am. Bird Bander 12:58–60.

Stewart, P. A., and H. A. Conner. 1980. Fixation of wintering Palm Warblers to a specific site. J. Field Ornithol. 51:365–367.

Stewart, R. E. 1949. Ecology of a nesting Red-shouldered Hawk population. Wilson Bull. 61:26–35.

Stewart, R. E. 1953. A life history study of the Yellow-throat. Wilson Bull. 65:99–115.

Stewart, R. M. 1973. Breeding behavior and life history of the Wilson's Warbler. Wilson Bull. 85:21–30.

Stewart, R. M., R. P. Henderson, and K. Darling. 1977. Breeding ecology of the Wilson's Warbler in the high Sierra Nevada, California. Living Bird 16:83–102.

Stiles, F. G. 1980. Evolutionary implications of habitat relations between permanent and winter resident landbirds in Costa Rica. Pages 421–435 *in* A. Keast and E. S. Morton, eds., Migrant birds in the Neotropics. Smithsonian Institution Press, Washington, D.C.

Stiles, F. G., and A. F. Skutch. 1989. A guide to the birds of Costa Rica. Cornell University Press, Ithaca, N.Y. 511 pp.

Stokes, A. W. 1950. Breeding behavior of the Goldfinch. Wilson Bull. 62:107–127.

Stoudt, J. H. 1982. Habitat use and productivity of Canvasbacks in southwestern Manitoba, 1961–72. Spec. Sci. Rep. Wildl. 248. U.S. Dept. of the Interior, Fish and Wildl. Serv., Washington, D.C. 31 pp.

Strohmeyer, D. L. 1977. Common Gallinule (*Gallinula chloropus*). Pages 110–117 *in* G. C. Sanderson, ed., Management of migratory shore and upland game birds in North America. International Association of Fish and Wildlife Agencies, Washington, D.C.

Stutchbury, B. J. 1991. Coloniality and breeding biology of Purple Martins (*Progne subis hesperia*) in saguaro cacti. Condor 93:666–675.

Sutton, G. M., and D. F. Parmelee. 1954. Survival problems of the Water Pipit in Baffin Island. Arctic 7:81–92.

Sutton, G. M., and D. F. Parmelee. 1955. Breeding of the Semipalmated Plover on Baffin Island. Bird Banding 26:137–147.

Sweeney, J. M., and W. D. Dijak. 1985. Ovenbird habitat capability model for an oak-hickory forest. Proc. Annu. Conf. Southeast. Assoc. Fish Wildl. Agencies 39:430–438.

Szaro, R. C., and R. P. Balda. 1982. Selection and monitoring of avian indicator species: an example from a ponderosa pine forest in the Southwest. Gen. Tech. Rep. RM-89. U.S. Dept. of Agric., Forest Serv., Rocky Mountain Forest and Range Exp. Stn., Fort Collins, Colo. 8 pp.

Taber, R. D. 1947. The Dickcissel in Wisconsin. Passenger Pigeon 9:39–46.

Tamm, S. 1989. Display behavior of male Calliope Hummingbirds during the breeding season. Condor 91:272–279.

Tate, G. R. 1992. Short-eared Owl (Asio flammeus). Pages 171–190 in K. J. Schneider and D. M. Pence, eds., Migratory nongame birds of management concern in the Northeast. U.S. Fish and Wildl. Serv., Region 5, Newton Corner, Mass.

Tate, J., Jr. 1973. Methods and annual sequence of foraging by the sapsucker. Auk 90:840–856.

Tate, J., Jr., and D. J. Tate. 1982. The blue list for 1982. Am. Birds 36:126–135.

Tatschl, J. L. 1967. Breeding birds of the Sandia Mountains and their ecological distribution. Condor 69:479–490.

Taylor, D. M., and C. D. Littlefield. 1986. Willow Flycatcher and Yellow Warbler response to cattle grazing. Am. Birds 40:1169–1173.

Taylor, R. J. 1984. Predation. Chapman and Hall, New York.

Taylor, W. K., and H. Hanson. 1970. Observations on the breeding biology of the Vermilion Flycatcher in Arizona. Wilson Bull. 82:315–319.

Temple, S. A., and J. R. Cary. 1988. Modelling dynamics of habitat-interior bird populations in fragmented landscapes. Conserv. Biol. 2:340–347.

Temple, S. A., and B. L. Temple. 1976. Avian population trends in central New York State, 1935–1973. Bird Banding 47:238–257.

Terborgh, J. W. 1974. Preservation of natural diversity: the problem of extinction prone species. BioScience 24:715–722.

Terborgh, J. W. 1989. Where have all the birds gone? Princeton University Press, Princeton, N.J.

Terborgh, J. W. 1992. Perspectives on the conservation of Neotropical migrant landbirds. Pages 7–12 in J. M. Hagan III and D. W. Johnston, eds., Ecology and conservation of Neotropical migrant landbirds. Smithsonian Institution Press, Washington, D.C. 609 pp.

Terres, J. K. 1980. The Audubon Society encyclopedia of North American birds. Knopf, New York. 1109 pp.

Thiollay, J. M. 1977. Autumn migration along the eastern coast of Mexico. Alauda 45:344–346.

Thiollay, J. M. 1979. Importance of an axis of migration along the east coast of Mexico. Alauda 47:235–245.

Thomas, J. W., tech. ed. 1979. Wildlife habitats in managed forests: the Blue Mountains of Oregon and Washington. Agric. Handb. 533. U.S. Dept. of Agric., Washington, D.C.

Thomas, J. W., R. Anderson, C. Maser, and E. Bull. 1979. Snags. Pages 60–77 in J. W. Thomas, tech. ed., Wildlife habitats in managed forests: the Blue Mountains of Oregon and Washington. Agric. Handb. 533. U.S. Dept. of Agric., Washington, D.C.

Literature Cited

Thomas, R. H. 1946. A study of Eastern Bluebirds in Arkansas. Wilson Bull. 58:143–183.

Thompson, C. F., and V. Nolan, Jr. 1973. Population biology of the Yellow-breasted Chat (*Icteria virens* L.) in southern Indiana. Ecol. Monogr. 43:145–171.

Thompson, W. L., R. H. Yahner, and G. L. Storm. 1990. Winter use and habitat characteristics of vulture communal roosts. J. Wildl. Manage. 54:77–83.

Titterington, R. W., H. S. Crawford, and B. N. Burgason. 1979. Songbird responses to commercial clearcutting in Maine spruce-fir forests. J. Wildl. Manage. 43:602–609.

Titus, K. 1990. Trends in counts of Scissor-tailed Flycatchers based on a nonparametric rank-trend analysis. Pages 164–166 *in* J. R. Sauer and S. Droege, eds., Survey designs and statistical methods for the estimation of avian population trends. Biol. Rep 90 (1). U.S. Dept. of the Interior, Fish and Wildl. Serv., Washington, D.C.

Titus, K., M. R. Fuller, and D. Jacobs. 1990. Detecting trends in hawk migration count data. Pages 105–113 *in* J. R. Sauer and S. Droege, eds., Survey designs and statistical methods for the estimation of avian population trends. Biol. Rep 90(1). U.S. Dept. of the Interior, Fish and Wildl. Serv., Washington, D.C.

Tobalske, B. W. 1992. Evaluating habitat suitability using relative abundance and fledging success of Red-naped Sapsuckers. Condor 94:550–553.

Todd, R. L. 1977. Black Rail, little black rail, black crane, farallon rail (*Laterallus jamaicensis*). Pages 71–83 *in* G. C. Sanderson, ed., Management of migratory shore and upland game birds in North America. International Association of Fish and Wildlife Agencies, Washington, D.C.

Tompkins, I. R. 1959. Life history notes on the Least Tern. Wilson Bull. 71:313–322.

Tramer, E. J., and T. R. Kemp. 1980. Foraging ecology of migrants and resident warblers and vireos in the highlands of Costa Rica. Pages 285–296 *in* A. Keast and E. S. Morton, eds., Migrant birds in the Neotropics. Smithsonian Institution Press, Washington, D.C.

Trimble, S. A. 1975. Habitat management series for unique or endangered species. Rep. 15. Merlin, *Falco columbarius*. Tech. Note 271. U.S. Dept. of the Interior, Bureau of Land Management, Washington, D.C. 41 pp.

Tuck, L. M. 1972. The snipes: a study of the genus *Capella*. Monogr. Ser. 5. Can. Wildl. Serv., Ottawa. 428 pp.

Twedt, D. J., W. J. Bleier, and G. M. Linz. 1991. Geographic and temporal variation in the diet of Yellow-headed Blackbirds. Condor 93:975–986.

Udvardy, M. D. F. 1977. The Audubon Society field guide to North American birds (west). Knopf, New York. 854 pp.

United Nations Food and Agricultural Organization. 1991. Second interim report on the state of tropical forests. Forest Resources Assessment 1990 Project, 10th World Forestry Congress, Paris, September 1991.

U.S. Congress. 1918. Migratory Bird Treaty Act. P.L. 65-186, ch. 128, 40 stat. 755; 16 U.S.C. 703-712.

U.S. Fish and Wildlife Service (USFWS). 1987. Migratory nongame birds of management concern in the United States. U.S. Dept. of the Interior, Fish and Wildl. Serv., Washington, D.C.

U.S. Fish and Wildlife Service (USFWS). 1991. Endangered and threatened wildlife and plants. 50 CFR 17.11 and 17.12. U.S. Dept. of the Interior, Fish and Wildl. Serv., Washington, D.C.

664

U.S. Fish and Wildlife Service (USFWS). 1992. Notice of petition finding and status review for the Southwestern Willow Flycatcher. Fed. Reg. 57(170):39664–39668.

Vahle, J. R., N. L. Dodd, and S. Nagiller. 1988. Osprey. Pages 37–47 *in* R. L. Glinski, B. G. Pendleton, M. B. Moss, M. N. LeFranc, Jr., B. A. Millsap, and S. W. Hoffman, eds., Proceedings of the Southwest Raptor Management Symposium and Workshop. Sci. and Tech. Ser. 11. Natl. Wildl. Fed., Washington, D.C. 395 pp.

Van Daele, L. J., and H. A. Van Daele. 1982. Factors affecting the productivity of Ospreys nesting in west-central Idaho. Condor 84:292–299.

Van Tyne, J. 1936. The discovery of the nest of the Colima Warbler (*Vermivora crissalis*). Misc. Publ. 33. University of Michigan Museum of Zoology, Ann Arbor. 11 pp.

Vega-Rivera, J. H. 1991. Dynamics of the avian community in a brush management area of south Texas. M.S. thesis, Texas A&I University, Kingsville.

Vega-Rivera, J. H., and J. H. Rappole. 1994. Composition and phenology of an avian community in the Rio Grande Plain of Texas. Wilson Bull. 106:366–380.

Verbeek, N. A. M. 1967. Breeding biology and ecology of the Horned Lark in alpine tundra. Wilson Bull. 79:208–218.

Verbeek, N. A. M. 1970. Breeding ecology of the Water Pipit. Auk 87:425–451.

Verbeek, N. A. M. 1975a. Northern wintering of flycatchers and residency of Black Phoebes in California. Auk 92:737–749.

Verbeek, N. A. M. 1975b. Comparative feeding behavior of three coexisting tyrannid flycatchers. Wilson Bull. 87:231–240.

Vermeer, K. 1970. Breeding biology of California and Ring-billed Gulls. Rep. Ser. 12. Can. Wildl. Serv., Ottawa. 52 pp.

Verner, J. 1965. Breeding biology of the Long-billed Marsh Wren. Condor 67:6–30.

Verner, J., and A. S. Boss, tech. coord. 1980. California wildlife and their habitats: western Sierra Nevada. Gen. Tech. Rep. PSW-37. U.S. Dept. of Agric., Forest Serv., Pacific Southwest Forest and Range Exp. Stn., Berkeley, Calif. 439 pp.

Verner, J., and G. H. Engelsen. 1970. Territories, multiple nest building, and polygyny in the Long-billed Marsh Wren. Auk 87:557–567.

Via, J., and D. C. Duffy. 1992. Northern Harrier (*Circus cyaneus*). Pages 89–118 *in* K. J. Schneider and D. M. Pence, eds., Migratory nongame birds of management concern in the Northeast. U.S. Fish and Wildl. Serv., Region 5, Newton Corner, Mass.

Vidal-Rodriguez, R. M. 1992. Abundance and seasonal distribution of Neotropical migrants during autumn in a Mexican cloud forest. Pages 370–376 *in* J. M. Hagan III and D. W. Johnston, eds., Ecology and conservation of Neotropical migrant landbirds. Smithsonian Institution Press, Washington, D.C. 609 pp.

Vogt, W. 1970. The avifauna in a changing ecosystem. Smithson. Contrib. Zool. 26.

Walkinshaw, L. H. 1935. Studies of the Short-billed Marsh Wren, *Cistothorus stellaris*, in Michigan. Auk 52:362–369.

Walkinshaw, L. H. 1937. The Yellow Rail in Michigan. Auk 56:227–237.

Walkinshaw, L. H. 1940. Summer life of the Sora Rail. Auk 57:153–168.

Walkinshaw, L. H. 1944. The Eastern Chipping Sparrow in Michigan. Wilson Bull. 56: 193–205.

Walkinshaw, L. H. 1949. The Sandhill Cranes. Bull. 29. Cranbrook Institute of Science, Bloomfield Hills, Mich. 202 pp.

Walkinshaw, L. H. 1953. Life-history of the Prothonotary Warbler. Wilson Bull. 65: 152–168.

Literature Cited

Walkinshaw, L. H. 1966. Summer biology of Traill's Flycatcher. Wilson Bull. 78:31–46.

Walkinshaw, L. H. 1983. Kirtland's Warbler. Bull. 58. Cranbrook Institute of Sciences, Bloomfield Hills, Mich. 207 pp.

Walkinshaw, L. H., and W. A. Dyer. 1961. The Connecticut Warbler in Michigan. Auk 78:379–388.

Wallace, J. 1986. Where have all the songbirds gone? Sierra (Mar.–Apr.):44–47.

Warren, R. J. 1991. Ecological justification for controlling deer populations in eastern national parks. Trans. N. Am. Wildl. Nat. Res. Conf. 56:56–66.

Waser, N. M. 1976. Food supply and nest timing of Broad-tailed Hummingbirds in the Rocky Mountains. Condor 78:133–135.

Wauer, R. H., and D. G. Davis. 1972. Cave Swallows in Big Bend National Park, Texas. Condor 74:482.

Webb, W. L., D. F. Behrend, and B. Saisorn. 1977. Effect of logging on songbird populations in a northern hardwood forest. Wildl. Monogr. 55:6–36.

Webster, C. G. 1964. Fall foods of Soras from two habitats in Connecticut. J. Wildl. Manage. 28:163–165.

Weeks, H. P., Jr. 1978. Nesting ecology of the Eastern Phoebe in southern Indiana. Wilson Bull. 91:441–454.

Weller, M. W. 1961. Breeding biology of the Least Bittern. Wilson Bull. 73:11–35.

Weller, M. W. 1964. Distribution and migration of the Redhead. J. Wild. Manage. 28:64–103.

Welsh, D. A. 1975. Savannah Sparrow breeding and territoriality on a Nova Scotia dune beach. Auk 92:235–251.

Welter, W. A. 1935. The natural history of the Long-billed Marsh Wren. Wilson Bull. 47:3–34.

Werschler, C. R. 1986a. The Piping Plover in Alberta. Prov. Mus. Alberta Nat. Hist. Occas. Pap. 9, pp. 235–238.

Werschler, C. R. 1986b. The Mountain Plover in Canada. Prov. Mus. Alberta Nat. Hist. Occas. Pap. 9, pp. 259–261.

Weston, H. G., Jr. 1947. Breeding behavior of the Black-headed Grosbeak. Condor 49:54–73.

Weston, J. B. 1969. Nesting ecology of the Ferruginous Hawk (*Buteo regalis*). Sci. Bull. Biol. Ser. 10. Brigham Young University, Salt Lake City. Pp. 25–36.

Wetmore, A. 1924. Food and economic relations of North American grebes. Bull. 1196. U.S. Dept. of Agric., Washington, D.C. 23 pp.

Wetmore, A. 1943. The birds of southern Veracruz, Mexico. Proc. U.S. Natl. Mus. 93:215–340.

Wetmore, A. 1972. Passeriformes: Dendrocolaptidae (woodcreepers) to Oxyruncidae (sharpbills). The birds of the Republic of Panama, part 3. Smiths. Misc. Coll. vol. 150, Washington, D.C. 631 pp.

Wheelwright, N. T. 1986. The diet of American Robins: an analysis of U.S. Biological Survey records. Auk 103:710–725.

Whitcomb, R. F. 1977. Island biogeography and "habitat islands" of eastern forest. Am. Birds 31:3–5.

Whitcomb, R. F., C. S. Robbins, J. F. Lynch, B. L. Whitcomb, M. K. Klimkiewicz, and D. Bystrak. 1981. Effects of forest fragmentation on the avifauna of the eastern deciduous forest. Pages 125–205 *in* R. L. Burgess and D. M. Sharpe, eds., Forest island dynamics in man-dominated landscapes. Springer-Verlag, New York.

White, H. C. 1953. The Eastern Belted Kingfisher in the Maritime Provinces. Bull. 97. Fisheries Resource Board of Canada, Ottawa. 44 pp.

Wiedenfeld, D. A. 1992. Foraging in temperate- and tropical-breeding and wintering male Yellow Warblers. Pages 321–328 *in* J. M. Hagan III and D. W. Johnston, eds., Ecology and conservation of Neotropical migrant landbirds. Smithsonian Institution Press, Washington, D.C. 609 pp.

Wiens, J. A., B. Van Horne, and J. T. Rotenberry. 1987. Temporal and spatial variations in the behavior of shrubsteppe birds. Oecologia 73:60–70.

Wiens, J. A., B. Van Horne, and J. T. Rotenberry. 1990. Comparisons of the behavior of Sage and Brewer's Sparrows in shrubsteppe habitats. Condor 92:264–266.

Wilbur, S. R., and J. A. Jackson. 1983. Vulture biology and management. University of California Press, Berkeley. 550 pp.

Wilcove, D. S. 1983. Population changes in the Neotropical migrants of the Great Smoky Mountains: 1947–1982. Unpubl. report to the World Wildl. Fund, U.S., Washington, D.C.

Wilcove, D. S. 1985. Nest predation in forest tracts and the decline of migratory songbirds. Ecology 66:1211–1214.

Wilcove, D. S. 1988. Changes in the avifauna of the Great Smoky Mountains: 1947–1983. Wilson Bull. 100:256–271.

Wilcove, D. S., and J. W. Terborgh. 1984. Patterns of population decline. Am. Birds 38:10–13.

Wilcox, L. A. 1959. A twenty year banding study of the Piping Plover. Auk 76:129–152.

Wilcox, L. 1980. Observations of the life history of Willets on Long Island, New York. Wilson Bull. 92:253–258.

Wiley, J. W., and B. N. Wiley. 1979. The biology of the White-crowned Pigeon. Wildl. Monogr. 64:1–54.

Williams, G. G. 1958. Evolutionary aspects of bird migration. Lida Scott Brown Lectures in Ornithology, University of California, Los Angeles.

Williams, L. 1952. Breeding behavior of the Brewer Blackbird. Condor 54:3–47.

Williams, T. C., and J. M. Williams. 1978. Orientation of trans-Atlantic migrants. Pages 239–251 *in* K. Schmidt-Koenig and W. T. Keeton, eds., Symposium on animal migration navigation and homing. Springer-Verlag, New York.

Williamson, P. 1971. Feeding ecology of the Red-eyed Vireo (*Vireo olivaceus*) and associated foliage-gleaning birds. Ecol. Monogr. 41:129–152.

Willimont, L. A., S. E. Senner, and L. Goodrich. 1988. Fall migration of Ruby-throated Hummingbirds in the northeastern United States. Wilson Bull. 100:482–488.

Willis, E. O. 1972. The behavior of Spotted Antbirds. Ornithol. Monogr. 10.

Willis, E. O. 1980. Ecological roles of migratory and resident birds on Barro Colorado Island, Panama. Pages 205–225 *in* A. Keast and E. S. Morton, eds., Migrant birds in the Neotropics. Smithsonian Institution Press, Washington, D.C.

Willson, M. F. 1966. Breeding ecology of the Yellow-headed Blackbird. Ecol. Monogr. 36:51–77.

Winker, K., J. H. Rappole, and M. A. Ramos. 1990. Population dynamics of the Wood Thrush in southern Veracruz, Mexico. Condor 92:444–460.

Literature Cited

Winker, K., D. W. Warner, and A. R. Weisbrod. 1992. The Northern Waterthrush and Swainson's Thrush as transients at a temperate inland stopover site. Pages 384–402 in J. M. Hagan III and D. W. Johnston, eds., Ecology and conservation of Neotropical migrant landbirds. Smithsonian Institution Press, Washington, D.C. 609 pp.

Witham, J. W., and M. L. Hunter, Jr. 1992. Population trends of Neotropical migrant landbirds in northern coastal New England. Pages 85–95 in J. M. Hagan III and D. W. Johnston, eds., Ecology and conservation of Neotropical migrant landbirds. Smithsonian Institution Press, Washington, D.C. 609 pp.

Withers, P. C. 1977. Energetic aspects of reproduction by the Cliff Swallow. Auk 94: 718–725.

Woffinden, N. D., and J. R. Murphy. 1983. Ferruginous Hawk nest site selection. J. Wildl. Manage. 47:216–219.

Wolf, L., R. M. Lejnieks, and C. R. Brown. 1985. Temperature fluctuations and nesting behavior of Rock Wrens in a high-altitude environment. Wilson Bull. 97:385–387.

Wray, T., II, and R. C. Whitmore. 1979. Effects of vegetation on nesting success of Vesper Sparrows. Auk 96:802–805.

Wunderle, J. M. 1992. Sexual habitat segregation in wintering Black-throated Blue Warblers in Puerto Rico. Pages 299–307 in J. M. Hagan III and D. W. Johnston, eds., Ecology and conservation of Neotropical migrant landbirds. Smithsonian Institution Press, Washington, D.C. 609 pp.

Wunderle, J. M., D. J. Lodge, and R. B. Waide. 1992. Short-term effects of Hurricane Gilbert on terrestrial bird populations on Jamaica. Auk 109:148–166.

Yahner, R. H. 1991. Avian nesting ecology in small even-aged aspen stands. J. Wildl. Manage. 55:155–159.

Young, H. 1955. Breeding behavior and nesting of the Eastern Robin. Am. Midl. Nat. 53:329–352.

Zach, R., and J. B. Falls. 1975. Response of the Ovenbird (Aves: Parulidae) to an outbreak of the spruce budworm. Can. J. Zool. 53:1669–1672.

Zarn, M. 1974a. Osprey, Pandion haliaetus carolinensis. Habitat management series for unique or endangered species. Tech. Note 254. U.S. Dept. of the Interior, Bureau of Land Management, Washington, D.C. 41 pp.

Zarn, M. 1974b. Burrowing Owl, Speotyto cunicularia hypugaea. Habitat management series for unique or endangered species. Tech. Note 250. U.S. Dept. of the Interior, Bureau of Land Management, Washington, D.C. 41 pp.

Zeranski, J. D., and T. R. Baptist. 1990. Connecticut birds. University Press of New England, Hanover, N.H. 328 pp.

Zimmerman, D. A., and S. H. Levy. 1960. Violet-crowned Hummingbird nesting in Arizona and New Mexico. Auk 77:470–471.

Zimmerman, J. L. 1977. Virginia Rail, Rallus limicola. Pages 46–56 in G. C. Sanderson, ed., Management of migratory shore and upland game birds in North America. International Association of Fish and Wildlife Agencies, Washington, D.C.

Zimmerman, J. L. 1982. Nesting success of Dickcissels (Spiza americana) in preferred and less preferred habitats. Auk 99:292–298.

Index

Where two or more page references to the text occur, the number in **boldface** indicates the main description.

Index

Index

Index